Springer Proceedings in Mathematics & Statistics

Volume 418

This book series features volumes composed of selected contributions from workshops and conferences in all areas of current research in mathematics and statistics, including data science, operations research and optimization. In addition to an overall evaluation of the interest, scientific quality, and timeliness of each proposal at the hands of the publisher, individual contributions are all refereed to the high quality standards of leading journals in the field. Thus, this series provides the research community with well-edited, authoritative reports on developments in the most exciting areas of mathematical and statistical research today.

Generalized Neighborhood Systems Approach for Information Retrieval Systems

A. S. Salama and Radwan Abu Gdairi

Abstract We proposed in this paper a new decision information retrieval model using rough sets that are generated by general binary relations. This model depends on generalized rough sets to deal with the relevance among users' queries and documents of the information retrieval systems. The research problem of this paper is how we close the interesting categories to the relevant terms of the interested categories. In the classical information retrieval approach, there are only two cases namely relevant documents and irrelevant documents. In the traditional rough set theory, document stream is separated to some regions—positive, boundary, and the negative region. In this paper, we used generalized membership relations to more classifications on information retrieval documents that enable to divide the document stream into 16 different regions.

Keywords Information retrieval · Neighborhood systems · Rough sets · Document classification · Memberships relations

MSC 54A05 · 54B05 · 54D35 · 03B70

1 Introduction

Information retrieval problem happens after the board information does not exist. In this paper, we are trading with classifying the most revealing portion of data on a group of documents with the intention of obtaining the greatest outcome on a latter uncertain bunching phase [1–4]. The aim is to find comparisons between the documents and a position board, and to find relations related to a non-literal nose. We suggest putting on the famous entropy system and then displaying the latter dissimilar

A. S. Salama (✉)
Department of Mathematics, Faculty of Science, Tanta University, Tanta, Egypt
e-mail: asalama@science.tanta.edu.eg

R. A. Gdairi
Department of Mathematics, Faculty of Science, Zarqa University, Zarqa, Jordan
e-mail: rgdairi@zu.edu.jo

D. Zeidan et al. (eds.), *Mathematics and Computation*, Springer Proceedings in Mathematics & Statistics 418, https://doi.org/10.1007/978-981-99-0447-1_1

actions to the correct selection of the notice data [5, 6]. This process carries the main quantity of information inside the minimum quantity of data. Spread over an exact collection process for a collection of words springs additional information to distinguish and distinct the forms afterward using the entropy allowance. This revenues significant outcome on the dispensation time and the right uncertain gathering of the documents group [7–20].

Information retrieval (IR) applications can barely be assessed based on the definitive test-gathering pattern; consequently, there is a need for new estimation approaches. The evaluation process of IR includes user modeling, criteria, measures, methodology, and new trends in IR evaluation [21, 22, 24].

There are many different definitions of IR measures, but some of them are in a very special way, essentially the definition of a new metric must consist of some basic stages:

- Beginning from the selected norm, suppose a fixed user attitude (e.g., interception after a confirmed number of relevant documents).
- Define the objective (e.g., the least number of documents visible is the better).
- Know the basic metric that conforms to the preferences (e.g., accuracy).
- Moreover, one can suppose a collection of users and count a metric mean of these metric values of this collection (e.g., for middle accuracy, it is presumed that at every relevant document, the same number of users discontinue).
- Finally, for bringing the same result for a set of queries or meetings, a gathering technique has to be picked (e.g., mathematical average).

2 Rough Set Theory

In this section, we give some facts about Pawlak rough sets that are needed in this paper. In addition, we introduce the generalized neighborhood systems and we generate some approximations that are used in information retrieval applications.

Pawlak in [23] defined the approximation space $App = (U, R)$, where U is a non-empty finite set and R is an equivalence binary relation on U. The lower and upper approximations of a subset $A \subseteq U$ are defined respectively as follows:

$$\underline{R}(A) = \{x \in U \ |[x]_R \subseteq A\},$$

$$\overline{R}(A) = \{x \in U|[x]_R \cap A \neq \emptyset\}.$$

The subsets $[x]_R$ form a partition of the universe U for all $x \in U$. The elements surely belong to A are called the positive region of A and are denoted by $POS(A) = \underline{R}(A)$. The elements surely not belong to A are called the negative region of A and are denoted by $NEG(A) = U - \overline{R}(A)$. The elements that possibly belong to A are called the boundary region and are denoted by $B(A) = \overline{R}(A) - \underline{R}(A)$.

Dia Zeidan · Juan C. Cortés · Aliaa Burqan ·
Ahmad Qazza · Jochen Merker · Gharib Gharib
Editors

Mathematics and Computation

IACMC 2022, Zarqa, Jordan, May 11–13

Springer

Editors
Dia Zeidan
School of Basic Sciences and Humanities
German Jordanian University
Amman, Jordan

Juan C. Cortés
Department of Applied Mathematics
Universitat Politècnica de València
Valencia, Spain

Aliaa Burqan
Department of Mathematics
Zarqa University
Zarqa, Jordan

Ahmad Qazza
Department of Mathematics
Zarqa University
Zarqa, Jordan

Jochen Merker
MNZ
Leipzig University of Applied Sciences
Leipzig, Germany

Gharib Gharib
Department of Mathematics
Zarqa University
Zarqa, Jordan

ISSN 2194-1009 ISSN 2194-1017 (electronic)
Springer Proceedings in Mathematics & Statistics
ISBN 978-981-99-0446-4 ISBN 978-981-99-0447-1 (eBook)
https://doi.org/10.1007/978-981-99-0447-1

Mathematics Subject Classification: 08A99, 30A99, 35A99, 44A99, 65Z99, 68M99, 76A99

This Springer imprint is published by the registered company Springer Nature Singapore Pte Ltd.
The registered company address is: 152 Beach Road, #21-01/04 Gateway East, Singapore 189721, Singapore

Contents

actions to the correct selection of the notice data [5, 6]. This process carries the main quantity of information inside the minimum quantity of data. Spread over an exact collection process for a collection of words springs additional information to distinguish and distinct the forms afterward using the entropy allowance. This revenues significant outcome on the dispensation time and the right uncertain gathering of the documents group [7–20].

Information retrieval (IR) applications can barely be assessed based on the definitive test-gathering pattern; consequently, there is a need for new estimation approaches. The evaluation process of IR includes user modeling, criteria, measures, methodology, and new trends in IR evaluation [21, 22, 24].

There are many different definitions of IR measures, but some of them are in a very special way, essentially the definition of a new metric must consist of some basic stages:

- Beginning from the selected norm, suppose a fixed user attitude (e.g., interception after a confirmed number of relevant documents).
- Define the objective (e.g., the least number of documents visible is the better).
- Know the basic metric that conforms to the preferences (e.g., accuracy).
- Moreover, one can suppose a collection of users and count a metric mean of these metric values of this collection (e.g., for middle accuracy, it is presumed that at every relevant document, the same number of users discontinue).
- Finally, for bringing the same result for a set of queries or meetings, a gathering technique has to be picked (e.g., mathematical average).

2 Rough Set Theory

In this section, we give some facts about Pawlak rough sets that are needed in this paper. In addition, we introduce the generalized neighborhood systems and we generate some approximations that are used in information retrieval applications.

Pawlak in [23] defined the approximation space $App = (U, R)$, where U is a non-empty finite set and R is an equivalence binary relation on U. The lower and upper approximations of a subset $A \subseteq U$ are defined respectively as follows:

$$\underline{R}(A) = \{x \in U \ |[x]_R \subseteq A\},$$

$$\overline{R}(A) = \{x \in U|[x]_R \cap A \neq \emptyset\}.$$

The subsets $[x]_R$ form a partition of the universe U for all $x \in U$. The elements surely belong to A are called the positive region of A and are denoted by $POS(A) = \underline{R}(A)$. The elements surely not belong to A are called the negative region of A and are denoted by $NEG(A) = U - \overline{R}(A)$. The elements that possibly belong to A are called the boundary region and are denoted by $B(A) = \overline{R}(A) - \underline{R}(A)$.

Generalized Neighborhood Systems Approach for Information Retrieval Systems

A. S. Salama and Radwan Abu Gdairi

Abstract We proposed in this paper a new decision information retrieval model using rough sets that are generated by general binary relations. This model depends on generalized rough sets to deal with the relevance among users' queries and documents of the information retrieval systems. The research problem of this paper is how we close the interesting categories to the relevant terms of the interested categories. In the classical information retrieval approach, there are only two cases namely relevant documents and irrelevant documents. In the traditional rough set theory, document stream is separated to some regions—positive, boundary, and the negative region. In this paper, we used generalized membership relations to more classifications on information retrieval documents that enable to divide the document stream into 16 different regions.

Keywords Information retrieval · Neighborhood systems · Rough sets · Document classification · Memberships relations

MSC 54A05 · 54B05 · 54D35 · 03B70

1 Introduction

Information retrieval problem happens after the board information does not exist. In this paper, we are trading with classifying the most revealing portion of data on a group of documents with the intention of obtaining the greatest outcome on a latter uncertain bunching phase [1–4]. The aim is to find comparisons between the documents and a position board, and to find relations related to a non-literal nose. We suggest putting on the famous entropy system and then displaying the latter dissimilar

A. S. Salama (✉)
Department of Mathematics, Faculty of Science, Tanta University, Tanta, Egypt
e-mail: asalama@science.tanta.edu.eg

R. A. Gdairi
Department of Mathematics, Faculty of Science, Zarqa University, Zarqa, Jordan
e-mail: rgdairi@zu.edu.jo

D. Zeidan et al. (eds.), *Mathematics and Computation*, Springer Proceedings in Mathematics & Statistics 418, https://doi.org/10.1007/978-981-99-0447-1_1

The accuracy measure of a subset $A \subseteq U$ in the approximation space $App = (U, R)$ is the division of the number of elements in the positive region of A by the number of elements in the upper approximation of A. Then the accuracy measure by symbols is given as follows:

$\alpha_R(A) = \frac{|POS(A)|}{|\overline{R}(A)|}$, where $|\overline{R}(A)| \neq 0$, $|A|$ is the cardinality of A.

Another accuracy measure of approximations in Pawlak approximation spaces is defined for any subset $A \subseteq U$ as follows:

$$\rho_R(A) = 1 - \frac{|B(A)|}{|U|}.$$

In this definition, it is obvious that $0 \leq \rho_R(A) \leq 1$. Moreover, if $\rho_R(A) = 1$ then A is called R-definable (or R-exact) set. Otherwise, it is called R-rough.

We believe that the second definition of accuracy is accurate than Pawlak definition since the second considers the negative region and Pawlak used the positive region. For representative, consider the following example.

Example 2.1 Let U=$\{d1, d2, d3, d4, d5\}$ be the universe of discourse and $R = \{(d1, d1), (d1, d4), (d2, d2), (d2, d3), (d3, d2), (d3, d3), (d4, d1), (d4, d4), (d5, d5)\}$ is an equivalence relation on U. The equivalence classes of R are given by: $[d1]_R = [d4]_R = \{d1, d4\}$, $[d2]_R = [d3]_R = \{d2, d3\}$ and $[d5]_R = \{d5\}$. Hence, the partition induced by R is $U/R = \{\{d1, d4\}, \{d2, d3\}, \{d5\}\}$. Let $A = \{d2, d4\}$ be any subset of U. Thus $\underline{R}(A) = \emptyset$ and $\overline{R}(A) = \{d1, d2, d3, d4\}$. So we have $\alpha_R(A) = 0$ and $\rho_R(A) = 1/5$. Obviously, $\rho_R(A)$ is accurate than $\alpha_R(A)$ since the element of the set $N(A) = \{d5\}$ is surely does not belong to A according to R. Further, let $B = \{d1, d5\}$. So $\underline{R}(B) = \{d5\}$ and $\overline{R}(B) = \{d1, d4, d5\}$. Hence $\alpha_R(B) = 1/3$ and $\rho_R(B) = 3/5$. Clearly, $\rho_R(B)$ is accurate than $\alpha_R(B)$ since the elements of the set $N(B) = \{d2, d3\}$ are surely do not belong to B with respect to R. Also, the element of $\underline{R}(B) = \{d5\}$ is surely belongs to B according to R. Consequently, we can decide with full certainty that $d5 \in B$ and $d2, d3 \notin B$. Accordingly, the accuracy should be equal to $3/5$.

Membership functions are another approach to approximate concepts in rough set theory. For any subset $A \subseteq U$, for all $x \in U$, Pawlak defined the membership function $\mu_A^R(x) : U \to [0, 1]$ as follows:

$\mu_A^R(x) = \frac{|[x]_R \cap A|}{|[x]_R|}$, where $|[x]_R| \neq 0$, $|[x]_R|$ is the cardinality of $[x]_R$.

New rough membership functions are defined when the general binary relations are used instead of equivalence relations in approximations as follows:

For any subset $A \subseteq U$, and for all $x \in U$, we define the general membership function $\mu_A^R(x) : U \to [0, 1]$ as follows:

$\mu_A^R(x) = \frac{|xR \cap A|}{|[x]_R|}$ or $\mu_A^R(x) = \frac{|Rx \cap A|}{|[x]_R|}$, where $xR = \{y \in U \mid xRy\}$ and $Rx = \{y \in U | yRx\}$.

3 Generalized Rough Set Theory

Now we can generalize the equivalence relation to be non-equivalence by dropping one of the three conditions on it (reflexively, symmetry, and transitivity). Suppose that U is a non-empty finite set and let $\mathcal{R}$ be an arbitrary binary relation on U, then the pair $GApp = (U, \mathcal{R})$ is called a generalized approximation space.

For any generalized approximation space $GApp = (U, \mathcal{R})$ the right neighborhood and the left neighborhood of an element $x \in U$ are defined as follows:

$N_r(x) = \{y \in U \mid x\mathcal{R}y\}$, $N_l(x) = \{y \in U | y\mathcal{R}x\}$.

The class of all right neighborhoods of $x \in U$ is called the right neighborhood system and is denoted by $NS_r(x) = \{N_r(x) : x \in U\}$. Also, the class of all left neighborhoods of $x \in U$ is called the left neighborhood system and is denoted by $NS_l(x) = \{N_l(x) : x \in U\}$. The union of the right and left neighborhoods of $x \in U$ is called mixed neighborhood system and is given by $NS_m(x) = \{N_r(x) \cup N_l(x) : x \in U\}$. The mixed neighborhood of an element $x \in U$ is denoted by $N_m(x)$ such that $N_m(x) \in NS_m(x)$.

Example 3.1 Let $U = \{d1, d2, d3, d4, d5\}$ be the universe of discourse and let $\mathcal{R} = \{(d1, d1), (d1, d2), (d2, d3), (d2, d5), (d4, d3), (d4, d4), (d5, d2), (d5, d4), (d5, d5)\}$ be any binary general relation defined on U. Then we have $N_r(d1) = \{d1, d2\}$, $N_r(d2) = \{d3, d5\}$, $N_r(d3) = \emptyset$, $N_r(d4) = \{d3, d4\}$, $N_r(d5) = \{d2, d4, d5\}$, $NS_r(d1) = \{\{d1, d2\}\}$, $NS_r(d2) = \{\{d3, d5\}\}$, $NS_r(d3) = \{\emptyset\}$, $NS_r(d4) = \{\{d3, d4\}\}$, and $NS_r(d5) = \{\{d2, d4, d5\}\}$. Also we have $N_l(d1) = \{d1\}$, $N_l(d2) = \{d1, d5\}$, $N_l(d3) = \{d2, d4\}$, $N_l(d4) = \{d4, d5\}$, $N_l(d5) = \{d2, d5\}$, $NS_l(d1) = \{\{d1\}\}$, $NS_l(d2) = \{\{d1, d5\}\}$,$NS_l(d3) = \{\{d2, d4\}\}$,$NS_l(d4) = \{\{d4, d5\}\}$, and $NS_l(d5) = \{\{d2, d5\}\}$. Then the mixed neighborhood systems are given by $NS_m(d1) = \{\{d1, d2\}, \{d1\}\}$, $NS_m(d2) = \{\{d3, d5\}, \{d1, d5\}\}$, $NS_m(d3) = \{\emptyset, \{d2, d4\}\}$,

$NS_m(d4) = \{\{d3, d4\}, \{d4, d5\}\}$, and $NS_m(d5) = \{\{d2, d4, d5\}, \{d2, d5\}\}$.

More generalizations can be made using the right and the left neighborhoods of an element $x \in U$ as follows:

- $\cap_r$- Neighborhood of $x \in U$ is defined by $\cap_r(x) = \cap_{x \in N_r(y)} N_r(y)$.
- $\cap_l$- Neighborhood of $x \in U$ is defined by $\cap_l(x) = \cap_{x \in N_l(y)} N_l(y)$.
- $\cap_{rl}$- Neighborhood of $x \in U$ is defined by $\cap_{rl}(x) = N_r(x) \cap N_l(x)$.
- $\cup_{rl}$- Neighborhood of $x \in U$ is defined by $\cup_{rl}(x) = N_r(x) \cup N_l(x)$.
- $\cap_{\langle rl \rangle}$- Neighborhood of $x \in U$ is defined by $\cap_{\langle rl \rangle}(x) = \cap_r(x) \cap \cap_l(x)$.
- $\cup_{\langle rl \rangle}$- Neighborhood of $x \in U$ is defined by $\cup_{\langle rl \rangle}(x) = \cap_r(x) \cup \cap_l(x)$.

4 Generalized Neighborhood Systems

We develop a new series of definitions of the lower and upper approximation approximations according to the general neighborhood systems. These new definitions are based on right, left, and mixed neighborhood systems. In addition, we give suitable definitions of the accuracy measures of the given approximations.

For any subset $A \subseteq U$, in the generalized approximation space, $GApp = (U, \mathcal{R})$ we have.

- Right lower approximation of A is defined by
  $\underline{\mathcal{R}}_r(A) = \cup\{\ N_r(x) |\ N_r(x) \subseteq A, \forall x \in U\}$.
- Right upper approximation of A is defined by
  $\overline{\mathcal{R}}_r(A) = \cup\{N_r(x) | N_r(x) \cap A \neq \varnothing, \forall x \in U\}$.
- Left lower approximation of A is defined by
  $\underline{\mathcal{R}}_l(A) = \cup\{\ N_l(x) | N_l(x) \subseteq A, \forall x \in U\}$.
- Left upper approximation of A is defined by
  $\overline{\mathcal{R}}_l(A) = \cup\{\ N_l(x) | N_l(x) \cap A \neq \emptyset, \forall x \in U\}$.
- Mixed lower approximation of A is defined by
  $\underline{\mathcal{R}}_m(A) = \cup\{\ N_m(x) | N_m(x) \subseteq A, \forall x \in U\}$.
- Mixed upper approximation of A is defined by
  $\overline{\mathcal{R}}_m(A) = \cup\{\ N_m(x)\ | N_m(x) \cap A \neq \emptyset,\ \ \forall\ x \in U\}$.
- $\cap_r -$ lower approximation of A is defined by
  $\underline{\mathcal{R}}_{\cap r}(A) = \cup\{\cap_r(x) | \cap_r (x) \subseteq A,\ \forall x \in U\}$.
- $\cap_r -$ upper approximation of A is defined by
  $\overline{\mathcal{R}}_{\cap r}(A) = \cup\{\cap_r(x) | \cap_r (x) \cap A \neq \emptyset,\ \forall x \in U\}$.
- $\cap_l -$ lower approximation of A is defined by
  $\underline{\mathcal{R}}_{\cap l}(A) = \cup\{\cap_l(x) | \cap_l (x) \subseteq A,\ \forall x \in U\}$.
- $\cap_l -$ upper approximation of A is defined by
  $\overline{\mathcal{R}}_{\cap l}(A) = \cup\{\cap_l(x) | \cap_l (x) \cap A \neq \emptyset,\ \forall x \in U\}$.
- $\cap_{rl} -$ lower approximation of A is defined by
  $\underline{\mathcal{R}}_{\cap rl}(A) = \cup\{\cap_{rl}(x) | \cap_{rl} (x) \subseteq A,\ \forall x \in U\}$.
- $\cap_{rl} -$ upper approximation of A is defined by
  $\overline{\mathcal{R}}_{\cap rl}(A) = \cup\{\cap_{rl}(x) | \cap_{rl}(x) \cap A \neq \varnothing, \forall x \in U\}$..
- $\cap_{\langle rl \rangle} -$ lower approximation of A is defined by
  $\underline{\mathcal{R}}_{\cap \langle rl \rangle}(A) = \cup\left\{\cap_{\langle rl \rangle}(x) | \cap_{\langle rl \rangle} (x) \subseteq A,\ \forall x \in U\right\}$.
- $\cap_{\langle rl \rangle} -$ upper approximation of A is defined by
  $\overline{\mathcal{R}}_{\cap \langle rl \rangle}(A) = \cup\left\{\cap_{\langle rl \rangle}(x) | \cap_{\langle rl \rangle} (x) \cap A \neq \emptyset,\ \forall x \in U\right\}$.
- $\cup_{\langle rl \rangle} -$ lower approximation of A is defined by
  $\underline{\mathcal{R}}_{\cup \langle rl \rangle}(A) = \cup\left\{\cup_{\langle rl \rangle}(x) | \cup_{\langle rl \rangle} (x) \subseteq A,\ \forall x \in U\right\}$.
- $\cup_{\langle rl \rangle} -$ upper approximation of A is defined
  $\overline{\mathcal{R}}_{\cup \langle rl \rangle}(A) = \cup\left\{\cup_{\langle rl \rangle}(x) | \cup_{\langle rl \rangle} (x) \cap A \neq \emptyset, \forall x \in U\right\}$.

For any subset $A \subseteq U$, in the generalized approximation space, $GApp = (U, \mathcal{R})$ we define the boundary, positive and negative regions of the subset A as follows:

- The boundary, positive and negative regions of a subset A using right neighborhood are defined respectively by $\mathcal{B}_r(A) = \overline{\mathcal{R}}_r(A) - \underline{\mathcal{R}}_r(A)$, $POS_r(A) = \underline{\mathcal{R}}_r(A)$, $NEG_r(A) = U - \overline{\mathcal{R}}_r(A)$.

- The boundary, positive and negative regions of a subset A using left neighborhood are defined respectively by $\mathcal{B}_l(A) = \overline{\mathcal{R}}_l(A) - \underline{\mathcal{R}}_l(A)$, $POS_l(A) = \underline{\mathcal{R}}_l(A)$, $NEG_l(A) = U - \overline{\mathcal{R}}_l(A)$.
- The boundary, positive and negative regions of a subset A using mixed neighborhood are defined respectively by $\mathcal{B}_m(A) = \overline{\mathcal{R}}_m(A) - \underline{\mathcal{R}}_m(A)$, $POS_m(A) = \underline{\mathcal{R}}_m(A)$, $NEG_m(A) = U - \overline{\mathcal{R}}_m(A)$.
- The boundary, positive and negative regions of a subset A using $\cap_r$—neighborhood are defined respectively by $\mathcal{B}_{\cap_r}(A) = \overline{\mathcal{R}}_{\cap_r}(A) - \underline{\mathcal{R}}_{\cap_r}(A)$, $POS_{\cap_r}(A) = \underline{\mathcal{R}}_{\cap_r}(A)$, $NEG_{\cap_r}(A) = U - \overline{\mathcal{R}}_{\cap_r}(A)$.
- The boundary, positive and negative regions of a subset A using $\cap_l$—neighborhood are defined respectively by $\mathcal{B}_{\cap_l}(A) = \overline{\mathcal{R}}_{\cap_l}(A) - \underline{\mathcal{R}}_{\cap_l}(A)$, $POS_{\cap_l}(A) = \underline{\mathcal{R}}_{\cap_l}(A)$, $NEG_{\cap_l}(A) = U - \overline{\mathcal{R}}_{\cap_l}(A)$.
- The boundary, positive and negative regions of a subset A using $\cap_{rl}$- neighborhood are defined respectively by $\mathcal{B}_{\cap_{rl}}(A) = \overline{\mathcal{R}}_{\cap_{rl}}(A) - \underline{\mathcal{R}}_{\cap_{rl}}(A)$, $POS_{\cap_{rl}}(A) = \underline{\mathcal{R}}_{\cap_{rl}}(A)$, $NEG_{\cap_{rl}}(A) = U - \overline{\mathcal{R}}_{\cap_{rl}}(A)$.
- The boundary, positive and negative regions of a subset A using $\cup_{rl}$- neighborhood are defined respectively by $\mathcal{B}_{\cup_{rl}}(A) = \overline{\mathcal{R}}_{\cup_{rl}}(A) - \underline{\mathcal{R}}_{\cup_{rl}}(A)$, $POS_{\cup_{rl}}(A) = \underline{\mathcal{R}}_{\cup_{rl}}(A)$, $NEG_{\cup_{rl}}(A) = U - \overline{\mathcal{R}}_{\cup_{rl}}(A)$.
- The boundary, positive and negative regions of a subset A using $\cap_{\langle rl\rangle}$- neighborhood are defined respectively by $\mathcal{B}_{\cap_{\langle rl\rangle}}(A) = \overline{\mathcal{R}}_{\cap_{\langle rl\rangle}}(A) - \underline{\mathcal{R}}_{\cap_{\langle rl\rangle}}(A)$, $POS_{\cap_{\langle rl\rangle}}(A) = \underline{\mathcal{R}}_{\cap_{\langle rl\rangle}}(A)$, $NEG_{\cap_{\langle rl\rangle}}(A) = U - \overline{\mathcal{R}}_{\cap_{\langle rl\rangle}}(A)$.
- The boundary, positive and negative regions of a subset A using $\cup_{\langle rl\rangle}$- neighborhood are defined respectively by $\mathcal{B}_{\cup_{\langle rl\rangle}}(A) = \overline{\mathcal{R}}_{\cup_{\langle rl\rangle}}(A) - \underline{\mathcal{R}}_{\cup_{\langle rl\rangle}}(A)$, $POS_{\cup_{\langle rl\rangle}}(A) = \underline{\mathcal{R}}_{\cup_{\langle rl\rangle}}(A)$, $NEG_{\cup_{\langle rl\rangle}}(A) = U - \overline{\mathcal{R}}_{\cup_{\langle rl\rangle}}(A)$.

For any subset $A \subseteq U$, in the generalized approximation space, $GApp = (U, \mathcal{R})$ the accuracy measures are defined as follows:

- $\sigma_r(A) = 1 - \frac{|\mathcal{B}_r(A)|}{|U|}$,
- $\sigma_l(A) = 1 - \frac{|\mathcal{B}_l(A)|}{|U|}$,
- $\sigma_m(A) = 1 - \frac{|\mathcal{B}_m(A)|}{|U|}$,
- $\sigma_{\cap_r}(A) = 1 - \frac{|\mathcal{B}_{\cap_r}(A)|}{|U|}$,
- $\sigma_{\cap_l}(A) = 1 - \frac{|\mathcal{B}_{\cap_l}(A)|}{|U|}$,
- $\sigma_{\cap_{rl}}(A) = 1 - \frac{|\mathcal{B}_{\cap_{rl}}(A)|}{|U|}$,
- $\sigma_{\cup_{rl}}(A) = 1 - \frac{|\mathcal{B}_{\cup_{rl}}(A)|}{|U|}$,
- $\sigma_{\cap_{\langle rl\rangle}}(A) = 1 - \frac{|\mathcal{B}_{\cap_{\langle rl\rangle}}(A)|}{|U|}$,
- $\sigma_{\cup_{\langle rl\rangle}}(A) = 1 - \frac{|\mathcal{B}_{\cup_{\langle rl\rangle}}(A)|}{|U|}$.

In all the above measures we have that: $0 \leq \sigma_\nabla(A) \leq 1$, for $\nabla \in \{r, l, m, \cap_r, \cap_l, \cap_{rl}, \cup_{rl}, \cap_{\langle rl\rangle}, \cup_{\langle rl\rangle}\}$. Moreover, if $\sigma_\nabla(A) = 1$ then A is the exact set otherwise, it is called rough.

The following results are related to the 16 lower and upper approximations above.

Theorem 4.1 For any subset $A \subseteq U$, in the generalized approximation space, $GApp = (U, \mathcal{R})$ we have.

(i) $\underline{\mathcal{R}}_m(A) = \underline{\mathcal{R}}_r(A) \cup \underline{\mathcal{R}}_l(A)$,
(ii) $\overline{\mathcal{R}}_m(A) = \overline{\mathcal{R}}_r(A) \cap \overline{\mathcal{R}}_l(A)$,
(iii) $\mathcal{B}_m(A) = \mathcal{B}_r(A) \cap \mathcal{B}_l(A)$.

Proof Suppose an element $x \in \left(\underline{\mathcal{R}}_r(A) \cup \underline{\mathcal{R}}_l(A)\right)$ then $x \in \underline{\mathcal{R}}_r(A) \vee x \in \underline{\mathcal{R}}_l(A)$. Then, $N_r(x) \subseteq A \vee N_l(x) \subseteq A$ then $\exists N_m(x)$, such that $N_m(x) \subseteq A$, then $x \in \underline{\mathcal{R}}_m(A)$. then we have $\underline{\mathcal{R}}_m(A) = \underline{\mathcal{R}}_r(A) \cup \underline{\mathcal{R}}_l(A)$. On the other hand, let $x \in \overline{\mathcal{R}}_m(A)$, then there are two cases:

First case: $x \in A$ yields $x \in \overline{\mathcal{R}}_r(A) \wedge x \in \overline{\mathcal{R}}_l(A)$, hence $x \in \left(\overline{\mathcal{R}}_r(A) \cap \overline{\mathcal{R}}_l(A)\right)$.

Second case: $x \in U - A$. Then $x \in \overline{\mathcal{R}}_m(A)$, then $\forall N_m(x)$, $N_m(x) \cap A \neq \emptyset$. Hence, $(N_r(x) \cap A \neq \emptyset) \wedge (N_l(x) \cap A \neq \emptyset)$, then $x \in \overline{\mathcal{R}}_r(A) \wedge x \in \overline{\mathcal{R}}_l(A)$. Then we have, $x \in \left(\overline{\mathcal{R}}_r(A) \cap \overline{\mathcal{R}}_l(A)\right)$. Suppose that $x \in \left(\overline{\mathcal{R}}_r(A) \cap \overline{\mathcal{R}}_l(A)\right)$, then we have.

1- $x \in A$ yield to $x \in \overline{\mathcal{R}}_m(A)$.

2- $x \in U - A$. Then $x \in \left(\overline{\mathcal{R}}_r(A) \cap \overline{\mathcal{R}}_l(A)\right)$, then $(N_r(x) \cap A \neq \emptyset) \wedge (N_l(x) \cap A \neq \emptyset)$, then $\forall N_m(x)$, $N_m(p) \cap A \neq \emptyset$, hence $x \in \overline{\mathcal{R}}_m(A)$. Hence, $\overline{\mathcal{R}}_m(A) = \overline{\mathcal{R}}_r(A) \cap \overline{\mathcal{R}}_l(A)$. Finally, let $x \in \mathcal{R}_m(A)$, hence $x \in \left(\overline{\mathcal{R}}_m(A) - \underline{\mathcal{R}}_m(A)\right)$, then $x \in \overline{\mathcal{R}}_m(A) \wedge x \notin \underline{\mathcal{R}}_m(A)$. Since $\overline{\mathcal{R}}_m(A) \subseteq \overline{\mathcal{R}}_r(A) \cap \overline{\mathcal{R}}_l(A)$ and $\underline{\mathcal{R}}_m(A) \supseteq \underline{\mathcal{R}}_r(A) \cup \underline{\mathcal{R}}_l(A)$. Then we have $x \in \left(\overline{\mathcal{R}}_r(A) \cap \overline{\mathcal{R}}_l(A)\right) \wedge x \notin \left(\underline{\mathcal{R}}_r(A) \cup \underline{\mathcal{R}}_l(A)\right)$

$$\Rightarrow \left(x \in \overline{\mathcal{R}}_r(A) \wedge x \in \overline{\mathcal{R}}_l(A)\right) \wedge \left(x \notin \underline{\mathcal{R}}_r(A) \wedge x \notin \underline{\mathcal{R}}_l(A)\right)$$
$$\Rightarrow \left(x \in \overline{\mathcal{R}}_r(A) \wedge x \notin \underline{\mathcal{R}}_r(A)\right) \wedge \left(x \in \overline{\mathcal{R}}_l(A) \wedge x \notin \underline{\mathcal{R}}_l(A)\right)$$
$$\Rightarrow x \in \left(\overline{\mathcal{R}}_r(A) - \underline{\mathcal{R}}_r(A)\right) \wedge x \in \left(\overline{\mathcal{R}}_l(A) - \underline{\mathcal{R}}_l(A)\right)$$
$$\Rightarrow x \in \mathcal{R}_r(A) \wedge x \in \mathcal{R}_l(A) \Rightarrow x \in (\mathcal{R}_r(A) \cap \mathcal{R}_l(A)).$$

Therefore, $\mathcal{R}_m(A) \subseteq \mathcal{R}_r(A) \cap \mathcal{R}_l(A)$.

Proposition 4.1 For any two subsets $A, B \subseteq U$, in the generalized approximation space, $GApp = (U, \mathcal{R})$ and for $\nabla \in \{r, l, m, \cap_r, \cap_l, \cap_{rl}, \cup_{rl}, \cap_{\langle rl \rangle}, \cup_{\langle rl \rangle}\}$, we have

(1) $\underline{\mathcal{R}}_\nabla(A) \subseteq A$.
(2) $\underline{\mathcal{R}}_\nabla(U) = U$.
(3) $\underline{\mathcal{R}}_\nabla(\emptyset) = \emptyset$.
(4) $A \subseteq B \Rightarrow \underline{\mathcal{R}}_\nabla(A) \subseteq \underline{\mathcal{R}}_\nabla(B)$.
(5) $\underline{\mathcal{R}}_\nabla(A \cap B) \subseteq \underline{\mathcal{R}}_\nabla(A) \cap \underline{\mathcal{R}}_\nabla(B)$.
(6) $\underline{\mathcal{R}}_\nabla(A \cup B) \supseteq \underline{\mathcal{R}}_\nabla(A) \cup \underline{\mathcal{R}}_\nabla(B)$.
(7) $\underline{\mathcal{R}}_\nabla(A) = U - \overline{\mathcal{R}}_\nabla(U - A)$.
(8) $A \subseteq \overline{\mathcal{R}}_\nabla(A)$.
(9) $\overline{\mathcal{R}}_\nabla(U) = U$.
(10) $\overline{\mathcal{R}}_\nabla(\emptyset) = \emptyset$.
(11) $A \subseteq B \Rightarrow \overline{\mathcal{R}}_\nabla(A) \subseteq \overline{\mathcal{R}}_\nabla(B)$.
(12) $\overline{\mathcal{R}}_\nabla(A \cap B) \subseteq \overline{\mathcal{R}}_\nabla(A) \cap \overline{\mathcal{R}}_\nabla(B)$.

(13) $\overline{\mathcal{R}}_\nabla(A \cup B) \supseteq \overline{\mathcal{R}}_\nabla(A) \cup \overline{\mathcal{R}}_\nabla(B)$.
(14) $\overline{\mathcal{R}}_\nabla(A) = U - \underline{\mathcal{R}}_\nabla(U - A)$.
(15) $\underline{\mathcal{R}}_\nabla(A) \subseteq \overline{\mathcal{R}}_\nabla(A)$.

Proof The proof of (1), (2), (3), (6), (7) and (8) follows directly from definitions.

(4) Let $A \subseteq B$ and $x \in \underline{\mathcal{R}}_\nabla(A)$, then $\exists\, N_\nabla(x)$ such that $N_\nabla(x) \subseteq A$. So $x \in \underline{\mathcal{R}}_\nabla(A) \subseteq A \subseteq B$. Thus we have $x \in B$ and there exist $N_\nabla(x)$ such that $N_\nabla(x) \subseteq A \subseteq B$. Hence $x \in \underline{\mathcal{R}}_\nabla(B)$ and so $\underline{\mathcal{R}}_\nabla(A) \subseteq \underline{\mathcal{R}}_m(B)$. Therefore, $A \subseteq B \Rightarrow \underline{\mathcal{R}}_\nabla(A) \subseteq \underline{\mathcal{R}}_\nabla(B)$.

(11) Let $A \subseteq B$ and $x \in \overline{\mathcal{R}}_\nabla(A)$, then we have.

(1) $x \in A \Rightarrow x \in A \subseteq B \Rightarrow \in B \subseteq \overline{\mathcal{R}}_\nabla(B) \Rightarrow x \in \overline{\mathcal{R}}_\nabla(B)$
(2) $x \in U - A$. Then $x \in \overline{\mathcal{R}}_\nabla(A) \Rightarrow \forall\ N_\nabla(x),\ N_\nabla(x) \cap A \neq \emptyset$ and since $A \subseteq B$ thus we have $\forall\, N_\nabla(x),\ N_\nabla(x) \cap B \neq \emptyset$ and hence we have
(a) $x \in B - A \Rightarrow x \in B \Rightarrow x \in \overline{\mathcal{R}}_\nabla(B)$.
(b) $x \in U - B$. So $\forall\ N_\nabla(x),\ N_\nabla(x) \cap B \neq \emptyset \Rightarrow x \in \overline{\mathcal{R}}_\nabla(B)$. Hence, by (1) and (2), we have $A \subseteq B \Rightarrow \overline{\mathcal{R}}_\nabla(A) \subseteq \overline{\mathcal{R}}_\nabla(B)$.

(5) Let $x \in \underline{\mathcal{R}}_\nabla(A \cap B) \Rightarrow x \in (A \cap B),\ \exists\ N_\nabla(x),\ N_\nabla(x) \subseteq (A \cap B) \Rightarrow x \in A,\ \exists\ N_\nabla(x), N_\nabla(x) \subseteq A\ \wedge\ x \in B,\ \exists\ N_\nabla(x),\ N_\nabla(x) \subseteq B\ \Rightarrow x \in \underline{\mathcal{R}}_\nabla(A) \wedge x \in \underline{\mathcal{R}}_\nabla(B)\ \Rightarrow\ x \in \underline{\mathcal{R}}_\nabla(A) \cap \underline{\mathcal{R}}_\nabla(B)$.

(6) $(A \cap B) \subseteq A \Longrightarrow \overline{\mathcal{R}}_\nabla(A \cap B) \subseteq \overline{\mathcal{R}}_\nabla(A)$ and $(A \cap B) \subseteq B \Longrightarrow \overline{\mathcal{R}}_\nabla(A \cap B) \subseteq \overline{\mathcal{R}}_\nabla(B)$. So $\overline{\mathcal{R}}_\nabla(A \cap B) \subseteq \overline{\mathcal{R}}_\nabla(A) \cap \overline{\mathcal{R}}_\nabla(B)$.

(12) $A \subseteq (A \cup B) \Longrightarrow \underline{\mathcal{R}}_\nabla(A) \subseteq \underline{\mathcal{R}}_\nabla(A \cup B)$ and $B \subseteq (A \cup B) \Longrightarrow \underline{\mathcal{R}}_\nabla(B) \subseteq \underline{\mathcal{R}}_\nabla(A \cup B)$. Hence $\underline{\mathcal{R}}_\nabla(A \cup B) \supseteq \underline{\mathcal{R}}_\nabla(A) \cup \underline{\mathcal{R}}_\nabla(B)$.

(13) Let $x \notin \overline{\mathcal{R}}_\nabla(A \cup B)$, then $x \notin (A \cup B)$ and $x \in (A \cup B)^C, \exists\, N_\nabla(x),\ N_\nabla(p) \cap (A \cup B) = \emptyset$. So

$x \in \left(A^C \cap B^C\right),\ \exists\ N_\nabla(x),\ (N_\nabla(x) \cap A) \cup (N_\nabla(x) \cap A) = \emptyset$. Thus.

$x \in U - A \exists\, N_\nabla(x),\ N_\nabla(x) \cap A = \emptyset\ \wedge\ x \in B^C,\ \exists\, N_\nabla(x),\ N_\nabla(x) \cap B = \emptyset$.

$\Rightarrow x \notin \overline{\mathcal{R}}_\nabla(A) \wedge x \notin \overline{\mathcal{R}}_\nabla(B) \Rightarrow x \notin \left(\overline{\mathcal{R}}_\nabla(A) \cup \overline{\mathcal{R}}_\nabla(B)\right)$. Hence, we have $\overline{\mathcal{R}}_\nabla(A \cup B) \supseteq \overline{\mathcal{R}}_\nabla(A) \cup \overline{\mathcal{R}}_\nabla(B)$.

(7) Let $x \in \underline{\mathcal{R}}_\nabla(A) \Longleftrightarrow x \in A, \exists\, N_\nabla(x), N_\nabla(x) \subseteq A \Longleftrightarrow x \in \left(A^C\right)^C,\ \exists\ N_\nabla(x), N_\nabla(x) \cap A^C = \emptyset \Longleftrightarrow x \notin \overline{\mathcal{R}}_\nabla\left(A^C\right) \Longleftrightarrow x \in \left(U - \overline{\mathcal{R}}_\nabla(U - A)\right)$. Hence $\underline{\mathcal{R}}_\nabla(A) = \left(\overline{\mathcal{R}}_\nabla\left(A^C\right)\right)^C$.

(14) Putting U − A for A in (7) we have $\overline{\mathcal{R}}_\nabla(A) = \left(\underline{\mathcal{R}}_\nabla\left(A^C\right)\right)^C$.

(15) Obviously, by (1) and (7) we get $\underline{\mathcal{R}}_\nabla(A) \subseteq \overline{\mathcal{R}}_\nabla(A)$.

Remark 4.1 For any two subsets $A, B \subseteq U$, in the generalized approximation space, $GApp = (U, \mathcal{R})$ and for $\nabla \in \{r, l, m, \cap_r, \cap_l, \cap_{rl}, \cup_{rl}, \cap_{\langle rl\rangle}, \cup_{\langle rl\rangle}\}$ the following properties are not necessarily true:

(1) $\underline{\mathcal{R}}_\nabla(A) = \underline{\mathcal{R}}_\nabla(\underline{\mathcal{R}}_\nabla(A))$.
(2) $\underline{\mathcal{R}}_\nabla(A) = \overline{\mathcal{R}}_\nabla(\underline{\mathcal{R}}_\nabla(A))$.
(3) $A \subseteq \underline{\mathcal{R}}_\nabla(\overline{\mathcal{R}}_\nabla(A))$.
(4) $\underline{\mathcal{R}}_\nabla(A) \subseteq \underline{\mathcal{R}}_\nabla(\underline{\mathcal{R}}_\nabla(A))$.
(5) $\underline{\mathcal{R}}_\nabla(A \cap B) = \underline{\mathcal{R}}_\nabla(A) \cap \underline{\mathcal{R}}_\nabla(B)$.
(6) $\overline{\mathcal{R}}_\nabla(A) = \overline{\mathcal{R}}_\nabla(\overline{\mathcal{R}}_\nabla(A))$.
(7) $\overline{\mathcal{R}}_\nabla(A) = \underline{\mathcal{R}}_\nabla(\overline{\mathcal{R}}_\nabla(A))$.
(8) $A \supseteq \overline{\mathcal{R}}_\nabla(\underline{\mathcal{R}}_\nabla(A))$.
(9) $\overline{\mathcal{R}}_\nabla(A) \supseteq \overline{\mathcal{R}}_\nabla(\overline{\mathcal{R}}_\nabla(A))$.
(10) $\overline{\mathcal{R}}_\nabla(A \cup B) = \overline{\mathcal{R}}_\nabla(A) \cup \overline{\mathcal{R}}_\nabla(B)$.

The following example illustrates the meaning of Remark 1 for $\nabla = m$.

Example 4.1 by recalling Example 2, we have.

(1) For the subset $A = \{d2, d5\}$, thus $\underline{\mathcal{R}}_m(A) = \{d5\}$ and $\underline{\mathcal{R}}_m(\underline{\mathcal{R}}_m A) = \emptyset$. Clearly, $\underline{\mathcal{R}}_m(A) \neq \underline{\mathcal{R}}_m(\underline{\mathcal{R}}_m(A))$.
(2) For the subset $A = \{d1, d3\}$, then $\underline{\mathcal{R}}_m(A) = \{d1, d3\}$ and $\overline{\mathcal{R}}_m(\underline{\mathcal{R}}_m(A)) = \{d1, d2, d3, d4\}$. So, $\underline{\mathcal{R}}_m(A) \neq \overline{\mathcal{R}}_m(\underline{\mathcal{R}}_m(A))$.
(3) For the subset $A = \{d4, d5\}$, therefore, $\overline{\mathcal{R}}_m(A) = \{d2, d4, d5\}$ and $\underline{\mathcal{R}}_m(\overline{\mathcal{R}}_m(A)) = \{d5\}$. Obviously, $A \not\subseteq \underline{\mathcal{R}}_m(\overline{\mathcal{R}}_m(A))$.
(4) For the subset $A = \{d1, d2, d4\}$, then $\underline{\mathcal{R}}_m(A) = \{d1, d4\}$ and $\underline{\mathcal{R}}_m(\underline{\mathcal{R}}_m(A)) = \{d1\}$. Thus, $\underline{\mathcal{R}}_m(A) \not\subseteq \underline{\mathcal{R}}_m(\underline{\mathcal{R}}_m(A))$.
(5) For the subset $A = \{d3, d4, d5\}$, and $B = \{d1, d2, d3, d4\}$. Then $\underline{\mathcal{R}}_m(A) = \{d3, d4, d5\}$, $\underline{\mathcal{R}}_m(B) = \{d1, d2, d3, d4\}$ and $\underline{\mathcal{R}}_m(A) \cap \underline{\mathcal{R}}_m(B) = \{d3, d4\}$. But $\underline{\mathcal{R}}_m(A \cap B) = \underline{\mathcal{R}}_m\{d3, d4\} = \{d3\}$. Therefore, $\underline{\mathcal{R}}_m(A \cap B) \neq \underline{\mathcal{R}}_m(A) \cap \underline{\mathcal{R}}_m(B)$.
(6) For the subset $A = \{d1, d4\}$ then $\overline{\mathcal{R}}_m(A) = \{d1, d2, d4\}$ and $\overline{\mathcal{R}}_m(\overline{\mathcal{R}}_m(A)) = \{d1, d2, d3, d5\}$. Thus, $\overline{\mathcal{R}}_m(A) \neq \overline{\mathcal{R}}_m(\overline{\mathcal{R}}_m(A))$.
(7) For the subset $A = \{d2, d5\}$ then $\overline{\mathcal{R}}_m(A) = \{d2, d4, d5\}$ and $\underline{\mathcal{R}}_m(\overline{\mathcal{R}}_m(A)) = \{d5\}$. So, $\overline{\mathcal{R}}_m(A) \neq \underline{\mathcal{R}}_m(\overline{\mathcal{R}}_m(A))$.
(8) For the subset $A = \{d3, d4, d5\}$ then $\underline{\mathcal{R}}_m(A) = \{d3, d4, d5\}$ and $\overline{\mathcal{R}}_m(\underline{\mathcal{R}}_m(A)) = \{d2, d3, d4, d5\}$. Clearly, $A \not\supseteq \overline{\mathcal{R}}_m(\underline{\mathcal{R}}_m(A))$.
(9) For the subset $A = \{d1, d5\}$ then $\overline{\mathcal{R}}_m(A) = \{d1, d4, d5\}$ and $\overline{\mathcal{R}}_m(\overline{\mathcal{R}}_m(A)) = \{d1, d2, d4, d5\}$. Thus, $\overline{\mathcal{R}}_m(A) \not\supseteq \overline{\mathcal{R}}_m(\overline{\mathcal{R}}_m(A))$.
(10) For the subset $A = \{d1, d4\}$ and $B = \{d2, d3\}$. Then $\overline{\mathcal{R}}_m(A) = \{d1, d2, d4\}$ $\overline{\mathcal{R}}_m(B) = \{d2, d3, d4\}$ and $\overline{\mathcal{R}}_m(A) \cup \overline{\mathcal{R}}_m(B) = \{a, b, c, d\}$. But $\overline{\mathcal{R}}_m(A \cup B) = \overline{\mathcal{R}}_m\{a, b, c, d\} = U$. Therefore, $\overline{\mathcal{R}}_m(A \cup B) \neq \overline{\mathcal{R}}_m(A) \cup \overline{\mathcal{R}}_m(B)$.

For any subset $A \subseteq U$, in the generalized approximation space, $GApp = (U, \mathcal{R})$ and for $\nabla \in \{r, l, m, \cap_r, \cap_l, \cap_{rl}, \cup_{rl}, \cap_{\langle rl\rangle}, \cup_{\langle rl\rangle}\}$, the general membership function is defined as the following way:

- x is $\nabla-$ surely belongs to A, written $x\underline{\in}_\nabla A$ if and only if $x \in \underline{\mathcal{R}}_\nabla(A)$.
- x is $\nabla-$ possibly belongs to A, written $x\overline{\in}_\nabla A$ if and only if $x \in \overline{\mathcal{R}}_\nabla(A)$.

According to this definition of general membership function, we can deduce that.

- If $x\underline{\in}_\nabla A$, then $x \in A$,
- If $x \in A$, then $x\overline{\in}_\nabla A$.

Using general membership function definition we can redefine the lower and upper approximations for any $\nabla \in \{r, l, m, \cap_r, \cap_l, \cap_{rl}, \cup_{rl}, \cap_{\langle rl\rangle}, \cup_{\langle rl\rangle}\}$ as follows:
$\underline{\mathcal{R}}_\nabla(\mathrm{A}) = \{x \in U : x\underline{\in}_\nabla A\}$, $\overline{\mathcal{R}}_\nabla(A) = \{x \in U : x\overline{\in}_\nabla A\}$.

Theorem 4.2 For any subset $A \subseteq U$, in the generalized approximation space, $GApp = (U, \mathcal{R})$, the following properties of the membership function hold:

1. $x\underline{\in}_{\cap_{rl}} A \Rightarrow x\underline{\in}_r A \Rightarrow x\underline{\in}_{\cup_{rl}} A,$
2. $x\underline{\in}_{\cap_{rl}} A \Rightarrow x\underline{\in}_l A \Rightarrow x\underline{\in}_{\cup_{rl}} A,$
3. $x\overline{\in}_{\cup_{rl}} A \Rightarrow x\overline{\in}_r A \Rightarrow x\overline{\in}_{\cap_{rl}} A,$
4. $x\overline{\in}_{\cup_{rl}} A \Rightarrow x\overline{\in}_l A \Rightarrow x\overline{\in}_{\cap_{rl}} A,$
5. $x\underline{\in}_{\cap_{<rl>}} A \Rightarrow x\underline{\in}_{\cap r} A \Rightarrow x\underline{\in}_{\cup_{<rl>}} A,$
6. $x\underline{\in}_{\cap_{<rl>}} A \Rightarrow \underline{\in}_{\cap l} A \Rightarrow x\underline{\in}_{\cup_{<rl>}} A,$
7. $x\overline{\in}_{\cup_{<rl>}} A \Rightarrow x\overline{\in}_{\cap r} A \Rightarrow x\overline{\in}_{\cap_{<rl>}} A,$
8. $x\overline{\in}_{\cup_{<rl>}} A \Rightarrow x\overline{\in}_{\cap l} A \Rightarrow x\overline{\in}_{\cap_{<rl>}} A.$

Proof

1. Since $x\underline{\in}_{\cap_{rl}} A \Rightarrow x \in \underline{\mathcal{R}}_{\cap_{rl}}(\mathrm{A})$, then $x \in \underline{\mathcal{R}}_r(A)$, hence $x\underline{\in}_r A$. Also, $x\underline{\in}_r A$, then $x \in \underline{\mathcal{R}}_r(A) \Rightarrow x \in \underline{\mathcal{R}}_{\cap_{rl}}(\mathrm{A})$, then $x\underline{\in}_{\cup_{rl}} A$.

By the same manner the rest of the theorem.

5 Construction of Information Retrieval System

The basic objective of developing an information retrieval system is to reduce the stress of a user for obtaining needed information. This stress can be spoken as the time a user applies in the entire stepladders primarily to interpret an article covering the needed information. The achievement of an information system is actually particular, based upon what information is desirable and the readiness of a user to accept results.

In the information retrieval system, the expression "is relevant to" is used to appear the results containing the query of the user. In fact, the expression of "is relevant to" is not a binary relation but it is a continuous function. From a system point of view, the information must be suitable for the criteria of the seeking query.

Information retrieval is classically two basic stages are as follows:

- In the first, we identify the possibly relevant documents.
- Second, we ranked the initiated documents.

Each information retrieval system has a numerous mechanisms as follows:

Indexing: A pre-process called indexing is required for documents from a corpus to match a given query. To make searching more effective, a retrieval system stores documents in a mental representation.

The greatest general data construction working by information retrieval systems is called the inverted file (IF). Every entrance in the inverted file covers information about only one term in the document group.

The indexing procedure includes numerous steps, which are defined as follows:

Coding: In this stage, documents text is analyzed and index words named codes are produced. Furthermore, at this phase, all characters controlled in the signs are often lower-cased and all punctuations are detached.

Removal of stop words: There are several common terms (e.g., "the", "an", "on", "of", … and so on) that seem in almost all documents of a corpus. Removing the stop words lets also decrease of the size of the produced document index. We eliminate only non-relational stop words to achieve relation comprehensive searching.

Stemming: It would be useful for retrieval if documents comprising alternatives of the query stretch were contained in an applicable document. Plurals, gerund forms, and past tense suffixes are instances of syntactical differences that avert a faultless competition between a query tenure and an individual document tenure.

Index Data Structure: The most generally used data structure is the overturned index, which is a word-oriented instrument. Overall, the reversed index construction covers two mechanisms: language and situation list. The terminology is a set of all dissimilar terms removed from the corpus by the overhead steps.

Query Parser: It achieves codes, stemming, and stop words elimination processes on the query so that it would be informal to achieve corresponding on indexed documents for these query footings.

Matching: Several information retrieval models, such as the Boolean model, vector space model, and probabilistic model, can approximate the significance of a document to a given enquiry.

Ranking: Completely the retrieved documents are ranked rendering to their implication groove using the produced educated ranking meaning.

User Interface: Interface achieves communication with the user by the attractive query as effort and showing documents rendering to their relevance notch as production (Fig. 1).

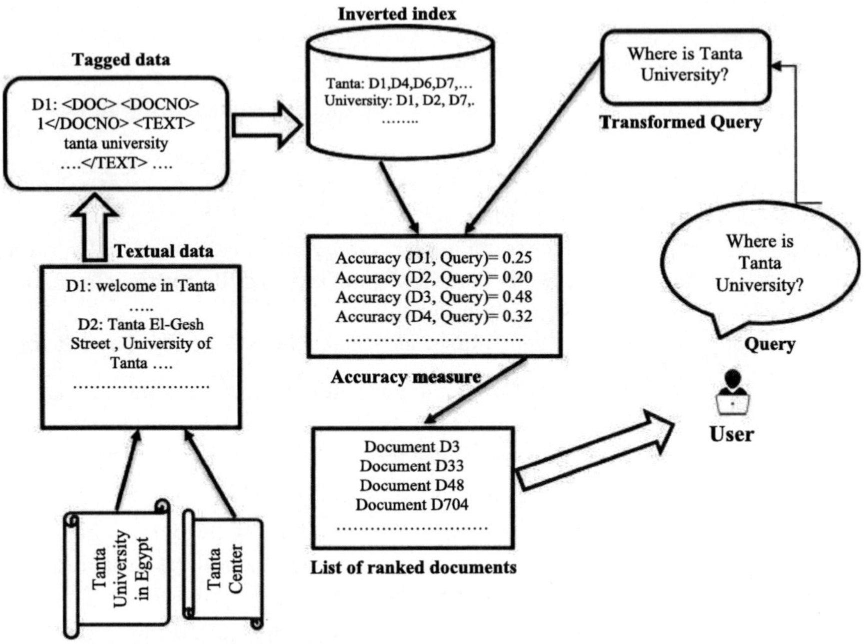

Fig. 1 Construction of information retrieval system

6 Query Expansion Techniques for Information Retrieval

In this section, we used the queries as keywords for searching in the corpus of documents. We need first to illustrate some basic definitions of the term frequency (tf) and the inverse document frequency (idf).

Tf-idf stances for *termfrequency-inverse documentfrequency* and the tf-idf weight is a heaviness frequently used in information retrieval and text removal. This weight is a mathematical amount used to assess how significant a word is to a document in an assortment or quantity. The rank surges proportionately to the number of times a word seems in the document but is offset by the incidence of the word in the corpus. Search engines often use differences of the tf-idf weighting structure as a vital implement in counting and positioning a document's significance agreed in a user query.

Some of the humblest position purposes are computed by adding the tf-idf for each query period; many more classy position functions are alternatives to this modest model.

Tf-idf can be positively used for stop words sifting in numerous subject fields with text summarization and organization.

Naturally, the tf-idf mass is collected by binary relationships: the first calculates the regularized Term Frequency (TF) (the amount of iterations of a word that appears in a text) that is divided by the entire number of words in that text. The second part is the Inverse Document Frequency (IDF), which is intended as the logarithm of the number of documents in the quantity separated by the number of documents where the exact term seems.

For instance, suppose that a document containing 1000 words where the word "university" appears 30 times. The term frequency (i.e., tf) for "university" is then (30/1000) = 0.03. Currently, suppose that we have 1,000,000 documents, and the word "university" seems in 10 thousands of these. Then, the inverse document frequency (i.e., idf) is considered as log(1,000,000 / 10,000) = 2. Therefore, the Tf-idf weight is the multiply of these numbers: 0.03 * 2 = 0.06.

7 Conclusion

In this paper, we proved that the approximations based on mixed neighborhood systems are accurate than the approximations based on either right neighborhood systems or left neighborhood systems. Furthermore, we believe that our definition of the accuracy measure is accurate than Pawlak's definition since our definition considers the negative region and Pawlak's definition does not consider it.

References

1. Abu-Donia, H.M.: Comparison between different kinds of approximations by using a family of binary relations. Knowl.-Based Syst. **21**, 911–919 (2008)
2. Abu-Donia, H.M., Nasef, A.A., Marei, E.A.: Finite information systems. Appl. Math. Inf. Sci. **1**, 13–21 (2007)
3. Abu-Donia, H.M., Salama, A.S.: Generalization of Pawlak's rough approximation spaces by using δβ-open sets. Int. J. Approx. Reas. **53**(7), 1094–1105 (2012)
4. Abu-Donia, H.M., Salama, A.S.: Fuzzy simple expansion. J. King Saud Univ. - Sci **22**(4), 223–227 (2010)
5. Al-shami, T.M.: Somewhere dense sets and ST1-spaces, Punjab university. J. Math. **49**(2), 101–111 (2017)
6. Al-shami, T.M., Noiri, T.: More notions and mappings via somewhere dense sets. Afr. Mat. **30**(7), 1011–1024 (2019)
7. Lashin, E,F., Kozae, A.M., Abo Khadra, A.A., Medhat, T.: Rough set theory for topological spaces. Int. J. Approx. Reason. **40**, 35–43 (2005)
8. Salama, A.S.: Topological solution of missing attribute values problem in incomplete information tables. Inf. Sci. **180**, 631–639 (2010)
9. Salama, A.S.: Generalizations of rough sets using two topological spaces with medical applications. Information **19**(7A), 2425–2440 (2016)
10. Salama, A.S.: Sequences of topological near open and near closed sets with rough applications. Filomat **34**(1), 51–58 (2020)
11. Salama, A.S.: Bitopological approximation space with application to data reduction in multi-valued information systems. Filomat **34**(1), 99–110 (2020)
12. Salama, A.S., Abd El-Monsef, M.M.E.: New topological approach of rough set generalizations. Int. J. Comput. Math. **88**(7), 1347–1357 (2011)
13. Salama, A.S., Abd El-Monsef, M.M.E.: Generalizations of rough set concepts. J. King Saud Univ. – Sci. **23**(1), 17–21 (2011)

14. Salama, A.S., El Barbary, O.G.: Future applications of topology by computer programming. Life Sci. J. **11**(4), 168–172, 25 (2014)
15. Salama, A.S., El Barbary, O.G.: New topological approach to vocabulary mining and document classification. Life Sci. J. **11**(5), 84–91 (2014)
16. Salama, A.S., El Barbary, O.G.: Topological approach to retrieve missing values in incomplete information systems. J. Egyptian Math. Soc. 1–5 (2017)
17. Salama, A.S.: Two new topological rough operators. J. Interdiscip. Math. **11**(1), 1–10 (2008)
18. Salama, A.S.: Topologies induced by relations with applications. J. Comput. Sci. **4**(10), 877–887 (2008)
19. Salama, A.S., Abd El-Monsef, M.M.E.: Generalizations of rough set concepts. J. King Saud Univ. – Sci. **23**(1), 17–21 (2011)
20. Salama, A.S., Abd El-Monsef, M.M.E.: New topological approach of rough set generalizations. Int. J. Comput. Math. **88**(7), 1347–1357 (2011)
21. El Barbary, O.G., Salama, A.S.: Feature selection for document classification based on topology. Egyptian Inf. J. article in press (2019)
22. El Barbary, O.G., Salama, A.S., Sayed Atlam, E.l.: Granular information retrieval using neighborhood systems. Math. Meth. Appl. Sci. 1–17 (2017)
23. Pawlak, Z.: Rough sets. Int. J. Comput. Inform. Sci. **11**, 341–356 (1982)
24. Wu, H., Luk, R., Wong, K., Kwok, K.: Interpreting TF-IDF term weights as making relevance decisions. ACM Trans. Inf. Syst. **26**(3) (2008)

Applying "Emad-Sara" Transform on Partial Differential Equations

Emad A. Kuffi, Elaf Sabah Abbas, and Sara Falih Maktoof

Abstract This work demonstrates the "Emad-Sara (ES)" integral transform, where its basic properties and its capability to find a particular solution for partial differential equations have been presented and proven via the solution of multiple fundamental physical partial differential equations.

Keywords Emad-Sara (ES) transform · First-order differential equations · Second-order differential equations · Five fundamental mathematical physics equations · Wave · Heat · Laplace's · Telegraph · Klein-Gordan

1 Introduction

Partial differential equations (PDE) represent a special case of ordinary differential equations, with multiple partial derivatives of unknown variables. PDE degree is identified by the highest derivative that appears in the equation. Applying a mathematical method that could solve PDE concludes a function converts to identity when substituted into the equation. PDEs have been used in various scientific fields, which yield from their ability to express physical problems in a mathematical formula that can be manipulated and solved via some mathematical method [1–3].

The significance of PDEs necessitated using the most effective mathematical methods for their solution [4–7]. Integral transforms' ability to transform problems from one domain to another to simplify their solution has positioned them as a priority in the domain of PDE solution. Mathematicians have proposed numerous

E. A. Kuffi (✉)
Department of Material Engineering, College of Engineering, Al-Qadisiyah University, Al Diwaniyah, Iraq
e-mail: emad.abbas@qu.edu.iq

E. Sabah Abbas
Communication Engineering Department, Al-Mansour University College, Baghdad, Iraq
e-mail: elaf.abbas@muc.edu.iq

S. F. Maktoof
Department of Mathematics, Faculty for Girls, University of Kufa, Kufa, Iraq

D. Zeidan et al. (eds.), *Mathematics and Computation*, Springer Proceedings in Mathematics & Statistics 418, https://doi.org/10.1007/978-981-99-0447-1_2

integral transforms to solve PDEs; each proposed transform has particular cases where it shines [8–13]. The substantial field of partial differential equations, on the other hand, has not yet benefited from the revolutionary Emad-Sara (ES) integral transform.

The ES integral transform is used in this work to solve first- and second-order differential equations, as well as several practical applications of differential equations, which are regarded basic in the mathematical physical area.

2 Fundamental Properties of Emad-Sara Transform

The Emad-Sara (ES) transform is defined for a function $f(t)$ as [14]:

$$ES[f(t)] = T(\alpha) = \frac{1}{\alpha^2}\int_0^{\infty} f(t)e^{-\alpha t}dt, \tag{1}$$

2.1 Emad-Sara Transform Existence [14]

ES transform is considered to exist for sufficiently large v, providing the integral:

$$\frac{1}{v^2}\int_{t=0}^{\infty} f(t)e^{-vt}dt = \lim_{p\to\infty}\int_{t=0}^{p} f(t)e^{-vt}dt.$$

Criteria for Convergence (I)

ES transform for the function $f(t)$ exist, if it has exponential order and $\int_0^p |f(t)|dt$ exist for any $p > 0$.

Since the convergence is needed to be shown only for sufficiently large v, then it is going to be assumed that $v > c and v > 0$.

$$\frac{1}{v^2}\int_0^{\infty} \left|f(t)e^{-vt}\right|dt = \frac{1}{v^2}\left[\int_0^{n} \left|f(t)e^{-vt}\right|dt + \int_n^{\infty} \left|f(t)e^{-vt}\right|dt\right],$$

$$\leq \frac{1}{v^2}\left[\int_0^{n} |f(t)|dt + \int_n^{\infty} e^{-vt}|f(t)|dt\right]$$

For: $\left[0 < \frac{1}{v2}e^{-vt} \leq 1\right]$

$$\leq \frac{1}{v^2}\left[\int_0^n |f(t)|dt + \int_n^\infty e^{-vt} M e^{ct} dt\right], \text{ (exponential order)}.$$

$$= \frac{1}{v^2}\left[\int_0^n |f(t)|dt + M\left[\frac{e^{(c-v)t}}{c-v}\Big|_n^\infty\right]\right].$$

For $v > c$

$$= \frac{1}{v^2}\left[\int_0^n |f(t)|dt + \mathrm{M}\frac{e^{(c-v)n}}{v-c}\right].$$

The first integral exists by assumption, and the second term is finite $v > c$.

The integral $\frac{1}{v^2}\int_0^\infty f(t)e^{-vt}dt$, converges absolutely and $ES\{f(t)\}$ exists.

Criteria for Convergence (II)

To satisfy criterion (I), $ES\{f(t)\}$ exists if:

- $f(t)$is of exponential order and on the closed interval $[0, p]$.
- $f(t)$ is bounded, piecewise, continuous and has a finite number of discontinuous requirements implying that $\int_b^0 |f(t)|dt$.

where $F(v) \to 0$ as $v \to \infty$.

Assuming $f(t)$ satisfy criterion (I), which implies $F(v) = ES\{f(t)\}$ will exist if $v \geq m$ for some m.

$|F(v)| = \left|\frac{1}{v^2}\int_0^\infty f(t)e^{-vt}dt\right| \leq \int_0^\infty \left|f(t)e^{-vt}\right|dt = G(v)v \to \infty, \frac{1}{v^2}e^{-vt} \to 0\ for\ t \geq 0.$

2.2 *Emad-Sara Transform Uniqueness [14]*

Suppose that the functions f and g are exponential type b, piecewise and continuous on the interval $[0, \infty)$. If $ES\{f(t)\} = ES\{g(t)\}$ when $s > b$, then $f(t) = g(t)$ for all t greater than or equal to zero.

2.3 *Derivation of the Emad-Sara Transform of Derivatives [14]*

Integration by parts is used to obtain the ES integral transform for partial derivatives, as follows:

$$ES\left[\frac{\partial f}{\partial t}(x,t)\right] = \int_0^\infty \frac{1}{\alpha^2}\frac{\partial f}{\partial t} e^{-\alpha t} dt = \lim_{p\to\infty} \int_0^p \frac{1}{\alpha^2}\frac{\partial f}{\partial t} e^{-\alpha t} dt$$

$$= \lim_{p\to\infty}\left\{\left[\frac{1}{\alpha^2}e^{-\alpha t}f(x,t)\right]_0^p + \frac{1}{\alpha}\int_0^p f(x,t)e^{-\alpha t}dt\right\}$$

$$= \alpha T(x,\alpha) - \frac{f(x,0)}{\alpha^2}.$$

$$ES\left[\frac{\partial f}{\partial t}(x,t)\right] = \alpha T(x,\alpha) - \frac{1}{\alpha^2}f(x,0). \tag{2}$$

Assuming the function f is a continuous and of exponential order, then:

$$ES\left[\frac{\partial f}{\partial x}\right] = \int_0^\infty \frac{1}{\alpha^2}e^{-\alpha t}\frac{\partial f}{\partial x}(x,t)dt = \frac{\partial}{\partial x}\int_0^\infty \frac{1}{\alpha^2}e^{-\alpha t}f(x,t)dt,$$

Using the Leibnitz rule:

$$ES\left[\frac{\partial f}{\partial x}\right] = \frac{\partial}{\partial x}[T(x,\alpha)],$$

$$ES\left[\frac{\partial f}{\partial x}\right] = \frac{d}{dx}[T(x,\alpha)]. \tag{3}$$

It is also possible to find:

$$ES\left[\frac{\partial^2 f}{\partial x^2}\right] = \frac{d^2}{dx^2}[T(x,\alpha)]. \tag{4}$$

To find $ES\left[\frac{\partial^2 f}{\partial t^2}(x,t)\right]$

Let $\frac{\partial f}{\partial t} = g$, then, by using Eq. (2):

$$ES\left[\frac{\partial^2 f}{\partial t^2}(x,t)\right] = ES\left[\frac{\partial g}{\partial t}(x,t)\right] = \alpha ES[g(x,t)] - \frac{1}{\alpha^2}g(x,0).$$

$$\therefore ES\left[\frac{\partial^2 f}{\partial t^2}(x,t)\right] = \alpha^2 T(x,\alpha) - \frac{1}{\alpha^2}\frac{\partial f}{\partial t}(x,0) - \frac{f(x,0)}{\alpha}. \tag{5}$$

In the same way, it is possible to extend this result to the nth partial derivative using mathematical induction.

3 Solving Some Partial Differential Equations Using ES Transform

The solutions to some first and second-order differential equations, as well as the five fundamental mathematical physics equations: wave, heat, Laplace's, telegraph, and Klein-Gordan, are demonstrated in this section.

Problem 1

Consider the first-order initial value problem (IVP):

$$\left.\begin{array}{r} u_x = 2u_t + u, \\ with\ u(x,0) = 6e^{-3x} \end{array}\right\} \quad (6)$$

And u is bounded for $x, t > 0$.

Solution:

Let T be the Emad-Sara (ES) transform of u.

Taking ES transform to Eq. (6), gives.

$\frac{dT(x,\alpha)}{dx} - 2\alpha T(x,\alpha) + \frac{2}{\alpha^2}u(x,0) = T(x,\alpha)$, this equation is a linear first-order ordinary differential equation.

$$\frac{dT(x,\alpha)}{dx} - (2\alpha+1)T(x,\alpha) = \frac{-12}{\alpha^2}e^{-3x},$$

The integral factor is $P = e^{-\int(2\alpha+1)dx} = e^{-(2\alpha+1)x}$,

Therefore, $T(x,\alpha) = \frac{1}{P}\int P.Qdx$,

$$\text{Then, } T(x,\alpha) = e^{(2\alpha+1)x}\int e^{-(2\alpha+1)x}\left(\frac{-12}{\alpha^2}\right)e^{-3x}dx$$

$$T(x,\alpha) = e^{(2\alpha+1)x}\left[\frac{-12}{\alpha^2}\int e^{-2(\alpha+2)x}dx\right]$$

$$T(x,\alpha) = e^{(2\alpha+1)x}\left[\frac{-12}{\alpha^2(-2\alpha-4)}e^{-2(\alpha+2)x} + C\right]$$

$$T(x,\alpha) = e^{(2\alpha+1)x}\left[\frac{6}{\alpha^3+2\alpha^2}e^{-2(\alpha+2)x} + C\right]$$

$$\therefore T(x,\alpha) = \frac{6}{\alpha^2(\alpha+2)}e^{-3x} + Ce^{(2\alpha+1)x},$$

Since T is bounded, then C should be equal to zero.

Taking inverse ES transform gives.

$$T(x,t) = 6e^{-3x}\cdot e^{-2t} \Longrightarrow T(x,t) = 6e^{-(3x+2t)}.$$

Problem 2

Consider the Laplace equation:

$$\left.\begin{aligned} u_{xx} + u_{tt} = 0, u(x,0) = 0 \\ u_t(x,0) = \cos(x), x, t > 0 \end{aligned}\right\} \tag{7}$$

Solution:

Let $T(\alpha)$ be the ES transform of u.

Taking ES transform to Eq. (7), gives.

$$T''(x,\alpha) + \alpha^2 T(x,\alpha) - \frac{1}{\alpha^2}\frac{\partial u}{\partial t}(x,0) - \frac{u(x,0)}{\alpha} = 0,$$
$$\frac{1}{\alpha^2}T''(x,\alpha) + T(x,\alpha) = \frac{\cos(x)}{\alpha^4}.$$

The concluded equation is a second-order nonhomogeneous ordinary differential equation that has a particular solution in the following form:

$$\begin{aligned} T(x,\alpha) &= \frac{\frac{1}{\alpha^4}\cos(x)}{\frac{1}{\alpha^2}D^2 + 1} = \frac{\frac{1}{\alpha^4}\cos(x)}{\frac{1}{\alpha^2}(-1) + 1}, \\ &= \frac{\frac{1}{\alpha^4}\cos(x)}{1 - \frac{1}{\alpha^2}} = \frac{1}{\alpha^2(\alpha^2 - 1)}\cos(x). \end{aligned} \tag{8}$$

where, $D^2 \equiv \frac{d^2}{dx^2}$.

Applying the inverse ES transform to Eq. (8) produces the solution to Eq. (7) in the form:

$$u(x,t) = \sinh(t)\cos(x).$$

Or

$$u(x,t) = \frac{1}{2}\left(e^t - e^{-t}\right)\cos(x) = \frac{1}{2}e^t\cos(x) - \frac{1}{2}e^{-t}\cos(x).$$

Problem 3

Consider the wave equation:

$$\left.\begin{aligned} u_{xx} - 4u_{tt} = 0, u(x,0) = \sin(\pi x) \\ u_t(x,0) = 0, x, t > 0 \end{aligned}\right\} \tag{9}$$

Solution:

Applying the ES transform to Eq. (9) and using the conditions provided, obtains

$$T''(x,\alpha) - 4\left[\alpha^2 T(x,\alpha) - \frac{1}{\alpha^2}u_t(x,0) - \frac{u(x,0)}{\alpha}\right] = 0,$$

$$\frac{1}{4\alpha^2}T''(x,\alpha) - T(x,\alpha) = \frac{-\sin(\pi x)}{\alpha^3},$$

$$T(x,\alpha) = \frac{\frac{-1}{\alpha^3}\sin(\pi x)}{\frac{1}{4\alpha^2}D^2 - 1} = \frac{\frac{-1}{\alpha^3}\sin(\pi x)}{\frac{1}{4\alpha^2}(-\pi)^2 - 1},$$

$$T(x,\alpha) = \frac{\frac{-1}{\alpha^3}\sin(\pi x)}{\frac{-\pi^2}{4\alpha^2} - 1} = \frac{1}{\alpha^3}\frac{4\alpha^2}{\pi^2 + 4\alpha^2}\sin(\pi x),$$

$$T(x,\alpha) = \frac{\alpha}{\alpha^2\left[\alpha^2 + \left(\frac{\pi}{2}\right)^2\right]}\sin(\pi x).$$

Applying the inverse ES transform produce the particular solution of Eq. (9) in the form:

$$u(x,t) = \cos\left(\frac{\pi}{2}t\right)\sin(\pi x).$$

Problem 4

Consider the homogeneous heat equation:

$$\left.\begin{array}{c} 4\frac{\partial u}{\partial t} = \frac{\partial^2 u}{\partial x^2},\ u(x,0) = \sin\left(\frac{\pi}{2}x\right) \\ x,t > 0 \end{array}\right\} \tag{10}$$

Solution:

Applying ES transform on Eq. (10), gives

$$T''(x,\alpha) - 4\left[\alpha T(x,\alpha) - \frac{u(x,0)}{\alpha^2}\right] = 0,$$

$$T''(x,\alpha) - 4\alpha T(x,\alpha) = \frac{-4\sin\left(\frac{\pi}{2}x\right)}{\alpha^2},$$

Now,

$$T(x,\alpha) = \frac{\frac{-4\sin\left(\frac{\pi}{2}x\right)}{\alpha^2}}{D^2 - 4\alpha} = \frac{\frac{-4}{\alpha^2}\sin\left(\frac{\pi}{2}x\right)}{-\left(\frac{\pi}{2}\right)^2 - 4\alpha},$$

$$T(x,\alpha)=\frac{\frac{-4}{\alpha^2}\sin\left(\frac{\pi}{2}x\right)}{\frac{-\pi^2-16\alpha}{4}}=\frac{16}{\alpha^2\left(\pi^2+16\alpha\right)}\sin\left(\frac{\pi}{2}x\right)=\frac{1}{\alpha^2}\left(\frac{1}{\alpha+\left(\frac{\pi}{4}\right)^2}\right)\sin\left(\frac{\pi}{2}x\right). \tag{11}$$

Applying the inverse ES transform to Eq. (11) produces the solution to Eq. (10) in the form:

$$u(x,t)=e^{\frac{-\pi^2}{16}t}.\sin\left(\frac{\pi}{2}x\right).$$

Problem 5

Consider the linear telegraph equation:

$$\left.\begin{array}{c} u_{xx}=u_{tt}+2u_t+u \\ \text{Subject to the inetial conditions} \\ u(x,0)=e^x,\, u_t(x,0)=-2e^x \end{array}\right\} \tag{12}$$

Solution:

Applying the ES transform to Eq. (12), obtains the following:

$T''(x,\alpha)-\alpha^2T(x,\alpha)+\frac{1}{\alpha^2}u_t(x,0)+\frac{1}{\alpha}u(x,0)-2\alpha T(x,\alpha)+\frac{2}{\alpha^2}u(x,0)-T(x,\alpha)=0,$

Providing the initial conditions to the concluded equation gives

$$T''(x,\alpha)-\alpha^2T(x,\alpha)+\frac{1}{\alpha^2}\left(-2e^x\right)+\frac{e^x}{\alpha}-2\alpha T(x,\alpha)+\frac{2e^x}{\alpha^2}-T(x,\alpha)=0,$$

$$\Rightarrow T''(x,\alpha)-\left(\alpha^2+2\alpha+1\right)T(x,\alpha)=\frac{-e^x}{\alpha},$$

$$\Rightarrow \frac{1}{\alpha^2+2\alpha+1}T''(x,\alpha)-T(x,\alpha)=\frac{-e^x}{\alpha\left(\alpha^2+2\alpha+1\right)},$$

$$\Rightarrow \frac{1}{(\alpha+1)^2}T''(x,\alpha)-T(x,\alpha)=\frac{-e^x}{\alpha(\alpha+1)^2},$$

$$\Rightarrow T(x,\alpha)=\frac{\frac{-e^x}{\alpha(\alpha+1)^2}}{\frac{1}{(\alpha+1)^2}D^2-1},$$

$$T(x,\alpha)=\frac{e^x}{\alpha^2(\alpha+2)}. \tag{13}$$

Applying the inverse ES transform to Eq. (13) produces the solution to Eq. (12) in the form:

$$u(x,t)=e^x.e^{-2t}=e^{x-2t}.$$

Problem 6

Consider the second-order linear homogeneous Klein-Gordan equation:

$$\left.\begin{array}{c} u_{tt} = u_{xx} + u_x + 2u, -\infty < x < \infty, t > 0 \\ \text{Subject to the inetial conditions} \\ u(x,0) = e^x, u_t(x,0) = 0 \end{array}\right\} \quad (14)$$

Solution:

Using ES transform on Eq. (14), gives

$$\alpha^2 T(x,\alpha) - \frac{1}{\alpha^2} u_t(x,0) - \frac{u(x,0)}{\alpha} - T''(x,\alpha) - T'(x,\alpha) - 2T(x,\alpha) = 0,$$

$$\alpha^2 T(x,\alpha) - \frac{e^x}{\alpha} - T''(x,\alpha) - T'(x,\alpha) - 2T(x,\alpha) = 0,$$

$$T''(x,\alpha) + T'(x,\alpha) - \left(v^2 - 2\right)T(x,\alpha) = \frac{-e^x}{\alpha},$$

$$\frac{1}{\left(\alpha^2 - 2\right)} T''(x,\alpha) + \frac{1}{\left(\alpha^2 - 2\right)} T'(x,\alpha) - T(x,\alpha) = \frac{-e^x}{\alpha\left(\alpha^2 - 2\right)},$$

$$T(x,\alpha) = \frac{\frac{-e^x}{\alpha\left(\alpha^2 - 2\right)}}{\frac{1}{\left(\alpha^2 - 2\right)} D^2 + \frac{1}{\left(\alpha^2 - 2\right)} D - 1},$$

$$T(x,\alpha) = \frac{-e^x}{\alpha\left[1 + 1 - \left(\alpha^2 - 2\right)\right]},$$

$$T(x,\alpha) = \frac{-e^x}{\alpha\left(2 - \alpha^2 + 2\right)} = \frac{-e^x}{\alpha\left(-\alpha^2 + 4\right)},$$

$$\therefore T(x,\alpha) = \frac{\alpha e^x}{\alpha^2\left(\alpha^2 - 4\right)}. \quad (15)$$

Applying the inverse ES to Eq. (15) gives the solution to Eq. (14) in the form:

$$u(x,\alpha) = \cosh(2t)e^x.$$

Or

$$u(x,\alpha) = \frac{1}{2}\left[e^{2t} + e^{-2t}\right]e^x = \frac{1}{2}e^{2t+x} + \frac{1}{2}e^{-2t+x}.$$

4 Conclusion

The novel integral Sara-Emad (ES) integral transform has been applied to solve partial differential equations. The proofs that accompanied applying the SE transform to partial differential equations and the solution of a practical example solidify the SE integral transform's ability to efficiently handle and provide the solution to the PDEs, making it a strong competitor to other integral transforms in solving partial differential equations.

References

1. Le Dret, H., Lucquin, B.: Partial Differential Equations: Modeling, Analysis and Numerical Approximation, 1st edn. Birkhäuser (2016)
2. Hillen, T., Leonard, I.E., van Roessel, H.: Partial Differential Equations: Theory and Completely Solved Problems, 1st edn. Wiley (2012)
3. Xie, W-C.: Differential Equations for Engineers. Cambridge University Press (2010)
4. Tatari, M., Dehghan, M.: A method for solving partial differential equations via radial basis functions: application to the heat equation. Eng. Anal. Boundary Elem. **34**(3), 206–212 (2010)
5. Bhatia, G.S., Arora, G.: Radial basis function methods for solving partial differential equations-a review. Indian J. Sci. Technol. **9**(45)
6. Abdel-Hassan, I.H.: Differential transformation technique for solving higher-order initial value problem. Appl. Math. Comput. **154**(2), 299–311 (2004)
7. Kilicman, A., Gadain, H.E.: A note on integral transforms and partial differential equations. Malaysian J. Math. Sci. **4**(1), 109–118 (2010)
8. Ahmed, S.A., Elzaki, T.M., Elbadri, M., Mohamed, M.Z.: Solution of partial differential equations by new double integral transform (Laplace - Sumudu transform). Ain Shams Eng. J. (2021)
9. Atangana, A., Noutchie, S.C.O.: On Multi-Laplace Transform for Solving Nonlinear Partial Differential Equations with Mixed Derivatives, Hindawi: Mathematical Problems in Engineering (2014)
10. Zhou, Z., Gao, X.: Laplace Transform Methods for a Free Boundary Problem of Time-Fractional Partial Differential Equation System. Discrete Dynamics in Nature and Society, Hindawi (2017)
11. Poonia, S.: Solution of differential equation using by Sumudu transform. Int. J. Math. Comput. Res. **2**(1), 316–323 (2013)
12. Elzaki, T.M., Ezaki, S.M.: Application of new transform "Elzaki Transform" to partial differential equations. Glob. J. Pure Appl. Math. **7**(1), 65–70 (2011)
13. Gupta, A.R., Aggarwal, S., Agrawal, D.: Solution of linear partial integro-differential equations using Kamal transform. Int. J. Latest Technol. Eng. Manage. Appl. Sci. **7**(7), 88–91 (2018)
14. Maktoof, S.F., Kuffi, E., Abbas, E.S.: "Emad-Sara Transform" a new integral transform. J. Interdiscip. Math. **24**(3), 2021 (2021)

Estimations of the Bounds for the Zeros of Polynomials Using Matrices

Ahmad Al-Swaftah, Aliaa Burqan, and Mona Khandaqji

Abstract Let $p(z) = z^n + \alpha_n z^{n-1} + \alpha_{n-2} z^{n-2} + \cdots + \alpha_2 z + \alpha_1$ be a monic polynomial of degree $n \geq 7$ with complex coefficients $\alpha_n, \alpha_{n-1}, \ldots.\alpha_1$, where $\alpha_1 \neq 0$. This paper investigates and estimates the upper bounds for the moduli of the zeros of p depending on the spectral norms, spectral radii, and the fifth power of the Frobenius companion. These upper bounds allow us to locate all the zeros of p in smaller annuli in the complex plane.

Keywords Bounds for the zeros of polynomials · Companion matrix · Spectral radius

2000 Mathematics Subject Classification 26A33 · 41A58

1 Introduction

Locating the zeros of polynomials is essential in many fields of study, including signal processing, control theory, communication theory, coding theory, and cryptography. Beginning with Cauchy, this classic problem drew a large number of mathematicians across time. Recently, several famous classical upper bounds for the moduli of the zeros of the monic complex polynomials have been established using the Frobenius companion matrix, which is a key connection between matrix theory and polynomial geometry. These bounds include Cauchy's bound [2], Carmichael and Mason's bound, Montel's bound [2] and Fujii and Kubo's bound [3]. In this paper, we will give a new estimate for the zeros of polynomials using the spectral norm and the spectral radius for the fifth power of the Frobenius companion matrix.

A. Al-Swaftah · A. Burqan (✉)
Department of Mathematics, Zarqa University, Zarqa, Jordan
e-mail: aliaaburqan@zu.edu.jo

M. Khandaqji
Department of Mathematics, Applied Science Private University, Amman, Jordan

D. Zeidan et al. (eds.), *Mathematics and Computation*, Springer Proceedings in Mathematics & Statistics 418, https://doi.org/10.1007/978-981-99-0447-1_3

Let $M_n(\mathbb{C})$ stands for the algebra of all $n \times n$ complex matrices. For $A \in M_n(\mathbb{C})$, the eigenvalues of A are denoted by $\lambda_i(A)$, for $i = 1, 2, \ldots, n$, arranged in such a way that

$$|\lambda_1(A)| \geq |\lambda_2(A)| \geq \cdots \geq |\lambda_n(A)|.$$

The singular values of A, (the eigenvalues of $|A| = (A^*A)^{\frac{1}{2}}$) are denoted by $s_i(A)$, $(1 \leq i \leq n)$, where they are arranged in such a way that

$$s_1(A) \geq s_2(A) \geq \cdots \geq s_n(A).$$

Recall that $s_i^2(A) = \lambda_j(A^*A) = \lambda_j(AA^*)$, for $j = 1, 2, \ldots, n$ and $s_1(A) = \|A\|$, where $\|A\|$ represent the spectral norm of A. For $A \in M_n(\mathbb{C})$, if λ is the eigenvalue of A and $r(A)$ represents the spectral radius of A, then for any matrix norm $|||\cdot|||$, we have

$$|\lambda| \leq r(A) \leq |||A|||.$$

Let $p(z) = z^n + \alpha_n z^{n-1} + \alpha_{n-2} z^{n-2} + \cdots + \alpha_2 z + \alpha_1$ be a monic polynomial of degree $n \geq 7$ with complex coefficients $\alpha_n, \alpha_{n-1}, \ldots .\alpha_1$, where $\alpha_1 \neq 0$. The following matrix

$$C = \begin{bmatrix} -\alpha_n & -\alpha_{n-1} & \cdots & -\alpha_2 & -\alpha_1 \\ 1 & 0 & \cdots & 0 & 0 \\ 0 & 1 & \cdots & 0 & 0 \\ \vdots & \vdots & \ddots & \vdots & \vdots \\ 0 & 0 & \cdots & 1 & 0 \end{bmatrix}_{n \times n}$$

is called the Frobenius companion matrix for p. It is well known that the characteristic polynomial of C is p itself, and so the eigenvalues of C are the zeros of p, see [4]. Using the fact that the eigenvalues of C are the roots of $p(z) = 0$, then for any matrix norm $|||\cdot|||$, $|z| \leq |||C|||$, where z is the zero of the monic polynomial p. Many mathematicians in the area have used Frobenius companion matrix C to derive bounds for the moduli of the zeros of the polynomial p; we list some of them below. Let z be any zero of p, then we note that some bounds are obtained by the classical approach.

Cauchy [2], proved that

$$|z| \leq 1 + \max\{|\alpha_1|, |\alpha_1|, \ldots, |\alpha_n|\},$$

Montal [2], proved that

$$|z| \leq 1 + |\alpha_1| + |\alpha_1| + \ldots + |\alpha_n|,$$

Cramichael and Mason [2], proved that

$$|z| \leq \left(1 + |\alpha_1|^2 + |\alpha_2|^2 + \ldots + |\alpha_n|^2\right)^{\frac{1}{2}}.$$

Others have provided bounds for zeros of polynomials based on matrix inequalities using the Frobenius companion matrix, such as

Fujii and Kubi [3], proved that

$$|z| \le \cos\left(\frac{\pi}{n+1}\right) + \frac{1}{2}\left(|\alpha_n| + \left(\sum_{j=1}^{n} |\alpha_j|^2\right)^{\frac{1}{2}}\right),$$

Linden [7], proved that

$$|z| \le \frac{|\alpha_n|}{2} + \left(\frac{n-1}{n}\left(n-1+\left|\sum_{j=1}^{n} |\alpha_j|^2 - \frac{|\alpha_n|^2}{2}\right|\right)\right)^{\frac{1}{2}},$$

Kittaneh [5], proved that

$$|z| \le \frac{1}{2}\left(|\alpha_n| + 1 + \sqrt{(|\alpha_n|-1)^2 + 4\sqrt{\sum_{j=1}^{n-1} |\alpha_j|^2}}\right).$$

Based on certain estimates for spectral norms and spectral radii of the square of the Frobenius companion matrices Kittaneh and Shebrawi, [6] obtained new bounds for the zeros of p as follows:

$$|z| \le \left(1 + \left(\sum_{j=1}^{n} |\alpha_j|^2 + |b_j|^2\right)\right)^{\frac{1}{4}}, \text{ where } b_j = \alpha_n\alpha_j - \alpha_{j-1}.$$

Also

$$|z| \le \left(\frac{1}{2}\left(|b_n| + \beta + \sqrt{(|b_n| - \beta)^2 + 4\gamma\sqrt{1 + |\alpha_n|^2}}\right)\right)^{\frac{1}{2}},$$

where

$$\gamma = \left(\sum_{j=1}^{n-1} |b_j|^2\right)^{\frac{1}{2}}$$

and

$$\beta = \sqrt{\frac{1}{2}\left(1 + \sum_{j=1}^{n-1} |\alpha_j|^2 + \sqrt{1 + \sum_{j=1}^{n-1} |\alpha_j|^2 - 4\left(|\alpha_1|^2 + |\alpha_2|^2\right)}\right)}.$$

They also obtained new bounds based on the spectral norms and the spectral radii of the cube of the Frobenius companion matrix.

Recently, Al Sawaftah and Burqan [1] have given another bound for the zeros of polynomials depending on the spectral norm of fourth of the Frobenius companion matrix. In this paper we will present more accurate bounds depending on the spectral norm and the spectral radii of C^5. In this paper let $N = C^5$. Thus,

$$N = \begin{bmatrix} e_n & e_{n-1} & \cdots & e_6 & e_5 & \cdots & e_1 \\ d_n & d_{n-1} & \cdots & d_6 & d_5 & \cdots & d_1 \\ c_n & c_{n-1} & \cdots & c_6 & c_5 & \cdots & c_1 \\ b_n & b_{n-1} & \cdots & b_6 & b_5 & \cdots & b_1 \\ -\alpha_n & -\alpha_{n-1} & \cdots & -\alpha_6 & -\alpha_5 & \cdots & -\alpha_1 \\ 1 & 0 & \cdots & 0 & 0 & \cdots & 0 \\ 0 & 1 & \cdots & 0 & 0 & \cdots & 0 \\ \vdots & \vdots & \ddots & \vdots & \vdots & \vdots & \vdots \\ 0 & 0 & \cdots & 1 & 0 & \cdots & 0 \end{bmatrix},$$

where

$$\begin{aligned} b_j &= \alpha_n \alpha_j - \alpha_{j-1}, c_j = -\alpha_n b_j + \alpha_{n-1}\alpha_j - \alpha_{j-2}, \\ d_j &= -\alpha_n c_j - \alpha_{n-1} b_j + \alpha_{n-2}\alpha_j - \alpha_{j-3}, \\ e_j &= -\alpha_n d_j - \alpha_{n-2} c_j - \alpha_{n-2} b_j + \alpha_{n-3}\alpha_j - \alpha_{n-4}, \end{aligned}$$

for $j = 1, 2, \ldots, n$.

2 Main Results

In this section, we obtain bounds for the spectral norm and the spectral radius of the matrix N, which we use it to estimate the zeros of polynomials.

Theorem 1 *Let z be a zero of $p(z) = z^n + \alpha_n z^{n-1} + \alpha_{n-2} z^{n-2} + \cdots + \alpha_2 z + \alpha_1$, with degree $n \geqslant 7$, then*

$$|z| \leq \left(1 + \sum_{j=1}^{n} |\alpha_j|^2 + |b_j|^2 + |c_j|^2 + |d_j|^2 + |e_j|^2\right)^{\frac{1}{10}}.$$

Proof Consider the following matrices

$$G_1 = \begin{bmatrix} e_n & e_{n-1} & \cdots & e_2 & e_1 \\ 0 & 0 & \cdots & 0 & 0 \\ 0 & 0 & \cdots & 0 & 0 \\ \vdots & \vdots & \ddots & \vdots & \vdots \\ 0 & 0 & \cdots & 0 & 0 \end{bmatrix}_{n\times n}, \quad G_2 = \begin{bmatrix} 0 & 0 & \cdots & 0 & 0 \\ d_n & d_{n-1} & \cdots & d_2 & d_1 \\ 0 & 0 & \cdots & 0 & 0 \\ \vdots & \vdots & \ddots & \vdots & \vdots \\ 0 & 0 & \cdots & 0 & 0 \end{bmatrix}_{n\times n},$$

$$G_3 = \begin{bmatrix} 0 & 0 & \cdots & 0 & 0 \\ 0 & 0 & \cdots & 0 & 0 \\ c_n & c_{n-1} & \cdots & c_2 & c_1 \\ \vdots & \vdots & \ddots & \vdots & \vdots \\ 0 & 0 & \cdots & 0 & 0 \end{bmatrix}_{n\times n}, \quad G_4 = \begin{bmatrix} 0 & 0 & \cdots & 0 & 0 \\ 0 & 0 & \cdots & 0 & 0 \\ 0 & 0 & \cdots & 0 & 0 \\ b_n & b_{n-1} & \cdots & b_2 & b_1 \\ 0 & 0 & \cdots & 0 & 0 \\ \vdots & \vdots & \ddots & \vdots & \vdots \\ 0 & 0 & \cdots & 0 & 0 \end{bmatrix}_{n\times n},$$

$$G_5 = \begin{bmatrix} 0 & 0 & \cdots & 0 & 0 \\ 0 & 0 & \cdots & 0 & 0 \\ 0 & 0 & \cdots & 0 & 0 \\ 0 & 0 & \cdots & 0 & 0 \\ -\alpha_n & -\alpha_{n-1} & \cdots & -\alpha_2 & -\alpha_1 \\ \vdots & \vdots & \ddots & \vdots & \vdots \\ 0 & 0 & \cdots & 0 & 0 \end{bmatrix}_{n\times n}$$

and the block matrix $G_6 = \begin{bmatrix} \mathbf{0} & \mathbf{0} \\ I_{n-5} & \mathbf{0} \end{bmatrix}_{n\times n}$, where I_{n-5} is the identity of order $n-5$. Then $\sum_{l=1}^{6} G_l = N$ with $G_l^* G_m = 0,\ 1 \le l, m \le 6,\ \ l \ne m$. Thus by the triangle inequality, and using the fact that $\|A\|^2 = \|A^*A\|$, for any matrix $A \in M_n(\mathbb{C})$, we get

$$\begin{aligned}
\|N\|^2 = \|N^*N\| &= \left\| \sum_{l=1}^{6} G_l^* G_l \right\| \\
&\le \sum_{l=1}^{6} \left\| G_l^* G_l \right\| = \sum_{l=1}^{6} \|G_l\|^2 \\
&= \sum_{j=1}^{n} \left(|e_j|^2 + |d_j|^2 + |c_j|^2 + |b_j|^2 + |\alpha_j|^2 \right) + 1.
\end{aligned}$$

Since

$$\|G_1\|^2 = \max\left\{\lambda : \lambda \in \sigma\left(G_1^* G_1\right)\right\} = \sum_{j=1}^{n} \left|e_j\right|^2,$$

Also

$$\|G_2\|^2 = \sum_{j=1}^{n} \left|d_j\right|^2, \|G_3\|^2 = \sum_{j=1}^{n} \left|c_j\right|^2, \|G_4\|^2 = \sum_{j=1}^{n} \left|b_j\right|^2, \|G_5\|^2 = \sum_{j=1}^{n} \left|\alpha_j\right|^2$$

$$\left\|G_6^* G_6\right\| = 1.$$

Therefore,

$$\left\|C^5\right\| = \|N\| \le \left(1 + \sum_{j=1}^{n} \left|e_j\right|^2 + \left|d_j\right|^2 + \left|c_j\right|^2 + \left|b_j\right|^2 + \left|\alpha_j\right|^2\right)^{\frac{1}{2}}.$$

Using the fact that $|z| \le \left\|C^5\right\|^{\frac{1}{5}}$, we get

$$|z| \le \left(1 + \sum_{j=1}^{n} \left|e_j\right|^2 + \left|d_j\right|^2 + \left|c_j\right|^2 + \left|b_j\right|^2 + \left|\alpha_j\right|^2\right)^{\frac{1}{10}}.$$

■

Let us recall some important Lemmas which are essential to establish our next results in this paper. These Lemmas can be found in [4].

Lemma 2 *If* $A = \begin{bmatrix} a & b \\ c & d \end{bmatrix}$, *then the spectral radius of* A,

$$r(A) = \frac{1}{2}\left(a + d + \sqrt{(a-d)^2 + 4bc}\right).$$

Lemma 3 *Let* $A \in M_n(\mathbb{C})$ *be partitioned as* $A = \begin{bmatrix} A_{11} & A_{12} \\ A_{21} & A_{22} \end{bmatrix}$, *where* A_{ij} *is an* $n_i \times n_j$ *matrix for* $i, j = 1, 2$ *with* $n_1 + n_2 = n$. *If* $\tilde{A} = \begin{bmatrix} \|A_{11}\| & \|A_{12}\| \\ \|A_{21}\| & \|A_{22}\| \end{bmatrix}$, *then* $r(A) \le r\left(\tilde{A}\right)$ *and* $\|A\| \le \left\|\tilde{A}\right\|$.

Lemma 4 *Let* $A = \begin{bmatrix} a & b \\ c & d \end{bmatrix}$, *then the spectral norm of* A *is*

$$\|A\| = \left(\frac{1}{2}\left(|a|^2 + |b|^2 + |c|^2 + |d|^2 + \gamma\right)\right)^{\frac{1}{2}},$$

where $\gamma = \sqrt{\left(|a|^2 + |c|^2 - |b|^2 - |d|^2\right)^2 + 4\left|a\bar{b} + c\bar{d}\right|^2}$.

Lemma 5 *Let*

$$B = \begin{bmatrix} -\alpha_{n-1} & -\alpha_{n-2} & \cdots & -\alpha_6 & -\alpha_5 & \cdots & -\alpha_1 \\ 0 & 0 & \cdots & 0 & 0 & \cdots & 0 \\ 0 & 0 & \cdots & 0 & 0 & \cdots & 0 \\ 0 & 0 & \cdots & 0 & 0 & \cdots & 0 \\ 0 & 0 & \cdots & 0 & 0 & \cdots & 0 \\ 1 & 0 & \cdots & 0 & 0 & \cdots & 0 \\ 0 & 1 & \cdots & 0 & 0 & \cdots & 0 \\ \vdots & \vdots & \ddots & \vdots & \vdots & \cdots & \vdots \\ 0 & 0 & \cdots & 1 & 0 & \cdots & 0 \end{bmatrix}_{n \times n},$$

with $n \geq 7$, *then*

$$\|B\| = \frac{1}{2}\left(1 + \mu + \sqrt{(1+\mu)^2 - 4\sum_{j=1}^{5}\left|\alpha_j\right|^2}\right),$$

where $\mu = \sum_{j=1}^{n-4}\left|\alpha_j\right|^2$.

The following partition matrix is needed to obtain the next result. For the matrix

$$N = \begin{bmatrix} e_n & e_{n-1} & \cdots & e_6 & e_5 & \cdots & e_1 \\ d_n & d_{n-1} & \cdots & d_6 & d_5 & \cdots & d_1 \\ c_n & c_{n-1} & \cdots & c_6 & c_5 & \cdots & c_1 \\ b_n & b_{n-1} & \cdots & b_6 & b_5 & \cdots & b_1 \\ -\alpha_n & -\alpha_{n-1} & \cdots & -\alpha_6 & -\alpha_5 & \cdots & -\alpha_1 \\ 1 & 0 & \cdots & 0 & 0 & \cdots & 0 \\ 0 & 1 & \cdots & 0 & 0 & \cdots & 0 \\ \vdots & \vdots & \ddots & \vdots & \vdots & \vdots & \vdots \\ 0 & 0 & \cdots & 1 & 0 & \cdots & 0 \end{bmatrix},$$

partition the matrix as $N = \begin{bmatrix} N_{11} & N_{12} \\ N_{21} & N_{22} \end{bmatrix}$, where

$$N_{11} = \begin{bmatrix} e_n & e_{n-1} & e_{n-2} & e_{n-3} \\ d_n & d_{n-1} & d_{n-2} & d_{n-3} \\ c_n & c_{n-1} & c_{n-2} & c_{n-3} \\ b_n & b_{n-1} & b_{n-2} & b_{n-3} \end{bmatrix}_{4\times 4},$$

$$N_{12} = \begin{bmatrix} e_{n-4} & \cdots & e_6 & e_5 & \cdots & e_1 \\ d_{n-4} & \cdots & d_6 & d_5 & \cdots & d_1 \\ c_{n-4} & \cdots & c_6 & c_5 & \cdots & c_1 \\ b_{n-4} & \cdots & b_6 & b_5 & \cdots & b_1 \end{bmatrix}_{4\times (n-4)},$$

$$N_{21} = \begin{bmatrix} -\alpha_n & -\alpha_{n-1} & -\alpha_{n-2} & -\alpha_{n-3} \\ 1 & 0 & 0 & 0 \\ 0 & 1 & 0 & 0 \\ 0 & 0 & 1 & 0 \\ 0 & 0 & 0 & 1 \\ 0 & 0 & 0 & 0 \\ \vdots & \vdots & \vdots & \vdots \\ 0 & 0 & 0 & 0 \end{bmatrix}_{(n-4)\times 4},$$

$$N_{22} = \begin{bmatrix} -\alpha_{n-4} & -\alpha_{n-5} & \cdots & -\alpha_6 & -\alpha_5 & -\alpha_4 & -\alpha_3 & -\alpha_2 & -\alpha_1 \\ 0 & 0 & \cdots & 0 & 0 & 0 & 0 & 0 & 0 \\ 0 & 0 & \cdots & 0 & 0 & 0 & 0 & 0 & 0 \\ 0 & 0 & \cdots & 0 & 0 & 0 & 0 & 0 & 0 \\ 0 & 0 & \cdots & 0 & 0 & 0 & 0 & 0 & 0 \\ 1 & 0 & \cdots & 0 & 0 & 0 & 0 & 0 & 0 \\ 0 & 1 & \cdots & 0 & 0 & 0 & 0 & 0 & 0 \\ \vdots & \vdots & \ddots & \vdots & \vdots & \vdots & \vdots & \vdots & \vdots \\ 0 & 0 & \cdots & 1 & 0 & 0 & 0 & 0 & 0 \end{bmatrix}_{(n-4)\times (n-4)}.$$

Now, as a result, we get the following:

Theorem 6 *Let z be a zero of $p(z) = z^n + \alpha_n z^{n-1} + \alpha_{n-2} z^{n-2} + \cdots + \alpha_2 z + \alpha_1$, with degree $n \geqslant 7$, then*

$$|z| \leq \left[\|N_{11}\| + \|N_{22}\| + \sqrt{(\|N_{11}\| - \|N_{22}\|)^2 + 4\,\|N_{12}\|\,\|N_{21}\|} \right]^{\frac{1}{5}}.$$

Proof Since N is partitioned as $N = \begin{bmatrix} N_{11} & N_{12} \\ N_{21} & N_{22} \end{bmatrix}$, applying Lemma 3, we have

$$r(N) \leq r\left(\begin{bmatrix} \|N_{11}\| & \|N_{12}\| \\ \|N_{21}\| & \|N_{22}\| \end{bmatrix} \right).$$

To find $\|N_{11}\|$, we partition N_{11} as $N_{11} = \begin{bmatrix} S_{11} & S_{12} \\ S_{21} & S_{22} \end{bmatrix}$, where

$$S_{11} = \begin{bmatrix} e_n & e_{n-1} \\ d_n & d_{n-1} \end{bmatrix}, \; S_{12} = \begin{bmatrix} e_{n-2} & e_{n-3} \\ d_{n-2} & d_{n-3} \end{bmatrix}, \; S_{21} = \begin{bmatrix} c_n & c_{n-1} \\ b_n & b_{n-1} \end{bmatrix}$$

and

$$S_{22} = \begin{bmatrix} c_{n-2} & c_{n-3} \\ b_{n-2} & b_{n-3} \end{bmatrix}.$$

Now, find the spectral norm for each S_{ij}, $i, j = 1, 2$, by using Lemma 4 as follows:

$$\alpha = \|S_{11}\| = \left(\frac{1}{2} \left(\sum_{j=n-1}^{n} |e_j|^2 + |d_j|^2 \right.\right.$$
$$\left.\left. + \sqrt{\left(|e_n|^2 + |d_n|^2 - |e_{n-1}|^2 - |d_{n-1}|^2\right)^2 + 4\left|e_n\overline{e_{n-1}} + d_n\overline{d_{n-1}}\right|^2} \right)\right)^{\frac{1}{2}},$$

$$\beta = \|S_{12}\| = \left(\frac{1}{2} \left(\sum_{j=n-3}^{n} |e_j|^2 + |d_j|^2 \right.\right.$$
$$\left.\left. + \sqrt{\left(|e_{n-2}|^2 + |d_{n-2}|^2 - |e_{n-3}|^2 - |d_{n-3}|^2\right)^2 + 4\left|e_{n-2}\overline{e_{n-3}} + d_{n-2}\overline{d_{n-3}}\right|^2} \right)\right)^{\frac{1}{2}},$$

$$\gamma = \|S_{21}\| = \left(\frac{1}{2} \left(\sum_{j=n-1}^{n} |c_j|^2 + |bd_j|^2 \right.\right.$$
$$\left.\left. + \sqrt{\left(|c_n|^2 + |b_n|^2 - |c_{n-1}|^2 - |b_{n-1}|^2\right)^2 + 4\left|c_n\overline{c_{n-1}} + b_n\overline{b_{n-1}}\right|^2} \right)\right)^{\frac{1}{2}},$$

and

$$\delta = \|S_{22}\| = \left(\frac{1}{2} \left(\sum_{j=n-3}^{n-2} |c_j|^2 + |b_j|^2 \right.\right.$$
$$\left.\left. + \sqrt{\left(|c_{n-2}|^2 + |b_{n-2}|^2 - |c_{n-3}|^2 - |b_{n-3}|^2\right)^2 + 4\left|c_{n-2}\overline{c_{n-3}} + b_{n-2}\overline{b_{n-3}}\right|^2} \right)\right)^{\frac{1}{2}},$$

Also, by Lemma 3, we have

$$\|N_{11}\| \leq \left\| \begin{bmatrix} \|S_{11}\| & \|S_{12}\| \\ \|S_{21}\| & \|S_{22}\| \end{bmatrix} \right\| = \left\| \begin{bmatrix} \alpha & \beta \\ \gamma & \delta \end{bmatrix} \right\|.$$

Again using Lemma 3 to get

$$\|N_{11}\| \leq \left(\frac{1}{2} \left(\alpha^2 + \beta^2 + \gamma^2 + \delta^2 + \sqrt{\left(\alpha^2 + \beta^2 - \gamma^2 - \delta^2\right)^2 + 4\left|\alpha\beta + \gamma\delta\right|^2} \right) \right)^{\frac{1}{2}}.$$

Now, $\|N_{12}\| = \left(r\left(N_{12}N_{12}^*\right)\right)^{\frac{1}{2}}$, where

$$N_{12}N_{12}^* = \begin{bmatrix} \sum_{j=1}^{n-4} |e_j|^2 & \sum_{j=1}^{n-4} e_j\overline{d_j} & \sum_{j=1}^{n-4} e_j\overline{c_j} & \sum_{j=1}^{n-4} e_j\overline{b_j} \\ \sum_{j=1}^{n-4} d_j\overline{e_j} & \sum_{j=1}^{n-4} |d_j|^2 & \sum_{j=1}^{n-4} d_j\overline{c_j} & \sum_{j=1}^{n-4} d_j\overline{b_j} \\ \sum_{j=1}^{n-4} c_j\overline{e_j} & \sum_{j=1}^{n-4} c_j\overline{d_j} & \sum_{j=1}^{n-4} |c_j|^2 & \sum_{j=1}^{n-4} c_j\overline{b_j} \\ \sum_{j=1}^{n-4} b_j\overline{e_j} & \sum_{j=1}^{n-4} b_j\overline{d_j} & \sum_{j=1}^{n-4} b_j\overline{c_j} & \sum_{j=1}^{n-4} |b_j|^2 \end{bmatrix}.$$

To find $\|N_{12}\|$, we partition $N_{12}N_{12}^*$ as $\begin{bmatrix} W_{11} & W_{12} \\ W_{21} & W_{22} \end{bmatrix}$, where

$$W_{11} = \begin{bmatrix} \sum_{j=1}^{n-4} |e_j|^2 & \sum_{j=1}^{n-4} e_j\overline{d_j} \\ \sum_{j=1}^{n-4} d_j\overline{e_j} & \sum_{j=1}^{n-4} |d_j|^2 \end{bmatrix}, \quad W_{12} = \begin{bmatrix} \sum_{j=1}^{n-4} e_j\overline{c_j} & \sum_{j=1}^{n-4} e_j\overline{b_j} \\ \sum_{j=1}^{n-4} d_j\overline{c_j} & \sum_{j=1}^{n-4} d_j\overline{b_j} \end{bmatrix}, .$$

$$W_{21} = \begin{bmatrix} \sum_{j=1}^{n-4} c_j\overline{e_j} & \sum_{j=1}^{n-4} c_j\overline{d_j} \\ \sum_{j=1}^{n-4} b_j\overline{e_j} & \sum_{j=1}^{n-4} b_j\overline{d_j} \end{bmatrix}, \quad W_{22} = \begin{bmatrix} \sum_{j=1}^{n-4} |c_j|^2 & \sum_{j=1}^{n-4} c_j\overline{b_j} \\ \sum_{j=1}^{n-4} b_j\overline{c_j} & \sum_{j=1}^{n-4} |b_j|^2 \end{bmatrix}.$$

Using Lemma 4 to get the spectral norm for each $W_{ij},\ i, j = 1, 2$ as follows:

$$\|W_{11}\| = \left(\frac{1}{2} \left(\left| \sum_{j=1}^{n-4} |e_j|^2 \right|^2 + \left| \sum_{j=1}^{n-4} |d_j|^2 \right|^2 + \left| \sum_{j=1}^{n-4} e_j\overline{d_j} \right|^2 + \left| \sum_{j=1}^{n-4} d_j\overline{e_j} \right|^2 + \sqrt{a+b} \right) \right)^{\frac{1}{2}},$$

$$\|W_{12}\| = \left(\frac{1}{2}\left(\left|\sum_{j=1}^{n-4} e_j\overline{c_j}\right|^2 + \left|\sum_{j=1}^{n-4} e_j\overline{b_j}\right|^2 + \left|\sum_{j=1}^{n-4} d_j\overline{c_j}\right|^2 + \left|\sum_{j=1}^{n-4} d_j\overline{b_j}\right|^2 + \sqrt{c+d}\right)\right)^{\frac{1}{2}},$$

$$\|W_{21}\| = \left(\frac{1}{2}\left(\left|\sum_{j=1}^{n-4} c_j\overline{e_j}\right|^2 + \left|\sum_{j=1}^{n-4} c_j\overline{d_j}\right|^2 + \left|\sum_{j=1}^{n-4} b_j\overline{e_j}\right|^2 + \left|\sum_{j=1}^{n-4} b_j\overline{d_j}\right|^2 + \sqrt{e+f}\right)\right)^{\frac{1}{2}},$$

$$\|W_{22}\| = \left(\frac{1}{2}\left(\left|\sum_{j=1}^{n-4} |c_j|^2\right|^2 + \left|\sum_{j=1}^{n-4} |b_j|^2\right|^2 + \left|\sum_{j=1}^{n-4} b_j\overline{c_j}\right|^2 + \left|\sum_{j=1}^{n-4} c_j\overline{b_j}\right|^2 + \sqrt{g+h}\right)\right)^{\frac{1}{2}},$$

where

$$a = \left(\left|\sum_{j=1}^{n-4} |e_j|^2\right|^2 + \left|\sum_{j=1}^{n-4} d_j\overline{e_j}\right|^2 - \left|\sum_{j=1}^{n-4} e_j\overline{d_j}\right|^2 - \left|\sum_{j=1}^{n-4} |d_j|^2\right|^2\right)^2,$$

$$b = 4\left|\left(\sum_{j=1}^{n-4} |e_j|^2\right)\left(\sum_{j=1}^{n-4} e_j\overline{d_j}\right) + \left(\sum_{j=1}^{n-4} d_j\overline{e_j}\right)\left(\sum_{j=1}^{n-4} |d_j|^2\right)\right|^2,$$

$$c = \left(\left|\sum_{j=1}^{n-4} e_j\overline{c_j}\right|^2 + \left|\sum_{j=1}^{n-4} d_j\overline{c_j}\right|^2 - \left|\sum_{j=1}^{n-4} e_j\overline{b_j}\right|^2 - \left|\sum_{j=1}^{n-4} d_j\overline{b_j}\right|^2\right)^2,$$

$$d = 4\left|\left(\sum_{j=1}^{n-4} e_j\overline{c_j}\right)\left(\sum_{j=1}^{n-4} e_j\overline{b_j}\right) + \left(\sum_{j=1}^{n-4} d_j\overline{c_j}\right)\left(\sum_{j=1}^{n-4} d_j\overline{b_j}\right)\right|^2,$$

$$e = \left(\left|\sum_{j=1}^{n-4} c_j\overline{e_j}\right|^2 + \left|\sum_{j=1}^{n-4} b_j\overline{e_j}\right|^2 - \left|\sum_{j=1}^{n-4} c_j\overline{d_j}\right|^2 - \left|\sum_{j=1}^{n-4} b_j\overline{d_j}\right|^2\right)^2,$$

$$f = 4\left|\left(\sum_{j=1}^{n-4} c_j\overline{e_j}\right)\left(\sum_{j=1}^{n-4} c_j\overline{d_j}\right) + \left(\sum_{j=1}^{n-4} b_j\overline{d_j}\right)\left(\sum_{j=1}^{n-4} d_j\overline{b_j}\right)\right|^2,$$

$$g = \left(\left|\sum_{j=1}^{n-4} |c_j|^2\right|^2 + \left|\sum_{j=1}^{n-4} b_j\overline{c_j}\right|^2 - \left|\sum_{j=1}^{n-4} c_j\overline{b_j}\right|^2 - \left|\sum_{j=1}^{n-4} |b_j|^2\right|^2\right)^2$$

$$h = 4\left|\left(\sum_{j=1}^{n-4} |c_j|^2\right)\left(\sum_{j=1}^{n-4} c_j\overline{b_j}\right) + \left(\sum_{j=1}^{n-4} b_j\overline{c_j}\right)\left(\sum_{j=1}^{n-4} b_j\overline{c_j}\right)\right|^2.$$

By Lemmas 2 and 3, we have

$$\|N_{12}\| = \left(r\left(N_{12}N_{12}^*\right)\right)^{\frac{1}{2}} \le \left(r\left(\begin{bmatrix} \|w_{11}\| & \|w_{12}\| \\ \|w_{21}\| & \|w_{22}\| \end{bmatrix}\right)\right)^{\frac{1}{2}}$$
$$= \left(\frac{1}{2}\left(\|w_{11}\| + \|w_{11}\| + \sqrt{(\|w_{11}\| - \|w_{22}\|)^2 + 4\|w_{12}\| \, \|w_{21}\|}\right)\right)^{\frac{1}{2}}.$$

Now,

$$\|N_{21}\| = \sqrt{|\alpha_n|^2 + |\alpha_{n-1}|^2 + |\alpha_{n-2}|^2 + |\alpha_{n-3}|^2 + 1},$$

and Lemma 5 yields

$$\|N_{22}\| = \left(\frac{1}{2}\left(1 + \mu + \sqrt{(1+\mu)^2 - 4\left(|\alpha_1|^2 + |\alpha_2|^2 + |\alpha_3|^2 + |\alpha_4|^2 + |\alpha_5|^2\right)}\right)\right)^{\frac{1}{2}},$$

where $\mu = \sum_{j=1}^{n-4} \left|\alpha_j\right|^2$ .Thus,

$$r(N) \le r\left(\begin{bmatrix} \|N_{11}\| & \|N_{12}\| \\ \|N_{21}\| & \|N_{22}\| \end{bmatrix}\right)$$
$$= \frac{1}{2}\left(\|N_{11}\| + \|N_{22}\| + \sqrt{(\|N_{11}\| - \|N_{22}\|)^2 + 4\|N_{12}\| \, \|N_{21}\|}\right).$$

Since $|z| \le r(C) = \left(r\left(C^5\right)\right)^{\frac{1}{5}} = (r(N))^{\frac{1}{5}}$, we have

$$|z| \le \left(\frac{1}{2}\left(\|N_{11}\| + \|N_{22}\| + \sqrt{(\|N_{11}\| - \|N_{22}\|)^2 + 4\|N_{12}\| \, \|N_{21}\|}\right)\right)^{\frac{1}{5}}.$$

This completes the proof. ■

References

1. Al Sawafteh, A., Burqan, A.: Bounds for the Zeros of Polynomials, Master thesis, Zarqa University (2021)
2. Fujii, M., Kubo, F.: Operator norms as bounds for roots of algebraic equations. Proc. Jpn. Acad. **49**(10), 805–808 (1973)
3. Fujii, M., Kubo, F.: Buzano's inequality and bounds for roots of algebraic equations. Proc. Am. Math. Soc. **117**(2), 359–361 (1993)
4. Horn, R. A., Johnson, C. R.: Matrix Analysis. Cambridge University Press (2012)

5. Kittaneh, F.: Bounds for the zeros of polynomials from matrix inequalities. Archiv der Mathematik **81**(5), 601–608 (2003)
6. Kittaneh, F., Shebrawi, K.: Bounds for the zeros of polynomials from matrix inequalities - II. Linear Multilinear Algebra **55**(2), 147–158 (2007)
7. Linden, H.: Bounds for zeros of polynomials using traces and determinants. Seminarberichte Fachbereich Mathematik FeU Hagen **69**, 127–146 (2000)

Applications on Formable Transform in Solving Integral Equations

Rania Saadeh, Bayan Ghazal, and Gharib Gharib

Abstract Mathematics is a powerful tool for global understanding and communication that organizes our lives and encourages the ability to solve problems. One of the most important aspects of mathematics is differential and integral equations, the real power of equations is that they provide a very precise way to describe various features of the world. In this article, we introduce an effective method to solve integral equations and integro-differential equations. We use the new transform called the formable integral transform for solving the Volterra integral equations of the second kind and integro-differential equations. To show the simplicity and applicability of the method, we introduce some examples and apply the transform to get the exact solutions.

Keywords Integral equations · Integro-differential equations · Formable transform

AMS Subject Classification: 45D05 · 45E10

1 Introduction

Mathematics is one of the most important fields in our lives. Mathematics and its applications promote innovation in order to reach solutions, and mathematical skills are also included in many jobs and fields, as they contribute to solving most physical, engineering, technological, and other problems [1].

Integral transforms are valuable because they are able to simplify differential, integral, and partial equations subject to certain boundary conditions, where the equation is converted by the transform, it gives an algebraic equation that can be easily solved.

R. Saadeh (✉) · B. Ghazal · G. Gharib
Zarqa University, Zarqa 13132, Jordan
e-mail: rsaadeh@zu.edu.jo

G. Gharib
e-mail: ggharib@zu.edu.jo

D. Zeidan et al. (eds.), *Mathematics and Computation*, Springer Proceedings in Mathematics & Statistics 418, https://doi.org/10.1007/978-981-99-0447-1_4

The integral transform of the function $g(x)$ where $x \in (-\infty, \infty)$ can be obtained by computing the improper integral

$$\pounds[g(x)](s) = \int_{-\infty}^{\infty} q(s,x)g(x)dx, \tag{1}$$

where $q(s, x)$ is called the kernel of the integral transform and s is the variable of the transform which might be real or complex number and independent of the variable x. The theory of integral transforms goes back to the work of P.S. Laplace in 1780 [2, 3] and Fourier in 1822.

Also, with what we have mentioned about the importance of integral transform, so many researchers have contributed to the existence of new integral transforms, such as the z-transform [4], Mellin integral transform [5], Laplace Carson transform [6], Hankel?s transform [7], Sumudu integral transform [8], ARA transform [9], and recently, in 2021, formable transform was introduced [10].

When scientists began studying natural phenomena, whether physical, chemical, biological, or engineering [11?19], the integral equations had an important role in explaining these phenomena and finding different solutions to them.

The integral equation for the function $g(x)$ and the kernel of integral equation $q(x, t)$ is defined by

$$\varphi(x)y(x) = g(x) + \lambda \int_{a}^{x} q(x,t)y(t)dt \tag{2}$$

while $y(x)$ is an unknown function that will be determined, ? is a non-zero, real or complex, parameter. The function $\varphi(x)$ determines the kind of integral equation.

As there are several types of integral equations, including Volterra integral equations and integro-differential equations [20].

2 Basic Definitions

In this section, we introduce some basic definitions that are essential to our research.

Definition 1 The second kind of Volterra integral equation of the function $g(x)$ is defined by [21, 22?]:

$$y(x) = g(x) + \lambda \int_{a}^{x} q(x,t)y(t)dt \tag{3}$$

where the kernel of the integral equation is $q(x, t)$.

Definition 2 Integro-differential equations of first order is defined by

$$y'(x) + y(x) + \int_{x_0}^{x} g(t, y(t))dt = f(x, y(x)), \quad y(x_0) = y_0, \quad x_0 \geq 0 \tag{4}$$

where $f(x, y(x))$ and $g(t, y(t))$ are given, such that $f(x, y(x))$ and $g(t, y(t))$ are generally nonlinear in $y(x)$ which is the variable of the integro-differential boundary value problems that need to be determined.

Definition 3 The convolution of the functions $f(x)$ and $g(x)$, which denoted by $(f * g)(x)$ is defined by the relation

$$(f * g)(x) = \int_{0}^{x} f(t)g(x - t)dt. \tag{5}$$

Definition 4 A unit step function or Heaviside step function is a piecewise function defined as follows:

$$u(x) = \begin{cases} 1, & x > 0 \\ 0, & x \leq 0 \end{cases}. \tag{6}$$

Definition 5 A function $g(x)$ is called a function of exponential order, a if there exist constants a, M and $b > 0$ such that $|g(x)| \leq Me^{ax}$ for all $x > b$.

3 Formable Integral Transform (FIT)

Through section three, we present the definition of the FT and some properties that are needed in our work. To see more about the Formable transform, see Ref. [10].

Definition 1 Assume that the function $g(x)$ is a piecewise continuous function of exponential order defined over the set

$$W = \left\{ g(x) : \exists N, \tau_1, \tau_2 > 0, |g(x)| < N \exp\left(\frac{|x|}{\tau_i}\right), \text{ if } x \in (-1)^i \times [0, \infty) \right\},$$

then, the Formable integral transform of a as the following form:

$$R[g(x)] = B(s, u) = s \int_{0}^{\infty} \exp(-sx)g(ux)dx, \tag{7}$$

which is equivalent to

$$R[g(x)] = \frac{s}{u}\int_0^{\infty} \exp\left(\frac{-sx}{u}\right) g(x)dx. \tag{8}$$

$$R[g(x)] = \frac{s}{u}\lim_{\tau\to\infty}\int_0^{\tau} \exp\left(\frac{-sx}{u}\right) g(x)dx, \quad s > 0, u > 0,$$

where $s\&u$ are the Formable transform?s variables, t is a real number and the integral is taken along the line $x = \tau$.

Definition 2 The inverse Formable transform of a function $g(x)$ is given by

$$R^{-1}[B(s, u)] = g(x) = \frac{1}{2\pi i}\int_{c-i\infty}^{c+i\infty} \frac{1}{s}\exp\left(\frac{sx}{u}\right) B(s, u)ds. \tag{9}$$

Properties of Formable Transform

We present some important properties and theorems of FT that this paper is based on

Property 1
(Linearity property)
Let $\alpha g_1(x)$ and $\beta g_2(x)$ be two functions in which the Formable transform exists for them, where α and β are non-zero arbitrary constants, then

$$R[\alpha g_1(x) + \beta g_2(x)] = \alpha R[g_1(x)] + \beta R[g_2(x)]. \tag{10}$$

Property 2
(Shifting in s-domain)
If the function $g(x)$ in which the Formable transform exists is multiplied with shift function x^n then

$$R\left[x^n g(x)\right] = (-u)^n s \frac{\partial^n}{\partial s^n}\left[\frac{R[g(x)]}{s}\right] \tag{11}$$

Property 3
(FT of the Derivatives)
Let $g^{(n)}(x)$ be the nth derivative of the function $g(x)$ such that $g^{(n)}(x) \in W$, then

$$R\left[g^{(n)}(x)\right] = \frac{s^n}{u^n} B(s, u) - \sum\nolimits_{k=0}^{n-1} \left(\frac{s}{u}\right)^{n-k} g^{(k)}(0) \tag{12}$$

Property 4
(FT of the Convolution)

If $F(s,u)$ and $G(s,u)$ are the Formable transforms of the functions $f(x)$ and $g(x)$ respectively, then

$$R[f(x) * g(x)] = \frac{u}{s} F(s,u) G(s,u). \tag{12}$$

Property 5
(FT of Derivatives)

If the function $g^{(n)}(x)$ is the n-th derivative of the function $g(x)$ where $g^{(n)}(x) \in W$, then

$$R\left[g^{(n)}(x)\right] = \frac{s^n}{u^n} B(s,u) - \sum_{k=0}^{n-1} \left(\frac{s}{u}\right)^{n-k} g^{(k)}(0) \tag{13}$$

Property 6
The FT of the unit step function $u(x)$ is given by

$$R[u(x)] = 1 \tag{14}$$

Proof

$$R[u(x)] = \frac{s}{u} \int_0^\infty \exp\left(\frac{-sx}{u}\right) dx = \frac{s}{u} \lim_{\alpha \to \infty} \left[\frac{-u}{s} \exp\left(\frac{-sx}{u}\right)\right]_0^\alpha = 1.$$

4 Main Results

In this section, we introduce two important theorems for solving integral equations of both types, Volterra integral equation and integro-differential equation using the Formable transform.

Theorem 1 Let $g(x)$ be a continues function defined on the interval $[a, b]$ and consider the Volterra integral equation of the second kind.

$$y(x) = g(x) + \lambda \int_a^x k(x,t) y(t) dt. \tag{15}$$

Assume that the kernel $k(x,t)$ is a difference kernel, that is, it has the property $k(s,t) = q(x-t)$, $k(x,t)$ depends on the difference *x-t*. Then, Eq. (15) can be written as

$$y(x) = g(x) + \lambda \int_a^x q(x-t)y(t)dt, \tag{16}$$

and has the solution

$$y(x) = R^{-1}\left[\frac{sG(s,u)}{s - \lambda u Q(s,u)}\right],$$

where

$$Q(s,u) = R[q(x)], \qquad G(s,u) = R[g(x)].$$

Proof Running the FT to Eq. (16), we get after using Property 4.

$$R[y(x)] = R[g(x)] + \lambda R\left[\int_a^x k(x,t)y(t)dt\right]$$

$$Y(s,u) = G(s,u) + \frac{\lambda u}{s} Q(s,u)Y(s,u), \tag{17}$$

where $Y(s,u) = R[y(x)]$.

Solving Eq. (17) for $Y(s,u)$, we have

$$Y(s,u) = \frac{sG(s,u)}{s - \lambda u Q(s,u)}, \qquad \lambda u Q(s,u) \neq s \tag{18}$$

The solution $y(x)$ is obtained by applying the inverse Formable transform to Eq. (18), to get

$$y(x) = R^{-1}\left[\frac{s\,G(s,u)}{s - \lambda u Q(s,u)}\right] \blacksquare \tag{19}$$

Theorem 2 Consider the integro-differential equation of the first order.

$$L\,y'(x) + P\,y(x) + \frac{1}{C}\int_0^x y(t)dt = E(x), \tag{20}$$

with the initial condition,

$$y(0) = a, \tag{21}$$

then, the solution of Eq. (20) and the initial condition (21) is given by

$$y(x) = \left[\frac{1}{L}e^{\frac{-P}{2L}x}\left(\cos\left(\sqrt{\frac{4L-P^2c}{4CL^2}}x\right) - \frac{P\sqrt{C}}{L\sqrt{4L-P^2C}}\sin\left(\sqrt{\frac{4L-P^2c}{4CL^2}}x\right)\right)\right.$$
$$\left.*\left(E(x) - \frac{aP}{2}\right)\right] + ae^{\frac{-P}{2L}x}\cos\left(\sqrt{\frac{4L-P^2c}{4CL^2}}x\right) \tag{22}$$

while L, P, and C are constant with $C \neq 0$.

Proof Firstly, we apply the Formable transform on both sides of Eq. (20) and use Property 4, we have

$$R\left[L\,y'(x)\right] + R[P\,y(x)] + \frac{1}{C}R\left[\int_0^x y(t)dt\right] = R[E(x)],$$

$$L\left(\frac{s}{u}Y(s,u) - \frac{s}{u}y(0)\right) + PY(s,u) + \frac{1}{C}\frac{u}{s}Y(s,u) = E(s,u), \tag{23}$$

where $Y(s,u) = R[y(x)]$ and $E(s,u) = R[E(x)]$.

Substituting the initial condition (21) and simplifying Eq. (23), we get

$$Y(s,u)\left[L\frac{s}{u} + P + \frac{u}{Cs}\right] = E(s,u) + aL\frac{s}{u}. \tag{24}$$

Solving Eq. (24) for $Y(s,u)$, gives

$$\begin{aligned}
Y(s,u) &= \frac{E(s,u) + aL\frac{s}{u}}{L\frac{s}{u} + P + \frac{u}{Cs}} \\
&= \frac{\frac{1}{L}E(s,u)su + as^2}{\left(s + \frac{P}{2L}u\right)^2 + \frac{4L-P^2C}{4CL^2}u^2} \\
&= \frac{\frac{su}{L}\left(E(s,u) - \frac{aP}{2}\right)}{\left(s + \frac{P}{2L}u\right)^2 + \frac{4L-P^2C}{4CL^2}u^2} + \frac{as\left(s + \frac{P}{2L}u\right)}{\left(s + \frac{P}{2L}u\right)^2 + \frac{4L-P^2C}{4CL^2}u^2} \\
&= \frac{u}{s}\left(E(s,u) - \frac{aP}{2}\right)\frac{\frac{s}{L}\left(s + \frac{P}{2L}u\right) - \frac{P}{2L^2}su}{\left(s + \frac{P}{2L}u\right)^2 + \frac{4L-P^2C}{4CL^2}u^2} + \frac{as\left(s + \frac{P}{2L}u\right)}{\left(s + \frac{P}{2L}u\right)^2 + \frac{4L-P^2C}{4CL^2}u^2}
\end{aligned}$$

$$Y(s,u) = \frac{u}{s}\left(E(s,u) - \frac{aP}{2}\right)\left(\frac{\frac{s}{L}\left(s + \frac{P}{2L}u\right)}{\left(s + \frac{P}{2L}u\right)^2 + \frac{4L-P^2C}{4CL^2}u^2} - \frac{P}{2L^2}\frac{su}{\left(s + \frac{P}{2L}u\right)^2 + \frac{4L-P^2C}{4CL^2}u^2}\right)$$
$$+ \frac{as\left(s + \frac{P}{2L}u\right)}{\left(s + \frac{P}{2L}u\right)^2 + \frac{4L-P^2C}{4CL^2}u^2} \tag{25}$$

Now, applying the inverse FT to both sides of Eq. (25), then running the convolution property on Eq. (13), we get the result

$$y(x) = \left[\frac{1}{L}e^{\frac{-P}{2L}x}\left(\cos\left(\sqrt{\frac{4L-P^2c}{4CL^2}}x\right) - \frac{P\sqrt{C}}{L\sqrt{4L-P^2C}}\sin\left(\sqrt{\frac{4L-P^2c}{4CL^2}}x\right)\right)\right.$$
$$\left. *\left(E(x) - \frac{aP}{2}\right)\right] + ae^{\frac{-P}{2L}x}\cos\left(\sqrt{\frac{4L-P^2c}{4CL^2}}x\right) \blacksquare$$

It is also worth mentioning that Eq. (20), which is an integro-differential equation, is an application to one of the most important physical phenomena, which is circuit analysis. According to Kirchhoff's second law, the net voltage drop across a closed loop is equal to the voltage $E(x)$,where $y(x)$ is a current function of time, L is the inductance, P is the resistance and C is the capacitance.

5 Applications and Examples

In this section, we introduce some examples of a set of integral equations and integro-differential equations and apply the proposed FT to get their solutions.

Example 1 Consider the Volterra integral equation.

$$y(x) = 1 + \int_0^x y(t)dt \tag{26}$$

Solution

As a first step, we apply the FT to Eq. (26), to get

$$Y(s,u) = 1 + \frac{u}{s}Y(s,u) \tag{27}$$

Simplifying Eq. (27), we get

$$Y(s,u) = \frac{s}{s-u} \tag{28}$$

Finally, we apply the inverse FT on Eq. (28), to get the

$$y(x) = e^x \tag{29}$$

Example 2 Consider the Volterra integral equation.

$$y(x) = 1 - \int_0^x (x - t)y(t)dt \tag{30}$$

Solution

Applying the FT to Eq. (30), we have

$$\begin{aligned} Y(s, u) &= 1 - \frac{u}{s}\frac{u}{s}Y(s, u) \\ &= 1 - \frac{u^2}{s^2}Y(s, u) \end{aligned} \tag{31}$$

Simplifying Eq. (31), we get

$$Y(s, u) = \frac{s^2}{s^2 + u^2} \tag{32}$$

Now, taking the inverse FT on both sides of Eq. (32), we get the solution

$$y(x) = \cos x \tag{33}$$

Example 3 Consider the Volterra integral equation.

$$y(x) = \sin x + \cos x + 2\int_0^x \sin(x - t)y(t)dt \tag{34}$$

Solution

We apply the FT on Eq. (34), to get

$$\begin{aligned} Y(s, u) &= \frac{su}{s^2 + u^2} + \frac{s^2}{s^2 + u^2} + 2\frac{u}{s}\frac{su}{s^2 + u^2}Y(s, u) \\ &= \frac{su}{s^2 + u^2} + \frac{s^2}{s^2 + u^2} + 2\frac{u^2}{s^2 + u^2}Y(s, u) \end{aligned} \tag{35}$$

Simplifying Eq. (35), we get

$$\begin{aligned} Y(s, u) &= \frac{su + s^2}{s^2 - u^2} \\ &= \frac{s}{s - u} \end{aligned} \tag{36}$$

Now, applying the inverse FT on Eq. (36), we have the solution

$$y(x) = e^x \tag{37}$$

Example 4 Consider the Volterra integral equation.

$$y(x) = \frac{x^3}{6} - \int_0^x (x - t)y(t)dt \tag{38}$$

Solution

We apply the FT on Eq. (38), to get

$$\begin{aligned} Y(s, u) &= \frac{u^3}{s^3} - \frac{u}{s}\frac{u}{s}Y(s, u) \\ &= \frac{u^3}{s^3} - \frac{u^2}{s^2}Y(s, u). \end{aligned} \tag{39}$$

Simplifying Eq. (39), we get

$$Y(s, u) = \frac{u}{s} - \frac{su}{s^2 + u^2} \tag{40}$$

Now, applying the inverse FT on Eq. (40), we have the solution

$$y(x) = x - \sin x \tag{41}$$

Example 5 Consider the integro-differential equation.

$$y' + 2y + 5\int_0^x y(t)dt = u(x) \tag{42}$$

with the initial conditions,

$$y(0) = 0 \tag{43}$$

Solution

We apply the FT on Eq. (42), and we get

$$\frac{s}{u}Y(s, u) - \frac{s}{u}y(0) + 2Y(s, u) + 5\frac{u}{s}Y(s, u) = 1. \tag{44}$$

Now, substituting the initial conditions (43) in Eq. (44),

$$\frac{s}{u}Y(s,u) + 2Y(s,u) + 5\frac{u}{s}Y(s,u) = 1 \tag{45}$$

Simplifying Eq. (45), we get

$$Y(s,u) = \frac{2su}{(s+u)^2 + 4u^2} \tag{46}$$

Now, we take the inverse FT on both sides of Eq. (46), and we get the solution

$$y(x) = \frac{1}{2}e^{-x}\sin 2x \tag{47}$$

Example 6 Consider the integro-differential equation.

$$y' + 3y + 2\int_0^x y(t)dt = 2e^{-3x} \tag{48}$$

with the initial conditions,

$$y(0) = 0 \tag{49}$$

Solution

Appling the formable transform to both sides of Eq. (48), we have

$$\frac{s}{u}Y(s,u) - \frac{s}{u}y(0) + 3Y(s,u) + 2\frac{u}{s}Y(s,u) = 2\frac{s}{s+3u}. \tag{50}$$

Now, substituting the initial conditions (49) in Eq. (50),

$$\frac{s}{u}Y(s,u) + 3Y(s,u) + 2\frac{u}{s}Y(s,u) = 2\frac{s}{s+3u}. \tag{51}$$

Simplifying Eq. (51), we get

$$\left[\frac{s^2 + 3su + 2u^2}{su}\right]Y(s,u) = 2\frac{s}{s+3u} \tag{52}$$

Solving Eq. (52) for $Y(s,u)$ gives

$$Y(s,u) = \frac{2s^2u}{(s+u)(s+2u)(s+3u)}$$
$$= \frac{-s}{(s+u)} + \frac{4s}{(s+2u)} + \frac{-3s}{(s+3u)} \tag{53}$$

Now, applying the inverse FT on Eq. (53), we get the solution

$$y(x) = -e^{-x} + 4e^{-2x} - 3e^{-3x} \tag{54}$$

Example 7 Consider the integro-differential equation.

$$y' - \frac{1}{2}\int_0^x (x-t)^2 y(t)dt = -x \tag{55}$$

with the initial conditions,

$$y(0) = 1 \tag{56}$$

Solution

We apply the FT on Eq. (55), and we have

$$\frac{s}{u}Y(s,u) - \frac{s}{u}y(0) - \frac{u}{s}\frac{u^2}{s^2}Y(s,u) = -\frac{u}{s}. \tag{57}$$

Now, we substitute the initial conditions (56) in Eq. (57),

$$\frac{s}{u}Y(s,u) - \frac{s}{u} - \frac{u^3}{s^3}Y(s,u) = -\frac{u}{s}. \tag{58}$$

Simplifying Eq. (58), we get

$$\left[\frac{s^4 - u^4}{s^3u}\right]Y(s,u) = \frac{s^2 - u^2}{su}. \tag{59}$$

Solving Eq. (59) for $Y(s,u)$ gives

$$Y(s,u) = \frac{s^2(s^2 - u^2)}{s^4 - u^4}$$
$$= \frac{s^2}{s^2 + u^2} \tag{60}$$

Now, we take the inverse FT on Eq. (60), and we get the solution

$$y(x) = \cos x \tag{61}$$

6 Conclusion

In this paper, we presented two basic theorems concerning the formable transform to solve the Volterra integral equations and the integro-differential equations with special kernels. We use the proposed integral transform to present the exact solutions of the target integral equations. Seven interesting examples were introduced and solved by the formable transform. As a future work, we attend to solve linear and nonlinear fractional differential equations by the formable transform.

References

1. Kusmaryono, I.: The importance of mathematical power in mathematics learning. International Conference on Mathematics. Science and Education, (2014).?
2. Widder, D.V.: The Laplace Transform. Princeton University Press, London, UK (1946)
3. Spiegel, M.R.: Theory and Problems of Laplace Transforms; Schaums Outline series. McGraw-Hill. New York (1965)
4. Sullivan, D.M.: Z-transform theory and the FDTD method. IEEE Trans. Anten. Propagat. **44**(1), 28?34 (1996)
5. Butzer, P.L, Jansche, S.A.: Direct approach to the Mellin transform. J. Fourier Anal. Appl. **3**(4), 325?76 (1997)
6. Makarov, A.M.: Application of the Laplace-Carson method of integral transformation to the theory of unsteady visco-plastic flows. J. Engrg. Phys. Thermophys. **19**, 94?99 (1970)
7. Yu, L., Huang. M., Chen, M., Chenm, W., Huang, W., Zhu, Z.: Quasi-discrete Hankel transform. Opt. Lett. **23**(6), 409?11(1998)
8. Watugala, G.: Sumudu transform: a new integral transform to solve differential equations and control engineering problems. Integrat. Educat. **24**(1), 35?43 (1993)
9. Saadeh, R., Qazza, A., Burqan, A.: A new integral transform: ARA transform and its properties and applications. Symmetry **12**(6), 925 (2020)
10. Saadeh, R., Ghazal, B.: A new approach on transforms: formable integral transform and its applications. Axioms **10**(4), 332 (2021)
11. Saadeh, R., Alaroud, M., Al-Smadi, M., Ahmad, R., Din, U.: Application of fractional residual power series algorithm to solve Newell-whitehead-Segel equation of fractional order. Symmetry **11**(12), 1431?486 27 (2019)
12. Saadeh, R., et al.: Numerical investigation for solving two-point fuzzy boundary value problems by reproducing kernel approach. Appl. Math. Inf. Sci. **10**(6), 1?13 470 21 (2016)
13. Gharib, G., Saadeh, R.: Reduction of the Self-dual Yang-Mills equations to Sinh-Poisson equation and exact solutions. 471 WSEAS Interact. Math. (20), 540?554, 472 22 (2021). https://doi.org/10.37394/23206.2021.20.57
14. Burqan, A., El-Ajou, A., Saadeh, R., Al-Smadi, M.: A new efficient technique using Laplace transforms and smooth expansions 473 to construct a series solutions to the time-fractional Navier-Stokes equations. Alex. Eng. J. **61**(2), 1069?1077 (2022)

15. Saadeh, R.: Numerical solutions of fractional convection-diffusion equation using finite-difference and finite-volume schemes. J. Math. Comput. Sci. **11**(6), 7872?7891 (2021)
16. Qazza, A., Burqan, A., Saadeh, R.: A new attractive method in solving families of fractional differential equations by a new transform. Mathematics. **9**(23), 3039 (2021)
17. Edwan, R. et al.: Solving time-space-fractional Cauchy problem with constant coefficients by finite-difference method. In: Computational Mathematics and Applications, pp. 25?46. Springer. Singapore (2020).?
18. Burqan, A., Saadeh, R., Qazza, A.: A novel numerical approach in solving fractional neutral pantograph equations via the ARA integral transform. Symmetry. **14**(1), 50 (2022)
19. Saadeh, R.: Numerical algorithm to solve a coupled system of fractional order using a novel reproducing kernel method. Alex. Eng. J. **60**(5), 4583?4591 (2021)
20. Prüss, J.: Evolutionary Integral Equations and Applications, vol. 87. Birkhäuser (2013).
21. Aggarwal, S., Sharma, N.: Laplace transform for the solution of first kind linear Volterra integral equation. J. Adv. Res. Appl. Matt. Stat. **4**(3&4), 16?23 (2019)
22. Aggarwal, S., Sharma, N., Chauhan, R.: Solution of linear Volterra integral equations of second kind using Mohand transform. Int. J. Res. Advent Technol. **6**(11), 3098?3102 (2018)

A New Authentication Scheme Based on Chaotic Maps and Factoring Problems

Nedal Tahat, Obaida M. Al-hazaimeh, and Safaa Shatnawi

Abstract Users of public key cryptography systems reveal their public keys, but they keep their private keys private. A public key directory is where all of your public keys are kept. Public key cryptosystems place a high value on preventing their keys from being forgery or otherwise tampered with. A key authentication mechanism is therefore required to confirm that an intrusion has not happened. A key authentication technique is developed in this study by solving numerous issues, including chaotic maps and factoring. When compared to schemes based on a single problem, the proposed scheme has been mathematically demonstrated to be safer. An alternative to conventional key authentication systems, the proposed scheme can help to design a cryptography system that addresses a variety of issues. In contrast, the newly created authentication technique involves only minimal and low-complexity computations, making it incredibly efficient.

Keywords Key authentication · Chaotic maps · Factoring · Cryptanalysis · Attacks

1 Introduction

In public key cryptosystems, the most important issue is to safeguard public keys from being hacked by adversaries. On a specific problem, such as discrete logarithm, a number of authentication systems have been proposed in the past. Using

---

N. Tahat (✉)
Department of Mathematics, Faculty of Science, The Hashemite University, 330127, Zarqa 13133, Jordan
e-mail: nedal@hu.edu.jo

O. M. Al-hazaimeh
Department of Information Technology, Al-Balqa Applied University, Salt, Jordan
e-mail: dr_obaida@bau.edu.jo

S. Shatnawi
Tabareyah Secondary School for Girls, Directorate of Education of the District of Qasbah Irbid, The Ministry of Education, Salt, Jordan

D. Zeidan et al. (eds.), *Mathematics and Computation*, Springer Proceedings in Mathematics & Statistics 418, https://doi.org/10.1007/978-981-99-0447-1_5

discrete logarithms, Horng and Yang [1] came up with a new approach for public key cryptosystems in 1996. Although this approach is comparable to the standard certificate-based scheme, it does not require any authority to authenticate keys, unlike most others. Using the password guessing attack, Zhan et al. [2] demonstrated in 1999 that Horng-approach Yang's is not secure.

Many academics, such as [3–9], presented an improved key authentication system based on a similar problem to improve security. However, due to current technological advancements, intruders may be able to simply solve the authentication technique based on a single problem. As a result, this research is being carried out in order to design a key authentication technique based on multiple problems. A new secure key authentication technique based on discrete logarithm and factoring issues has been suggested recently [10]. However, their scheme is somewhat time-consuming. As early as 1989, an algorithmic design for image encryption based on a chaotic map was introduced [5, 11]. There has been an increase in work on this subject recently, as a few methods have been given in the research art [5, 12–17]. In comparison to public cryptosystems that use modular exponential computing or scalar multiplication on elliptic curves, chaotic map-based systems require the least computational complexity. Using a combination of chaotic maps and factorization problems, we present a novel approach to secure key authentication that improves security and reduces the number of operations required for both user registration and key authentication. Using chaotic maps and factoring to authenticate keys has not yet been proposed to our knowledge.

Here are the rest of the sections of the paper's structure: Sect. 2 has a few introductions. Section 3 explains a new key authentication method. In Sect. 4, we describe the security and performance of our proposed key authentication method, followed by numerical simulation in Sect. 5. Section 6 concluded this research work.

2 Preliminaries

In this section, we will briefly review the fundamental concept of the Chebyshev chaotic map [12, 17–21].

2.1 Chebyshev Chaotic Map

Authentication of the Chebyshev chaotic maps is provided [22–24]. The structure of the Chebyshev polynomials is reviewed in Fig. 1 [25].

To make it clear, consider the following example: A variable called x has an interval $[-1, 1]$, and n is an integer number. The Chebyshev polynomial $T_n(x) : [-1, 1] \to [-1, 1]$ is defined as $T_n(x) = \cos\left(n \cos^{-1}(x)\right)$, and the Chebyshev polynomial map $T_n(x) : \nu \to \nu$ of degree n is defined by

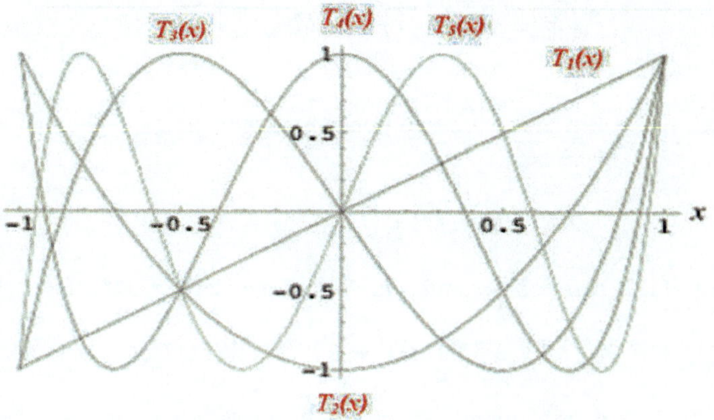

Fig. 1 Chebyshev polynomials structure

$$T_n(x) = 2xT_{n-1}(x) - T_{n-2}(x)\,; \quad n \geq 2 \tag{1}$$

where $T_0(x) = 1$, $T_1(x) = x$. Some Chebyshev polynomials are $T_2(x) = 2x^2 - 1$, $T_3(x) = 4x^3 - 3x$, $T_4(x) = 8x^4 - 8x^2 + 1$ and $T_5(x) = 16x^5 - 20x^3 + 5x$. The semi-group feature of Chebyshev polynomials is one of the most important properties of Chebyshev polynomials which is given by

$$T_r(T_s(x)) = T_{rs}(x) \tag{2}$$

This property has the direct consequence of ensuring that Chebyshev polynomials commute when used in the composition.

$$T_r(T_s(x)) = T_s(T_r(x))$$

Specifically, Zhang [15] shows that the semi-group property applies to Chebyshev polynomials defined on the interval $(-\infty, \infty)$, which has the effect of increasing the security of Chebyshev polynomials. It is possible to express the enhanced Chebyshev polynomials by

$$T_n(x) = (2xT_{n-1}(x) - T_{n-2}(x))(\text{mod } p) \tag{3}$$

where $n \geq 2$, $x \in (-\infty, \infty)$, and the large prime number is p, we obtain that

$$T_{rs}(x) = T_r(T_s(x)) = T_s(T_r(x))$$

Theorem 1 ([12]) **Let** $f(M) = t^2 - 2Mt + 1$ and α, β be two roots of $f(M)$. If $M = \frac{1}{2}(\alpha + \beta)$, in this case, the number of possible solutions is met by.

$$T_a(M) = \frac{\left(M + \sqrt{M^2 - 1}\right)^a + \left(M - \sqrt{M^2 - 1}\right)^a}{2} (\mathrm{mod}\ p) \quad (4)$$

Theorem 2 ([12]) If a and b are two positive integers and $a > b$, then we obtain that

$$2T_a(M).T_b(M) = T_{a+b}(M) + T_{a-b}(M) \quad (5)$$

Theorem 3 ([12]) **If** $a = b + c$ and p is a prime (i.e., large number), we obtain that

$$[T_a(M)]^2 + [T_b(M)]^2 + [T_c(M)]^2 = 2T_a(M)T_b(M)T_c(M) + 1(\mathrm{mod}\ p) \quad (6)$$

Lemma 1 ([12]) Let the elements of a finite field are g and h, i.e., if $g+g^{-1} = h+h^{-1}$ then $g = h$ or $g = h^{-1}$.

Lemma 2 ([12]) For any $g \in GF(p)$ and $y = g^t$ for some integer t, we can find an integer $M \in GF(p)$ and then construct a chaotic maps sequence $\{T_a(M)\}$, in polynomial time such that.

$$\frac{1}{2}\left(y + y^{-1}\right) = T_t(M) \in T_a(M) \quad (7)$$

Lemma 3 ([12]) Let p, n, and α are the same as earlier; and G is the group formed by the combination of these three. To obtain the value of v such that $a = T_{v^2(mod\ n)}(\alpha) mod\ p$, where a is given and $a \in G$, one must solve both the chaotic maps problem in G and the factorization of n.

Theorem 4 The discrete logarithm problem over $GF(p)$ can be solved in polynomial time if a method AL can be used to solve the chaotic mapping problem over $GF(p)$.

2.2 Computational Problem

To demonstrate the security of our proposed cryptosystem, we give the following important mathematical features of Chebyshev chaotic maps:

(a) Semi-group property: Given $x \in [-1, 1]$,

$$\begin{aligned} T_r(T_s(x)) &= \cos\left(r\cos^{-1}\left(s\cos^{-1}(x)\right)\right) = \cos\left(rs\cos^{-1}(x)\right) \\ &= T_{sr}(x) = T_s(T_r(x)) \end{aligned}$$

(b) Chaotic maps problem: If two items x and y are given, the discrete logarithm problem's task is to find integers s, such that $T_s(x) = y$.

Table 1 Initialization settings

Notations/parameters description	
$\overline{p}, \overline{q}$ and p	Three large strong primes, $n = \overline{pq}$
β	A primitive element, in $\{1, 2, \ldots p - 1\}$ and satisfying $\beta^n \equiv 1 \bmod p$
$g(\alpha) = T_\alpha(\beta) \bmod p$	As a public function, the chaotic map function of g is used
K_{pub}	Public key
K_{priv}	Private key

(c) If three elements x, $T_r(x)$, and $T_s(x)$, are given, Computing elements $T_{rs}(x)$. is the goal of the Diffie-Hellman problem.

Table 1 lists certain initialization settings (i.e., notations, and parameters) that are used in the developed approach.

3 Key Authentication Scheme

Phase one of the proposed system is user registration, and phase two is key authentication. It's safe to assume that Abu will log into the system manually. After that, a server will house all of Abu's publicly available data as the third level of authority. According to this plan, the server, the third authority, can be trusted. Once she has verified the public key of her recipient, Mimi can send her message to Abu. Because of this, Mimi as a sender will execute key authentication computations in order to overcome this issue. The proposed scheme of key authentication is as follows.

3.1 User Registration Phase

The phase of user registration involves the following steps:

- For each of the three numbers, pick a huge prime number p and two separate prime integers $\overline{p}$ and $\overline{q}$.
- Compute Eqs. 8 and 9, respectively.

$$n = \overline{pq} \text{ where } \quad n/(p-1), \tag{8}$$

$$\varphi(n) = (\overline{p} - 1)(\overline{q} - 1) \tag{9}$$

- Choose randomly any integer, e where $\gcd(\varphi(n), e) = 1$

- Find d, where

$$ed \equiv 1 \bmod \varphi(n) \tag{10}$$

- Randomly choose any integer, x where $x \in \{1, 2, \ldots, p-1\}$
- Compute the following equations:

$$g(x) = T_x(\beta) \bmod p \tag{11}$$

$$K_{priv} = (xg(x) + d) \bmod n \tag{12}$$

$$K_{pub} = T_{K_{priv}^2 (\bmod n)}(\beta) \bmod p \tag{13}$$

- Choose randomly any integer, r where $r \in Z_p^*$ and a password, pwd where $pwd \in Z_p^*$, then calculate the following:

$$g(pwd) = T_{pwd}(\beta) \bmod p \tag{14}$$

$$Y = T_r(\beta) \bmod p \tag{15}$$

$$V = g(pwd + r) \equiv T_{(pwd+r)}(\beta) \bmod p \tag{16}$$

- The encrypted password is $g(pwd)$. The user will send $g(pwd)$, Y and V to a server secretly.
- According to Eq. 17, a server will determine whether or not the user is authentic. The value of V will be stored in an encrypted password table if the equation is correct. $h(Y)$ will be calculated by a server and stored in an encrypted password database.

$$[v]^2 + \left[T_{pwd}(\beta)\right]^2 + [Y]^2 = \left(2VYT_{pwd}(\beta) + 1\right) \tag{17}$$

- Finally, compute the certificate (i.e., C) by the following equations, and then store the obtained parameters (i.e., K_{pub}, e and C) in the directory of the public key.

$$W = \left(pwd + r + K_{priv}^{2}\right)(\bmod n) \tag{18}$$

$$C = T_d(W) \bmod p \tag{19}$$

3.2 Key Authentication Phase

The process of authentication is used to build confidence between two or more interactive entities. This phase involves the following steps:

- Mimi will receive the following information from Abu: X, Y, and Z. She will then compute

$$m = T_e(C) \bmod p \tag{20}$$

- In order to make sure that Abu's public key has not been tampered with, Mimi will check to see whether the following equation is correct. A cryptosystem message will be encrypted using the public key of Mimi, if it is valid.

$$[T_m(\beta)]^2 + [V]^2 + \left[K_{pub}\right]^2 = \left(2V.K_{pub}.T_m(\beta) + 1\right)(\bmod p) \tag{21}$$

4 Security and Performance Analysis

It is in this section that the verification, security analysis, and efficiency analysis of the proposed key authentication technique are presented in the following sub-sections.

4.1 Security Analysis

The proposed method for discrete logarithm and chaotic maps (i.e., Chebyshev) is evaluated based on its computational complexity. A thorough study of the proposed scheme's advantages over some cryptanalysis issues has also been carried out to demonstrate its efficiency.

Theorem 5 If the user registration step goes successfully, the key authentication phase will go smoothly as well.

$$\begin{aligned}
[T_m(\beta)]^2 + [V]^2 + [K_{pub}]^2 &= [T_{T_e(C)}(\beta)]^2 + [T_{(pwd+r)}(\beta)]^2 + \left[T_{K^2_{priv}(\mathrm{mod}\,n)}(\beta)\right]^2 (\mathrm{mod}\,p) \\
&= [T_{T_e(C)}(\beta)]^2 + [T_{(pwd+r)}(\beta)]^2 + \left[T_{K^2_{priv}(\mathrm{mod}\,n)}(\beta)\right]^2 (\mathrm{mod}\,p) \\
&= [T_{T_e(T_d(W))}(\beta)]^2 + [T_{(pwd+r)}(\beta)]^2 + \left[T_{K^2_{priv}(\mathrm{mod}\,n)}(\beta)\right]^2 (\mathrm{mod}\,p) \\
&= [T_{T_{ed(\mathrm{mod}\,\varphi(n))}(W)}(\beta)]^2 + [T_{(pwd+r)}(\beta)]^2 + \left[T_{K^2_{priv}(\mathrm{mod}\,n)}(\beta)\right]^2 (\mathrm{mod}\,p) \\
&= [T_W(\beta)]^2 + [T_{(pwd+r)}(\beta)]^2 + \left[T_{K^2_{priv}(\mathrm{mod}\,n)}(\beta)\right]^2 (\mathrm{mod}\,p) \\
&= \left[T_{\left(pwd+r+K^2_{priv}\right)}(\beta)\right]^2 + [T_{(pwd+r)}(\beta)]^2 + \left[T_{K^2_{priv}(\mathrm{mod}\,n)}(\beta)\right]^2 (\mathrm{mod}\,p) \\
&= 2T_{\left(pwd+r+K^2_{priv}\right)}(\beta) T_{(pwd+r)}(\beta) T_{K^2_{priv}(\mathrm{mod}\,n)}(\beta) + 1(\mathrm{mod}\,p) \\
&= 2T_W(\beta)\, V\, K_{pub} + 1(\mathrm{mod}\,p) \\
&= 2T_{T_{ed(\mathrm{mod}\,\varphi(n))}(W)}(\beta)\, V\, K_{pub} + 1(\mathrm{mod}\,p) \\
&= 2T_{T_e(C)}(\beta)\, V\, K_{pub} + 1(\mathrm{mod}\,p) \\
&= 2T_m(\beta)\, V\, K_{pub} + 1\ (\mathrm{mod}\,p)
\end{aligned}$$

Factoring attack: If the adversary (i.e., Adv) is able to solve the factoring problem, then Eq. 12 provides the value of d. To obtain K_{priv}, Adv requires knowledge of the value of x in advance. However, obtaining the value of x is difficult due to the computational infeasibility of the discrete logarithm problem [1]. Moreover, Based on Eqs. 18 and 19, $(pwd + r)$ must be known in advance in order to get the K_{priv} value. Because the chaotic map problem is difficult to solve, it is difficult to get $(pwd + r)$.

Chaotic maps attack: If Adv is able to solve a chaotic map, the value of x can be found by referencing Eq. 11. Since d is known from Eq. 12, the value of K_{priv} can also be determined. The factoring issues in Eq. 10 make it difficult to obtain d value. K^2_{priv} can be determined if Eq. 13 can be figured out. But, factoring problems are notoriously difficult to solve, so this is impossible.

4.2 Performance Analysis

In comparison to other key authentication schemes, the computation of Chebyshev polynomial problem allows for smaller faster computation, key sizes, and significant savings in terms of memory, bandwidth, and energy. The ECC encryption algorithm has a high computational complexity, but the chaotic maps algorithm avoids scalar modular and multiplication exponentiation calculations, allowing for greater efficiency than the ECC algorithm. To make it easier to calculate the cost of computation, we use the following notations as listed in Table 2 [22, 24, 25].

Table 2 Notations

Notation	Description	Value
T_h	Hash function computation time	$T_h \approx 0.0005s$
T_{ch}	Extended chaotic function computation time	$T_{ch} \approx 0.172s$
T_{exp}	Exponentiation function computation time	$T_{exp} \approx 5.37s$
T_{mul}	Multiplication function computation time	$T_{mul} \approx 0.00207$

Table 3 Comparative analysis of the proposed scheme and existing scheme

	Phase	Criterion	State of the art scheme	Time complexity (Total)	Total (s)
State of the art scheme [10]	User registration	Time complexity	$6T_{exp} + 2T_{mul} + T_h + 2T_{qrt}$	$8T_{exp} + 3T_{mul} + T_h + 2T_{qrt}$	42.972
	Key authentication	Time complexity	$2T_{exp} + T_{mul}$		
Proposed scheme	User registration	Time complexity	$6T_{ch} + 2T_{mul} + T_h + 2T_{qrt}$	$9T_{ch} + 4T_{mul} + 5T_{qrt} + T_h$	1.567
	Key authentication	Time complexity	$36T_{ch} + 3T_{qrt} + 2T_{mul}$		

It is shown in Table 3 that the proposed scheme has a lower time complexity than that in [10]. The proposed scheme is more efficient than the one in [10]. While their scheme requires 42.972 s to complete, ours only takes 1.567 s to complete.

5 Numerical Simulation of the Proposed Scheme

5.1 User Registration Phase

- Abu choose randomly a large prime number, $p = 1427$ and two distinct prime numbers, $\overline{p} = 23$ and $\overline{q} = 31$. Compute $n = 23 \times 31 = 713$ and $\varphi(n) = 660$. $\beta = 12$ with order 713such that $12^{713} \equiv 1 \bmod 1427$.
- Choose $e = 113$ such that $\gcd(113, 660) = 1$. Compute an integer d such that $ed \equiv 1 \bmod 660$. Then $d = 257$. Randomly choose any integer, $x = 231$ and then calculate the following:

$$g(x) = T_{231}(12) \bmod 1427 = 100$$

$$K_{priv} = (231(100) + 257) \bmod 713 = 541$$

$$K_{pub} = T_{541^2(mod\,713)}(12) mod\ 1427 = 1010$$

- Choose randomly any integer, $r = 173$ where $r \in Z_p^*$ and a password, $pwd = 141$ where $pwd \in Z_p^*$, and we obtain that

$$g(pwd) = T_{141}(12)\text{mod }\ 1427 = 395$$

$$Y = T_{173}(12)\text{mod } 1427 = 598$$

$$V = g(pwd + r) \equiv T_{(141+173)}(12)\text{mod } 1427 = 1039$$

- To determine whether a user is authentic, the following equation must be checked by the server:

$$[v]^2 + \left[T_{pwd}(\beta)\right]^2 + [Y]^2 = 2VYT_{pwd}(\beta) + 1\ (\text{mod p})$$

$$[v]^2 + \left[T_{pwd}(\beta)\right]^2 + [Y]^2 = \left(1039^2 + 598^2 + 395^2\right)\text{mod } 1427 = 618$$

$$2VYT_{pwd}(\beta) + 1\ (\text{mod } p) = 2(1039)(395)(598) + 1(mod\ 1427 = 618)$$

- Compute the certificate (i.e., C) by the following equations, and then store the obtained value in the directory of the public key.

$$W = (141 + 173 + 351)(\text{mod} 713) = 665$$

$$C = T_d(W)\text{mod } p = T_{257}(665) = 797$$

$$K_{pub} = 1010,\ e = 113 \text{ and } C = 797$$

5.2 Key Authentication Phase

- Before making a calculation, Mimi will gather Abu's data (i.e., K_{pub}, e. and C) and then calculate the following:

$$m = T_{113}(797) \bmod 1427 = 585$$

- Mimi will validate whether or not the following equation is true to ensure that Abu's public key has not been altered:

$$[T_m(\beta)]^2 + [V]^2 + \left[K_{pub}\right]^2 = 2V.K_{pub}.T_m(\beta) + 1(\bmod p)$$

$$[T_m(\beta)]^2 + [V]^2 + \left[K_{pub}\right]^2 = 47^2 + 1039^2 + 1010^2(\bmod 1427) = 1286$$

$$2V.K_{pub}.T_m(\beta) + 1(\bmod p) = 2(1039)(1010)(47) + 1(\bmod 1427) = 1286$$

- Then we have

$$[T_m(\beta)]^2 + [V]^2 + \left[K_{pub}\right]^2 = 2V.K_{pub}.T_m(\beta) + 1(\bmod p)$$

- Finally, Mimi will encrypt the message before sending it to Abu using a cryptosystem using the public key.

6 Conclusion

Using chaotic maps and factoring problems, this paper proposes a key authentication scheme that is both secure and efficient. When compared to other key authentication schemes, such as elliptic curves, ElGamal, and RSA, the scheme takes advantage of the inherent advantages of chaotic map cryptosystems such as computationally less intensive and smaller key size. Additionally, the comparison of efficiency has been discussed. We conclude from the results that our scheme is superior to the Suparlan et al. schemes. Although this scheme is said to be more secure, it is less efficient than the existing single problem authentication scheme.

References

1. Horng, G., Yang, C.-S.: Key authentication scheme for cryptosystems based on discrete logarithms. Comput. Commun. **19**, 848–850 (1996)
2. Zhan, B., Li, Z., Yang, Y., Hu, Z.: On the security of HY-key authentication scheme. Comput. Commun. **22**, 739–741 (1999)
3. Tahat, N., Alomari, A., Al–Freedi, A., Al-Hazaimeh, O.M., Al–Jamal, M.F.: An efficient identity-based cryptographic model for Chebyhev chaotic map and integer factoring based cryptosystem. J. Appl. Secur. Res. **14**, 257–269 (2019)

4. Lee, C.-C., Hwang, M.-S., Li, L.-H.: A new key authentication scheme based on discrete logarithms. Appl. Math. Comput. **139**, 343–349 (2003)
5. Al-Hazaimeh, O.M.: A new dynamic speech encryption algorithm based on Lorenz chaotic map over internet protocol. Int. J. Electr. Comput. Eng. **10**, 4824 (2020)
6. Wang, M., Liu, D., Zhu, L., Xu, Y., Wang, F.: LESPP: lightweight and efficient strong privacy preserving authentication scheme for secure VANET communication. Computing **98**, 685–708 (2016)
7. Tahat, N., Alomari, A., Al-Hazaimeh, O.M., Al-Jamal, M.F.: An efficient self-certified multi-proxy signature scheme based on elliptic curve discrete logarithm problem. J. Discret. Math. Sci. Cryptography **23**, 935–948 (2020)
8. Yoon, E.-J., Yoo, K.-Y.: Secure key authentication scheme based on discrete logarithms. In: Third International Conference on Next Generation Web Services Practices (NWeSP'07), pp. 73–78 (2007)
9. Al-Hazaimeh, O.M.: Combining audio samples and image frames for enhancing video security. Indian J. Sci. Technol. **8**, 940 (2015)
10. Suparlan, A., Abd Nassir, A., Ismail, N., Shohaimay, F., Ismail, E.S.: Secure key authentication scheme based on discrete logarithm and factoring problems. In: Regional Conference on Science, Technology and Social Sciences (RCSTSS 2014), pp. 221–229 (2016)
11. Boneh, D., Franklin, M.: Identity-based encryption from the Weil pairing. In: Annual International Cryptology Conference, pp. 213–229 (2001)
12. Chain, K., Kuo, W.-C.: A new digital signature scheme based on chaotic maps. Nonlinear Dyn. **74**, 1003–1012 (2013)
13. Chen, W., Quan, C., Tay, C.: Optical color image encryption based on Arnold transform and interference method. Opt. Commun. **282**, 3680–3685 (2009)
14. Li, X., Zhao, D.: Optical color image encryption with redefined fractional Hartley transform. Optik **121**, 673–677 (2010)
15. Martin, K., Lukac, R., Plataniotis, K.N.: Efficient encryption of wavelet-based coded color images. Pattern Recogn. **38**, 1111–1115 (2005)
16. Obaida, M.A.-H.: Combining audio samples and image frames for enhancing video security. Indian J. Sci. Technol. **8**, 940 (2015)
17. Tay, C.J., Quan, C., Chen, W., Fu, Y.: Color image encryption based on interference and virtual optics. Opt. Laser Technol. **42**, 409–415 (2010)
18. Liu, Y., Xue, K.: An improved secure and efficient password and chaos-based two-party key agreement protocol. Nonlinear Dyn. **84**, 549–557 (2016)
19. Yoon, E.-J.: Efficiency and security problems of anonymous key agreement protocol based on chaotic maps. Commun. Nonlinear Sci. Numer. Simul. **17**, 2735–2740 (2012)
20. Tahat, N., Tahat, A.A., Albadarneh, R.B., Edwan, T.A.: Design of identity-based blind signature scheme upon chaotic maps. Int. J. Online Biomed. Eng. **16** (2020)
21. Al-Hazaimeh, O.M., Abu-Ein, A.A., Nahar, K.M., Al-Qasrawi, I.S.: Chaotic elliptic map for speech encryption. Indonesian J. Electr. Eng. Comput. Sci. **25**, 1103–1114 (2022)
22. Tahat, N., Ismail, E.S., Aljammal, A.H.: A cryptosystem based on chaotic maps and factoring problems. Int. J. Math. Operat. Res. **15**, 55–64 (2019)
23. Zhang, F., Chen, X.: Cryptanalysis of Huang–Chang partially blind signature scheme. J. Syst. Softw. **76**, 323–325 (2005)
24. Gura, N., Patel, A., Wander, A., Eberle, H., Shantz, S.C.: Comparing elliptic curve cryptography and RSA on 8-bit CPUs. In: International Workshop on Cryptographic Hardware and Embedded Systems, pp. 119–132 (2004)
25. El Bakrawy, L.M., Ghali, N.I., Hassanien, A.E., Kim, T.H: A fast and secure one-way hash function. In: International Conference on Security Technology, pp. 85–93 (2011)

A Pro Rata Definition of the Fractional-Order Derivative

Ramzi B. Albadarneh, Ahmad M. Adawi, Sa'ud Al-Sa'di, Iqbal M. Batiha, and Shaher Momani

Abstract In this paper a novel definition of the fractional-order derivative operator will be introduced. This operator will be called "pro rata" due to its ratio form as well as its geometric behavior that it is proportional to the fractional-order value. Some properties and theorems will be investigated. As an inverse of the fractional-order derivative operator, the integral of fractional order will be introduced. Some illustrative examples will be given.

Keywords Fractional derivative · Fractional integral

1 Introduction

Non-integer calculus is the calculus of differentiation and integration of arbitrary orders (real or complex), called fractional-order derivatives and integrals, which generalizes the concept of differentiation and fold integration of integer-order [19, 21]. The history of non-integer calculus set out approximately in the meanwhile when the traditional calculus was recognized. It was early reported in a letter between the mathematical geniuses Leibniz and L'Hôpital in 1695, where the semiderivative's idea was proposed. From that time forward, a lot of well-known physicists and mathematicians have mainly investigated fractional-order derivatives and integrals in a purely mathematical context, without its real applications, the basic concepts being connected with the names of Grunwald, Letnikov, Riemann, Abel, Liouville, and many more. But over the past few decades, it was turned out that the non-integer calculus has gained much attention as a result of its common appearance in different

R. B. Albadarneh (✉) · A. M. Adawi · S. Al-Sa'di
Department of Mathematics, Faculty of Science, The Hashemite University, P.O Box 330127, Zarqa 13133, Jordan
e-mail: rbadarneh@hu.edu.jo

I. M. Batiha
Department of Mathematics, Al Zaytoonah University of Jordan, Amman 11733, Jordan

I. M. Batiha · S. Momani
Nonlinear Dynamics Research Center (NDRC), Ajman University, Ajman, UAE

D. Zeidan et al. (eds.), *Mathematics and Computation*, Springer Proceedings in Mathematics & Statistics 418, https://doi.org/10.1007/978-981-99-0447-1_6

implementations in the scopes of engineering, electrical networks, fluid mechanics, diffusive transport, control theory, optics and signal processing, etc. [1–5, 7, 8, 16, 26, 27]. It should be noted that in current literature the terms "derivative" is used for positive orders and "integral" (for negative orders).

In the scope of mathematics, there exist several definitions of fractional-order differentiation and integration in the literature presently, involving Caputo [9, 10], Riemann-Liouville [11, 20], Crünwald-Letnikov [14], Riesz [24, 25], Weyl [6, 24, 25], Jumarie [15], Hadamard [24, 25]. The most famous definition that has been popularized is due to Riemann and Liouville, which depends in its construction on the nth-Cauchy's integral formula that relies only on a straightforward integration. The definition is obtained as follow: Let $a, T, \alpha \in \mathbb{R}$ such that $a < T, n = \max\{0, [\alpha] + 1\}$ and $f(t)$ be an integrable function on (a, T). For $n > 0$, if $f(t)$ is n-times differentiable function on (a, T) except on a set of measure zero, then for $t \in (a, T)$

$$ {}^{R}D_{a}^{\alpha} f(t) = \frac{1}{\Gamma(n-\alpha)} \frac{d^n}{dt^n} \int_a^t \frac{f(x)}{(t-x)^{\alpha-n+1}}\, dx = \frac{d^n}{dt^n}\left[{}^{R}I_{a}^{n-\alpha} f(t)\right] \tag{1} $$

where I_a^{α} is the fractional integral operator of order $\alpha > 0$. In particular, this operator can be outlined as the convolution integral of the function $t^{\alpha-1}$ and the function f itself, i.e.,

$$ {}^{R}I_{a}^{\alpha} f(t) = \frac{1}{\Gamma(\alpha)} \int_a^t (t-x)^{\alpha-1} f(x)\, dx \tag{2} $$

where $\Gamma(\cdot)$ is the gamma function, which is defined by

$$ \Gamma(z) = \int_0^{\infty} e^{-t} t^{z-1} dt, \quad \Re e(z) > 0. \tag{3} $$

In fact, formula (2) is a generalized formulation of the following Cauchy's formula for repeated integration of a continuous function f on $\mathbb{R}$, if $\alpha \in \mathbb{N}$ and $(n-1)!$ is replaced by its generalization $\Gamma(\alpha)$, see [22]:

$$ \int_a^s \int_a^{s_1} \int_a^{s_2} \dots \int_a^{s_{n-1}} f(s_n) ds_n ds_{n-1} \dots ds_2 ds_1 = \frac{1}{(n-1)!} \int_a^s (s-t)^{n-1} f(t) dt, \tag{4} $$

for $n \in \mathbb{N}, a, s \in \mathbb{R}, s > a$.

On the other hand, if $\alpha = k$ with $k \in \mathbb{N}$, then we have $n = k + 1$ and also we get

$$ {}^{R}D_{a}^{k} f(t) = \frac{1}{\Gamma(1)} \frac{d^{k+1}}{dt^{k+1}} \int_a^t f(x)\, dx = \frac{d^k f(t)}{dt^k}. \tag{5} $$

A useful alternative operator for the fractional-order Riemann-Liouville derivative operator was introduced originally by Caputo in 1967. This operator was then approved by Mainardi and Caputo in 1971 to be called later on by the fractional-order Caputo derivative operator. The definition of this operator can be defined as

$$ {}^{C}D_a^{\alpha} f(t) = \frac{1}{\Gamma(n-\alpha)} \int_a^t \frac{f^{(n)}(x)}{(t-x)^{\alpha-n+1}}\, dx, \quad n-1 < \alpha < n \tag{6} $$

It is clear that the fractional-order Caputo derivative operator is more limiting than the fractional-order Riemann-Liouville derivative operator. This is because it needs the existence of the nth-derivative of the function under consideration. At the same time, it is worth noting that the functions that not having the 1st-order derivative could have, in view of Riemann-Liouville sense, derivatives of fractional-order values less than one. In addition, it should be also noted that the fractional-order derivative of an arbitrary function does not need to be a continuous function at the origin and it does not need to be differentiable too.

However, the Caputo operator has confirmed its ability to greatly match with observational data that is typically used to describe the performance of several engineering and applied science problems. It is very important to point out that the Riemann-Liouville definition has certain drawbacks and limitations, especially in describing several real-life applications. This actually backs the fact that asserts these applications need certain definitions of the fractional-order derivative that can allow the usage of initial conditions that are physically interpretable. For instance, the fractional-order Riemann-Liouville derivative operator of a constant function does not equal zero. Besides, the fractional-order derivative of a given function will have a singularity at the origin, whenever it is constant at the origin. In this regard, it has been shown that the Caputo operator is highly advantageous for such tasks. In particular, such operator has an ability of using the initial conditions reported for the problems formulated by using certain differential equations of fractional order. Moreover, the fractional-order derivative of a constant function is zero by using this operator.

To this point, we have introduced two expressions of the fractional-order derivative operators. Actually, the existence of several expressions of the identical notion raises the query, are these definitions equivalent? The brief reply to this query in general is no, although the differentiation and integration operators are interchanged in the corresponding definitions of the Caputo fractional derivative and Riemann-Liouville fractional derivative. More particularly, it can be noted that, with the help of using Riemann-Liouville operator, the function at hand is first integrated $n-\alpha$-times and then differentiated n-times. On the other hand, with the help of using the Caputo operator, the same function is first differentiated n-times and then integrated $n-\alpha$-times. In general, the two aforesaid definitions cannot be coincided. That is

$$ {}^{R}D^{\alpha} f \neq {}^{C}D^{\alpha} f. \tag{7} $$

However, it was shown in [13] that the above two definitions can be coincided if and only if the function $f(x)$ together with its first $n-1$-derivatives vanish at $x=0$. More precisely, for $t>0$, $n-1<\alpha<n$, and $n \in \mathbb{N}$, we have

$$^{C}D^{\alpha}f(t) = {}^{R}D^{\alpha}f(t) - \sum_{k=0}^{n-1} \frac{t^{k-\alpha}}{\Gamma(k+1-\alpha)} f^{(k)}(0). \tag{8}$$

Proposition 1 *Let $n-1 < \alpha < n$, $n \in \mathbb{N}$, $\alpha \in \mathbb{R}$ and $f(t)$ be a function such that $^{C}D^{\alpha}f(t)$ exists. Then the following properties for the Caputo fractional derivatives hold:*

$$\lim_{\alpha \to n^-} {}^{C}D^{\alpha}f(t) = f^{(n)}(t), \tag{9}$$

$$\lim_{\alpha \to n-1^+} {}^{C}D^{\alpha}f(t) = f^{(n-1)}(t) - f^{(n-1)}(0). \tag{10}$$

In the same regard, it should be mentioned here that there is another operator for computing the fractional-order derivatives. This operator is called the Crünwald-Letnikov operator. It can be obtained under the assumption that assumes the function $f(t)$ must be n-times continuously differentiable on $[a, t]$. However, the Crünwald-Letnikov operator can be defined as follows:

$$^{G}D_a^{\alpha}f(t) = \sum_{k=0}^{n-1} \frac{f^{(k)}(a)}{\Gamma(-\alpha+k+1)}(t-a)^{-\alpha+k} + \frac{1}{\Gamma(n-\alpha)} \int_a^t \frac{f^{(n)}(x)}{(t-x)^{\alpha-n+1}}\, dx \tag{11}$$

Therefore, by considering a category of functions $f(t)$, possessing n-continuous derivatives for $t \geq 0$, as well as by means of carrying out some differentiations and frequent integrations by parts, the Riemann-Liouville operator can be inferred by the Crünwald-Letnikov operator.

It should be mentioned that all the definitions of fractional derivatives above satisfy the linearity property, that is

$$D^{\alpha}(\mu f(x) + \lambda g(x)) = \mu D^{\alpha}f(x) + \lambda D^{\alpha}g(x). \tag{12}$$

Recently in 2014 in [18], a novel straightforward fractional-order derivative definition called the *conformable fractional derivative* was proposed. This definition agrees with the traditional definitions of Riemann-Liouville and Caputo in dealing with polynomials. In particular, if $f : [0, \infty) \to \mathbb{R}$, then the conformable fractional derivative of order α of the function f can be outlined as follows:

$$T_{\alpha}f(t) := \lim_{\epsilon \to 0} \frac{f(t+\epsilon t^{1-\alpha}) - f(t)}{\epsilon}, \quad \text{for all } t > 0,\ \alpha \in (0, 1). \tag{13}$$

If f is α-differentiable in some $(0, a)$, $a > 0$, and $\lim_{t \to 0^+} f^{(\alpha)}(t)$ exists, then the fractional derivative at 0 is defined by $f^{(\alpha)}(0) = \lim_{t \to 0^+} f^{(\alpha)}(t)$. The authors in [18] proved some properties for the above definition. For example, they proved that if f is differentiable, then $T_{\alpha}(f)(t) = t^{1-\alpha} f'(t)$. However, the zero-order derivative of a function does not return the function, i.e., $T_0 f(t) \neq f(t)$, see [17, 23]. Besides, the derivative reported in (13) does not verify the index law; $T_\alpha T_\beta f(t) \neq T_{\alpha+\beta} f(t)$ for general α and β, and it does not verify the generalized Leibniz rule. However, it verifies the product rule,

$$T_\alpha(fg) = fT_\alpha(g) + gT_\alpha(f). \tag{14}$$

Furthermore, the definition given in (13) satisfies the interpolation property. In other words, for $0 < \alpha < 1$, we have

$$\lim_{\alpha \to 1^-} T_\alpha f(t) = \frac{df}{dt}, \quad \lim_{\alpha \to 0^+} T_\alpha f(t) = t\frac{df}{dt}. \tag{15}$$

For $n-1 < \alpha < n$:

$$\lim_{\alpha \to n^-} T_\alpha f(t) = \frac{d^n f}{dt^n}, \quad \lim_{\alpha \to n-1^+} T_\alpha f(t) = t\frac{d^n f}{dt^n}, \tag{16}$$

In addition, the author in [18] proved similar results to the classical Mean Value Theorem and Rolle's Theorem.

The organization of this article is arranged in the following manner: In the next section, we propose a new definition of the derivative of order α where $\alpha \in [0, 1]$, and a general definition of derivatives of higher order, then we prove some important properties. In Sect. 3, we introduce a generalized definition of the integral of order α. We conclude the paper with some remarks in Sect. 4.

2 Fractional Derivative

In this section, we give our new definition of the derivative of order α of a continuous function f at a point x and prove several results that are close to those found in classical calculus.

Definition 1 Let $f(x)$ be a continuous function on the interval $(x-\epsilon, x+\epsilon)$ where $\epsilon > 0$. The derivative of order $\alpha \in [0, 1]$ is defined by

$$D^\alpha f(x) = \frac{d^\alpha f}{dx^\alpha} = \lim_{h \to 0} \frac{\alpha f(x+h) + [(1-\alpha)h - \alpha]f(x)}{h} \tag{17}$$

If the derivative of f of order α exists, then we will say that f is α-differentiable. Notice that if a function $f(x)$ is differentiable on an interval $[a, b]$ then the derivative $D^\alpha f$ is defined. This leads to the following theorem.

Theorem 1 *Let $f(x)$ be a continuous function on the interval $[a, b]$. If $0 < \alpha \leq 1$ and $x \in [a, b]$, then $D^\alpha f(x)$ exists if and only if $f'(x)$ exists. Consequently,*

$$D^\alpha f(x) = (1-\alpha)f(x) + \alpha f'(x), \quad 0 < \alpha < 1. \tag{18}$$

Proof Let $x \in [a, b]$. First assume that $D^\alpha f(x)$ exists, hence the limit in (17) exists, and

$$\alpha \lim_{h \to 0} \frac{f(x+h) - f(x)}{h} + (1-\alpha) f(x) = D^\alpha f(x) \tag{19}$$

hence,

$$\lim_{h \to 0} \frac{f(x+h) - f(x)}{h} = \frac{\alpha - 1}{\alpha} f(x) + \frac{1}{\alpha} D^\alpha f(x). \tag{20}$$

This implies that

$$f'(x) = \frac{\alpha - 1}{\alpha} f(x) + \frac{1}{\alpha} D^\alpha f(x) \tag{21}$$

exist. Conversely, assume that

$$f'(x) = \lim_{h \to 0} \frac{f(x+h) - f(x)}{h} \tag{22}$$

exists, then

$$\lim_{h \to 0} \frac{f(x+h) - f(x)}{h} + (1-\alpha) f(x) = \lim_{h \to 0} \frac{\alpha f(x+h) + [(1-\alpha)h - \alpha] f(x)}{h} \tag{23}$$

exists.

It follows that, if a function f is α-differentiable, $\alpha \in (0, 1]$ at $x_o \in (a, b)$, then f is continuous at x_o.

Example 1 Using formula (18) we can compute the derivative of order α of some functions, for example:

1. Let $f(x) = c$, the constant function. Then $D^\alpha(f) = (1-\alpha)c$.
2. Let $f(x) = Ax + B$. Then $D^\alpha(f) = (1-\alpha)(Ax + B) + \alpha A$.
3. Let $f(x) = x^p$. Then $D^\alpha(f) = (1-\alpha)x^p + \alpha p x^{p-1}$.
4. Let $f(x) = e^x$. Then $D^\alpha(f) = e^x$.

For higher order derivatives case we can generalize the definition to the following:

Definition 2 Let n be a positive integer and $\alpha \in [n, n+1]$. If $f(x)$ is an $n+1$ differentiable on $[a, b]$, then

$$\begin{aligned} D^\alpha f = \lim_{h \to 0} \frac{1}{h} \Big((n+1-\alpha) \left(f^{(n-1)}(x+h) - f^{(n-1)}(x) \right) \\ + (\alpha - n) \left(f^{(n)}(x+h) - f^{(n)}(x) \right) \Big). \end{aligned} \tag{24}$$

Similar argument used in Theorem 1 shows that

$$D^\alpha f = \frac{d^\alpha f}{dx^\alpha} = (n+1-\alpha)\frac{d^n f}{dx^n} + (\alpha - n)\frac{d^{n+1} f}{dx^{n+1}}. \tag{25}$$

We notice from Definition 2 of the fractional derivative of order α that when the parameter α varies from the integer n to the integer $n+1$ then the fractional derivative of order α varies continuously from the $n^t h$ derivative to the $(n)^{th}$ derivative with

$$\lim_{\alpha \to n^+} D^\alpha f = \frac{d^n f}{dx^n} \quad \text{and} \quad \lim_{\alpha \to n+1^-} D^\alpha f = \frac{d^{n+1} f}{dx^{n+1}}$$

and hence

$$\lim_{\alpha \to n} D^\alpha f = \frac{d^n f}{dx^n}.$$

This desired property is not confirmed in many other definitions of the fractional derivative. For the illustration of this property see Fig. 1.

It can be shown that in the case of α is an integer, this definition reduces to the standard definition of the nth-derivative of $f(x)$. This shows that our definition of the derivative of order α is a generalization of the integer-order derivative. Now we are going to obtain some general properties of our new definition of the derivative of order α. First, the linearity of the differential operator D^α is ensured by the following theorem:

Theorem 2 *Let n be a non-negative integer and $\alpha \in [n, n+1]$. If $f(x)$ and $g(x)$ are two functions such that both $D^\alpha f$ and $D^\alpha g$ exists. Then the derivative of order α is a linear operator, i.e.,*

$$D^\alpha (\lambda f(x) + \mu g(x)) = \lambda D^\alpha f(x) + \mu D^\alpha g(x), \tag{26}$$

for any constants λ, μ.

Proof The proof follows directly from Eq. (25) and the linearity of the limit.

This definition of the derivatives of order α carries with it some important properties, that will show importance when solving equations involving integrals and derivatives of general order.

Proposition 2 (The Product Rule of Fractional Derivative)

1. *If $0 \le \alpha \le 1$, and f, g are two differentiable functions, then*

$$D^\alpha (fg) = \alpha \left(fg' + f'g\right) + (1-\alpha) fg. \tag{27}$$

2. *Let n be a non-negative integer. If $n \le \alpha \le n+1$, and f, g are two $(n+1)$-differentiable functions, then*

$$D^{\alpha}(fg) = (\alpha - n)\sum_{k=0}^{n+1}\left(D^k f\right)\left(D^{n+1-k} g\right) + (n+1-\alpha)\sum_{k=0}^{n}\left(D^k f\right)\left(D^{n-k} g\right). \tag{28}$$

Proposition 3 (The Iterated Fractional Derivative) *Let n be a non-negative integer, and $n \leq \alpha \leq n+1$. We denote the 2nd-iterated fractional derivative $D^{\alpha}D^{\alpha}f$ by $(D^{\alpha})^2 f$. Then*

$$(D^{\alpha})^2 f = (\alpha - n)^2 f^{(2n+2)} + 2(\alpha - n)(n+1-\alpha) f^{(2n+1)} + (n+1-\alpha)^2 f^{(2n)}. \tag{29}$$

In general, for a positive integer k, we write $(D^{\alpha})^k f = D^{\alpha}D^{\alpha}\ldots D^{\alpha} f$, k times, then the kth-iterated derivative of f is given by

$$(D^{\alpha})^k f = \sum_{j=0}^{k}\binom{k}{j}(\alpha - n)^j (n+1-\alpha)^{k-j} f^{(kn+j)}. \tag{30}$$

The following theorem shows that the derivative of order α defined in (24) is commutative.

Theorem 3 *Let n, m be two non-negative integers, and f be an $(n+m+2)$ differentiable function on (a, b). If $\alpha \in [n, n+1]$ and $\beta \in [m, m+1]$, then*

$$D^{\alpha}D^{\beta} f(x) = D^{\beta}D^{\alpha} f(x), \quad \forall x \in (a, b). \tag{31}$$

Proof To begin with, let $x \in (a, b)$, if we apply (25) twice we get

$$\begin{aligned} D^{\alpha}\left(D^{\beta} f(x)\right) &= D^{\alpha}\left((\beta - m) f^{(m+1)}(x) + (m+1-\beta) f^{(m)(x)}\right) \\ &= (\alpha - n)(\beta - m) f^{(n+m+2)}(x) \\ &\quad + (\alpha - n)(m+1-\beta) f^{(n+m+1)}(x) \\ &\quad + (n+1-\alpha)(\beta - m) f^{(n+m+1)}(x) \\ &\quad + (n+1-\alpha)(m+1-\beta) f^{(n+m)}(x). \end{aligned}$$

Similarly,

$$\begin{aligned} D^{\beta}\left(D^{\alpha} f(x)\right) &= D^{\beta}\left((\alpha - n) f^{(n+1)}(x) + (n+1-\alpha) f^{(n)(x)}\right) \\ &= (\beta - m)(\alpha - n) f^{(n+m+2)}(x) \\ &\quad + (\beta - m)(n+1-\alpha) f^{(n+m+1)}(x) \\ &\quad + (m+1-\beta)(\alpha - n) f^{(n+m+1)}(x) \\ &\quad + (m+1-\beta)(n+1-\alpha) f^{(n+m)}(x). \end{aligned}$$

Hence, $D^{\alpha}D^{\beta} f(x) = D^{\beta}D^{\alpha} f(x)$, for any $x \in (a, b)$.

We noticed that the definition (18) is equivalent to the classical definition of the first derivative of a function f. This suggests that there are corresponding results similar to the classical Rolle's theorem and the Mean Value theorem for the derivative of order α definition, as we prove in the next theorems.

Theorem 4 (Rolle's Theorem for Derivative of Order α) *Let $f : [a, b] \to \mathbb{R}$ be a continuous function on $[a, b]$ and α-differentiable on (a, b) for some $\alpha \in [0, 1]$. If $f(a) = f(b)$, then there exists $c \in (a, b)$ such that*

$$D^{\alpha} f(c) = (1 - \alpha) f(c). \tag{32}$$

Proof Since f is α-differentiable for $\alpha \in [0, 1]$ then f is differentiable on (a, b). Hence, by the classical Rolle's theorem, there exists $c \in (a, b)$ such that $f'(c) = 0$. Consequently, by Eq. (18),

$$D^{\alpha} f(c) = \alpha f'(c) + (1 - \alpha) f(c) = (1 - \alpha) f(c), \quad 0 < \alpha < 1 \tag{33}$$

completing the proof.

The following result is a direct consequence of the ordinary mean value theorem.

Theorem 5 (Mean Value Theorem for Derivative of order α) *Let $f : [a, b] \to \mathbb{R}$ be a continuous function on $[a, b]$ and differentiable on (a, b). Then there exists $c \in (a, b)$ such that*

$$D^{\alpha} f(c) = \alpha \left(\frac{f(b) - f(a)}{b - a} \right) + (1 - \alpha) f(c), \quad 0 < \alpha < 1. \tag{34}$$

Proof The asserted conclusion follows directly by applying Eq. (18).

The following theorem is a general version of Rolle's theorem for the derivative of order α.

Theorem 6 (Generalized Rolle's Theorem for Derivative) *Let $f : [a, b] \to \mathbb{R}$ be a continuous function on $[a, b]$ and n times differentiable on (a, b). If $f(x) = 0$ at the $n + 1$ distinct points $\{x_i\}_{i=0}^{n}$ such that $a \leq x_0 < x_1 < \cdots < x_n \leq b$, then there exists $c \in (x_0, x_n)$, and hence in (a, b), such that*

$$D^{\alpha} f(c) = (n - \alpha) f^{(n-1)}(c), \tag{35}$$

for any $\alpha \in [n - 1, n]$.

Proof Since f is n-times differentiable on (a, b) then by the ordinary generalized Rolle's theorem [12, p. 549], there exists $c \in (a, b)$ with $f^{(n)}(c) = 0$. If $n - 1 < \alpha < n$, then by (25) we have

$$D^{\alpha}f(c) = (\alpha - n + 1)f^{(n)}(c) + (n - \alpha)f^{(n-1)}(c), \quad n - 1 < \alpha < n \tag{36}$$

simple computations leads to (35).

Proposition 4 *If f is n-differentiable at a point $x = c$, for some positive integer n, and*

$$f(c) = f'(c) = f''(c) = \cdots = f^{(n)}(c), \tag{37}$$

then $D^{\alpha}f(c) = f(c)$ for any $\alpha \in [0, n]$.

Proposition 5 *If f is differentiable on $[a, b]$, and $f(x) \in [c, d]$, $f'(x) \in [c, d]$, for all $x \in [a, b]$, then $D^{\alpha}f(x) \in [c, d]$ for all $x \in [a, b]$ and $\alpha \in [0, 1]$.*

The following propositions give some geometric representations of the derivative of order α defined in the Definition 1.

Proposition 6 *Let f be differentiable on $[a, b]$.*

(a) If $f(x) \leq f'(x)$ for all $x \in [a, b]$, then $f(x) \leq D^{\alpha}f(x) \leq f'(x)$ for all $x \in [a, b]$, $\alpha \in [0, 1]$,
(b) If $f'(x) \leq f(x)$ for all $x \in [a, b]$, then $f'(x) \leq D^{\alpha}f(x) \leq f(x)$ for all $x \in [a, b]$, $\alpha \in [0, 1]$.

We note that proposition 6 stated that the graph of $D^{\alpha}f(t)$ for $0 < \alpha < 1$ always lies between the functions $f(t)$ and $f'(t)$, for example, the fractional derivative using our definition for $f(t) = \frac{1}{2}\sin(t^2)$ for different values of α is shown in Fig. 1.

Proposition 7 *If f is twice differentiable on $[a, b]$ and f and f' are increasing (decreasing) on $[a, b]$, then the graph of $D^{\alpha}f$ is increasing (decreasing) on $[a, b]$, for all $\alpha \in [0, 1]$.*

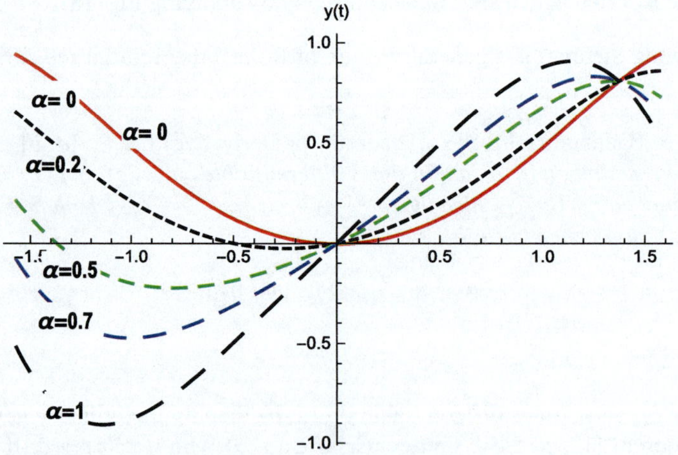

Fig. 1 The graph of $D^{\alpha}f(t)$ where $f(t) = \frac{1}{2}\sin(t^2)$ for $\alpha = 0, 0.2, 0.5, 0.7$, and 1

Proposition 8 *If f is 3rd-differentiable on $[a, b]$ and f and f' are concave up (concave down) on $[a, b]$, then the graph of $D^\alpha f$ is concave up (concave down) on $[a, b]$, for all $\alpha \in (0, 1)$.*

Proposition 9 *For $\alpha \in [0, 1]$,*

$$(1-\alpha)\Big(\textit{Area between } f(x) \textit{ and } D^\alpha f(x)\Big) = \alpha\Big(\textit{area between } f'(x) \textit{ and } D^\alpha f(x)\Big). \tag{38}$$

Proof The area between $f(x)$ and $D^\alpha f(x)$ over an interval I is given by

$$\int_I |D^\alpha f(x) - f(x)|\, dx = \int_I |((1-\alpha)f(x) + \alpha f'(x)) - f(x)|\, dx = \alpha \int_I |f(x) - f'(x)|\, dx. \tag{39}$$

Similar computations gives that the area between $f'(x)$ and $D^\alpha f(x)$ over the interval I is $(1-\alpha)\int\limits_I |f(x) - f'(x)|\, dx$, the conclusion follows immediately.

Recall that a continuously differentiable function is monotone in some interval $[a, b]$ if and only if its first derivative does not change its sign there. We now state general results involving derivatives of order α holds, the proofs are based on the standard definition of limits and the fact that $f'(x) = \lim\limits_{\alpha \to 1^-} D^\alpha f(x)$.

Proposition 10 *Let $f \in C^1[a, b]$.*

1. *If there exists $\alpha_o \in [0, 1)$ such that $D^\alpha f(x) > 0$ for all $x \in [a, b]$ and every $\alpha \in (\alpha_0, 1)$, then f is increasing on $[a, b]$.,*
2. *If there exists $\alpha_o \in [0, 1)$ such that $D^\alpha f(x) < 0$ for all $x \in [a, b]$ and every $\alpha \in (\alpha_0, 1)$, then f is decreasing on $[a, b]$.*

Proposition 11 *Let $f \in C^1[a, b]$.*

1. *If f is increasing on $[a, b]$ then for $x \in [a, b]$ there exists $\alpha_x \in (0, 1)$ such that $D^\alpha f(x) \geq 0$ for all $\alpha \in (\alpha_x, 1)$.*
2. *If f is decreasing on $[a, b]$ then for $x \in [a, b]$ there exists $\alpha_x \in (0, 1)$ such that $D^\alpha f(x) \leq 0$ for all $\alpha \in (\alpha_x, 1)$.*

Proposition 12 *If for all $x \in [a, b]$, $D^\alpha f(x) = 0$ and $D^\beta f(x) = 0$ for some $\alpha, \beta \in [0, 1]$ with $\alpha \neq \beta$, then $f(x) = 0$ for all $x \in [a, b]$.*

Proposition 13 *If $\alpha \in (0, 1)$ then the Laplace transform of $f(x)$ is given by*

$$\mathcal{L}\{D^\alpha f\} = \mathcal{L}\{(1-\alpha)f + \alpha f'\} \tag{40}$$

$$= (\alpha(s-1)+1)F(s) - f(0), \tag{41}$$

where $F(s)$ is the Laplace transform of f, and for $\alpha \in (n, n+1)$, the Laplace transform of $f(x)$ is given by

$$\mathcal{L}\{D^\alpha f\} = (n+1-\alpha)\left(s^n F(s) - \sum_{k=0}^{n-1} s^{n-k} f^{(k)}(0)\right)$$
$$+(\alpha-n)\left(s^{n+1} F(s) - \sum_{k=0}^{n} s^{n+1-k} f^{(k)}(0)\right).$$

3 Integral of Order α

Now we introduce a generalized definition of the integral of order α as follows:

Definition 3 Let $f(t)$ be a function defined on $[a, x]$. If $0 < \alpha \leq 1$, then the integral of order α of f is defined by

$$I_a^\alpha f(x) = \frac{1}{\alpha}\int_a^x \exp\left[\left(\frac{1-\alpha}{\alpha}\right)(t-x)\right] f(t)\,dt. \tag{42}$$

Observe that the integral of a continuous function f is an anti-derivative of f. We prove this property in the next theorem.

Theorem 7 *Let f be a continuous function such that $I_a^\alpha f(x)$ exists $\alpha \in (0, 1]$. Then*

$$D^\alpha I_a^\alpha f(x) = f(x), \quad x \geq a.$$

Proof Since f is continuous, then $I_a^\alpha f$ is differentiable, hence, By (18) we have

$$D^\alpha(I_a^\alpha f) = \alpha\frac{d(I_a^\alpha f)}{dx} + (1-\alpha)(I_a^\alpha f) = f(x), \tag{43}$$

solving we get

$$I_a^\alpha f = \frac{1}{\alpha}\int_a^x \exp\left[\left(\frac{1-\alpha}{\alpha}\right)(t-x)\right] f(t)\,dt + c\exp\left[\left(\frac{1-\alpha}{\alpha}\right)x\right], \tag{44}$$

where c is an arbitrary constant. Setting the constant c to be zero we get (42), with $D^\alpha I^\alpha f(x) = f(x)$.

Definition 3 can be generalized for the integral of higher order as follows:

Definition 4 Let $f(t)$ be a function defined on $[a, x]$. If $n < \alpha \leq n+1$, then the integral of order α of f is defined by

$$I_a^\alpha f(x) = \frac{1}{(n-1)!(\alpha-n)}\int_a^x (x-s)^{n-1}\int_a^s \exp\left[\left(\frac{n+1-\alpha}{\alpha-n}\right)(t-s)\right] f(t)\,dt\,ds. \tag{45}$$

Similar argument used in the proof of Theorem (7) can be used to show that the definition above satisfies the property $D^\alpha I_a^\alpha f(x) = f(x)$ for all $x \geq a$. Indeed, consider

$$(\alpha - n)\frac{d^{n+1}(I_a^\alpha f)}{dx^{n+1}} + (n + 1 - \alpha)\frac{d^n(I_a^\alpha f)}{dx^n} = f(x), \tag{46}$$

equivalently,

$$(\alpha - n)\frac{d}{dx}\left(\frac{d^n(I_a^\alpha f)}{dx^n}\right) + (n + 1 - \alpha)\frac{d^n(I_a^\alpha f)}{dx^n} = f(x). \tag{47}$$

Solving, we get

$$\frac{d^n(I_a^\alpha f)}{dx^n} = \frac{1}{\alpha - n}\int_a^x \exp\left[\left(\frac{n+1-\alpha}{\alpha - n}\right)(t - x)\right] f(t)\,dt + c\exp\left(\frac{n+1-\alpha}{\alpha - n}x\right), \tag{48}$$

where c is an arbitrary constant. Integrating the last expression n-times, and setting the arbitrary constants to be zero, we obtain (45).

Theorem 8 *Let f be α-function for $\alpha \in (0, 1)$. Then for all $x > a$ we have*

$$I_a^\alpha (D^\alpha f(x)) = f(x) - f(a)\exp\left[\left(\frac{\alpha - 1}{\alpha}\right)(x - a)\right].$$

Note that Theorems 7 and 8 show that the derivative of order α and the integral of order α of a function f on $[a, b]$ are inverse of each other provided that $f(a) = 0$.

4 Illustrative Examples

Example 2 $D^\alpha(e^x) = e^x$ for any $\alpha > 0$. This desired property cannot be satisfied with other definitions.

Example 3 Consider the following initial value problem: $D^{1/2}y = 1$, $y(0) = 1$. This equation can be reduced to: $\frac{dy}{dx} + y = 2$, $y(0) = 1$, which has the solution: $y = 2 - e^x$.

Example 4 Consider the following differential equation: $D^{3/2}y - D^{1/2}y = 0$. This equation can be reduced to: $\frac{d^2y}{dx^2} - y = 0$. The initial value problem has the general solution: $y = c_1e^x + c_2e^{-x}$.

5 Conclusion

In this work, a novel definition of the fractional-order derivative operator has been presented. It has been found that the proposed definition is an extension of the classical operator. The following properties have been inferred:

1. For n is a non-negative integer and $n < \alpha < n + 1$ we have:
$D^{\alpha} f = (n + 1 - \alpha)\frac{d^n f}{dx^n} + (\alpha - n)\frac{d^{n+1} f}{dx^{n+1}}$
2. $\lim_{\alpha \to n} D^{\alpha} f = \frac{df^n}{dx^n}$
3. Sum rule.
4. Product rules.
5. Commutative rule.

It is clear that solving fractional differential equations with the pro rata definition is easier than solving such equations with some other definitions. The computing of Laplace transforms and other transforms is also easier than computing them with some other definitions. The fractional-order integral operator has been also defined.

References

1. Albadarneh, R.B., Alomari, A.B., Tahat, N., Batiha, I.: Analytic solution of nonlinear singular BVP with multi-order fractional derivatives in electrohydrodynamic flows. Int. J. Math. Comput. Sci. **11**(4), 1125 (2021). https://doi.org/10.11591/ijece.v11i6.pp5367-5378
2. Albadarneh, R.B., Batiha, I., Adwai, A., Tahat, N., Alomari, A.B.: Numerical Riemann-Liouville fractional derivative operator. Int. J. Electr. Comput. Eng. **11**(16), 5367–5378 (2021). https://doi.org/10.11591/ijece.v11i6.pp5367-5378
3. Albadarneh, R.B., Batiha, I., Alomari, A.B., Tahat, N.: Numerical approach for approximating the caputo fractional-order derivative operator. AIMS Math. **6**(11), 12743–12756 (2021). https://doi.org/10.3934/math.2021735
4. Albadarneh, R. B., Batiha, I., Tahat, N. and Alomari, A. B.: Analytical solutions of linear and non-linear incommensurate fractional-order coupled systems. Indones. J. Electr. Eng. Comput. Sci. **21**(2), 776–790 (2021). https://doi.org/10.11591/ijeecs.v21.i2.pp776-790
5. Alomari, A.K., Drabseh, G.A., Al-Jamal, M.F., Albadarneh, R.B.: Numerical simulation for fractional phi-4 equation using homotopy sumudu approach. Int. J. Simul. Process Model. **16**(1), 26–33 (2021). https://doi.org/10.1504/IJSPM.2021.113072
6. Atangana, A.: Numerical solution of space-time fractional derivative of groundwater flow equation. In: Proceedings of the International Conference of Algebra and Applied Analysis, vol. 2, p. 20 (2012)
7. Batiha, I.M., Albadarneh, R.B., Momani, S., Jebril, I.H.: Dynamics analysis of fractional-order hopfield neural networks. Int. J. Biomath. **13**(08) (2020). https://doi.org/10.1142/S1793524520500837
8. Beyer, H., Kempfle, S.: Definition of physically consistent damping laws with fractional derivatives. ZAMM-J. Appl. Math. Mech./Zeitschrift fur Angewandte Mathematik und Mechanik **75**(8), 623–635 (1995). https://doi.org/10.1002/zamm.19950750820
9. Caputo, M.: Elasticit'a e dissipazione (zanichelli bologna) (1969)
10. Caputo, M., Mainardi, F.: Linear models of dissipation in anelastic solids. La Rivista del Nuovo Cimento (1971-1977) **1**(2), 161–198 (1971). https://doi.org/10.1007/BF02820620

11. Das, S.: Functional Fractional Calculus. Springer, Berlin, Heidelberg (2011). https://doi.org/10.1007/978-3-642-20545-3
12. Epperson, F.J.: An Introduction to Numerical Methods and Analysis. Wiley (2013)
13. Gorenflo, R., Mainardi, F.: Essentials of Fractional Calculus (2000)
14. Grunwald, A.K.: Uber "begrente" derivationen und deren anwedung. Zangew Math und Phys **12**, 441–480 (1867)
15. Jumarie, G.: Modified Riemann-Liouville derivative and fractional Taylor series of nondifferentiable functions further results. Comput. Math. Appl. **51**(9–10), 1367–1376 (2006). https://doi.org/10.1016/j.camwa.2006.02.001
16. Jumarie, G.: Path probability of random fractional systems defined by white noises in coarse-grained time applications of fractional entropy. Frac. Diff. Eq **1**(1), 45–87 (2011). https://doi.org/10.7153/fdc-01-03
17. Katugampol, U.N.: Correction to "what is a fractional derivative?" by ortigueira and machado [journal of computational physics, vol. 293, 15 july 2015, pages 4-13. special issue on fractional pdes]. J. Comput. Phys. **321**, 1255–1257 (2016). https://doi.org/10.1016/j.jcp.2016.05.052
18. Khalil, R., Al Horani, M., Yousef, A., Sababheh, M.: A new definition of fractional derivative. J. Comput. Appl. Mathemat. **264**, 65–70 (2014). https://doi.org/10.1016/j.cam.2014.01.002
19. Mainardi, F.: Fractals and Fractional Calculus in Continuum Mechanics. In: Caepinteri, A., Mainardi, F. (eds.) CISM Courses and Lectures, vol. 378. Springer, Wie (1997). https://doi.org/10.1007/978-3-7091-2664-6_7
20. Mainardi, F.: Fractional Calculus and Waves in Linear Viscoelasticity: An Introduction to Mathematical Models. World Scientific (2010). https://doi.org/10.1142/p614
21. Miller, K.S., Ross, B.: An Introduction to the Fractional Calculus and Fractional Differential Equations (1993)
22. Oldham, K., Spanier, J.: The Fractional Calculus. Academic, New York, London (1974)
23. Ortigueira, M.D., Machado, J.T.: What is a fractional derivative? J. Comput. Phys. **293**, 4–13 (2015). https://doi.org/10.1016/j.jcp.2014.07.019
24. Podlubny, I.: Geometric and physical interpretation of fractional integration and fractional differentiation. Fract. Calc. Appl. Anal. **5**(4), 367–386 (2002)
25. Samko, S.G., Kilbas, A.A., Mariche, O.I.: Integrals and Derivatives of Fractional Order and some Applications of them (1987)
26. Silva, M., Machado, J., Lopes, A.: Fractional order control of a hexapod robot. Nonlinear Dyn. **38**(1–4), 417–433 (2004). https://doi.org/10.1007/s11071-004-3770-8
27. Yan, J., Li, C.: On chaos synchronization of fractional differential equations. Chaos, Solitons Fractals **32**(2), 725–735 (2007). https://doi.org/10.1016/j.chaos.2005.11.062

Investigating Multicollinearity in Factors Affecting Number of Born Children in Iraq

Salisu Ibrahim, Mowafaq Muhammed Al-Kassab, and Muhammed Qasim Al-Awjar

Abstract The occurrence of multiple multicollinearities in many multiple regression models leads to significant problems that can affect the results of the entire multiple regression model, and among the problems is the low accuracy of the estimated coefficients, which reduces the statistical power of the model. The effect of sensitivity on the estimated coefficients is due to a small swing in the model. This paper discusses the two basic approaches to defining a multilinear relationship. The first approach is the correlation coefficient (CC) and the second is the variance inflation factor (VIF). Hill regression, principal component regression, intention root regression, and weighted regression are advanced regression models to investigate the existence of multiple multicollinearities, and these results will process, reduce and stabilize the multiple multicollinearities between independent variables, and help predict the best-fit model. Finally, we came up with the best suitable model.

Keywords Multiple regression · Multicollinearity · Correlation coefficient · Variance inflation factor Smoking mother

1 Introduction

Multicollinearity occurs when explanatory variables in a regression model are correlated. This correlation is a problem because independent variables should be independent. If the degree of correlation between variables is high enough, it can cause

S. Ibrahim (✉) · M. M. Al-Kassab
Department of Mathematics Education, Tishk International University, Erbil, Kurdistan Region, Iraq
e-mail: salisu.ibrahim@tiu.edu.iq; ibrahimsalisu46@yahoo.com; ibrahimsalisu46@tiu.edu.iq

M. M. Al-Kassab
e-mail: mowafaq.muhammed@tiu.edu.iq

M. Q. Al-Awjar
Department of Statistics and Informatics, College of Computers and Mathematics, University of Mosul, Mosul, Iraq
e-mail: mqy.alawjar@uomosul.edu.iq

D. Zeidan et al. (eds.), *Mathematics and Computation*, Springer Proceedings in Mathematics & Statistics 418, https://doi.org/10.1007/978-981-99-0447-1_7

problems when you fit the model and interpret the results [1]. A key goal of regression analysis is to isolate the relationship between each independent variable and the dependent variable [2]. Multicollinearity makes it hard to interpret your coefficients, and it reduces the power of your model to identify independent variables that are statistically significant. These are serious problems. However, the good news is that you don't always have to find a way to fix multicollinearity [3]. Several studies examined and discussed the problems of multicollinearity for the regression model and also emphasized that the major problem related to multicollinearity comprises uneven and biased standard errors and impractical explanations of the results [4–6].

In this paper, we considered the correlation coefficient (CC) and the variance inflation factor (VIF) approaches for identifying the multicollinearity among the independent variables, in the year 2015. Multiple regression is considered for the prediction of the best models. Based on the results, we discovered that there is multicollinearity among the factors, these necessitate the use of CC and the VIF approaches to tackle, reduce, and fixed the multicollinearity among the independent variables. Lastly, we came up with the best-fitted model. This paper is scheduled as: Sect. 2 provides the methods for investigating multicollinearity. The results and diagnosed multicollinearity are presented in Sects. 3 and 4, respectively. The conclusion follows in Sect. 5.

2 Materials and Methods for Investigating Multicollinearity

In this section, we present the materials and methods used for investigating the multicollinearity within the independent variables. The dataset was selected at random from 100 women's records, moreover, the dataset used in this study is collected from the Babil Governorate health center [7]. The independent variables (IVs) are husband age, mother weight, the mother age, years of marriage, smoking mother, number of dead children, the mother age when married, number of sports hours per week, the mother with the thyroid gland, the mother sleeping hours per week, the mother taking medicine, breastfeeding duration per month, and mother job, while the dependent variable (DV) is the number of born children. Other factors like financial assistance, chronic illness (breast cancer), stress due to job, illegal drug, and house activities can be among the leading risk factors affecting the number of born children. These factors lead to serious health conditions that make one vulnerable to Covid 19, see [8]. When it comes to the application perspective, the authors in [8–10] make use of commutativity to study the relation and the sensitivity between systems, the idea can be extended to investigate the commutativity and sensitivity between the independent variables, The main aim of this research is to investigate multicollinearity using some techniques such as i) correlation coefficient and ii) variance inflation factor.

2.1 Correlation Coefficient

Pearson's correlation coefficient (also called Pearson's *R*) is a relationship coefficient regularly utilized in direct relapse. The formula of the Pearson correlation coefficient is given as

$$r = \frac{\sum_{i=1}^{n}(x_i - \overline{x})(y_i - \overline{y})}{\sqrt{\sum_{i=1}^{n}(x_i - \overline{x})^2 \sum_{i=1}^{n}(y_i - \overline{y})^2}}, \tag{1}$$

where n is a sample size, r is the correlation coefficient, y_i and x_i are dependent and independent variables indexed in i, respectively. If the correlation coefficient value is higher with the pairwise variables, it indicates the possibility of collinearity. In general, if the absolute value of the Pearson correlation coefficient is close to 0.8, collinearity is likely to exist [11].

2.2 Variance Inflation Factor (VIF)

Variance Inflation Factor (VIF) is a simple way to detect multicollinearity in a regression model, it is used to determine the correlation between independent variables. The VIF measures how much the variance is inflated. VIF is calculated as

$$VIF_j = \frac{1}{1 - R_{ij}^2} = \frac{1}{Tolerance}. \tag{2}$$

Please observe that the higher the tolerance, the lower the VIF, and the limited possibility for multicollinearity among the variables. The VIF with the value of 1 clearly shows that there is no correlation between the independent variables. But if the VIF has a value within 1 < VIF < 5, it suggests that there is a moderate correlation between the variables, with VIF between 5 ≤ VIF ≤ 10, it indicates multicollinearity that needs corrective action, and VIF > 10 are indications of severe correlation between the variables, with critical levels of the multicollinearity [12].

2.3 Multiple Linear Regression

The multiple linear regression model is given as

$$\sum_{j=1}^{13} \beta_0 + \beta_j x_{ij} + e_i. \tag{3}$$

where β_0, β_i are the unknown constants, x_i are the IVs, y is the DV and e_i is the error term that has a normal distribution with mean o and variance σ^2. The mother age (x_1), the mother age when married (x_2), mother weight (x_3), smoking mother (x_4), husband age (x_5), years of marriage (x_6), number of dead children (x_7), number of sports hours per week (x_8), the mother with the thyroid gland (x_9), mother sleeping hours per week (x_{10}), the mother taking marriage (x_{11}), breastfeeding duration per week (x_{12}), and mother job (x_{13}) are the IVs and also the number of born child (y) is the DV.

3 Results

The author in [13] discusses some primary techniques for detecting multicollinearity using the questionnaire survey data on customer satisfaction. In this section, we statistically detect the multicollinearity among the independent variables using the correlation coefficient method in Eq. (1), VIF in Eq. (2), and lastly with the help of multiple linear regression in Eq. (3).

3.1 *Investigating Multicollinearity Using Pairwise Scatterplot*

The scatterplot is one of the methods used for detecting multicollinearity by observing the relationship between the variables. The dots depicted in Fig. 1 represent the values of two variables.

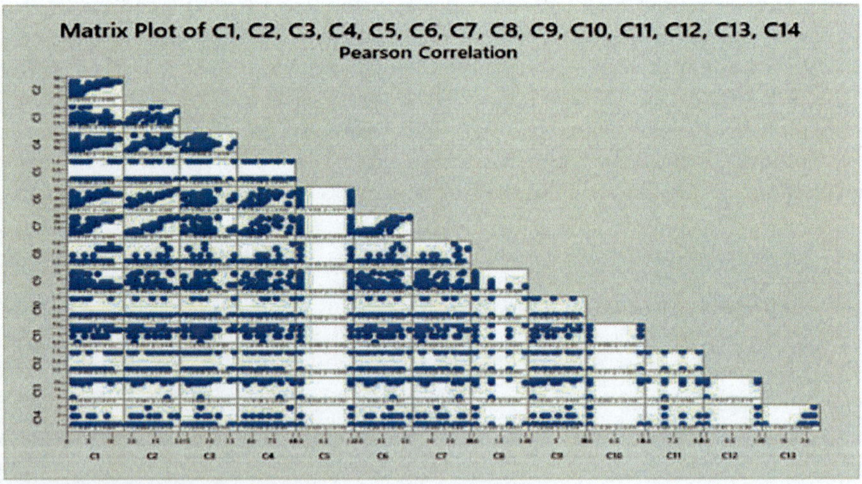

Fig. 1 Scatterplot of pairwise variables

3.2 *Investigating Multicollinearity Using Pearson's Correlations*

Pearson's correlations are a very important method used to investigate collinearity between the independent variables. Table 2 shows the relationships in terms of collinearity between the independent variables. The results obtained from the overall correlation detected the collinearity between the variables, the most highly correlated variables are (x_1), (x_3), (x_5), and (x_6). The mother age (x_1) versus mother weight (x_3) has [r = 0.638, c.f = (0.505, 0.741), p < 0.05], the mother age (x_1) versus husband age (x_5) has no logical relation, the mother age (x_1) versus years of marriage (x_6) has [r = 0.932, c.f = (0.9, 0.954), p < 0.05], the mother weight (x_3) versus husband age (x_5) has no logical relation, the mother weight (x_3) versus years of marriage (x_6) has [r = 0.597, c.f = (0.451, 0.7101), p < 0.05], and husband age (x_5) versus years of marriage (x_6) has [r = 0.850, c.f = (0.784, 0.897), p < 0.05]. The Pearson correlation coefficient is close to 0.8, this shows the existence of collinearity between the variables (Table 1).

The model is given as

$$y = -1.03 + 0.294x_1 - 0.338x_2 - 0.0614x_3 - 1.04x_4 - 0.0310x_5 + 1.38x_7 - 0.0904x_8 + 2.19x_9 + 0.165x_{10} - 1.75x_{11} + 0.246x_{12} + 0.521x_{13}$$

The overall significance of the model is given in Table 3.

Table 1 Descriptive statistics

Variable	N	Mean	SE Mean	Median	Mode
y	100	3.570	0.299	2.000	1
x_1	100	31.030	0.838	30.000	22
x_2	100	18.640	0.321	18.500	19
x_3	100	68.520	0.894	67.000	67
x_4	100	0.3800	0.0488	0.0000	0
x_5	100	34.170	0.829	33.500	42
x_6	100	12.390	0.883	9.500	3
x_7	100	0.3400	0.0655	0.0000	0
x_8	100	3.850	0.291	3.000	2
x_9	100	0.0900	0.0288	0.0000	0
x_{10}	100	8.0900	0.0740	8.0000	8
x_{11}	100	0.3000	0.0461	0.0000	0
x_{12}	100	23.210	0.276	24.000	24
x_{13}	100	1.1700	0.0428	1.0000	1

Table 2 Pearson's correlations coefficients

Variables	y	x_1	x_2	x_3	x_4	x_5	x_6	x_7	x_8	x_9	x_{10}	x_{11}	x_{12}
x_1	0.659												
x_2	−0.293	0.047											
x_3	0.241	0.638	0.024										
x_4	−0.240	0.084	0.211	0.174									
x_5	0.577	0.918	0.060	0.671	−0.016								
x_6	0.732	0.932	−0.319	0.597	0.003	0.850							
x_7	0.437	0.451	0.083	0.354	0.129	0.384	0.398						
x_8	−0.276	−0.198	0.130	−0.008	−0.052	−0.199	−0.235	−0.360					
x_9	−0.002	0.024	0.254	0.194	0.330	−0.019	−0.070	0.104	0.077				
x_{10}	0.063	0.083	0.133	−0.128	−0.040	0.067	0.030	−0.231	0.161	−0.086			
x_{11}	−0.089	0.317	0.074	0.499	0.566	0.291	0.274	0.328	−0.102	0.175	−0.139		
x_{12}	0.109	−0.047	0.090	−0.114	0.030	−0.081	−0.077	−0.118	0.207	−0.062	0.204	0.037	
x_{13}	−0.187	−0.066	0.052	0.156	0.026	0.166	−0.082	−0.208	0.094	−0.126	−0.049	0.149	−0.005

Table 3 Analysis of variance

Model	DF	Adj SS	Adj MS	F-value	P-value
Regression	12	697.9	58.154	27.40	0.000
Residual error	87	184.656	2.122		
Total	99	882.5			

3.3 *Investigating Multicollinearity Using Variance Inflation Factor (VIF)*

The variance inflation factor (VIF) identifies the correlation between independent variables and the strength of that correlation. The regression analysis illustrated in Table 4 detected multicollinearity by identifying variables with p-value > 0.05 and VIF > 5. These results show that the mother age (x_1), the mother age when married (x_2), mother weight (x_3), smoking mother (x_4), years of marriage (x_6), number of dead children (x_7), the mother with the thyroid gland (x_9), mother sleeping hours per week (x_{10}), and mother job (x_{13}) are statistically significant while husband age (x_5), number of sports hours per week (x_8), the mother taking marriage (x_{11}), breastfeeding duration per week (x_{12}) are not statistically significant. Moreover, the model indicates that the mother age (x_1) and husband age (x_5) has the highest VIFs of 10.8 and 11.7, respectively. This indicates serious multicollinearity that requires removal.

The R-square is 79%.

Table 4 Regression analysis

Predictor	Coef	SE Coef	T-value	P-value	VIF
Constant	−1.030	2.658	−0.39	0.699	
β_1	0.29400	0.05732	5.13	0.000	10.8
β_2	−0	0.05007	−6.76	0.000	1.2
β_3	−0.06140	0.02669	−2.30	0.024	2.7
β_4	−1.0369	0.4119	−2.52	0.014	1.9
β_5	−0.03101	0.06052	−0.51	0.610	11.7
β_7	1.3775	0.2919	4.72	0.000	1.7
β_8	−0.09044	0.05865	−1.54	0.127	1.4
β_9	2.1872	0.5818	3.76	0.000	1.3
β_{10}	0.1649	0.2215	0.74	0.459	1.3
β_{11}	−1.7490	0.4640	−3.77	0.000	2.1
β_{12}	0.24604	0.05639	4.36	0.000	1.1
β_{13}	0.5215	0.4549	1.15	0.255	1.8

Table 5 Analysis of variance table and overall significant of the model

Model	DF	Adj SS	Adj MS	F-Value	P-Value
Regression	11	697.296	63.391	30.12	0.000
Residual Error	88	185.214	2.105		
Total	99	882.510			

4 Diagnosed Multicollinearity

There are several methods to remove multicollinearity, the authors in [14, 15] studied the application of latent roots regression to multicollinear data, but in this research, we will consider i) removal of variables with high VIF and ii) removing non-significant variables.

4.1 *Diagnosed Multicollinearity by Removing High VIF*

In our model, the mother age (x_1) and husband age (x_5) has the highest VIFs of 10.8 and 11.7 respectively. The correlation between the mother age (x_1) and husband age (x_5) is significant with $r = 0.918$, see Table 2. So instead of removal both of them, we keep the mother age (x_1) with a VIF of 10.8 and remove the husband age (x_5) with VIF 11.7, we obtained a new model in Table 6. We can see that all the VIFs are down to satisfactory values with (VIFs < 5). The model is given as

$$\begin{aligned} y = &- 0.86 + 0.2678x_1 - 0.3415x_2 - 0.0648x_3 - 0.974x_4 + 1.379x_7 \\ &- 0.085x_8 + 2.186x_9 + 0.157x_{10} - 1.745x_{11} + 0.2477x_{12} + 0.394x_{13} \end{aligned}$$

The overall significance of the model is given in Table 5.
The R-square is 79%.

4.2 *Diagnosed Multicollinearity by Removing Non-significant Variables*

Removing the husband age (x_5) with VIF 11.7 is not enough to predict the best model, since we still have some variables such as the number of sports hours per week (x_8), mother sleeping hours per week (x_{10}), and mother job (x_{13}) that are not statistically significant in Table 7. This necessitates the removal of this variable. We can see that after removing the non-significant variables, the p-values of all the variables are down to satisfactory values with ($p < 0.05$) in Table 8. The model is given as

Table 6 Regression analysis

Term	Coef	SE Coef	T-value	P-value	VIF
Constant	−0.86	2.63	−0.33	0.745	
β_1	0.2678	0.0259	10.34	0.000	2.22
β_2	−0.3415	0.0495	−6.90	0.000	1.19
β_3	−0.0648	0.0257	−2.52	0.014	2.49
β_4	−0.974	0.391	−2.49	0.015	1.72
β_7	1.379	0.291	4.74	0.000	1.70
β_8	−0.0850	0.0574	−1.48	0.143	1.31
β_9	2.186	0.579	3.77	0.000	1.31
β_{10}	0.157	0.220	0.71	0.478	1.25
β_{11}	−1.745	0.462	−3.78	0.000	2.13
β_{12}	0.2477	0.0561	4.42	0.000	1.12
β_{13}	0.394	0.379	1.04	0.302	1.24

$$y = 0.89 + 0.2746x_1 - 0.3407x_2 - 0.0690x_3 - 0.950x_4 + 1.382x_7 + 1.985x_9 - 1.661x_{11} + 0.2349x_{12}$$

The overall significance of the model is given in Table 7.
The R-square is 78%.

Table 7 Analysis of variance

Source	DF	Adj SS	Adj MS	F-value	P-value
Regression	8	689.73	86.216	40.70	0.000
Residual error	91	192.78	2.118		
Total	99	882.51			

Table 8 Regression analysis

Term	Coef	SE Coef	T-value	Value	VIF
Constant	0.89	2.10	0.42	0.673	
β_1	0.2746	0.0242	11.32	0.000	1.93
β_2	−0.3407	0.0481	−7.08	0.000	1.12
β_3	−0.0690	0.0242	−2.85	0.005	2.18
β_4	−0.950	0.392	−2.42	0.017	1.71
β_7	1.382	0.260	5.32	0.000	1.35
β_9	1.985	0.566	3.50	0.001	1.24
β_{11}	−1.661	0.457	−3.63	0.000	2.07
β_{12}	0.2349	0.0545	4.31	0.000	1.06

5 Conclusion

This paper investigates the multicollinearity relation among the independent variables, the mother age, the mother age when married, husband age, mother weight, years of marriage, smoking mother, number of sports hours per week, number of dead children, the mother with the thyroid gland, the mother sleeping hours per week, the mother taking medicine, breastfeeding duration per month, and mother job, the model obtained proves to be not significant since some variables have p less than 0.05. These were a result of multicollinearity among the variables. The two methods correlation coefficient and variance inflation factor proposed in this work were used to detect the multicollinearity among the variables. Among several methods to remove collinearity, we consider two methods; removing variables with high VIF and removing variables that are not statistically significant ($p < 0.05$). Lastly, we obtained the best-fitted model that predicts the factors affecting the number of born children in Iraq. Moreover, the ANOVA obtained from Table 7 shows that the model is more fitted since we observed a monotone increment in the f-value, from 27.40 in Table 3 to 40.70 in Table 7. Furthermore, more advanced research techniques such as the ridge regression method, latent root regression, weighted regression method, and principal components regression can be used to detect collinearity [16, 17]. The results are validated with Minitab version 19.

Funding No funding.

References

1. Young, D.S.: Handbook of Regression Methods, pp. 109–136. CRC Press, Boca Raton (2017)
2. Frank, E.H., Jr.: Regression Modeling Strategies: With Applications to Linear Models, Logistic Regression, and Survival Analysis, pp. 121–142. Springer, New York (2001)
3. Hosmer, D.W., Lemeshow, S., Sturdivant, R.X.: Applied Logistic Regression. Wiley, New Jersey (2013)
4. Pedhajur, E.J.: Multiple Regression in Behavioral Research: Explanation and Prediction, 3rd edn. Thomson Learning, Wadsworth (1997)
5. Keith, T.Z.: Multiple Regression and Beyond: An Introduction to Multiple Regression and Structural Equation Modeling, 2nd edn. Taylor and Francis, New York (2015)
6. Aiken, L.S., West, S.G.: Multiple Regression: Testing and Interpreting Interactions. Sage, Newbury Park (1991)
7. Majid, S., Alsabah, S.: Parameters estimation of the multiple linear regression. The mode under Multicollinearity problem. Iraqi Acad. Sci. J. **12**(1), 1–28 (2020)
8. Ibrahim, S., Al-Kassab, M.M.: Using linear regression analysis to study the recovery cases of COVID 19 in Erbil, Kurdistan Region. Drugs Cell Therap. Hematol. **10**(1), 1226–1239 (2021)
9. Ibrahim, S., Koksal, M.E.: Commutativity of sixth-order time-varying linear systems. Circuits, Syst., Signal Process. **40**(10), 4799–4832 (2021).View at: Publisher Site I Google Scholar
10. Ibrahim, S., Koksal, M.E.: Realization of a fourth-order linear time-varying differential system with nonzero initial conditions by cascaded two second-order commutative pairs. Circuits, Syst. Signal Process. **40**(6), 3107–3123 (2021).View at: Publisher Site I Google Scholar

11. Ibrahim, S., Rababah, A.: Decomposition of fourth-order euler-type linear time-varying differential system into cascaded two second-order euler commutative pairs. Complexity ArticleID 3690019, 9 (2022). https://doi.org/10.1155/2022/3690019
12. Gunst, R.F., Webster, J.T.: Regression analysis and problems of multicollinearity. Commun. Stat. **4**(3), 277–292 (1975)
13. Belsley, D.A.: Conditioning Diagnostics: Collinearity and Weak Data in Regression. Wiley Inc., New York (1991)
14. Shrestha, N.: Detecting multicollinearity in regression analysis. Am. J. Appl. Math. Stat. **8**(2), 39–42 (2020)
15. Al-Kassab, M.M., Adnan, M.A., Dilnas, S.Y.: Studying the effect of some variables on the economic growth using latent roots method (11), 1–10 (2019)
16. Al-kassab, M.M., Dilnas, S.Y.: Application of latent roots regression to multicollinear data. J. Adv. Res. Comput. Sci. Eng. **4**(12), 1–11 (2017)
17. Ibrahim, S.: Numerical approximation method for solving differential equations. Eurasian J. Sci. Eng. **6**(2), 157–168 (2020). https://doi.org/10.23918/eajse.v6i2p157
18. Rababah, A., Ibrahim, S.: Weighted G^1-multi-degree reduction of Bézier curves. Int. J. Adv. Comput. Sci. Appl. **7**(2), 540–545 (2016)

Hilbert–Schmidt Numerical Radius Inequalities for Certain 2×2 Operator Matrices

Tasnim Alkharabsheh, Khalid Shebrawi, and Mohammed Abu-Saleem

Abstract We prove several Hilbert–Schmidt numerical radius inequalities for certain 2×2 operator matrices. Among other inequalities, it is shown that if $X, Y \in C_2$, then

$$w_2\left(\widetilde{T}_t\right) \leq \frac{1}{\sqrt{2}}\left(\left\| |Y|^t \left|X^*\right|^{1-t} \right\|_2 + \left\| |X|^t \left|Y^*\right|^{1-t} \right\|_2\right) \text{ for all } t \in [0, 1],$$

where $T = \begin{bmatrix} 0 & X \\ Y & 0 \end{bmatrix}$. Also, we introduce some applications of our inequalities.

Keywords Numerical radius · Hilbert–Schmidt numerical radius · Usual operator norm · Hilbert–Schmidt norm · Operator matrix and inequality.

Mathematics Subject Classification. 15A18 · 15B48 · 15A60 · 47A12 · 47A30 · 47A63 · 47B15.

1 Introduction

Let $\mathcal{H}$ be a complex Hilbert space with inner product $\langle \cdot, \cdot \rangle$. Let $\mathfrak{B}(\mathcal{H})$ be the space of all bounded linear operators on $\mathcal{H}$. For $X \in \mathfrak{B}(\mathcal{H})$, let $w(X)$, $w_2(X)$, $\|X\|$ and $\|X\|_2$ denote the numerical radius, the Hilbert–Schmidt numerical radius, the usual operator norm, and the Hilbert–Schmidt norm of X, respectively. Recall that the Hilbert–Schmidt norm of X is defined by

T. Alkharabsheh · K. Shebrawi (✉) · M. Abu-Saleem
Department of Mathematics, Al-Balqa Applied University, Salt, Jordan
e-mail: khalid@bau.edu.jo

M. Abu-Saleem
e-mail: m_abusaleem@bau.edu.jo

D. Zeidan et al. (eds.), *Mathematics and Computation*, Springer Proceedings in Mathematics & Statistics 418, https://doi.org/10.1007/978-981-99-0447-1_8

$$\|X\|_2 = \left(\mathrm{tr}X^*X\right)^{\frac{1}{2}} = \left(\sum_{j=1}^{\infty} s_j^2(X)\right)^{1/2}, \tag{1}$$

where $s_1(X) \geq s_2(X) \geq s_3(X) \geq \cdots$ are the singular values of X. Note that X belongs to the trace class C_1 if tr $|X|$ is finite, and X belongs to the Hilbert–Schmidt class C_2 if $\|X\|_2$ is finite. The Cauchy–Schwarz inequality (see, e.g., [4, p. 96]) says that if $X, Y \in C_2$, then $XY \in C_1$, and

$$|\mathrm{tr}XY| \leq \|X\|_2\|Y\|_2. \tag{2}$$

It is known that $\|X\|_2$ is unitarily invariant, in the sense that for $X \in C_2$ and unitary $U, V \in \mathfrak{B}(\mathcal{H})$, we have

$$\|\mathrm{UXV}\|_2 = \|X\|_2, \tag{3}$$

also, $\|X\|_2$ is self-adjoint, that is for every $X \in \mathfrak{B}(\mathcal{H})$ we have

$$\left\|X^*\right\|_2 = \|X\|_2. \tag{4}$$

It is also known that for $X \in \mathfrak{B}(\mathcal{H})$,

$$w(X) = \sup_{\theta\in\mathbb{R}} \left\|\mathrm{Re}\left(e^{i\theta}X\right)\right\| \tag{5}$$

(see, e.g.,[7]).

Let $N(.)$ be a norm on $\mathfrak{B}(\mathcal{H})$. A generalization of the numerical radius has been recently introduced in [1] by defining $w_N(\mathrm{X}) = \sup_{\theta\in\mathbb{R}} N\left(\mathrm{Re}\left(e^{i\theta}X\right)\right)$ for every $X \in \mathfrak{B}(\mathcal{H})$. Thus, $w_N(X) \geq N(\mathrm{Re}X)$ and $w_N(X) \geq N(\mathrm{Im}X)$. In particular, we have $w(X) \geq \|ReX\|$ and $w(X) \geq \|ImX\|$. In fact, $w_N(\mathrm{X})$ and, in particular, $w(\mathrm{X})$ define norms on $\mathfrak{B}(\mathcal{H})$.

An important property of $w_N(\mathrm{X})$ is that if $N(.)$ is weakly unitarily invariant, then so is $w_N(.)$, that is, for $X, U \in \mathfrak{B}(\mathcal{H})$ such that U is unitary, we have

$$w_N\left(\mathrm{UXU}^*\right) = w_N(X), \tag{6}$$

and self-adjoint, that is, $w_N(X) = w_N\left(X^*\right)$. Moreover, clearly if $X \in \mathfrak{B}(\mathcal{H})$ is self-adjoint, then $w(X) = \|X\|$. The triangle inequality for $w_N(.)$ is given by

$$w_N(X+Y) \leq w_N(X) + w_N(Y), \tag{7}$$

where $X, Y \in \mathfrak{B}(\mathcal{H})$. The following formula for $w_2(X)$ is recently proved in [1], say that if $X \in C_2$, then

$$w_2(X) = \sqrt{\frac{1}{2}\|X\|_2^2 + \frac{1}{2}\left|\mathrm{tr}X^2\right|}. \tag{8}$$

Also, for $X \in C_2$, we have

$$\frac{1}{\sqrt{2}}\|X\|_2 \leq w_2(X) \leq \|X\|_2. \tag{9}$$

A special case of 8 can be found in [3], which says if $X \in C_2$ and $X^2 = 0$, then

$$w_2(X) = \sqrt{\frac{1}{2}\|X\|_2^2 + \frac{1}{2}\left|trX^2\right|} = \frac{1}{\sqrt{2}}\|X\|_2. \tag{10}$$

For $X \in \mathfrak{B}(\mathcal{H})$ with a polar decomposition $X = U|X|$, the generalized Aluthge transform $\tilde{X}$ of X is given by

$$\tilde{X_t} = |X|^t U |X|^{1-t} \text{ for all } t \in [0, 1]. \tag{11}$$

Here U is a partial isometry and $|X| = (X^*X)^{\frac{1}{2}}$.

In Sect. 2 of this paper, we give several Hilbert–Schmidt numerical radius inequalities, including lower and upper bounds for certain 2×2 operator matrices. In Sect. 3, we give some applications of our results given in Sect. 2.

2 Inequalities for Certain 2×2 Operator Matrices

The aim of this section is to give bounds for the Hilbert–Schmidt numerical radius for $\begin{bmatrix} X & Y \\ -Y & -X \end{bmatrix}$ and $\begin{bmatrix} X & Y \\ Z & W \end{bmatrix}$.

A special assertion is given to the off-diagonal parts of 2×2 operator matrices. First, we need the following lemma, which has been recently proved by Aldalabih and Kittaneh [3].

Lemma 2.1 *Let $X, Y \in C_2$. Then*

(a) $w_2\left(\begin{bmatrix} 0 & X \\ e^{i\theta}Y & 0 \end{bmatrix}\right) = w_2\left(\begin{bmatrix} 0 & X \\ Y & 0 \end{bmatrix}\right)$ *for every $\theta \in \mathbb{R}$.*

(b) $w_2\left(\begin{bmatrix} 0 & X \\ Y & 0 \end{bmatrix}\right) = w_2\left(\begin{bmatrix} 0 & Y \\ X & 0 \end{bmatrix}\right)$.

(c) $w_2\left(\begin{bmatrix} 0 & Y \\ Y & 0 \end{bmatrix}\right) = \sqrt{2}w_2(Y)$.

(d) $w_2\left(\begin{bmatrix} X & Y \\ Y & X \end{bmatrix}\right) \leq \sqrt{w_2^2(X+Y) + w_2^2(X-Y)}$.

(e) $w_2\left(\begin{bmatrix} X & 0 \\ 0 & Y \end{bmatrix}\right) \leq \sqrt{w_2^2(X) + w_2^2(Y)}$.

(f) $w_2\left(\begin{bmatrix} X & Y \\ 0 & 0 \end{bmatrix}\right) = \sqrt{w_2^2(X) + \frac{1}{2}\|Y\|_2^2}$.

Lemma 2.2 *Let $X \in C_2$. Then*

$$w_2\left(\begin{bmatrix} X & 0 \\ 0 & -X \end{bmatrix}\right) = \sqrt{2}w_2(X). \tag{1}$$

Proof Using formula (8), we have

$$w_2^2\left(\begin{bmatrix} X & 0 \\ 0 & -X \end{bmatrix}\right) = \frac{1}{2}\left(\left\|\begin{bmatrix} X & 0 \\ 0 & -X \end{bmatrix}\right\|_2^2 + \left|\operatorname{tr}\begin{bmatrix} X & 0 \\ 0 & -X \end{bmatrix}^2\right|\right).$$

Since

$$\left\|\begin{bmatrix} X & 0 \\ 0 & -X \end{bmatrix}\right\|_2^2 = \|X\|_2^2 + \|-X\|_2^2 = 2\|X\|_2^2,$$

and

$$\left|\operatorname{tr}\begin{bmatrix} X & 0 \\ 0 & -X \end{bmatrix}^2\right| = \left|\operatorname{tr}X^2 + \operatorname{tr}X^2\right| = 2\left|\operatorname{tr}X^2\right|.$$

It follows that

$$w_2^2\left(\begin{bmatrix} X & 0 \\ 0 & -X \end{bmatrix}\right) = \|X\|_2^2 + \left|\operatorname{tr}X^2\right| = 2w_2^2(X).$$

And so,

$$w_2\left(\begin{bmatrix} X & 0 \\ 0 & -X \end{bmatrix}\right) = \sqrt{2}w_2(X).$$

□

Theorem 2.3 *Let $X, Y \in C_2$. Then*

$$\sqrt{2}\max\left(w_2(X), w_2(Y)\right) \le w_2\left(\begin{bmatrix} X & Y \\ -Y & -X \end{bmatrix}\right) \le \sqrt{2}\left(w_2(X) + w_2(Y)\right). \tag{2}$$

Proof By Lemma 2.2, we have

$$\begin{aligned} w_2\left(\begin{bmatrix} X & Y \\ -Y & -X \end{bmatrix}\right) &\ge w_2\left(\begin{bmatrix} X & 0 \\ 0 & -X \end{bmatrix}\right) \\ &= \sqrt{2}w_2(X). \end{aligned} \tag{3}$$

And by Lemma 2.1 (a,c), we have

$$\begin{aligned} w_2\left(\begin{bmatrix} X & Y \\ -Y & -X \end{bmatrix}\right) &\geq w_2\left(\begin{bmatrix} 0 & Y \\ -Y & 0 \end{bmatrix}\right) \\ &= w_2\left(\begin{bmatrix} 0 & Y \\ Y & 0 \end{bmatrix}\right) \\ &= \sqrt{2}w_2(Y). \end{aligned} \tag{4}$$

From inequalities (3) and (4), we have

$$w_2\left(\begin{bmatrix} X & Y \\ -Y & -X \end{bmatrix}\right) \geq \sqrt{2}\max\left(w_2(X), w_2(Y)\right). \tag{5}$$

This proves the first inequality of the theorem. To prove the second inequality, by using Inequality (7), Lemma 2.1 (a,c) and Lemma 2.2, we have

$$\begin{aligned} w_2\left(\begin{bmatrix} X & Y \\ -Y & -X \end{bmatrix}\right) &\leq w_2\left(\begin{bmatrix} X & 0 \\ 0 & -X \end{bmatrix}\right) + w_2\left(\begin{bmatrix} 0 & Y \\ -Y & 0 \end{bmatrix}\right) \\ &= w_2\left(\begin{bmatrix} X & 0 \\ 0 & -X \end{bmatrix}\right) + w_2\left(\begin{bmatrix} 0 & Y \\ Y & 0 \end{bmatrix}\right) \\ &= \sqrt{2}\left(w_2(X) + w_2(Y)\right). \end{aligned} \tag{6}$$

From inequalities (5) and (6), we have

$$\sqrt{2}\max\left(w_2(X), w_2(Y)\right) \leq w_2\left(\begin{bmatrix} X & Y \\ -Y & -X \end{bmatrix}\right) \leq \sqrt{2}\left(w_2(X) + w_2(Y)\right).$$

□

Remark 2.4 Let $X \in C_2$. Then

$$\left\|\begin{bmatrix} 0 & X \\ X^* & 0 \end{bmatrix}\right\|_2 = \sqrt{2}\|X\|_2. \tag{7}$$

Indeed if $T = \begin{bmatrix} 0 & X \\ X^* & 0 \end{bmatrix}$, then T is self-adjoint and

$$\begin{aligned} \|T\|_2^2 &= \left\|\begin{bmatrix} 0 & X \\ X^* & 0 \end{bmatrix}\right\|_2^2 \\ &= \operatorname{tr}\left(T^*T\right) \\ &= \operatorname{tr}\left(\begin{bmatrix} XX^* & 0 \\ 0 & X^*X \end{bmatrix}\right) \\ &= \operatorname{tr}\left(XX^*\right) + \operatorname{tr}\left(X^*X\right) \\ &= 2\operatorname{tr}\left(X^*X\right) \quad (\text{since } \operatorname{tr}(AB) = \operatorname{tr}(BA)) \\ &= 2\|X\|_2^2. \end{aligned}$$

Theorem 2.5 *Let $X \in C_2$. Then*

$$w_2\left(\begin{bmatrix} 0 & X \\ X^* & 0 \end{bmatrix}\right) = \sqrt{2}\|X\|_2. \tag{8}$$

Proof Let $T = \begin{bmatrix} 0 & X \\ X^* & 0 \end{bmatrix}$. Then, by formula (8), we have

$$w_2(T) = \sqrt{\frac{1}{2}\|T\|_2^2 + \frac{1}{2}\left|\mathrm{tr}T^2\right|}.$$

Indeed, by remark 2.4, we have

$$\|T\|_2^2 = \left\|\begin{bmatrix} 0 & X \\ X^* & 0 \end{bmatrix}\right\|_2^2 = 2\|X\|_2^2$$

and

$$\begin{aligned}\left|\mathrm{tr}T^2\right| &= \left|\mathrm{tr}\begin{bmatrix} 0 & X \\ X^* & 0 \end{bmatrix}^2\right| \\ &= \left|\mathrm{tr}\begin{bmatrix} XX^* & 0 \\ 0 & X^*X \end{bmatrix}\right| \\ &= \left|\mathrm{tr}\left(XX^*\right) + \mathrm{tr}\left(X^*X\right)\right| \\ &= 2\left|\mathrm{tr}\left(X^*X\right)\right|,\end{aligned}$$

it follows that

$$\begin{aligned} w_2(T) &= \sqrt{\|X\|_2^2 + |\mathrm{tr}\,(X^*X)|} \\ &= \sqrt{\|X\|_2^2 + \left|\|X\|_2^2\right|} \quad \text{(by (1))} \\ &= \sqrt{2\|X\|_2^2} \quad (\text{since} \|\cdot\|_2 \geq 0) \\ &= \sqrt{2}\|X\|_2. \end{aligned}$$

□

Lemma 2.6 *Let $X, Y \in C_2$. Then*

$$w_2\left(\begin{bmatrix} 0 & X \\ Y & 0 \end{bmatrix}\right) = \frac{1}{\sqrt{2}} \sup_{\theta \in \mathbb{R}} \left\|X + e^{i\theta}Y^*\right\|_2. \tag{9}$$

Proof

$$
\begin{aligned}
w_2\left(\begin{bmatrix} 0 & X \\ Y & 0 \end{bmatrix}\right) &= \sup_{\theta\in\mathbb{R}} \left\| \operatorname{Re}\left(\begin{bmatrix} 0 & e^{i\theta}X \\ e^{i\theta}Y & 0 \end{bmatrix}\right)\right\|_2 \\
&= \frac{1}{2}\sup_{\theta\in\mathbb{R}} \left\| \begin{bmatrix} 0 & e^{i\theta}\left(X+e^{-2i\theta}Y^*\right) \\ e^{i\theta}\left(Y+e^{-2i\theta}X^*\right) & 0 \end{bmatrix}\right\|_2 \\
&= \frac{1}{2}\sup_{\theta\in\mathbb{R}} \left\| \begin{bmatrix} 0 & e^{i\theta}\left(X+e^{-2i\theta}Y^*\right) \\ e^{-i\theta}\left(X+e^{-2i\theta}Y^*\right)^* & 0 \end{bmatrix}\right\|_2 \\
&= \frac{1}{\sqrt{2}}\sup_{\theta\in\mathbb{R}} \left\| e^{i\theta}\left(X+e^{-2i\theta}Y^*\right)\right\|_2 \text{ (by remark 2.4)} \\
&= \frac{1}{\sqrt{2}}\sup_{\theta\in\mathbb{R}} \left\| X+e^{-2i\theta}Y^*\right\|_2 \text{ (since } \left|e^{i\theta}\right| = 1) \\
&= \frac{1}{\sqrt{2}}\sup_{\theta\in\mathbb{R}} \left\| X+e^{i\theta}Y^*\right\|_2 .
\end{aligned}
$$

□

To prove the next theorem, we need the following lemma which has been given in [3].

Lemma 2.7 *Let $X, Y \in C_2$. Then*

$$
\frac{\max\left(w_2(X+Y), w_2(X-Y)\right)}{\sqrt{2}} \le w_2\left(\begin{bmatrix} 0 & X \\ Y & 0 \end{bmatrix}\right) \le \frac{w_2(X+Y)+w_2(X-Y)}{\sqrt{2}}. \tag{10}
$$

Theorem 2.8 *Let $A, B, X \in C_2$ such that A and B are self-adjoint. Then*

$$
\|X\|_2 \le w_2(X+A) + w_2(X+iB). \tag{11}
$$

Proof It follows from Lemma 2.6 that

$$
\frac{\|S+T^*\|_2}{\sqrt{2}} \le w_2\left(\begin{bmatrix} 0 & S \\ T & 0 \end{bmatrix}\right),
$$

and from Lemma 2.7

$$
w_2\left(\begin{bmatrix} 0 & S \\ T & 0 \end{bmatrix}\right) \le \frac{w_2(S+T)+w_2(S-T)}{\sqrt{2}}.
$$

Therefore,

$$
\frac{\|S+T^*\|_2}{\sqrt{2}} \le w_2\left(\begin{bmatrix} 0 & S \\ T & 0 \end{bmatrix}\right) \le \frac{w_2(S+T)+w_2(S-T)}{\sqrt{2}},
$$

and so

$$\|S+T^*\|_2 \le w_2(S+T)+w_2(S-T). \tag{12}$$

Now, letting

$$S = X + \frac{A+iB}{2} \text{ and } T = \frac{-A+iB}{2}.$$

Then,

$$T^* = \frac{-A-iB}{2},\ S+T = X+iB,\ S+T^* = X \text{ and } S-T = X+A. \tag{13}$$

Substituting relation (13) in Inequality (12), we have

$$\|X\|_2 \le w_2(X+A)+w_2(X+iB). \tag{14}$$

Notice that if we let $A = B = 0$ in Inequality (14), then we have

$$\|X\|_2 \le 2w_2(X),$$

and so

$$\frac{\|X\|_2}{2} \le w_2(X).$$

□

Theorem 2.9 *Let $X, Y, Z, W \in C_2$. Then*

$$w_2\left(\begin{bmatrix} X & Y \\ Z & W \end{bmatrix}\right) \le \sqrt{w_2^2(X)+\frac{1}{2}\|Y\|_2^2}+\sqrt{w_2^2(W)+\frac{1}{2}\|Z\|_2^2}. \tag{15}$$

Proof Let $U = \begin{bmatrix} 0 & I \\ I & 0 \end{bmatrix}$ be a unitary operator. Then by Inequality (7), we have

$$\begin{aligned} w_2\left(\begin{bmatrix} X & Y \\ Z & W \end{bmatrix}\right) &\le w_2\left(\begin{bmatrix} X & Y \\ 0 & 0 \end{bmatrix}\right)+w_2\left(\begin{bmatrix} 0 & 0 \\ Z & W \end{bmatrix}\right) \\ &= w_2\left(\begin{bmatrix} X & Y \\ 0 & 0 \end{bmatrix}\right)+w_2\left(U^*\begin{bmatrix} W & Z \\ 0 & 0 \end{bmatrix}U\right) \\ &= w_2\left(\begin{bmatrix} X & Y \\ 0 & 0 \end{bmatrix}\right)+w_2\left(\begin{bmatrix} W & Z \\ 0 & 0 \end{bmatrix}\right) \quad \text{(by Identity (6))} \\ &= \sqrt{w_2^2(X)+\frac{1}{2}\|Y\|_2^2}+\sqrt{w_2^2(W)+\frac{1}{2}\|Z\|_2^2}. \quad \text{(by Lemma 2.1 (f))} \end{aligned}$$

Hence,

$$w_2\left(\begin{bmatrix} X & Y \\ Z & W \end{bmatrix}\right) \leq \sqrt{w_2^2(X) + \frac{1}{2}\|Y\|_2^2} + \sqrt{w_2^2(W) + \frac{1}{2}\|Z\|_2^2}.$$

□

Now, we need the following lemma, which has been recently proved by Aldalabih and Kittaneh [3].

Lemma 2.10 *Let $T \in C_2$ have the Cartesian decomposition $T = X + iY$. Then*

$$\frac{w_2(T)}{2} \leq \frac{1}{\sqrt{2}} w_2\left(\begin{bmatrix} 0 & X \\ Y & 0 \end{bmatrix}\right) \leq w_2(T). \tag{16}$$

Theorem 2.11 *Let $T \in C_2$ have the Cartesian decomposition $T = X + iY$. Then*

$$w_2(T) \leq \sqrt{w_2^2(X) + w_2^2(Y) + \|X\|_2 \|Y\|_2}. \tag{17}$$

Proof Since X and Y are self-adjoint, then

$$\begin{aligned} \|T\|_2^2 &= \operatorname{tr}\left(T^*T\right) \\ &= \operatorname{tr}((X - iY)(X + iY)) \\ &= \operatorname{tr}\left(X^2 + Y^2\right) \\ &= \|X\|_2^2 + \|Y\|_2^2. \end{aligned}$$

Now, by formula (8), we have

$$\begin{aligned} w_2^2(T) &= \frac{1}{2}\|T\|_2^2 + \frac{1}{2}\left|\operatorname{tr} T^2\right| \\ &= \frac{1}{2}\left(\|X\|_2^2 + \|Y\|_2^2\right) + \frac{1}{2}\left|\operatorname{tr}(X + iY)^2\right| \\ &= \frac{1}{2}\left(\|X\|_2^2 + \|Y\|_2^2\right) + \frac{1}{2}\left|\operatorname{tr}\left(X^2 - Y^2 + 2iXY\right)\right| \\ &\leq \frac{1}{2}\left(\|X\|_2^2 + \|Y\|_2^2\right) + \frac{1}{2}\left|\operatorname{tr}\left(X^2\right)\right| + \frac{1}{2}\left|\operatorname{tr}\left(Y^2\right)\right| + |\operatorname{tr}(XY)| \quad \text{(by the Triangle inequality)} \\ &\leq \frac{1}{2}\left(\|X\|_2^2 + \left|\operatorname{tr}\left(X^2\right)\right|\right) + \frac{1}{2}\left(\|Y\|_2^2 + \left|\operatorname{tr}\left(Y^2\right)\right|\right) + \|X\|_2 \|Y\|_2 \quad \text{(by the inequality (2))} \\ &= w_2^2(X) + w_2^2(Y) + \|X\|_2 \|Y\|_2. \quad \text{(by the formula (8))} \end{aligned}$$

Therefore,

$$w_2(T) \leq \sqrt{w_2^2(X) + w_2^2(Y) + \|X\|_2 \|Y\|_2}.$$

□

Corollary 2.12 *Let $T \in C_2$ have the Cartesian decomposition $T = X + iY$. Then*

$$\frac{1}{\sqrt{2}} w_2\left(\begin{bmatrix} 0 & X \\ Y & 0 \end{bmatrix}\right) \leq \sqrt{w_2^2(X) + w_2^2(Y) + \|X\|_2 \|Y\|_2}. \tag{18}$$

Proof By Lemma 2.10, we have

$$\frac{1}{\sqrt{2}} w_2\left(\begin{bmatrix} 0 & X \\ Y & 0 \end{bmatrix}\right) \leq w_2(T)$$
$$\leq \sqrt{w_2^2(X) + w_2^2(Y) + \|X\|_2\|Y\|_2}. \quad \text{(by Theorem 2.11)}$$

□

Theorem 2.13 *Let $X, Y \in C_2$. Then*

$$w_2\left(\widetilde{T}_t\right) \leq \frac{1}{\sqrt{2}}\left(\left\| |Y|^t \left|X^*\right|^{1-t} \right\|_2 + \left\| |X|^t \left|Y^*\right|^{1-t} \right\|_2\right) \tag{19}$$

for all $t \in [0, 1]$, where $T = \begin{bmatrix} 0 & X \\ Y & 0 \end{bmatrix}$.

Proof Let $X = U|X|$ and $Y = V|Y|$ be the polar decomposition of the operators X and Y, and let $T = \begin{bmatrix} 0 & X \\ Y & 0 \end{bmatrix}$. Then

$$\begin{bmatrix} 0 & X \\ Y & 0 \end{bmatrix} = \begin{bmatrix} 0 & U \\ V & 0 \end{bmatrix}\begin{bmatrix} |Y| & 0 \\ 0 & |X| \end{bmatrix}$$

is the polar decomposition of T. Then by (11), we have

$$\begin{aligned}
\tilde{T} &= |T|^t \begin{bmatrix} 0 & U \\ V & 0 \end{bmatrix} |T|^{1-t} \\
&= \begin{bmatrix} |Y|^t & 0 \\ 0 & |X|^t \end{bmatrix}\begin{bmatrix} 0 & U \\ V & 0 \end{bmatrix}\begin{bmatrix} |Y|^{1-t} & 0 \\ 0 & |X|^{1-t} \end{bmatrix} \\
&= \begin{bmatrix} 0 & |Y|^t U |X|^{1-t} \\ |X|^t V |Y|^{1-t} & 0 \end{bmatrix},
\end{aligned}$$

where $|X| = (X^*X)^{\frac{1}{2}}$ and $|Y| = (Y^*Y)^{\frac{1}{2}}$.

Now, by Inequality (7), we have

$$\begin{aligned}
w_2\left(\widetilde{T}_t\right) &= w_2\left(\begin{bmatrix} 0 & |Y|^t U |X|^{1-t} \\ |X|^t V |Y|^{1-t} & 0 \end{bmatrix}\right) \\
&\leq w_2\left(\begin{bmatrix} 0 & |Y|^t U |X|^{1-t} \\ 0 & 0 \end{bmatrix}\right) + w_2\left(\begin{bmatrix} 0 & 0 \\ |X|^t V |Y|^{1-t} & 0 \end{bmatrix}\right) \\
&= \frac{1}{\sqrt{2}}\left(\left\| |Y|^t U |X|^{1-t} \right\|_2 + \left\| |X|^t V |Y|^{1-t} \right\|_2\right). \quad \text{(by Identity (10)}
\end{aligned}$$

Since

$$\left|X^*\right|^2 = XX^* = U|X|^2U^*$$

and

$$|X|^2 = U^* \left|X^*\right|^2 U,$$

then

$$|X|^{1-t} = U^* \left|X^*\right|^{1-t} U.$$

Therefore

$$\begin{aligned}\left\||Y|^t U |X|^{1-t}\right\|_2 &= \left\||Y|^t U U^* \left|X^*\right|^{1-t} U\right\|_2 \\ &= \left\||Y|^t \left|X^*\right|^{1-t}\right\|_2. \quad \text{(by the identity (3))}\end{aligned}$$

Similarly,

$$\begin{aligned}\left\||X|^t V |Y|^{1-t}\right\|_2 &= \left\||X|^t V V^* \left|Y^*\right|^{1-t} V\right\|_2 \\ &= \left\||X|^t \left|Y^*\right|^{1-t}\right\|_2.\end{aligned}$$

Thus,

$$w_2\left(\widetilde{T}_t\right) \le \frac{1}{\sqrt{2}}\left(\left\||Y|^t \left|X^*\right|^{1-t}\right\|_2 + \left\||X|^t \left|Y^*\right|^{1-t}\right\|_2\right).$$

□

Corollary 2.14 *Let $X, Y \in C_2$. Then*

$$w_2\left(\widetilde{T}_t\right) \le w_2\left(|Y|^t \left|X^*\right|^{1-t}\right) + w_2\left(|X|^t \left|Y^*\right|^{1-t}\right). \tag{20}$$

Proof By Theorem 2.13, we have

$$\begin{aligned}w_2\left(\widetilde{T}_t\right) &\le \frac{1}{\sqrt{2}}\left(\left\||Y|^t \left|X^*\right|^{1-t}\right\|_2 + \left\||X|^t \left|Y^*\right|^{1-t}\right\|_2\right) \\ &\le w_2\left(|Y|^t \left|X^*\right|^{1-t}\right) + w_2\left(|X|^t \left|Y^*\right|^{1-t}\right). \quad \text{(by Inequality (9))}\end{aligned}$$

□

3 Applications

In this section, we present some applications of some of our results given in Sect. 2. First, we need the observation that for any two real numbers a and b, we have

$$\frac{a+b}{2} = \max(a,b) - \frac{|a-b|}{2}. \tag{1}$$

Theorem 3.1 *Let $X, Y \in C_2$. Then*

$$w_2\left(\begin{bmatrix} 0 & X \\ Y & 0 \end{bmatrix}\right) + \frac{|w_2(X+Y) - w_2(X-Y)|}{\sqrt{2}} \leq \sqrt{2}\left(\|X\|_2 + \|Y\|_2\right). \tag{2}$$

In particular,

$$|w_2(ReX) - w_2(ImX)| \leq \|X\|_2. \tag{3}$$

Proof By Lemma 2.7 and Identity (1), we have

$$\begin{aligned}
w_2\left(\begin{bmatrix} 0 & X \\ Y & 0 \end{bmatrix}\right) &\leq \frac{w_2(X+Y) + w_2(X-Y)}{\sqrt{2}} \\
&= \frac{\sqrt{2}\left(w_2(X+Y) + w_2(X-Y)\right)}{2} \\
&= \sqrt{2}\max\left(w_2(X+Y), w_2(X-Y)\right) - \frac{\sqrt{2}\,|w_2(X+Y) - w_2(X-Y)|}{2} \\
&= \sqrt{2}\max\left(w_2(X+Y), w_2(X-Y)\right) - \frac{|w_2(X+Y) - w_2(X-Y)|}{\sqrt{2}} \\
&\leq 2w_2\left(\begin{bmatrix} 0 & X \\ Y & 0 \end{bmatrix}\right) - \frac{|w_2(X+Y) - w_2(X-Y)|}{\sqrt{2}} \quad \text{(by Lemma 2.7)} \\
&\leq 2\left(w_2\left(\begin{bmatrix} 0 & X \\ 0 & 0 \end{bmatrix}\right) + w_2\left(\begin{bmatrix} 0 & 0 \\ Y & 0 \end{bmatrix}\right)\right) - \frac{|w_2(X+Y) - w_2(X-Y)|}{\sqrt{2}} \\
&= 2\left(\frac{\|X\|_2}{\sqrt{2}} + \frac{\|Y\|_2}{\sqrt{2}}\right) - \frac{|w_2(X+Y) - w_2(X-Y)|}{\sqrt{2}} \text{(by Identity (10))} \\
&= \sqrt{2}\left(\|X\|_2 + \|Y\|_2\right) - \frac{|w_2(X+Y) - w_2(X-Y)|}{\sqrt{2}}.
\end{aligned}$$

Hence,

$$w_2\left(\begin{bmatrix} 0 & X \\ Y & 0 \end{bmatrix}\right) + \frac{|w_2(X+Y) - w_2(X-Y)|}{\sqrt{2}} \leq \sqrt{2}\left(\|X\|_2 + \|Y\|_2\right).$$

To prove Inequality (3), let $Y = X^*$ in Inequality (2) to get

$$\begin{aligned}
w_2\left(\begin{bmatrix} 0 & X \\ X^* & 0 \end{bmatrix}\right) &\leq \sqrt{2}\left(\|X\|_2 + \left\|X^*\right\|_2\right) - \frac{|w_2\left(X+X^*\right) - w_2\left(X-X^*\right)|}{\sqrt{2}} \\
\sqrt{2}\|X\|_2 &\leq \sqrt{2}\left(\|X\|_2 + \left\|X^*\right\|_2\right) - \frac{|w_2\left(X+X^*\right) - w_2\left(X-X^*\right)|}{\sqrt{2}} \quad \text{(by Theorem 2.5)} \\
\sqrt{2}\|X\|_2 &\leq 2\sqrt{2}\|X\|_2 - \sqrt{2}\,|w_2(\mathrm{Re}X) - w_2(\mathrm{Im}X)| \quad \text{(by the identity (4))} \\
\|X\|_2 &\leq 2\|X\|_2 - |w_2(\mathrm{Re}X) - w_2(\mathrm{Im}X)|.
\end{aligned}$$

Hence,

$$|w_2(\mathrm{Re}X) - w_2(\mathrm{Im}X)| \leq \|X\|_2.$$

□

Theorem 3.2 *Let $X, Y \in C_2$. Then*

$$w_2\left(\begin{bmatrix} 0 & X \\ Y & 0 \end{bmatrix}\right) + (\|X\|_2 + \|Y\|_2) + \frac{\left|\sqrt{2}w_2(X+Y) - (\|X\|_2 + \|Y\|_2)\right|}{2} + \frac{\left|\sqrt{2}w_2(X-Y) - (\|X\|_2 + \|Y\|_2)\right|}{2} \leq 2\sqrt{2}\left(w_2(X) + w_2(Y)\right). \quad (4)$$

In particular,

$$(2+\sqrt{2})\|X\|_2 \leq 4\sqrt{2}w_2(X) - \left|\sqrt{2}w_2(ReX) - \|X\|_2\right| - \left|\sqrt{2}w_2(ImX) - \|X\|_2\right|. \quad (5)$$

Proof By Lemma 2.7, we have

$$\begin{aligned}
&w_2\left(\begin{bmatrix} 0 & X \\ Y & 0 \end{bmatrix}\right) + (\|X\|_2 + \|Y\|_2) \\
&\leq \frac{w_2(X+Y) + w_2(X-Y)}{\sqrt{2}} + (\|X\|_2 + \|Y\|_2) \\
&= \frac{\sqrt{2}w_2(X+Y) + \sqrt{2}w_2(X-Y)}{2} + (\|X\|_2 + \|Y\|_2) \\
&= \frac{\sqrt{2}w_2(X+Y) + (\|X\|_2 + \|Y\|_2)}{2} + \frac{\sqrt{2}w_2(X-Y) + (\|X\|_2 + \|Y\|_2)}{2} \\
&= \max\left(\sqrt{2}w_2(X+Y), (\|X\|_2 + \|Y\|_2)\right) - \frac{\left|\sqrt{2}w_2(X+Y) - (\|X\|_2 + \|Y\|_2)\right|}{2} \\
&+ \max\left(\sqrt{2}w_2(X-Y), (\|X\|_2 + \|Y\|_2)\right) - \frac{\left|\sqrt{2}w_2(X-Y) - (\|X\|_2 + \|Y\|_2)\right|}{2} \quad \text{(by Identity (1))} \\
&\leq \max\left(\sqrt{2}\left(w_2(X) + w_2(Y)\right), (\|X\|_2 + \|Y\|_2)\right) - \frac{\left|\sqrt{2}w_2(X+Y) - (\|X\|_2 + \|Y\|_2)\right|}{2} \\
&+ \max\left(\sqrt{2}\left(w_2(X) + w_2(Y)\right), (\|X\|_2 + \|Y\|_2)\right) - \frac{\left|\sqrt{2}w_2(X-Y) - (\|X\|_2 + \|Y\|_2)\right|}{2} \quad \text{(by Inequality (7))} \\
&= 2\max\left(\sqrt{2}\left(w_2(X) + w_2(Y)\right), (\|X\|_2 + \|Y\|_2)\right) - \frac{\left|\sqrt{2}w_2(X+Y) - (\|X\|_2 + \|Y\|_2)\right|}{2} \\
&- \frac{\left|\sqrt{2}w_2(X-Y) - (\|X\|_2 + \|Y\|_2)\right|}{2} \\
&= 2\sqrt{2}\left(w_2(X) + w_2(Y)\right) - \frac{\left|\sqrt{2}w_2(X+Y) - (\|X\|_2 + \|Y\|_2)\right|}{2} \\
&- \frac{\left|\sqrt{2}w_2(X-Y) - (\|X\|_2 + \|Y\|_2)\right|}{2}, \quad \text{(by Inequality (9))}
\end{aligned}$$

and so

$$w_2\left(\begin{bmatrix} 0 & X \\ Y & 0 \end{bmatrix}\right) + (\|X\|_2 + \|Y\|_2) + \frac{\left|\sqrt{2}w_2(X+Y) - (\|X\|_2 + \|Y\|_2)\right|}{2}$$
$$+ \frac{\left|\sqrt{2}w_2(X-Y) - (\|X\|_2 + \|Y\|_2)\right|}{2} \leq 2\sqrt{2}\,(w_2(X) + w_2(Y))\,.$$

To prove Inequality (5), letting $Y = X^*$ in Inequality (4) to get

$$w_2\left(\begin{bmatrix} 0 & X \\ X^* & 0 \end{bmatrix}\right) + \left(\|X\|_2 + \left\|X^*\right\|_2\right)$$
$$\leq 2\sqrt{2}\left(w_2(X) + w_2\left(X^*\right)\right) - \frac{\left|\sqrt{2}w_2\left(X + X^*\right) - (\|X\|_2 + \|X^*\|_2)\right|}{2}$$
$$- \frac{\left|\sqrt{2}w_2\left(X - X^*\right) - (\|X\|_2 + \|X^*\|_2)\right|}{2}.$$

Then, by Theorem 2.5 and Identity (4), we have

$$(2+\sqrt{2})\|X\|_2 \leq 4\sqrt{2}w_2(X) - \frac{\left|\sqrt{2}w_2(2\mathrm{Re}X) - 2\|X\|_2\right|}{2} - \frac{\left|\sqrt{2}w_2(2\mathrm{Im}X) - 2\|X\|_2\right|}{2}$$
$$(2+\sqrt{2})\|X\|_2 \leq 4\sqrt{2}w_2(X) - \left|\sqrt{2}w_2(\mathrm{Re}X) - \|X\|_2\right| - \left|\sqrt{2}w_2(\mathrm{Im}X) - \|X\|_2\right|.$$

□

References

1. Abu-Omar, A., Kittaneh, F.: A generalization of the numerical radius. Linear Algebr. Appl. **569**, 323–334 (2019)
2. Abu-Omar, A., Kittaneh, F.: Numerical radius inequalities for n× n operator matrices. Linear Algebr. Appl. **468**, 18–26 (2015)
3. Aldalabih, A., Kittaneh, F.: Hilbert-Schmidt numerical radius inequalities for operator matrices. Linear Algebr. Appl. **581**, 72–84 (2019)
4. Bhatia, R.: Matrix Analysis. Springer, New York (1997)
5. Hirzallah, O., Kittaneh, F., Shebrawi, K.: Numerical radius inequalities for certain 2×2 operator matrices. Integr. Equ. Oper. Theory. **71**, 129–147 (2011)
6. Shebrawi, K.: Numerical radius inequalities for certain 2×2 operator matrices II. Linear Algebr. Appl. **523**, 1–12 (2017)
7. Yamazaki, T.: On upper and lower bounds of the numerical radius and an equality condition. Stud. Math. **178**, 83–89 (2007)

Model Reduction and Implicit–Explicit Runge–Kutta Schemes for Nonlinear Stiff Initial-Value Problems

Younis A. Sabawi, Mardan A. Pirdawood, Hemn M. Rasool, and Salisu Ibrahim

Abstract The main goal of this paper is the use of the implicit–explicit Runge–Kutta method for finding the numerical approximate solutions for chemical reaction problems that contain stiff and no stiff terms. The stiff part is treated by an implicit scheme, while the second part is treated by an explicit scheme. This combination results in an efficient numerical scheme that is able to solve stiff problems quickly and accurately. An important factor in our proposed method is to reduce the number of iterations, which, consequently, leads to a reduction in the computational cost of the scheme. Additionally, this method is a significant step forward in the field of solving stiff problems. The accuracy of the suggested scheme is computed through pointwise error. It offers a robust and efficient numerical approach that is able to achieve high levels of accuracy. Numerical experiments show that there is good agreement and accuracy between the original solution and reduction problems.

Keywords Model reduction · IMEX-RK methods · Stiff problems · miRNA model

Y. A. Sabawi (✉) · M. A. Pirdawood · H. M. Rasool
Department of Mathematics, Faculty of Science and Health, Koya University, Koya KOY45, Koy Sanjaq, Kurdistan Region - F.R. Iraq, Iraq
e-mail: younis.abid@koyauniversity.org

M. A. Pirdawood
e-mail: mardan.ameen@koyauniversity.org

H. M. Rasool
e-mail: hemn.mohammed@koyauniversity.org

Y. A. Sabawi · S. Ibrahim
Department of Mathematics Education, Faculty of Education, Tishk International University, Erbil, Iraq
e-mail: salisu.ibrahim@tiu.edu.iq

D. Zeidan et al. (eds.), *Mathematics and Computation*, Springer Proceedings in Mathematics & Statistics 418, https://doi.org/10.1007/978-981-99-0447-1_9

1 Introduction

Numerous areas of chemical reactions, ecological interactions, biological processes, enzymatic reaction models, and cell signalling pathway models can be modelled as a system of stiff ordinary differential equations. The key idea of stiff problems is to give a great role in understanding and identifying these effects on the model dynamics. Because of their difficulties, most of these problems do not have exact analytic solutions. Furthermore, these problems have very different time scales occurring simultaneously. Therefore, many research have attracted much interest in this field and many numerical schemes have been proposed over the years, such as the Euler method, Runge–kutta method, multistep methods [1–4], Finite difference method [5–7], Finite element methods [8–11], linear regression analysis [12].

The most popular methods for solving stiff problems are the Runge–Kutta method. The disadvantages of these methods are that they do not work well for stiff differential equations in spite of providing a good understanding of the model's dynamical behaviour. The aim of this work is to propose a method that will tackle the difficulties that appear during the modelling process, most especially the transformation models of miRNA to stiff nonlinear equations with an implicit method. This method is called Implicit–Explicit (IMEX) schemes [13–18]. Consider the numerical method of the following system of stiff ordinary differential equation:

$$\frac{d\boldsymbol{u}}{\partial t} = F(t, \boldsymbol{u}(t)) + G(t, \boldsymbol{u}(t)), \tag{1}$$

A key idea for the proposed method is to split the right-hand side of (1) into stiff F (t, $\mathbf{u}$(t)) and nonstiff G(t, $\mathbf{u}$(t)). . Note that an explicit Runge–Kutta (*ERK*) method is used to solve the nonstiff part F and a diagonally implicit Runge–Kutta (*DIRK*) method is employed to solve the stiff part G. The popular family of IMEX schemes for DIRK and ERK terms takes the following form [10, 11]:

$$\begin{array}{c|ccccc} c_1 & a_{11} & 0 & 0 & \cdots & 0 \\ c_2 & a_{21} & a_{22} & 0 & \cdots & 0 \\ c_3 & a_{31} & a_{32} & a_{33} & \cdots & 0 \\ \vdots & \vdots & \vdots & \vdots & \ddots & \vdots \\ c_s & a_{s1} & a_{s2} & a_{s3} & \cdots & a_{ss} \\ \hline & b_1 & b_2 & b_3 & \cdots & b_s \end{array}$$

$$\begin{array}{c|ccccc}
\hat{c}_1 & 0 & 0 & 0 & \cdots & 0 \\
\hat{c}_2 & \hat{a}_{21} & 0 & 0 & \cdots & 0 \\
\hat{c}_3 & \hat{a}_{31} & \hat{a}_{32} & 0 & \cdots & 0 \\
\vdots & \vdots & \vdots & \vdots & \ddots & \vdots \\
\hat{c}_s & \hat{a}_{s1} & \hat{a}_{s2} & \hat{a}_{s3} & \cdots & 0 \\
\hline
 & \hat{b}_1 & \hat{b}_2 & \hat{b}_3 & \cdots & \hat{b}_s
\end{array}$$

$$\begin{array}{c|c}
c & A \\
\hline
 & b^T
\end{array}
\qquad
\begin{array}{c|c}
\hat{c} & \hat{A} \\
\hline
 & \hat{b}^T
\end{array}$$

(Ex) and (Im) are denoted to the explicit and implicit components. The Implicit–Explicit scheme, defined by its Butcher coefficients $(A^{[Ex]}, A^{[Im]}, b^{[Ex]}, b^{[Im]}, c^{[Ex]}, c^{[Im]})$ is given by

$$\boldsymbol{u}^{n+1} = \boldsymbol{u}^n + \Delta t \sum_{i=1}^{s} \left(b_i^{[Im]} \boldsymbol{k}_i^{[Im]} + b_i^{[Ex]} \boldsymbol{k}_i^{[Ex]} \right), \tag{2}$$

where $\boldsymbol{k}_i^{[Im]}$ and $\boldsymbol{k}_i^{[Ex]}$ are the discrete counterparts of the stiff and nonstiff operators respectively in (2), $\mathrm{F_s}$ and $\mathrm{F_{ns}}$,

$$\boldsymbol{k}_i^{[Im]} = F(t_i + c_i \Delta t, \boldsymbol{u}_i(t)), \boldsymbol{k}_i^{[Ex]} = G(t_i + c_i \Delta t, \boldsymbol{u}_i(t)),$$

and the stage values are defined as

$$\boldsymbol{u}_i = \boldsymbol{u}^n + \nu t \sum_{j=1}^{s} \left(a_{ij} \boldsymbol{k}_i^{[Im]} + \hat{a}_{ij} \boldsymbol{k}_i^{[Ex]} \right). \tag{3}$$

Applying DIRK schemes for the implicit part, the above expression gives

$$\boldsymbol{u}_i = \boldsymbol{u}^n + \Delta t \sum_{j=1}^{i-1} \left(a_{ij} \boldsymbol{k}_i^{[Im]} + \hat{a}_{ij} \boldsymbol{k}_i^{[Ex]} \right) + \Delta t a_{ii} \boldsymbol{k}_i^{[Im]}. \tag{4}$$

To deal with the linearly implicit cases, we use

$$(\boldsymbol{I} - \Delta t a_{ii} \boldsymbol{K}) \boldsymbol{u}_i = \boldsymbol{u}^n + \Delta t \sum_{j=1}^{i-1} \left(a_{ij} \boldsymbol{k}_i^{[Im]} + \hat{a}_{ij} \boldsymbol{k}_i^{[Ex]} \right). \tag{5}$$

The rest of this work is structured as follows. In Sect. 2, the model problem is introduced with the reduction method. In Sect. 3, Implicit–Explicit Runge–Kutta methods are presented. Numerical experiments are shown with different types of examples in Sect. 4. Finally, conclusions are given in Sect. 5.

2 Model Equations of miRNA

The skin, muscles, and bones you see are built inside the cells. All these cells involve billions of proteins. Certainly, proteins are a crucial molecular segment for every living organism on this planet [19]. MicroRNAs (miRNAs) involve 20–22 nucleotide RNAs that accentuate the process of eukaryotic messenger RNAs and also have an essential role in phylogeny, carcinogenesis, stress responses, and virus infection [20]. The miRNAs are single-stranded RNA fragments of around 21–23 nucleotides in the length, which accentuate the organization and classification of genes and translational qualifications [21]. miRNAs function, at any rate in part, to prevent the formation of proteins by messenger RNAs and contribute to the progress mRNA deacetylation, decamping, and 5' to 3' reduce of the mRNA body [22]. They were first described in 1993 [23], the RISC effector complex and elaborate microRNAs are associated, which incorporates as a key part Argonaut protein? MicroRNAs character quality representation by conducting the RISC complex near particular target mRNAs. Till present, there is an enormous controversy in deciding the precise composition of this restraint [24]. See Fig. 1. through explanation explaining the process of protein translation is given for the mathematical modelling of miRNA. We simply examined the previous analysis of the translation of miRNA protein given in [25]. Then, we use conservation laws for model reduction and the system is reduced to three equations from seven equations; model reduction is an important way to be dimensionless and it is applied in many previously published papers [26–29].

A nonlinear variant for the translation model was introduced to express the impact of mRNA interference with translation initiation factors. The recycling for initiation factors ($eIF4F$) and ribosomal subunits (60 S and 40 S) is precisely taken into study. The model involves six chemical reactions of the 60 S, 40 S, F,$eIF4F$, R, and A varieties including four chemical reactions, all supposedly irreversible; see Fig. 2. The model reaction will be given as:

1 $eIF4F + 40S \to F$, , construction for the initiation complex (rate k_1).
2 $F \to A$, assembly of some developed and cap-independent initiation treads, such as investigating the UTR to start codon A (rate k_2).
3 $A \to R$, , assembly of protein translation and ribosomes (rate k_3).
4 $80S \to 60S + 40S$, , recycling regarding the ribosomal subunits (rate k_4).

By employing stoichiometric vectors, mass action law, and reaction rates, we describe the model equations moreover the model is represented by the next system of nonlinear ordinary differential equations:

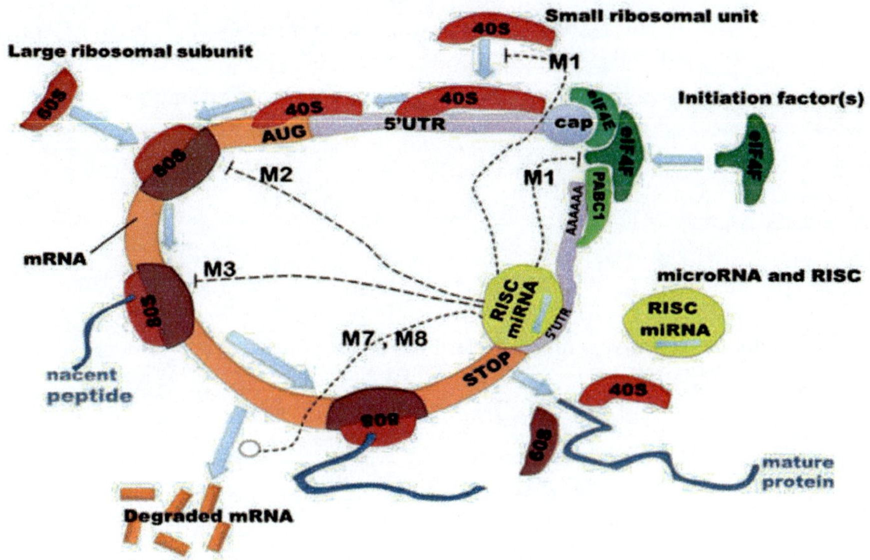

Fig. 1 The protein translation manner with microRNA tools

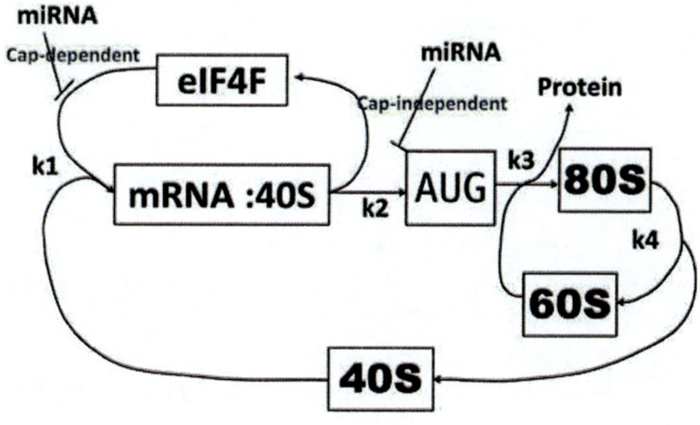

Fig. 2 Nonlinear protein's model pathways

$$\frac{d[40S](t)}{dt} = k_4[R] - k_1[eIF4F][40S],$$

$$\frac{d[eIF4F](t)}{dt} = k_2[F] - k_1[eIF4F][40S],$$

$$\frac{d[F](t)}{dt} = -k_2[F] + k_1[eIF4F][40S],$$

$$\frac{d[A](t)}{dt} = -k_3[60S][A] + k_2[F], \tag{6}$$

$$\frac{d[60S](t)}{dt} = k_4[R] - k_3[60S][A],$$

$$\frac{d[R](t)}{dt} = -k_4[R] + k_3[60S][A],$$

$$P_{synth}(t) = k_3[A](t).$$

The system (6) involves three independent of the conservations laws

$$[R] + [40S] + [F] + [A] = [40S]_0,$$

$$[eIF4F] + [F] = [eIF4F]_0, \tag{7}$$

$$[R] + [60S] = [60S]_0,$$

where $[eIF4F]_0$, $[60S]_0$, and $[40S]_0$ are total quantities of prepared initiation factor and big and small ribosomal subunits, respectively. By applying the conservation laws (7), the latter variables are eliminated [17]

$$[eIF4F] = -[F] + [eIF4F]_0,$$

$$[60S] = -[R] + [60S]_0, \tag{8}$$

$$[A] = -[R] + [40S]_0 - [F] - [40S].$$

Then, system (6) will have the form

$$\frac{d[40S](t)}{dt} = k_4[R] - k_1[eIf4F]_0[40S] + k_1[F][40S],$$

$$\frac{d[F](t)}{dt} = -k_1[F][40S] + k_1[eIF4F][40S] - k_2[F], \tag{9}$$

$$\frac{d[R](t)}{dt} = k_3[R][40S] + k_3[eIF4F]_0[40S] + k_3[R][F] - k_3[60S]_0[40S] + k_3([R])^2$$

$$-k_3[60S]_0[F] - (k_4 + [40S]_0 + k3[40S]_0)[R].$$

There are some assumptions on the initial variable states and model parameters

$$k_4 \ll k_1, k_2, k_3,$$

$$k_3 \gg k_1, k_2,$$

$$[eIF4F]_0 \ll [40S]_0, \tag{10}$$

$$[eIF4F]_0 < [60S]_0 < [40S]_0.$$

For more details about the mRNA model and assumptions, see [28–30]. Depending on the assumptions in Eq. (10) and introducing new variables

$$x_1 = \frac{[40S]}{[40S]_0}, x_2 = \frac{F}{[eIF4F]_0}, x_3 = \frac{R}{[eIF4F]_0}.$$

The system (9) takes the form

$$\frac{dx_1}{dt} = \alpha_1 x_1(x_2 - 1) + \alpha_2 x_3$$

$$\frac{dx_2}{dt} = \alpha_3 x_1(1 - x_2) - \alpha_4 x_2 \tag{11}$$

$$\frac{dx_3}{dt} = \alpha_5(1 - x_1) - \alpha_6 x_2 + \alpha_7 x_1 x_3 + \alpha_8 x_2 x_3 + x_3^2 - \alpha_9 x_3,$$

where $\alpha_1 = k_1[eIF4F]_0, \alpha_2 = k_4\frac{[eIF4F]_0}{[40S]_0}, \alpha_3 = k_1[40S]_0, \alpha_4 = k_2, \alpha_5 = \frac{k_3[40S]_0[60S]_0}{[eIF4F]_0}, \alpha_6 = k_3[60S]_0,$
$\alpha_7 = k_3[40S]_0, \alpha_8 = k_3[eIF4F]_0, \alpha_9 = k_3[60S]_0 + [40S]_0 + k_4.$

3 The Proposed Method

This section aims to use the high-order IMEX-RK scheme presented in Sect. 1 for solving the model equations of miRNA presented in Sect. 2. To do this, recalling (11), and for brevity, this can be written as

$$F_{Im}(\mathrm{t}, \mathrm{x(t)}) = \begin{bmatrix} -\alpha_1 x_1 + \alpha_2 x_3 \\ \alpha_3 x_1 - \alpha_4 x_2 \\ \alpha_5 - \alpha_5 x_1 - \alpha_6 x_2 - \alpha_9 x_3 \end{bmatrix},$$

$$F_{Ex}(\mathrm{t}, \mathrm{x(t)}) = \begin{bmatrix} \alpha_1 x_1 x_2 \\ -\alpha_3 x_1 x_2 \\ \alpha_7 x_1 x_3 + \alpha_8 x_2 x_3 + \alpha_8 x_2 \end{bmatrix}.$$

Go back to (11), and substituting the above equations in (11), leads to

$$\frac{du}{dt} = F_{Im}(t, \boldsymbol{u}(t)) + F_{Ex}(t, \boldsymbol{u}(t)), \tag{12}$$

where $\boldsymbol{u}(t) = [x_1(t)x_2(t)x_3(t)]^T$.

A key idea for the proposed method is to split the right-hand side of (12) into stiff $F_{Im}(t, \mathbf{u}(t))$ and nonstiff ($F_{Ex}(t, \mathbf{u}(t))$). Note that an explicit Runge–Kutta (*ERK*) method is used to solve the no stiff part (F_{Ex}) and a diagonally implicit Runge–Kutta (*DIRK*) method is employed to solve the stiff part (F_{Im}).

c_1	1/4	0	0	0	0
c_2	0.34114705729	1/4	0	0	0
c_3	0.80458720789	−0.07095262154	1/4	0	0
c_4	−0.52932607329	1.15137638494	−0.80248263237	1/4	0
c_5	0.11933093090	0.55125531344	−0.1216872844	0.20110104014	1/4
	0.11933093090	0.551255313447	−0.1216872844	0.20110104014943	¼

and

$\hat{c}_1$	0	0	0	0	0
$\hat{c}_2$	0.39098372	0	0	0	0
$\hat{c}_3$	1.09436646	0.33181504274	0	0	0
$\hat{c}_4$	0.14631668	0.69488738	0.46893381	0	0
$\hat{c}_5$	−1.33389883	2.90509214	−1.06511748	0.27210900509	0
	0.119330930	0.551255313	−0.12168728449	0.201101040	1/4

Algorithm 1 IMEX-RK(5, 4, 4)

1: Input $\boldsymbol{u}_0$,no of stages, no of iterations, Time
2: Put h = Time/no of iterations
3: The matrices $A^{[E]}, A^{[I]}, b^{[E]}$ and $b^{[I]}$ can be obtained in the Butcher Table.
4: **for** n = 0 : (no of iterations) − 1 **do**
5: **accum1** ⟵ $\mathbf{u}_n$
6: **for** i = 0 : (no of stages) − 1 **do**
7: **accum2** ⟵ $u_n + h \cdot (A^{[I]}_{ij} \cdot \mathbf{F}_{\mathbf{Im}}(:, \mathbf{u}_n))$.
8: **for** j = 0 : $(i-1)$ **do**
9: **accum2** ⟵ **accum2**+$h \cdot (A^{[Im]}_{ij} \cdot \mathbf{k}^{[Im]}_j + A^{[Ex]}_{ij} \cdot \mathbf{k}^{[Ex]}_j)$.
10: **end do**
11: $\mathbf{k}^{[Im]}_i$ ⟵ $\mathbf{F}_{\mathbf{Im}}$(:, **accum2**).
12: $\mathbf{k}^{[Ex]}_i$ ⟵ $\mathbf{F}_{\mathbf{Ex}}$(:, **accum2**).
13: **accum1**⟵ **accum1**+h $\cdot (b^{[Im]}_i \cdot \mathbf{k}^{[Im]}_i + b^{[Ex]}_i \cdot \mathbf{k}^{[Ex]}_i)$.
14: **end do**
15: u_{n+1} ⟵ **accum1**.
16: **end do**

4 Numerical Experiments

The goal of this section is to illustrate the performance of a presented method through an implementation based on Matlab programming. IMEX-RK (4, 5, 5) and classical Runge–Kutta method *ERK4* are used for solving (6) and (12), where the number 4 is the order of the scheme, 5 is the number of stages; implicit and explicit schemes. Some examples are utilized in this paper. We measure the error between exact and approximation by

$$L_{abc} = |y(t_i) - y_i|$$

a. Example 1

The protein translation pathways with microRNA model defined in (6), with$[40S]_0 = 100$,$[60S]_0 = 25$, $[eIF4F]_0 = 6$, $[40S](0) = [40S]_0$, $F(0) = 0$ and$R(0) = 0$. Also,$h = 0.01$,$k_1 = 0.04$, $k_2 = 0.05$, $k_3 = 1$, $k_4 = 0.0001$, and the initial conditions for system (12) become $x_1(0) = 1$, $x_2(0) = 0$ and$x_3(0) = 0$. Runge–Kutta method of order 4 ($ERK4$) is used as the exact solution of (12) with the stepsize $h = 0.001$

There are some effective results based on the approximate solutions, especially for more understanding of the global dynamics. From Fig. 3, the reduced models are very close to the original model. Figure 4 shows us the protein from the initial point till about 85 min is stable ßnd then the height rate of protein is produced until 370 min, then it remains stable, because of the homeostasis mechanism.

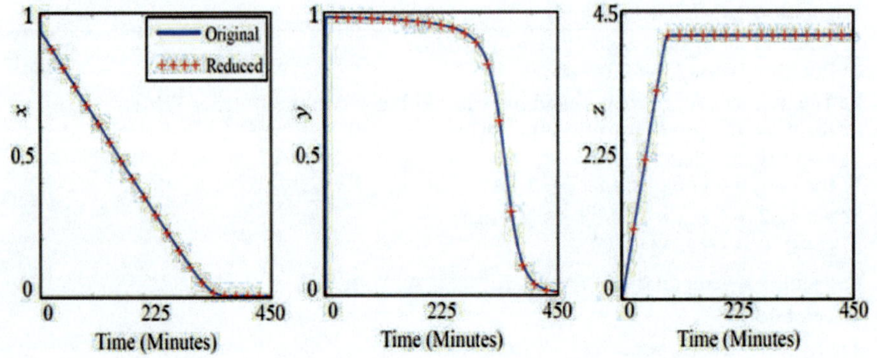

Fig. 3 Example 1. Numerical solutions by using IMEX-RK (4,5,5) method for original (6) and reduced (12) problems

b. Example 2 [28]. Consider the chemical reaction problem

$A + B \rightarrow 2A$, A + B $\rightarrow A + C$, A+C $\rightarrow$ $2C$, A $\rightarrow$ C. The numerical values of the rate constants are $k_1 = 0.01$, $k_2 = 0.001$, $k_3 = 0.001$, $k_4 = 0.1$. The mathematical model of the above chemical reaction can be transferred by a set of three ODEs as

$$\begin{pmatrix} y_1'(t) \\ y_2'(t) \\ y_3'(t) \end{pmatrix} = \begin{pmatrix} k_1 y_1(t) y_2(t) - k_3 y_1(t) y_3(t) - k_4 y_1(t) \\ (-k_1 + k_2 y_2(t)) y_1(t) y_3(t) \\ k_2 k_1 y_1(t) y_2(t) + k_3 y_1(t) y_3(t) + k_4 y_1(t) \end{pmatrix}.$$

Runge–Kutta method of order 4 ($ERK4$) is used as the exact solution of (12) with the stepsize $h = 0.00001$.

Tables 2 and 3 show the good accuracy of the IMEX – RK method using step size $h = 1, h = 0.001$, which are better than the classical Rung–Kutta, in which case $h = 0.0001$, h = 0.000001. This leads to the number of iterations IMEX much less than for the number of iterations ERK, and consequently leading to a reduction in the computational cost of the scheme IMEX.

c. Example 3 [29]. Consider the initial value problem

$$\frac{\mathrm{dy}_1}{\mathrm{dt}} = -1.001 y_1 + 0.999 y_2 + 2 y_1 y_2,$$

$$\frac{\mathrm{dy}_2}{\mathrm{dt}} = 0.999 y_1 -- 1.001 y_2 + y_1^2 + y_2^2,$$

$$y(0) = 0,\ y(0) = -1.$$

This problem can be written as stiff and no stiff terms as

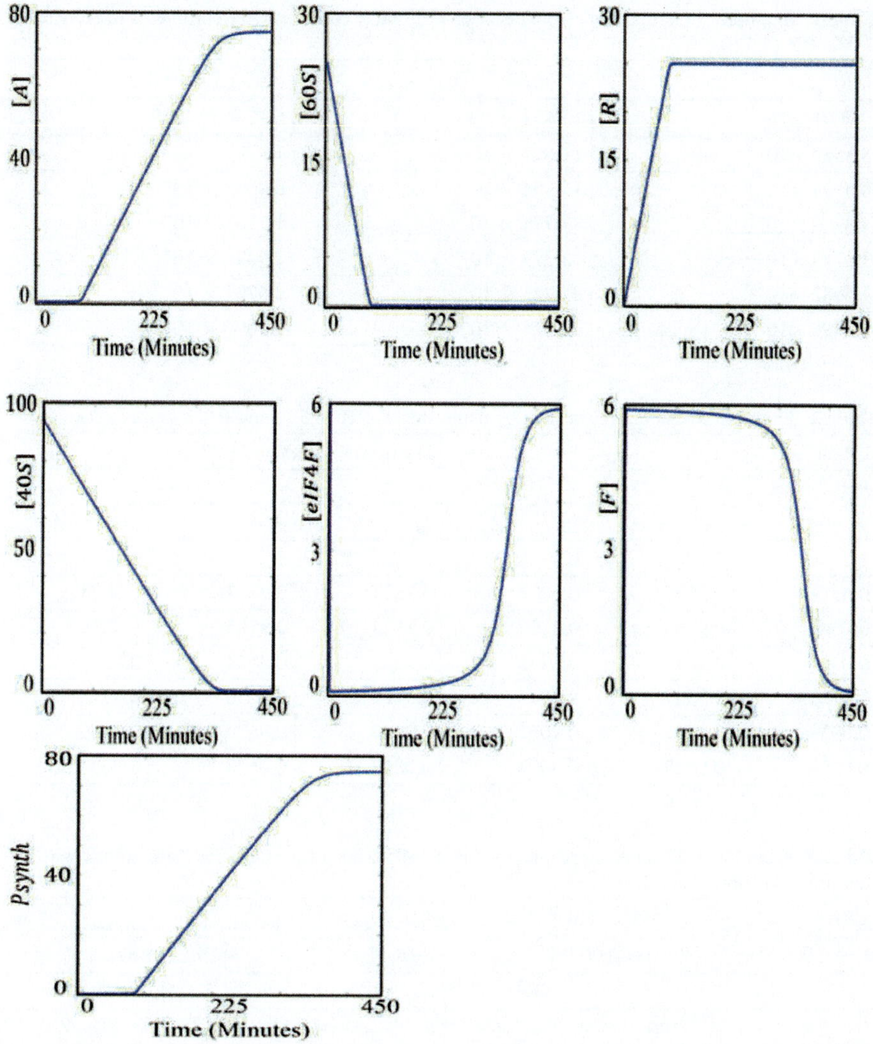

Fig. 4 Example 1. Numerical solutions by using IMEX-RK (4,5,5) method for original (6) problem

$$\frac{dy}{dt} = F_{Im}(t, y(t)) + F_{Ex}(t, y(t)),$$

$$y(t) = (y_1(t), y_2(t))^T, \mathrm{F_{Im}} = \begin{pmatrix} -1.001 & 0.999 \\ 0.999 & -1.001 \end{pmatrix} y, \mathrm{F_{Ex}} = \begin{pmatrix} 2y_1y_2 \\ y_1^2 + y_2^2 \end{pmatrix}.$$

The exact solution is

$$y_1(t) = \frac{1000}{2001e^{2000t} - 1} - \frac{1}{3e^{2t} - 1},$$

Table 1 Example 1. Comparing errors between original (6) and reduction problems (12) by using IMEX-RK (4,5,5)

$\lvert x_{1\text{Original}} - x_{1\text{Reduced}} \rvert$	$\lvert x_{2\text{Original}} -- x_{2\text{Reduced}} \rvert$	$\lvert x_{3\text{Original}} - x_{3\text{Reduced}} \rvert$
9.8091×10^{-8}	1.6583×10^{-6}	6.1240×10^{-6}
8.1589×10^{-8}	1.3857×10^{-6}	7.9795×10^{-10}
6.5194×10^{-8}	1.1154×10^{-6}	6.4068×10^{-8}
4.8905×10^{-8}	8.4774×10^{-7}	1.2703×10^{-7}
3.2773×10^{-8}	5.8377×10^{-7}	1.9478×10^{-7}
1.6982×10^{-8}	3.2723×10^{-7}	2.6707×10^{-7}
2.8490×10^{-9}	9.8524×10^{-8}	3.4094×10^{-7}

Table 2 Example 2. Comparing results by using IMEX-RK (4,5,5) and (ERK4) methods

ERK4			IMEX-RK (4,5,5)		
$h = 0.0001$			$h = 1$		
y_1	y_2	y_3	y_1	y_2	y_3
1	1	0	1	1	0
0.3319	0.9834	0.6078	0.3324	0.9833	0.6073
0.1099	0.9705	0.8102	0.1103	0.9705	0.8099
0.0364	0.9655	0.8773	0.0365	0.9654	0.8771
0.0120	0.9637	0.8995	0.0121	0.9636	0.8995
0.0039	0.9631	0.9069	0.0040	0.9630	0.9069
0.0013	0.9629	0.9093	0.0013	0.9628	0.9093

Table 3 Example 3. Comparing results by using IMEX-RK (4,5,5) and ERK4 methods with the measure of errors

ERK4		IMEX-RK (4,5,5)		$L_{abs}(y_i) = \lvert y_{iApproximate} - y_{iExact} \rvert$			
h = 0.000001		h = 0.001		ERK4		IMEX-RK(4,5,5)	
y_1	y_2	y_1	y_2	$L_{abs}(y_1)$	$L_{abs}(y_2)$	$L_{abs}(y_1)$	$L_{abs}(y_2)$
0	−1	0	−1	0	0	0	0
−0.0472	−0.0472	−0.0472	−0.0472	3.38e-14	3.38e-14	6.32e-8	6.32e-8
−0.0061	−0.0061	−0.00614	−0.0061	6.05e-15	6.05e-15	1.42e-8	1.42e-8
−0.0008	−0.0008	−0.00082	−0.0008	1.03e-15	1.03e-15	2.73e-9	2.73e-9
−0.0001	−0.0001	−0.00011	−0.0001	1.69e-16	1.69e-16	4.82e-10	4.82e-10
−0.0000	−0.0000	−0.00001	−0.0000	2.70e- 17	2.70e-17	8.03e-11	8.03e-11

$$y_2(t) = -\frac{1000}{2001e^{2000t} - 1} - \frac{1}{3e^{2t} - 1}.$$

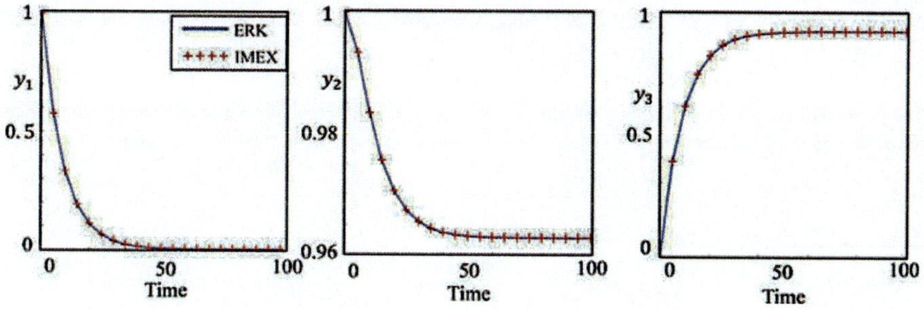

Fig. 5 Example 2. Numerical solutions by using IMEX-RK (4,5,5) and (ERK4) methods

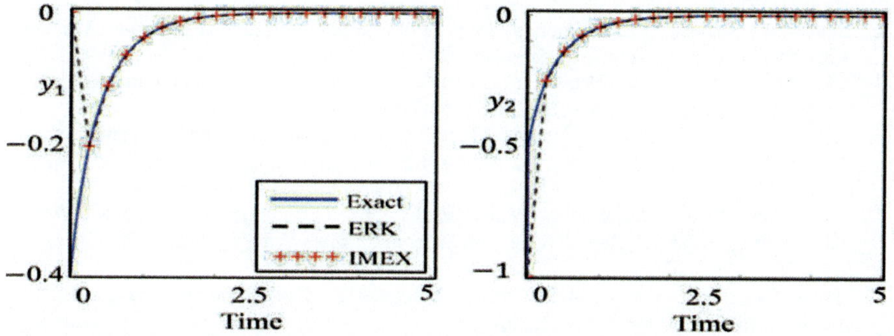

Fig. 6 Example 3. Numerical solutions by using IMEX-RK (4,5,5) and ERK4 methods

5 Conclusions

The miRNA protein translation nonlinear model has been studied, involving seven species and four parameters. For modelling the system, the classical chemical kinetics under constant rates and the law of mass action are applied. To minimize the number of model species and parameters, we have added some new variables. The main goal of this paper is the use of implicit–explicit Runge–Kutta method for finding the numerical approximate solutions for chemical reaction problems that contain stiff and no stiff terms. The stiff part is treated by an implicit scheme, while the second part is treated by an explicit scheme. An important factor in our proposed method is to reduce the number of iterations and consequently leading to a reduction in the computational cost of the scheme. To illustrate the performance of the presented method, three examples are used. It is clearly confirmed that the proposed method is much better than ERK in terms of computational cost and stability. In the future, these numerical methods can be used in complex cell signalling pathways and high dimensional nonlinear systems. This work can be extended to use finite element and discontinuous Galerkin methods for estimating this type of problem in terms of

$L_\infty(L^2)$ and $L_\infty(H^1)$[33–37]. Another interesting aspect of this work is the fact that commutativity and stability can be applicable to such problems [38–45].

Acknowledgements The authors would like to thank Tishk International and Koya Universities for their financial support.

References

1. Pareschi, L., & Russo, G.: Implicit–explicit Runge–Kutta schemes and app lications to hyperbolic systems with relaxation. J. Sci. Comput. 25(1), 129 (2005). https://doi.org/10.1007/s10915-004-4636-4
2. Lapidus, L., Seinfeld, J.H.: Numerical Solution of Ordinary Differential Equtions. Academic Press (1971). Mar 31. Another Reference
3. Atkinson, K., Han, W., Stewart, D.E.: Numerical Solution of Ordinary Differential Equations. Wiley (2011). https://doi.org/10.1002/9781118164495
4. Griffiths, D.F., Higham, D.J.: Numerical Methods for Ordinary Differential Equations Initial Value Problems. Springer Science Business Media (2010). https://doi.org/10.1007/978-0-85729-148-6
5. Islam, M.A.: A comparative study on numerical solutions of initial value problems (IVP) for ordinary differential equations (ODE) with Euler and Runge Kutta methods. Am. J. Comput. Math. **5**(03), 393 (2015)
6. Manaa, S.A., Moheemmeed, M.A., Hussien, Y.A.: A numerical solution for sine-gordon type system. Tikrit J. Pure Sci. **15**(3) (2010)
7. Sabawi, Y.A., Ahmed, S.B., Hamad, H.Q.: Numerical treatment of Allen's equation using semi implicit finite difference methods. Eurasian J. Sci. Eng. **8**, 90–100 (2022). https://doi.org/10.23918/eajse.v8i1p90
8. Martins, R.C., Fachada, N.: Finite element procedures for enzyme. In: Chemical Reaction and In- Silico'Genome Scale Networks. (2015). Aug 1. https://doi.org/10.48550/arXiv.1508.02506
9. Younis, A.S.: A posteriori error analysis in finite element approximation for fully discrete semilinear parabolic problems. In: Finite Element Methods and Their Applications, IntechOpen (2020)
10. Sabawi, Y.A.: A posteriori –error bounds in discontinuous Galerkin methods for semidiscrete semilinear parabolic interface problems. Baghdad Sci. J. **18**(3), 0522 (2021). https://doi.org/10.21123/bsj.2021.18.3.0522
11. Younis A.S.: A posteriori $L_\infty(H^1)$ error bound in finite element approximation of semdiscrete semilinear parabolic problems. In: 2019 First International Conference of Computer and Applied Sciences (CAS), pp. 102–10. IEEE, 2019. https://doi.org/10.1109/CAS47993.2019.9075699
12. Ibrahim, S., Muhammed Tawfeeq Al-kassab, M.: Using linear regression analysis to study the recovery cases of COVID 19 in Erbil, Kurdistan Region. Drugs Cell Therap. Hematol. **10**(1), 1226–1123 (2021). https://www.dcth.org/index.php/journal
13. Ascher, U.M., Ruuth, S.J. and Spiteri, R.J.: Implicit-explicit runge-kutta methods for time-dependent partial differential equations. Appl. Numer. Math. **25**(2–3), 151–167 (1997). https://doi.org/10.1016/S0168-9274(97)00056-1
14. Pareschi, L., & Russo, G.: Implicit–explicit Runge–Kutta schemes and applications to hyperbolic systems with relaxation. J. Sci. Comput. **25**(1), 129 (2005). https://doi.org/10.1007/s10915-004-4636-4
15. Sabawi, Y. A.: Combination between single diagonal implicit and explicit runge kutta (imex-rk) methods for solving stiff differential equations. Tikrit J. Pure Sci. **16**(1), 94–101 (2011)

16. Xiao, A., Zhang, G., Yi, X.: Two classes of implicit–explicit multistep methods for nonlinear stiff initial– value problems. Appl. Math. Comput. **15**(247), 47–60 (2014). https://doi.org/10.1016/j.amc.2014.08.066
17. Sabawi, Y.A., Pirdawood, M.A., Sadeeq, M.I.: A compact fourth-order implicit-explicit Runge-Kutta type method for solving diffusive Lotka–Volterra system. In: Journal of Physics: Conference Series 2021, vol. 1999, No. 1, p. 012103. IOP Pulishing. https://doi.org/10.1088/1742-6596/1999/1/012103
18. Pirdawood, M.A., Sabawi, Y.A.: High-order solution of generalized burgers–fisher equation using compact finite difference and DIRK methods. In: Journal of Physics: Conference Series, vol. 1999, No. 1, p. 012088. IOP Publishing (2021)
19. Xu, F., et al.: Dynamics of microRNA-mediated motifs. IET Syst. Biol. **3**(6), 490–504 (2009). https://doi.org/10.1088/1742-6596/1999/1/012088
20. Nissan, T., Parker, R.: Computational analysis of miRNA-mediated repression of translation implications for models of translation initiation inhibition. RNA **14**(13), 1480–1491 (2008). https://doi.org/10.1261/rna.1072808
21. Eulalio, A., Huntzinger, E., Izaurralde, E.: Getting to the root of miRNA-mediated silencing. Cell **132**(1), 9–14 (2008). https://doi.org/10.1016/j.cell.2007.12.024
22. Filipowicz, W., Bhattacharyya, S.N., Sonenberg, N.: Mechanisms of post- transcriptional regulation by microRNAs: Are the answers in sight? Nat. Rev. Genet. **9**(2), 102–114 (2008)
23. Jackson, R.J., Standart, N.: How do microRNAs regulate gene expression? Sci. Stke **2007**(367), re1 (2007). https://doi.org/10.1042/BST0361224
24. Valencia-Sanchez, M.A., et al.: Control of translation and mRNA degradation by miRNAs and si andsiRNAs. Genes Dev. **20**(5), 515–524 (2006). https://doi.org/10.1101/gad.1399806
25. Lee, R.C., Feinbaum, R.L., Ambros, V.: The C elegans heterochronic gene lin-4 encodes smal RNAs with antisense complementarity to lin-14. Cell **75**(5), 843–854 (1993).https://doi.org/10.1016/0092-8674(93)90529-Y
26. Glaser, V.: Tapping miRNA-regulated pathways: expression profiling ramps up to supportdiagonstics and drug discovery. Genet. Eng. Biotechnol. News **28**(5) (2008)
27. Zinovyev, A., et al.: Dynamical modeling of microRNA action on the protein translation process. BMC Syst. Biol. **4**(1), 13 (2010). https://doi.org/10.1186/1752-0509-4-13
28. Lewin, B., Cells.: Jones and Bartlett learning (2007)
29. Akgül, A., Khoshnaw, S.H., Rasool, H.M.: Minimizing cell signalling pathway elements using lumping parameters. Alexandria Eng. J. (2020).https://doi.org/10.1016/j.aej.2020.01.041
30. Khoshnaw, S.H., Rasool, H.M.: Mathematical modelling for complex biochemical networks and identification of fast and slow reactions. In: The International Conference, on Mathematical and Related Sciences. Springer (2019)
31. Boyce, W.E., DiPrima, R.C., Meade, D.B.: Elementary Differential Equations. Wiley (2017)
32. Shoufu, L.: Theory of Computational Methods for Stiff Differential Equations. Hunan Science and T andTechnology Publisher (1997)
33. Sabawi, Y.A.: Adaptive discontinuous Galerkin methods for interface problems, PhD Thesis, University of Leicester, Leicester, UK (2017)
34. Cangiani, A., Georgoulis, E.H., Sabawi, Y.A.: Adaptive discontinuous Galerkin methods for elliptic interface problems. Math. Comput. **87**, no. 314, 2675–2707 (2018). https://doi.org/10.1090/mcom/3322
35. Cangiani, A., Georgoulis, E.H., Sabawi, Y.A.: Convergence of an adaptive discontinuous Galerkin method for elliptic interface problems. J. Comput. Appl. Math. **367**. https://doi.org/10.1016/j.cam.2019.112397
36. Sabawi, Y.A.: Posteriori error bound for fullydiscrete semilinear parabolic integro-differential equations. In Journal of Physics: Conference Series, vol. 1999, No. 1, p. 012085. IOP Publishing. https://doi.org/10.1088/1742-6596/1999/1/012085
37. Khalaf, A.D., Zeb, A., Sabawi, Y.A., Djilali, S., Wang, X.: Optimal rates for the parameter prediction of a Gaussian Vasicek process. Europ. Phys. J. Plus. **136**(8), 1–7 (2021). https://doi.org/10.1140/epjp/s13360-021-01738-9

38. Ibrahim, S., Rababah, A.: Decomposition of fourth-order euler-type linear time-varying differential system into cascaded two second-order Euler commutative pairs. Complexity **2022**, ArticleID 3690019, 9 (2022). https://doi.org/10.1155/2022/3690019
39. Ibrahim, S., Koksal, M.E.: Commutativity of sixth-order time-varying linear systems. Circuits, Syst., Signal Process. **40**(10), 4799–4832 (2021).View at: https://doi.org/10.1007/s00034-021-01709-6
40. Ibrahim, S., Koksal, M. E.: Realization of a fourth-order linear time-varying differential system with nonzero initial conditions by cascaded two second-order commutative pairs. Circuits, Syst., Signal Process. **40**(6), 3107–3123. https://doi.org/10.1007/s00034-020-01617-1
41. Sabawi, Y.A., Pirdawood, M.A., Khalaf, A.D.: September. Semi-implicit and explicit runge kutta methods for stiff ordinary differential equations. In: Journal of Physics: Conference Series, vol. 1999, No. 1, p. 012100. IOP Publishing (2021). https://doi.org/10.1088/1742-6596/1999/1/012100
42. Sabawi, Y.A., Pirdawood, M.A., Khalaf, A.D.: Signal diagonally implicit Runge Kutta (SDIRK) methods for solving stiff ordinary problems. AIP Conf. Proc. **2457**, 020015 (2023). https://doi.org/10.1063/5.0118644
43. Rasool, H.M., Pirdawood, M.A., Sabawi, Y.A., Mahmood, R.H., Khalil, P.A.: Model reduction and analysis for ERK cell signalling pathway using implicit-explicit Rung-Kutta methods. *Passer Journal of Basic and Applied Sciences, 4*(Special issue), 160–178. https://doi.org/10.24271/psr.2022.161692
44. Pirdawood, M.A., Rasool, H.M., Sabawi, Y.A., Azeez, B.F.: Mathematical modeling and analysis for COVID-19 model by using implicit-explicit Rung-Kutta methods. *Academic Journal of Nawroz University (AJNU), 11*(3). https://doi.org/10.25007/ajnu.v11n3a1244
45. Sadeeq, M.I., Omar, F.M., Pirdawood, M.A.: Numerical solution of Hirota-Satsuma coupled Kdv system By Rbf-Ps method. *Journal of Duhok University, 25*(2), 164–175. https://doi.org/10.26682/sjuod.2022.25.2.15

Using Atomic Solution Method to Solve the Fractional Equations

Gharib M. Gharib, Maha S. Alsauodi, Ahlam Guiatni, Mohammad A. Al-Omari, and Abed Al-Rahman M. Malkawi

Abstract In this work, we will use a new method termed atomic solution to solve the fractional integral equation and ordinary and partial differential equations. When the separation of variables does not work, the tensor product of Banach spaces is utilized. Here, it is necessary to link the fractional derivatives according to the definitions adopted by various scientists who set theories and their applications to serve this method, using mathematical analysis in different spaces, referring to the concepts of fractional transformations, such as Laplace and Sumudu, to eventually produce accurate, effective, and predictable solutions that are applicable.

Keywords Atomic solution · Conformable derivative · Tensor product

1 Introduction

In [1], the definition conformable fractional derivative is defined

$$T_\alpha(f)(\mathrm{t}) = \lim_{\varepsilon \to 0} \frac{f(t + \varepsilon t^{1-\alpha}) - f(t)}{\varepsilon}$$

where $\alpha \in (0, 1)$,

G. M. Gharib (✉)
Department of Mathematics, Faculty of Science, Zarqa University, Zarqa, Jordan
e-mail: ggharib@zu.edu.jo

M. S. Alsauodi · M. A. Al-Omari
Khawarizmi University Technical College, Amman, Jordan

A. Guiatni
Department of Mathematics, Faculty ofExact Science and Informatics, The University of Jijel, Jijel, Algeria
e-mail: ahlam.guiatni@univ-jijel.dz

A. A.-R. M. Malkawi
Department of Mathematics, Faculty of Science, The University of Jordan, Amman, Jordan

D. Zeidan et al. (eds.), *Mathematics and Computation*, Springer Proceedings in Mathematics & Statistics 418, https://doi.org/10.1007/978-981-99-0447-1_10

If f is α − differentiable on (0, k), k > 0, and $\lim_{t\to 0^+} T_\alpha(f)(t)$ exists, then we define

$$T_\alpha(f)(0) = \lim_{t\to 0} T_\alpha(f)(t)$$

Properties:

1. $T_\alpha(af + bg) = aT_\alpha(f) + bT_\alpha(g)$, foralla, b $\in R$.
2. $T_\alpha(\lambda) = 0$, for constant, we have:
3. $T_\alpha(fg) = fT_\alpha(g) + gT_\alpha(f)$.
4. $T_\alpha\left(\frac{f}{g}\right) = \frac{gT_\alpha(f) - fT_\alpha(g)}{g^2}$, g(t) $\neq 0$.
5. $T_\alpha(f)(t) = t^{1-\alpha} f^{/}(t)$.

Proof Let h= $\varepsilon t^{1-\alpha}$ in definition, then $\varepsilon = ht^{\alpha-1}$ as $\varepsilon \to 0, h \to 0$

$$\begin{aligned} T_\alpha(f)(t) &= \lim_{\varepsilon\to 0} \frac{f\left(t + \varepsilon t^{1-\alpha}\right) - f(t)}{\varepsilon} \\ &= \lim_{h\to 0} \frac{f(t+h) - f(t)}{ht^{\alpha-1}} \\ &= t^{1-\alpha} \lim_{h\to 0} \frac{f(t+h) - f(t)}{h} \\ &= t^{1-\alpha} f^{/}(t). \end{aligned}$$

The fractional derivatives:

(i) $T_\alpha(t^p) = pt^{p-\alpha}$
(ii) $T_\alpha\left(\sin\frac{1}{\alpha}t^\alpha\right) = \cos\frac{1}{\alpha}t^\alpha$.
(iii) $T_\alpha\left(\cos\frac{1}{\alpha}t^\alpha\right) = -\sin\frac{1}{\alpha}t^\alpha$.
(iv) $T_\alpha\left(e^{\frac{1}{\alpha}t^\alpha}\right) = e^{\frac{1}{\alpha}t^\alpha}$.

If a function is differentiable, it must be α−conformable differentiable, but the converse is not true, for example, take $f(t) = 2\sqrt{t}$. Then, $T_{\frac{1}{2}}(f)(0) = 1$, but $T_1(f)(0)$ does not exist.

The separation of variables and Fourier series are the most common methods for solving partial differential equations [2]. But in many equations, the separation of variables is not applicable. In this case, the theory of tensor product of Banach spaces gives us some types of solutions called atomic solutions.

We recommend [1–9] for more information.

In this paper, we will solve some fractional partial differential equations named atomic solution by using the tensor product of Banach spaces.

In this paper we Using Atomic Solution Method to Solve the Fractional Equations

$$D_x^\alpha D_x^\alpha u - C^2 D_y^\beta D_y^\beta u = D_x^\alpha D_y^\beta u \tag{1}$$

With the following conditions:

$$u(0,0) = 1,\ D_x^{\alpha} D_y^{\beta} u(0,0) = 1$$

These conditions can be formulated in the form

$$P(0) = Q(0) = 1 \text{and} P^{\alpha}(0) = Q^{\alpha}(0) = 1 \quad (*)$$

This is a linear fractional partial differential equation. But the separation of variables is very difficult to work. Hence, we go for an atomic solution.

Procedure

Let

$$u(x,\ y) = P(x)Q(y) = P \otimes Q \quad (2)$$

be an atomic solution of (1), where P(x) and Q(y) are not constants.

Now, substitute (2) in (1) to get

$$P^{2\alpha}(x)Q(y) - C^2 Q^{2\beta}(y)P(x) = P^{\alpha}(x)Q^{\beta}(y) \quad (3)$$

Equation (3) has the tonsorial form

$$P^{2\alpha} \otimes Q - C^2 Q^{2\beta} \otimes P = P^{\alpha} \otimes Q^{\beta} \quad (4)$$

Since the sum of two atoms is an atom. we have two cases:

Case (i)

$$Q^{2\beta} = Q^{\beta} = Q$$

Let us discuss the case

$$Q^{2\beta} = Q^{\beta}$$

By using the property of Conformable fractional Laplace transform [10, 11]

$$L_{\beta}\{Q^{2\beta}(y)\} = L_{\beta}\{Q^{\beta}(y)\}$$

$$s^2 G_{\beta}(s) - sQ(0) - Q^{\beta}(0) = sG_{\beta}(s) - Q(0)$$

$$(s^2 - s)G_{\beta}(s) - s - 1 = -1$$

$$(s^2-s)G_\beta(s) = \text{s}$$

$$G_\beta(s) = \frac{\text{s}}{s^2 - s} = \frac{1}{s-1} \tag{5}$$

Applying the inverse Conformable Laplace Transform to both sides of Eq. (5), we obtained

$$L_\beta{}^{-1}\{G_\beta(s)\} = \text{Q(y)} = e^{\frac{y^\beta}{\beta}} \tag{6}$$

Similarly, $Q^\beta = Q$ with conditions in (*) gives the same solution in (6).
In the same way, $Q^{2\beta} = Q$ gives the same solution in (6).
Now to find P(x), we go back to (3) and substitute (6) in (3), to get

$$P^{2\alpha}(x) - C^2 P(x) - P^\alpha(x) = 0$$

By using the property of Conformable fractional Laplace transform [10, 11]

$$L_\alpha\{P^{2\alpha}(x)\} - C^2 L_\alpha\{\text{P(x)}\} - L_\alpha\{P^\alpha(x)\} = 0$$

$$s^2 F_\alpha(\text{s}) - s P(0) - P^\alpha(0) - c^2 F_\alpha(\text{s}) - s F_\alpha(\text{s}) + \text{P}(0) = 0$$

$$(s^2 - s - c^2) F_\alpha(\text{s}) = -1 + 1 + s$$

So

$$F_\alpha(\text{s}) = \frac{\text{s}}{s^2 - s - c^2} = \frac{\text{s}}{(s^2 - s + \frac{1}{4}) - \frac{1}{4} - c^2} = \frac{\text{s}}{(s-\frac{1}{2})^2 - (\frac{1}{4} + c^2)}$$

$$L_\alpha{}^{-1}\{F_\alpha(s)\} = L_\alpha{}^{-1}\left\{\frac{\text{s}}{(s-\frac{1}{2})^2 - (\frac{1}{4} + c^2)}\right\}$$

We get

$$P(x) = e^{\frac{1}{2}\frac{x^\alpha}{\alpha}} \cosh\sqrt{\frac{1}{4} + c^2}\frac{x^\alpha}{\alpha}$$

In this case, the atomic solution is

$$\text{u(x, y)} = \text{P(x)Q(y)} = (e^{\frac{1}{2}\frac{x^\alpha}{\alpha}} \cosh\sqrt{\frac{1}{4} + c^2}\frac{x^\alpha}{\alpha}) e^{\frac{y^\beta}{\beta}}$$

This ends the discussion of case (i).

Case (ii):

$$P^{2\alpha} = P = P^{\alpha}$$

Let us discuss the case

$$P^{\alpha} = P$$

By using the property of Conformable fractional Laplace transform [10]

$$L_{\alpha}\{P^{\alpha}(x)\} = L_{\alpha}\{P(x)\}$$

We get

$$P(x) = e^{\frac{x^{\alpha}}{\alpha}} \tag{7}$$

Similarly, $P^{2\alpha} = P^{\alpha}$ with conditions in (*) gives the same solution in (7).
In the same way, $P^{2\alpha} = P$ gives the same solution in (7).
Now to find Q(y),
Now, we go back to (3) and we substitute (7) in (3) to get

$$e^{\frac{x^{\alpha}}{\alpha}} Q(y) - c^2 Q^{2\beta}(y) e^{\frac{x^{\alpha}}{\alpha}} = e^{\frac{x^{\alpha}}{\alpha}} Q^{\beta}(y)$$

$$e^{\frac{x^{\alpha}}{\alpha}} [Q(y) - c^2 Q^{2\beta}(y) - Q^{\beta}(y)] = 0$$

$$e^{\frac{x^{\alpha}}{\alpha}} \neq 0, \rightarrow Q(y) - c^2 Q^{2\beta}(y) - Q^{\beta}(y) = 0$$

By using the property of Conformable fractional Laplace transform [8]

$$L_{\alpha}\{Q(y)\} - c^2 L_{\alpha}\{Q^{2\beta}(y)\} - L_{\alpha}\{Q^{\beta}(y)\} = 0$$

$$G_{\beta}(s) - c^2\left[s^2 G_{\beta}(s) - sQ(0) - Q^{\beta}(0)\right] - sG_{\beta}(s) + Q(0) = 0$$

$$G_{\beta}(s) - c^2 s^2 G_{\beta}(s) + sc^2 Q(0) + c^2 Q^{\beta}(0) - sG_{\beta}(s) + Q(0) = 0$$

$$G_{\beta}(s)\left[1 - c^2 s^2 - s\right] = \frac{-sc^2 - c^2 - 1}{1 - c^2 s^2 - s}$$

$$G_{\beta}(s) = \frac{sc^2 + c^2 + 1}{c^2 s^2 + s - 1} = \frac{sc^2 + c^2 + 1}{c^2[s^2 + \frac{s}{c^2}] - 1}$$

$$= \frac{sc^2 + c^2 + 1}{c^2[s^2 + \frac{s}{c^2} + \frac{1}{4c^4}] - \frac{1}{4c^4} - 1}$$

$$= \frac{sc^2 + c^2 + 1}{c^2\left[s + \frac{1}{2c^2}\right]^2 - \frac{1}{4c^4} - 1}$$

$$= \frac{sc^2 + c^2 + 1}{c^2\left[s + \frac{1}{2c^2}\right]^2 - \left[1 + \frac{1}{4c^4}\right]}$$

$$= \frac{sc^2 + c^2 + 1}{c^2[\left(s + \frac{1}{2c^2}\right)^2 - \frac{1}{c^2}\left(1 + \frac{1}{4c^4}\right)]}$$

$$= \frac{c^2\left[s + 1 + \frac{1}{c^2}\right]}{c^2[\left(s + \frac{1}{2c^2}\right)^2 - \frac{1}{c^2}\left(1 + \frac{1}{4c^4}\right)]}$$

$$= \frac{s + 1 + \frac{1}{c^2}}{\left(s + \frac{1}{2c^2}\right)^2 - \left(\frac{1}{c^2} + \frac{1}{4c^4}\right)}$$

Put

$$k^2 = \frac{1}{c^2} + \frac{1}{4c^4}$$

We get

$$G_\beta(s) = \frac{s + 1 + \frac{1}{c^2}}{(s + \frac{1}{2c^2})^2 - k^2}$$

$$G_\beta(s) = \frac{s}{(s + \frac{1}{2c^2})^2 - k^2} + \frac{1 + \frac{1}{c^2}}{(s + \frac{1}{2c^2})^2 - k^2} \tag{8}$$

Applying the inverse Conformable Laplace Transform to both sides of Eq. (8), we obtained

$$L_\beta{}^{-1}\{G_\beta(s)\} = \mathrm{Q(y)}$$

$$= L_\beta{}^{-1}\left\{\frac{s}{(s + \frac{1}{2c^2})^2 - k^2}\right\} + L_\beta{}^{-1}\left\{\frac{1 + \frac{1}{c^2}}{(s + \frac{1}{2c^2})^2 - k^2}\right\}$$

$$= e^{-\frac{1}{2c^2}\frac{y^\beta}{\beta}} \cos\sqrt{k^2}\frac{y^\beta}{\beta} + \frac{1 + \frac{1}{c^2}}{\mathrm{k}} e^{-\frac{1}{2c^2}\frac{y^\beta}{\beta}} \sin\sqrt{k^2}\frac{y^\beta}{\beta}$$

$$= e^{-\frac{1}{2c^2}\frac{y^\beta}{\beta}} \cos\mathrm{k}\frac{y^\beta}{\beta} + \frac{1 + \frac{1}{c^2}}{\mathrm{k}} e^{-\frac{1}{2c^2}\frac{y^\beta}{\beta}} \sin\mathrm{k}\frac{y^\beta}{\beta}$$

where

$$k = \frac{1}{c^2} + \frac{1}{4c^4}$$

So the general atomic solution is

$$\mathrm{u(x, y) = P(x)Q(y)} = \left(e^{-\frac{1}{2c^2}\frac{y^\beta}{\beta}} \cos k\frac{y^\beta}{\beta} + \frac{1+\frac{1}{c^2}}{k} e^{-\frac{1}{2c^2}\frac{y^\beta}{\beta}} \sin k\frac{y^\beta}{\beta}\right) e^{\frac{x^\alpha}{\alpha}}$$

This ends the discussion of case (ii).

References

1. Khalil, R., Al Horani, M., Yousef, A., Sababheh, M.: A new definition of fractional derivative. J. Comput. Appl. Math. **264** (2014)
2. Abu Hammad, I., Khalil, R.: Fractional fourier series with applications. Am. J. Comput. Appl. Math. **4** (2014)
3. Atangana, A., Baleanu, D., Alsaedi, A.: New properties of conformable derivative, Open, Mathematics, vol. 13 (2015)
4. Alhorani, M., Abu Hammad, M., Khalil, R.: Variation of parameters for local fractional non homogeneous linear-deferential equations. J. Math. Comput. Sci. **16** (2016)
5. Alhorani, M., Khalil, R.: Total fractional deferential with applications to exact fractional, deferential equations. Int. J. Comput. Math. **95** (2018)
6. Alhorani, M., Khalil, R., Aldarawi, I.: Fractional Cauchy Euler Deferential Equation, vol. 28 (2019)
7. Mhailan, M., Abu Hammad, M., Al Horani, M., Khalil, R.: On fractional vector analysis. J. Math. Comput. Sci. **10** (2020)
8. Abdeljawad, T.: On conformable fractional calculus. J. Comput. Appl. Math. **279** (2015)
9. Chung, W.: Fractional Newton mechanics with conformable fractional derivative. J. Comput. Appl. Math. **290** (2015)
10. Al-Zhour, Z., Alrawajeh, F., Al-Mutairi, N., Alkhasawneh, R.: New results on the conformable fractional Sumudu transform: theories and applications. Int. J. Anal. Appl. **17** (2019)
11. Kilbas, A., Srivastava, H., Trujillo, J.: Theory and Applications of Fractional Differential Equations, vol. 204. North-Holland Mathematics Studies. Elsevier, Amsterdam (2006)

Analysis in the Algebra $A(E)$

Sabra Ramadan

Abstract In this paper, we will define the linear and bilinear operations on the algebra $A(E)$ which is defined in (Ramadan in J Taibah Univ Sci 290–293, 2017, [1]). Also, we shall define the generalized numbers according to the given space $A(E)$ to study the initial and boundary differential equations and other mathematical models in this algebra.

Keywords Generalized functions · Locally convex algebra · Distribution · Algebra

1 Introduction

Following the general method of construction algebras of new distributions (generalized functions) [2] and used in [3, 4], the algebra $A(E)$ was defined in [1] as a factor algebras $A(E) = G_{\theta_1}(E) \backslash G_{\theta_2}(E)$, where

$$G_{\theta_1}(E) = \{(f_k) \in G(E) : \exists m \in \mathbb{N}, \forall \alpha \in I, \exists d_\alpha > 0,\ p_\alpha(f_k) \leq \theta_1 \forall k\}$$

$$G_{\theta_2}(E) = \{(f_k) \in G(E) : \forall m \in \mathbb{N}, \forall \alpha \in I, \exists d_\alpha > 0,\ p_\alpha(f_k) \leq \theta_2 \forall k\}$$

where θ_i be a multivariable functions

$$\theta_i\ :\ I \times I \times I \times .. \times I\ \times N \times N \times ..... \times N\ \times\ R \times R \times ..... \times R\ \ \rightarrow\ \ R^+ \cup \{0\}$$

and $G(E)$ be the set of all sequences in G, and E be separated by complete locally convex algebra [5].

The topology on E defined by P_α (collection of semi norms), where α belongs to the real algebra [interval I and satisfies the following conditions:

S. Ramadan (✉)
Faculty of Science, Jazan University, P.O. Box 2097, Jizan, Saudi Arabia
e-mail: rmsabra@jazanu.edu.sa

D. Zeidan et al. (eds.), *Mathematics and Computation*, Springer Proceedings in Mathematics & Statistics 418, https://doi.org/10.1007/978-981-99-0447-1_11

For each $\alpha \in I$, there exists $\beta \in I$ and a constant $C_\alpha > 0$ for which

$$p_\alpha\ (f.g) \leq C_\alpha\ p_\beta(f)\ p_\beta(g) \qquad \forall\, f,\ g \in E$$

The importance of the algebra A(E) appears when we study a special case when $E = S(\mathbb{R})$ the space of rapid decay functions with the usual topology defined by the family of semi norms

$$p_{n,l}(f) = \sup_{\substack{k \leq n \\ m \leq l}} q_{k,m}(f)$$

where

$q_{k,m}(f) = \sup_{x \in R} \left| x^k f^{(m)}(x) \right|,$

Now define the algebra.

$A(E) = G_{\theta_1}(S(\mathbb{R})) \backslash G_{\theta_2}(S(\mathbb{R})),$

where

$$\theta_1(\alpha, k, m) = d_\alpha e^{km}, \quad \theta_2(\alpha, k, m) = d_\alpha e^{-km},$$

The embedding of the dual space $S^*(\mathbb{R})$ which is called the space of tempered generalized functions (or sometimes it is called the space of tempered distributions) is defined in the following way:

Define the operator

$$B_f : S^*(R) \rightarrow A(S(R)),$$

where

$$B_f : u \in S^*(R) \rightarrow ((2\pi)^{-1}(f_k \,.\, u * g_k) + G_{\theta_2}(S(R)) \in A(S(R))$$

where

$f_k = f\left(\frac{x}{k}\right),\ f \in D(\mathbb{R}),\ f(x) = \begin{cases} 0 & |x| > 2 \\ 1 & |x| \leq 1 \end{cases},$

and

$$g_k = F\,(f_k(x)) = k\,F(f(kx))$$

where F is the Fourier transform.

Theorem. 1.1

(i) B_f is the injective linear operator,
(ii) If $f \in S^*(\mathbb{R})$, then $((2\pi)^{-1}(f_k.f * g_k) \in G(S(\mathbb{R})),$
(iii) If $g \in S^*(\mathbb{R})$, then $((2\pi)^{-1}(f_k.g * g_k) \in G_{\theta_1}(S(\mathbb{R})),$

(iv) If $|u_1(x) - u_2(x)| < \varepsilon$ for each $\varepsilon > 0$, and for each $x \in \mathbb{R}$, then $\left|B_f u_1 - B_f u_2\right| < \varepsilon$.

From this theorem, we conclude that

$$S(\mathbb{R}) \subset S^*(\mathbb{R}) \subset A(S(\mathbb{R}))$$

This means that we can define the associative multiplication of distributions as an element of algebra $A(S(\mathbb{R}))$, for example, if $\delta(x)$ be the Dirac delta function, then

$$B_f : \delta \in S^*(\mathbb{R}) \to ((2\pi)^{-1}(f_k.\delta * g_k) + G_{\theta_2}(S(\mathbb{R})) = (2\pi)^{-1} * g_k + G_{\theta_2}(S(\mathbb{R})) \in A(S(\mathbb{R}))$$

from which we can define the power of $\delta(x) \in S^*(\mathbb{R})$ by [see 1, 2].

$$\delta^n(x) = [(2\pi)^{-1} * g_k + G_{\theta_2}(S(R)]^n = [\frac{1}{(2\pi)^n} * (g_k(x))^n + G_{\theta_2}(S(R)]$$
$$\in A(S(R)).$$

2 Generalized Numbers According to the Algebra $A(E)$

Let $G(\mathbb{C})$ be the set of all complex sequences, that is, $G(\mathbb{C}) = \{(z_k) : z_k \in \mathbb{C}\}$, then we define the following subsets:

$$G_M(C) = \left\{(z_k) : z_k \in C : \exists m \in N, \text{ and } \exists\, c > 0 \ \forall k \ |z_k| < c\, e^{mk}\right\}$$

$$G_N(C) = \left\{(z_k) : z_k \in C : \forall m \in N, \forall k \ \exists\, c > 0 \ |z_k| < c\, e^{-mk}\right\}$$

Now if $z = (z_k)$, $w = (w_k)$ are two elements of the set $G_M(\mathbb{C})$, and $\mu = (\mu_k)$ be an element of the set $G_N(\mathbb{C})$, then

$$|zw| = |w_k z_k| \le d\, e^{(m_1+m_2)k}$$

for all k and for some m_1, m_2 from which we conclude that $G_M(\mathbb{C})$ is a sub algebra of the algebra $G(\mathbb{C})$.

Also, if we take

$$|z\mu| = |\mu_k z_k| \le d\, e^{(m_1-m)k}$$

for all m and for some m_1 which means that $G_N(\mathbb{C})$ be ideal in the space $G(\mathbb{C})$..

Now the set of generalized numbers we define as a factor algebra:

$$\mathbb{C}^* = G_M(\mathbb{C})/G_N(\mathbb{C})$$

Theorem 1.2 Let $f = (f_k) \in G_{\theta_1}(S(\mathbb{R}))$, and $g = (g_k) \in G_{\theta_2}(S(\mathbb{R}))$, and $z_0 \in \mathbb{C}$, then

1. $f(z_0) = (f_k(z_0)) \in G_M(C)$,

2. $g(z_0) = (g_k(z_0)) \in G_N(C)$

Proof The proof is trivial by using the properties of the spaces $G_{\theta_1}(S(\mathbb{R}))$, $G_{\theta_2}(S(\mathbb{R}))$, $G_M(\mathbb{C})$, and $G_N(\mathbb{C})$.

The properties of the algebra $A(E)$ and the definition of the generalized numbers $\mathbb{C}^*$ give us the opportunity to study the differential equations with initial and boundary conditions in the algebra $A(S(\mathbb{R}))$..

Moreover, the generalized numbers help us to define extended linear functional in our space (E), for example, the extended Lebesgue integral will be defined as we shall show in the latest section of this paper.

3 Linear and Bilinear Operations on Algebra $A(E)$

Let $T : E \to E$ be a continuous linear operator, then for each $u \in E$, and for each $\alpha \in A$, there is a constant $d_\alpha > 0$ and$\beta \in A$, such that

$$p_\alpha (T(u)) \leq d_\alpha \, p_\beta (u) \tag{3.1}$$

and if $B : ExE \to E$ be bilinear and continuous, then for each $\alpha \in A$ there is $\beta \in A$ and a constant $C_\alpha > 0$ such that for each $u, v \in E$

$$p_\alpha (B(u, v)) \leq C_\alpha \, p_\beta (u) \, p_\beta (v) \tag{3.2}$$

Theorem 1.3 The following embeddings are true:

1. $T(G_{\theta_1}(E)) \subset G_{\theta_1}(E)$,
2. $T(G_{\theta_2}(E)) \subset G_{\theta_2}(E)$,

Moreover, the operator T does not depend on the representative.

Proof Let $f = (f_k) \in G_{\theta_1}(S(\mathbb{R}))$, and $(f_k{}^*)$ be other representatives of f.

Consider

$$p_\alpha [T(f_k) - T(f_k^*)] = p_\alpha [T(f_k - f_k^*)] \leq C_\alpha \, p_\beta (f_k - f_k^*) \leq$$
$$\leq C_\alpha e^{-mk} \quad \forall \, m$$

that is

$$T(f_k) - T(f_k^*) \in G_{\theta_2}(E)$$

Now consider $p_\alpha[T(f_k)]$, and using (3.1), we conclude that

$$p_\alpha[T(f_k)] \leq C_\alpha p_\beta(f_k) \leq C_\alpha e^{mk}$$

that is

$$T(G_{\theta_1}(E)) \subset G_{\theta_1}(E).$$

Similarly if

$$g = (g_k) \in G_{\theta_2}(E),$$

then

$$T(g_k) \in G_{\theta_2}(E)$$

that is

$$T(G_{\theta_2}(E)) \subset G_{\theta_2}(E).$$

Now we can define the extended linear operator

$$\overline{T} : A(E) \to A(E)$$

Similarly, we can use inequality (3.2) to prove that.

$$B(G_{\theta_2}(E)xG_{\theta_2}(E)) \subset G_{\theta_2}(E),$$

and similarly, we define the extended bilinear map

$$\overline{B} : A(E)xA(E) \to A(E)$$

4 Analysis in the Algebra $A(E)$

In this section, we will give definitions of extended Linear functionals, extended Fourier transform, and extended differentiation as a special case of linear and bilinear operations on the algebra $A(S(\mathbb{R}))$.

Let $f : S(\mathbb{R}) \to \mathbb{C}$ be any linear functional defined on the space $S(\mathbb{R})$, then by previous results and definitions, we know that

$$f(G_{\theta_1}(S(R))) \subset G_M(C),$$

And

$$f(G_{\theta_2}(S(R))) \subset G_N(C)$$

So the extended linear functional

$$f : A(S(R)) \rightarrow C^*$$

will be correct and it is defined by

$$f(u) = f(u_k) + G_N(C) \in C^*.$$

For example, the Lebesgue integral over the compact K is defined by

$$\int_K : A(S(R)) \rightarrow C^*,$$

where

$$\int_K u = \int_K u_k + G_N(C) \in C^*.$$

Now since the differential operator $D : S(\mathbb{R}) \rightarrow S(\mathbb{R})$ is linear, then we can define the extended differential operator on the algebra $A(S(\mathbb{R}))$ in the following way:

$$\overline{D} : A(S(R)) \rightarrow A(S(R))$$

where

$$\overline{D}(u) = D(u_k) + G_{\theta_2}(S(R)) \in A(S(R)),$$

also, we define the extended Fourier transform by

$$\overline{F} : A(S(R)) \rightarrow A(S(R))$$

where

$$\overline{F}(u) = F(u_k) + G_{\theta_2}(S(R)) \in A(S(R)).$$

Now we can define the extended convolution on the algebra $A(S(\mathbb{R}))$ by

$$\bar{*} : A(S(R))xA(S(R)) \rightarrow A(S(R))$$

where

$$u \,\bar{*}\, v = (u_k * v_k) + G_{\theta_2}(S(R)) \in A(S(R)).$$

Remind that the associative multiplication is defined on the space $A(S(\mathbb{R}))$ by

$$\otimes : A(S(R))xA(S(R)) \rightarrow A(S(R))$$

$$u \otimes v = (u_k \,.\, v_k) + G_{\theta_2}(S(R)) \in A(S(R))$$

The following results show that the operations of differentiation $\overline{D}$, Fourier transformation $\overline{F}$, convolution $\bar{*}$, and multiplication $\otimes$ preserved many important properties:

Theorem 1.4 The operations of differentiation $\overline{D}$, Fourier transformation $\overline{F}$, convolution $\bar{*}$, and multiplication $\otimes$ satisfy the following properties:

(1) the Fourier transformation $\overline{F} : A(S(\mathbb{R})) \rightarrow A(S(\mathbb{R}))$ is an isomorphism;
(2) $\overline{F}[\overline{D}^n u] = (ix)^n (\overline{F}(u))$ for each $u \in A(S(R))$;
(3) $\overline{D}^n [\overline{F} u] = \overline{F}((ix)^n u)$ for each $u \in A(S(R))$;
(4) $\overline{F}(u \,\bar{*}\, v) = \overline{F}(u) \otimes \overline{F}(v)$ for each $u, v \in A(S(R))$;
(5) $\overline{F}(u \otimes v) = \overline{F}(u) \,\bar{*}\, \overline{F}(v)$ for each $u, v \in A(S(R))$;

Proof **(1)** It is known that [5] the usual Fourier transform $F : S(\mathbb{R}) \rightarrow S(\mathbb{R})$ is an isomorphism.

Now consider

$$\begin{aligned} \overline{F}(u + v) &= F(u_k + v_k) + G_{\theta_2}(S(R)) = F(u_k) + F(v_k) + G_{\theta_2}(S(R)) \\ &= \overline{F}(u) + \overline{F}(v), \end{aligned}$$

and

$$\overline{F}(\alpha u) = F(\alpha u_k) + G_{\theta_2}(S(R)) = \alpha F(u_k) + G_{\theta_2}(S(R)) = \alpha \overline{F}(u),$$

that is $\overline{F}$ is linear.

Now, let

$$\overline{F}(u) = \overline{F}(v) \rightarrow$$

$F(u_k) = F(v_k) \Rightarrow F(u_k - v_k) \in G_{\theta_2}(S(R)) \Rightarrow u \approx v$,, which means that $\overline{F}$ is injective.

Finally, if

$$v \in A(S(R)) \to v = (v_k) + G_{\theta_2}(S(R)),$$

then there is an element

$$u = F^{-1}(v_k) + G_{\theta_2}(S(R))$$

such that

$$\overline{F}u = FF^{-1}(v_k) + G_{\theta_2}(S(R)) = (v_k) + G_{\theta_2}(S(R)) = v,$$

that is the transform $\overline{F}$ is surjective.

(5) Consider.

$\overline{F}(u \otimes v) = F(u_k v_k) + G_{\theta_2}(S(R)) = F(u_k) * F(v_k) + G_{\theta_2}(S(R)) = \overline{F}(u) \overline{*} \overline{F}(v)$ Similarly, we can prove properties **(2–4)**.

5 Conclusion

The generalized numbers according to the given space $A(E)$ have been defined. Also, we have extended linear and bilinear operations on the algebra $A(E)$. These results will give us the opportunity to study the boundary differential equations and other mathematical models in this algebra.

Acknowledgements The author would like to acknowledge Jazan University for its support.

References

1. Ramadan, S.: Algebra A(E). J, Taibah University for Sci. 290–293 (2017)
2. Antonevich, A., Radyno, Y.: Dokl. Acad. Nauk. SSSR, 267–270 (1991)
3. Radyno, Y.: Ngo, Fu Tkan, Ramadan, Sabra. Russian Acad. Sci. Dokl, 20–24 (1993)
4. Ramadan, S.: One algebra of new generalized functions. J. Math. Stat. 15–16 (2007)
5. Antonevich A., Radyno, Y.: Functional analysis and integral equations, Minsk (2006)

Applications of Conformable Fractional Weibull Distribution

Sondos Rasem, Amer Dabahneh, and Ma'mon Abu Hammad

Abstract The aim of this research is to generate probability density functions of random variables of the Weibull distribution using fractional differential equations (FDE). And the second aims to find some basic concepts such as cumulative distribution, survival, and hazard functions. Expected values, rth moments, mean, variance, skewness, and kurtosis are all introduced as conformable fractional analogs. It also presents conformable fractional analogs of various entropy measures, such as Shannon, Renyi, and Tsallis entropy measures. Distributions have many applications in probability and other applied sciences.

Keywords Probability distribution · Conformable fractional · Conformable derivative · Entropy

1 Introduction

Fractional derivatives have shown to be extremely beneficial in a variety of fields. Conformable fractional derivative is introduced by Khalil et al. in 2014 [1–4]. In this work, we introduce a fractional distribution that can be employed in a variety of probability and applied sciences applications [1, 5, 6].

S. Rasem · A. Dabahneh (✉) · M. A. Hammad
Al-Zaytoonah University of Jordan, Queen Alia Airport St. 594, Amman 11942, Jordan
e-mail: dababneh.amer@zuj.edu.jo

M. A. Hammad
e-mail: m.abuhammad@zuj.edu.jo

D. Zeidan et al. (eds.), *Mathematics and Computation*, Springer Proceedings in Mathematics & Statistics 418, https://doi.org/10.1007/978-981-99-0447-1_12

Definition 1.1 Let $\mathfrak{w} : [0, \infty) \to \mathbb{R}$ and $t > 0$, then, given $\mathfrak{w}$ of order α, the "conformable fractional derivative" is defined by [7]

$$D_\alpha(\mathfrak{w})(t) = \lim_{\mathfrak{w}\to 0} \frac{\mathfrak{w}\left(t + \mathfrak{w}t^{1-\alpha}\right) - \mathfrak{w}(t)}{\mathfrak{w}}$$

For all $t > 0, \alpha \in (0, 1)$. If $\mathfrak{w}$ is $\alpha-$ differentiable in some $(0, a), a > 0$, and $\lim_{t\to 0^+} \mathfrak{w}^{(\alpha)}(t)$ exists, then specify $\mathfrak{w}^{(\alpha)}(0) = \lim_{t\to 0^+} \mathfrak{w}^{(\alpha)}(t)$

The conformable fractional derivatives of $\mathfrak{w}$ of order α are denoted by $\mathfrak{w}^{(\alpha)}(t)$ for $D_\alpha(\mathfrak{w})(t)$. Furthermore, we simply state $\mathfrak{w}$ is $\alpha-$ differentiable if the conformable fractional derivative of $\mathfrak{w}$ of order α exists. It's worth noting that $D_\alpha(t^p) = pt^{p-\alpha}$ [6].

Definition 1.2 $\int_c^t(\mathfrak{w})(x)d^\alpha x = \int_c^t \frac{\mathfrak{w}(x)}{x^{1-\alpha}}dx$, where the integral is the standard integral incorrect Riemann integral, and $\alpha \in (0, 1)$ [6].

All of the classical features of the ordinary first derivative are satisfied by the conformable derivative. In addition, the following propositions are correct based on this derivative:

(a) $D_\alpha(a\mathfrak{w} + b\varphi) = aD_\alpha(\mathfrak{w}) + bD_\alpha(\varphi)$
(b) $D_\alpha(t^p) = pt^{p-\alpha}$, for all $p \in \mathbb{R}$.
(c) $D_\alpha(\mathfrak{w}\varphi) = \mathfrak{w}D_\alpha(\varphi) + \varphi D_\alpha(\mathfrak{w})$
(d) $D_\alpha\left(\frac{\mathfrak{w}}{\varphi}\right) = \frac{\varphi D_\alpha(\mathfrak{w}) - \mathfrak{w}D_\alpha(\varphi)}{\varphi^2}$

2 Conformable Fractional Differential Equation

In this section, we introduce conformable Weibull distribution. Consider the following conformable differential equation [8–10]:

$x^\alpha D_\alpha y + \left(\alpha\beta\left(\frac{x^\alpha}{\theta}\right)^\beta - (\beta - 1)\right)y = 0$ Where $0 < \alpha < 1, \beta, \theta > 0, x > 0$

If y has an ordinary derivative y' with respect to x, then the above equation is equivalent to the ordinary differential equation.

$x^\alpha x^{1-\alpha} y' + \left(\alpha\beta\left(\frac{x^\alpha}{\theta}\right)^\beta - \beta - 1)\right)y = 0.$

Thus

$$\frac{y'}{y} = \frac{\beta - 1}{x} - \frac{\alpha\beta}{\theta\beta}x^{\alpha\beta-1}.$$

$$\text{Lny} = (\beta - 1)\ln x - \frac{\alpha\beta}{\theta\beta}\frac{x^{\alpha\beta}}{\alpha\beta} + c.$$

$y = Ax^{(\beta-1)}e^{-\left(\frac{x^\alpha}{\theta}\right)^\beta}$; where $A = e^c > 0$

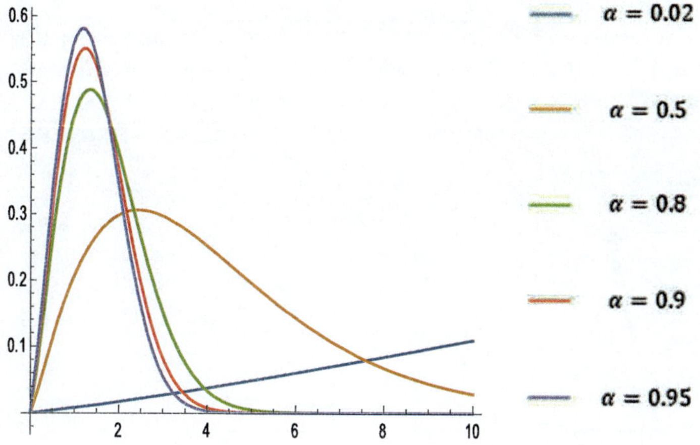

Fig. 1 Pdf $\theta = 1.5; \beta = 2.2;$

Set $g_\alpha(x) = Ax^{(\beta-1)}e^{-\left(\frac{x^\alpha}{\theta}\right)^\beta}$. For $g_\alpha(x)$ to be a conformable probability distribution function (CFPDF) with support $(0, \infty)$, we need $\int_0^\infty g_\alpha(x)d^\alpha x = 1$

Thus.

$A = \frac{\alpha\beta}{\Gamma\left(\frac{\alpha+\beta-1}{\alpha\beta}\right)\theta^{\left(\frac{\beta-1}{\alpha}+1\right)}}$.

Hence

$$g_\alpha(x) = \frac{\alpha\beta x^{\beta-1}e^{-\left(\frac{x^\alpha}{\theta}\right)^\beta}}{\theta^{\left(\frac{\beta-1}{\alpha}+1\right)}\Gamma\left(\frac{\alpha+\beta-1}{\alpha\beta}\right)} \tag{1}$$

Figure 2 There are no values in the $\alpha-$weibull distribution that are less than zero. The distribution becomes a Mesokurtosis distribution as the value rises [11, 12]

Consequently

$$\lim_{\alpha\to 1^-} g_\alpha(x) = \frac{\beta x^{\beta-1}e^{-\left(\frac{x}{\theta}\right)^\beta}}{\theta^\beta}, \beta, \theta > 0, x > 0$$

This is the PDF of a Weibull distribution that is denoted by Weibull(β, θ), so the CPDF $g_\alpha(x)$ is a generalization of the PDF of a Weibull distribution.

3 Applications to Conformable A−Weibull Distribution

In this section, we obtain the basic conformable fractional probabilistic properties of this distribution [1].

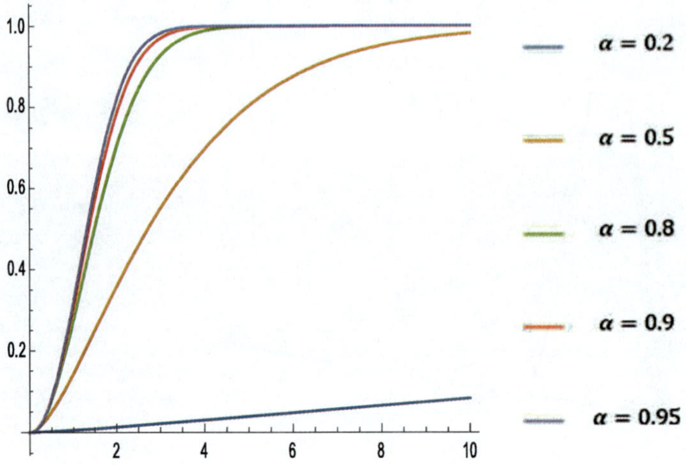

Fig. 2 CDF $\theta = 1.5$; $\beta = 2.2$;

A. **The conformable fractional cumulative distribution function (CFCDF) αCDF.**

When $x \sim \alpha weibull(\beta, \theta)$ so the CPDF is defined by

$$G_\alpha(x) = \int_0^x g_\alpha(t) d^\alpha t$$

To evaluate this integral, use the transformation $w = \left(\frac{x^\alpha}{\theta}\right)^\beta$ to get

$$G_\alpha(x) = 1 - \frac{\Gamma\left(\frac{\alpha+\beta-1}{\alpha\beta}, \left(\frac{x^\alpha}{\theta}\right)^\beta\right)}{\Gamma\left(\frac{\alpha+\beta-1}{\alpha\beta}\right)}. \quad (2)$$

2 The conformable fractional quantiles

Quantiles are points in a distribution that correspond to the rank order of its values.

By sorting it, you can find any quantile for a sample. The median is the sorted sample's middle value (the middle quantile, 50th percentile). The minimum and highest values are the limitations. Centiles or percentiles can be used to describe any other points between these two.

3 The conformable fractional survival function (CFSF) (S_α)

Conformable fractional survival function of X is given by

$$S_\alpha(X) = 1 - G_\alpha(X)$$

Table 1 Quantiles of the distribution classified by q in the rows and α in the columns. The parameters of the distribution are β =0.8 and θ = 0.9

α	q		
	0.25	0.5	0.75
0.1	0.001	0.001	0.003
0.2	0.001	0.001	0.016
0.3	0.001	0.001	0.047
0.4	0.001	0.002	0.111
0.5	0.001	0.006	0.227
0.6	0.001	0.027	0.435
0.7	0.001	0.111	0.817
0.8	0.001	0.473	1.589
0.9	0.002	2.645	3.58

$$S_{\alpha}(X) = 1 - \frac{\Gamma\left(\frac{\alpha+\beta-1}{\alpha\beta}\right) - \Gamma\left(\frac{\alpha+\beta-1}{\alpha\beta}, \left(\frac{x^{\alpha}}{\theta}\right)^{\beta}\right)}{\Gamma\left(\frac{\alpha+\beta-1}{\alpha\beta}\right)}. \tag{3}$$

4 Conformable fractional hazard function (CFHF)(H_{α})

Conformable fractional hazard function of X is given by

$$H_{\alpha} = \frac{S_{\alpha}(X)}{g_{\alpha}(x)}.$$

$$\mathrm{h}_{\alpha}(X) = \frac{\alpha\beta\left(\frac{1}{\theta}\right)^{\frac{\alpha+\beta-1}{\alpha}} x^{\beta-1} \mathrm{e}^{-\left(\frac{x^{\alpha}}{\theta}\right)^{\beta}}}{\Gamma\left(\frac{\alpha+\beta-1}{\alpha\beta}\right)\left(-\Gamma\left(\frac{\alpha+\beta-1}{\alpha\beta}, \left(\frac{x^{\alpha}}{\theta}\right)^{\beta}\right)\right)}. \tag{4}$$

5 Conformable Fractional expectation E_{α}

Definition D.1 Conformable fractional expectation E_{α} of a function $v(x)$ is given by $E_{\alpha}v(X) = \int v(x)g_{\alpha}(x)d^{\alpha}x$

So.

$$E_{\alpha}X^{r} = \int_{-\infty}^{\infty} \frac{\alpha\beta x^{r+\beta-1} e^{-\left(\frac{x^{\alpha}}{\theta}\right)^{\beta}}}{\theta^{\frac{\beta-1}{\alpha}+1}\Gamma\left(\frac{\alpha+\beta-1}{\alpha\beta}\right)x^{1-\alpha}} dx$$

$$= \frac{\left(\frac{1}{\theta}\right)^{\frac{r}{\alpha}}\Gamma\left(\frac{r+\alpha+\beta-1}{\alpha\beta}\right)}{\Gamma\left(\frac{\alpha+\beta-1}{\alpha\beta}\right)}.$$

The first four terms are given:

$$E_\alpha X = \mu_\alpha = \frac{\left(\frac{1}{\theta}\right)^{\frac{1}{\alpha}} \Gamma\left(\frac{\alpha+\beta}{\alpha\beta}\right)}{\Gamma\left(\frac{\alpha+\beta-1}{\alpha\beta}\right)} \tag{5}$$

$$E_\alpha X^2 = \frac{\left(\frac{1}{\theta}\right)^{\frac{2}{\alpha}} \Gamma\left(\frac{\alpha+\beta+1}{\alpha\beta}\right)}{\Gamma\left(\frac{\alpha+\beta-1}{\alpha\beta}\right)} \tag{6}$$

$$E_\alpha X^3 = \frac{\left(\frac{1}{\theta}\right)^{\frac{3}{\alpha}} \Gamma\left(\frac{\alpha+\beta+2}{\alpha\beta}\right)}{\Gamma\left(\frac{\alpha+\beta-1}{\alpha\beta}\right)} \tag{7}$$

$$E_\alpha X^4 = \frac{\left(\frac{1}{\theta}\right)^{\frac{4}{\alpha}} \Gamma\left(\frac{\alpha+\beta+3}{\alpha\beta}\right)}{\Gamma\left(\frac{\alpha+\beta-1}{\alpha\beta}\right)} \tag{8}$$

6 The conformable fractional variance

Ma'mon and others defined the conformable variance [1].

$$\alpha\sigma^2 = \mathrm{E}_\alpha\left(\mathrm{X}^2\right) - (\mu_\alpha)^2$$

So,

$$\alpha\sigma^2 = \frac{\left(\left(\frac{1}{\theta}\right)^{\beta}\right)^{-\frac{2}{\alpha\beta}} \left(\Gamma\left(\frac{\alpha+\beta-1}{\alpha\beta}\right)\Gamma\left(\frac{\alpha+\beta+1}{\alpha\beta}\right) - \Gamma\left(\frac{1}{\alpha} + \frac{1}{\beta}\right)^2\right)}{\Gamma\left(\frac{\alpha+\beta-1}{\alpha\beta}\right)^2} \tag{9}$$

7 The conformable fractional standard deviation

The standard deviation is calculated from the variance and is denoted by $\alpha\sigma$

$$\alpha\sigma = \sqrt{\frac{\left(\left(\frac{1}{\theta}\right)^{\beta}\right)^{-\frac{2}{\alpha\beta}} \left(\Gamma\left(\frac{\alpha+\beta-1}{\alpha\beta}\right)\Gamma\left(\frac{\alpha+\beta+1}{\alpha\beta}\right) - \Gamma\left(\frac{1}{\alpha} + \frac{1}{\beta}\right)^2\right)}{\Gamma\left(\frac{\alpha+\beta-1}{\alpha\beta}\right)^2}} \tag{10}$$

8 The conformable fractional Skewness is given by

$\alpha sk = \frac{E_\alpha(X-\mu)^3}{\alpha\sigma^3}$.
So,

by (5), (6), (7), and (9).

$$\alpha sk = \frac{\left(\frac{1}{\theta}\right)^{\frac{6}{\alpha}} \theta^{\frac{3}{\alpha}} \Gamma\left(\frac{\alpha+\beta-1}{\alpha\beta}\right)^{3}\left[-(\alpha+\beta)\Gamma\left(\frac{1}{\alpha}+\frac{1}{\beta}\right)^{3}+3\alpha\beta\Gamma\left(\frac{1}{\alpha}+\frac{1}{\beta}\right)^{2}\Gamma\left(\frac{1}{\alpha}+\frac{1}{\beta}+1\right)\right]}{(\alpha+\beta)\left(-\Gamma\left(\frac{1}{\alpha}+\frac{1}{\beta}\right)^{2}+\Gamma\left(\frac{\alpha+\beta-1}{\alpha\beta}\right)\Gamma\left(\frac{\alpha+\beta+1}{\alpha\beta}\right)\right)^{3}}$$

$$-\frac{\left(\frac{1}{\theta}\right)^{\frac{6}{\alpha}} \theta^{\frac{3}{\alpha}} \Gamma\left(\frac{\alpha+\beta-1}{\alpha\beta}\right)^{3}\left[3(\alpha+\beta)\Gamma\left(\frac{1}{\alpha}+\frac{1}{\beta}\right)\Gamma\left(\frac{\alpha+\beta-1}{\alpha\beta}\right)\Gamma\left(\frac{\alpha+\beta+1}{\alpha\beta}\right)\right]}{(\alpha+\beta)\left(-\Gamma\left(\frac{1}{\alpha}+\frac{1}{\beta}\right)^{2}+\Gamma\left(\frac{\alpha+\beta-1}{\alpha\beta}\right)\Gamma\left(\frac{\alpha+\beta+1}{\alpha\beta}\right)\right)^{3}}$$

$$+\frac{\left(\frac{1}{\theta}\right)^{\frac{6}{\alpha}} \theta^{\frac{3}{\alpha}} \Gamma\left(\frac{\alpha+\beta-1}{\alpha\beta}\right)^{3}\left[(\alpha+\beta)\Gamma\left(\frac{\alpha+\beta-1}{\alpha\beta}\right)^{2}\Gamma\left(\frac{\alpha+\beta+2}{\alpha\beta}\right)\right]}{(\alpha+\beta)\left(-\Gamma\left(\frac{1}{\alpha}+\frac{1}{\beta}\right)^{2}+\Gamma\left(\frac{\alpha+\beta-1}{\alpha\beta}\right)\Gamma\left(\frac{\alpha+\beta+1}{\alpha\beta}\right)\right)^{3}}$$

I. **The conformable fractional kurtosis is given by**

$$\alpha ku = \frac{E_{\alpha}(X-\mu)^{4}}{\alpha\sigma^{4}}$$

So,

by (5), (6), (7), and (9).

$$\alpha ku = \frac{\left(\frac{1}{\theta}\right)^{\frac{8}{\alpha}} \theta^{\frac{4}{\alpha}} \Gamma\left(\frac{\alpha+\beta-1}{\alpha\beta}\right)^{4}\left[(\alpha+\beta)\Gamma\left(\frac{1}{\alpha}+\frac{1}{\beta}\right)^{4}-4\alpha\beta\Gamma\left(\frac{1}{\alpha}+\frac{1}{\beta}\right)^{3}\Gamma\left(\frac{1}{\alpha}+\frac{1}{\beta}+1\right)\right]}{((\alpha+\beta)\left(-\Gamma\left(\frac{1}{\alpha}+\frac{1}{\beta}\right)^{2}+\Gamma\left(\frac{\alpha+\beta-1}{\alpha\beta}\right)\Gamma\left(\frac{\alpha+\beta+1}{\alpha\beta}\right)\right)^{4}}.$$

$$+\frac{\left(\frac{1}{\theta}\right)^{\frac{8}{\alpha}} \theta^{\frac{4}{\alpha}} \Gamma\left(\frac{\alpha+\beta-1}{\alpha\beta}\right)^{4}\left[6(\alpha+\beta)\Gamma\left(\frac{1}{\alpha}+\frac{1}{\beta}\right)^{2}\Gamma\left(\frac{\alpha+\beta-1}{\alpha\beta}\right)\Gamma\left(\frac{\alpha+\beta+1}{\alpha\beta}\right)\right]}{((\alpha+\beta)\left(-\Gamma\left(\frac{1}{\alpha}+\frac{1}{\beta}\right)^{2}+\Gamma\left(\frac{\alpha+\beta-1}{\alpha\beta}\right)\Gamma\left(\frac{\alpha+\beta+1}{\alpha\beta}\right)\right)^{4}}.$$

$$-\frac{\left(\frac{1}{\theta}\right)^{\frac{8}{\alpha}} \theta^{\frac{4}{\alpha}} \Gamma\left(\frac{\alpha+\beta-1}{\alpha\beta}\right)^{4}\left[4(\alpha+\beta)\Gamma\left(\frac{1}{\alpha}+\frac{1}{\beta}\right)\Gamma\left(\frac{\alpha+\beta-1}{\alpha\beta}\right)^{2}\Gamma\left(\frac{\alpha+\beta+2}{\alpha\beta}\right)\right]}{((\alpha+\beta)\left(-\Gamma\left(\frac{1}{\alpha}+\frac{1}{\beta}\right)^{2}+\Gamma\left(\frac{\alpha+\beta-1}{\alpha\beta}\right)\Gamma\left(\frac{\alpha+\beta+1}{\alpha\beta}\right)\right)^{4}}.$$

$$+\frac{\left(\frac{1}{\theta}\right)^{\frac{8}{\alpha}} \theta^{\frac{4}{\alpha}} \Gamma\left(\frac{\alpha+\beta-1}{\alpha\beta}\right)^{4}\left[(\alpha+\beta)\Gamma\left(\frac{+\alpha+\beta-1}{\alpha\beta}\right)^{3}\Gamma\left(\frac{\alpha+\beta+3}{\alpha\beta}\right)\right]}{((\alpha+\beta)\left(-\Gamma\left(\frac{1}{\alpha}+\frac{1}{\beta}\right)^{2}+\Gamma\left(\frac{\alpha+\beta-1}{\alpha\beta}\right)\Gamma\left(\frac{\alpha+\beta+1}{\alpha\beta}\right)\right)^{4}}$$

10 Conformable Entropy Measures

F.1. Shannon Conformable Fractional Entropy

Definition F.1 Conformable fractional Shannon entropy [13, 14] of a random variable x whose $g_{\alpha}(x)$ is defined $SH_{\alpha}(x) = -E_{\alpha}\log g_{\alpha}(X)$

$$SH_\alpha(x) = \frac{\Gamma\left(\frac{\alpha+\beta+\alpha\beta-1}{\alpha\beta}\right)}{\Gamma\left(\frac{\alpha+\beta-1}{\alpha\beta}\right)} - Log(\alpha) - Log(\beta) + \frac{\beta+\alpha-1}{\alpha}Log(\theta)$$

$$+Log\left(\Gamma\left(\frac{\alpha+\beta-1}{\alpha\beta}\right)\right) - \frac{(\alpha+\beta-2)}{\alpha}\Psi\left(\frac{\alpha+\beta-1}{\alpha\beta}\right)$$

where $\Psi(z)$ the digamma function.

F.2. Tsallis Conformable Fractional entropy.

Definition F.2 Conformable fractional Tsallis entropy of a random variable x whose $g_\alpha(x)$ is defined $SH_{T,\xi}(x) = \frac{1}{1-\xi}\log(E_\alpha(g_\alpha(X))^{\xi-1} - 1)$

$$SH_{T,\alpha,\xi} = \frac{\alpha^{\xi-1}\beta^{\xi-1}\theta^{\frac{1-\xi}{\alpha}}\xi^{-\frac{(\alpha+\beta-2)\xi+1}{\alpha\beta}}\Gamma\left(\frac{\alpha+\beta-1}{\alpha\beta}\right)^{-\xi}\Gamma\left(\frac{(\alpha+\beta-2)\xi+1}{\alpha\beta}\right) - 1}{1-\xi}.$$

Hence, $\lim_{\xi\to1} SH_{T,\alpha,\xi} = SH_\alpha(x)$. So the limit of conformable fractional Tsallis entropy is equal to the conformable fractional Shannon entropy.

F.3. Conformable Fractional Renyi entropy.

Definition F.3 Conformable fractional Renyi entropy of a random variable x whose $g_\alpha(x)$ is defined $SH_{R,\alpha,\xi} = \frac{1}{1-\xi}\log(E_\alpha(g_\alpha(X))^{\xi-1})$

$$SH_{R,\alpha,\xi} = \frac{1}{\alpha\beta(-1+\xi)}$$

$$\left(\begin{array}{c} \beta\xi Log(\theta) + Log(\xi) - 2\xi Log(\xi) + \alpha\xi Log(\xi) \\ +\beta\xi Log(\xi) + \alpha\beta\xi Log\left(\Gamma\left(\frac{\alpha+\beta-1}{\alpha\beta}\right)\right) - \alpha\beta Log\left(\Gamma\left(\frac{1+(\alpha+\beta-2)\xi}{\alpha\beta}\right)\right) \\ -\alpha\beta(-1+\xi)Log(\alpha) - \alpha\beta(-1+\xi)Log(\beta) - \beta Log(\theta) \end{array}\right).$$

Hence, $\lim_{\xi\to1} SH_{R,\alpha,\xi} = SH_\alpha(x)$. So the limit of conformable fractional Renyi entropy is equal to the conformable fractional Shannon entropy.

References

1. Abu Hammad, M., Awad, A., Khalil R.: Properties of conformable fractional chi-square probability distribution. J. Math. Comput. Sci. 1239–1250 (2020). (Khalil, R., AlHorani, M., Yousef, A., Sababheh, M.: A new definition of fractional derivative. J. Comput. Appl. Math. **264** (2014), 65–70, 2020)
2. Khalil, R., AlHorani, M., Abu Hammad, M.: Geometric meaning of conformable derivative via fractional cords. J. Math. Comput. Sci. 241–245 (2019)

3. Erden, S.: Weighted inequalities involving conformable integrals and its applications for random variable and numerical integration. Filomat **34**(8), 2785–2796 (2020)
4. Abdeljawad, T.: On conformable fractional calculus. J. Comput. Appl. Math. **279**, 57–66 (2015)
5. Batiha, I.M., Oudetallah, J., Ouannas, A., Al-Nana, A.A., Jebril, I.H.: Tuning the fractional-order PID-controller for blood glucose level of diabetic patients. Int. J. Adv. Soft Comput. Appl. **13**(2), 1–10 (2021)
6. Ouannas, A., batiha, I.M., khennaoui, A.-A., Jebril, I.H., On the 0-1 test for chaos Applied to the Generalized Fractional order Arnold Map. In: 2021 International Conference on Information Technology ICIT 2021-Proceedings, 2021, pp. 242–245, 9491633
7. Khalil, R., AlHorani, M., Yousef, A., Sababheh, M.: A new definition of fractional derivative. J. Comput. Appl. Math. 65–70 (2014)
8. Dababneh, A., Albarmawi, B., Abu Hammad, M., Zraiqat, A., Hamadneh, T.: Conformable Fractional Bernoulli Differential Equation with Applications. (JEEIT), pp. 7–19 (2019)
9. Bezziou, M., Dahmani, Z., Sarikaya, M.Z., Jebril, I.: New mixed operators for fractional integrations with some applications. Math. Eng. Sci. Aerosp. 95–108 (2021)
10. Abu Hammad, M., Awad, A., Khalil, R., Aldabbas. E.: Fractional distributions and probability density functions of random variables generated using FDE. J. Math. Comput. Sci. 522–534 (2020)
11. Jebril, I., Abu Hammad, M., Nouh, E., Hamidi, R., Dalahmeh, Y., Almutlak, S.: Properties of conformable fractional gamma with two parameters probability distribution. In: ICIT2021-Proceeding, pp. 16–18 (2021)
12. Abu Hammad, M., Jebril, I., AbuJudeh, D., Dalahmeh, Y., Abrikah, S.A.: Properties of conformable fractional Rayleigh probability distribution. ICIT. art.no. 9491658, pp. 13–15 (2021).
13. Awad, A.M.: A statistical information measure. Dirasat (Science), pp. 7–20 (1987)
14. Abu Hammad, M., Awad, A.M.: Distribution of Shannon statistic from normal sample. Metron 259–275 (2007)

Stable Second-Order Explicit Runge-Kutta Finite Difference Time Domain Formulations for Modeling Graphene Nano-Material Structures

Omar Ramadan

Abstract In this paper, stable second-order Runge-Kutta finite difference time domain (RK-FDTD) formulations are introduced for modeling graphene nano-material structures. In this respect, a differencing scheme in which the electric field and the associated current density are collocated in time and space is used for incorporating graphene's dispersion into the FDTD algorithm. The stability of the formulations is studied by using the root-locus method, and it is shown that the given formulations maintain the conventional Courant-Friedrichs-Lewy (CFL) time-step stability limit. The stability and the accuracy of the formulations are validated through a numerical test that investigates the tunneling phenomena of electromagnetic wave propagation through an infinite free-standing graphene layer.

Keywords Explicit finite difference time domain (FDTD) · Auxiliary differential equation (ADE) · Root-locus stability analysis · Second-order Runge-Kutta (RK) scheme · Graphene nano-material

1 Introduction

Graphene nano-material has attracted tremendous attention due to its exceptional electrical, optical, and mechanical properties [1], and this increases the interest of developing accurate and efficient numerical techniques for modeling graphene structures. In the last two decades, the finite difference time domain (FDTD) method [2], which is known to be one of the most popular electromagnetic time domain numerical techniques, has been widely used in graphene simulations [3–5]. In these approaches, the auxiliary differential equation method is used for incorporating graphene's dispersion into the FDTD algorithm.

In this paper, alternative formulations based on combining the second- order Runge-Kutta scheme [6] with the FDTD algorithm are introduced for model graphene

O. Ramadan (✉)
Computer Engineering Department, Eastern Mediterranean University, Gazimagusa, Mersin 10, Turkey
e-mail: omar.ramadan@emu.edu.tr

D. Zeidan et al. (eds.), *Mathematics and Computation*, Springer Proceedings in Mathematics & Statistics 418, https://doi.org/10.1007/978-981-99-0447-1_13

structures. In this respect, a differencing scheme in which the electric field and the associated current density are collocated in time and space is used for incorporating graphene's dispersion into the FDTD algorithm. The stability of the formulations is studied by using the root-locus method [7], and it is shown that the given formulations maintain the conventional Courant-Friedrichs-Lewy (CFL) time-step stability limit. The stability and accuracy of the formulations are validated through a numerical test that investigates the tunneling phenomena of electromagnetic wave propagation through an infinite free-standing graphene layer.

2 Formulations

At microwave and THz frequency regimes, the surface conductivity of graphene can be written as [8]

$$\sigma_{\mathtt{intra}} = \frac{\sigma_0}{\mathbf{j}\omega + v} \tag{1}$$

where v is the scattering rate, and $\sigma_0 = e^2 k_B T \left(\frac{\mu_c}{k_B T} + 2\ln(e^{-\mu_c/k_B T} + 1\right)/\pi\hbar^2$ is the static conductivity, where $-e$ is the electron charge, k_B is Boltzmann's constant, T is the temperature, μ_c is the chemical potential, and $\hbar$ is the reduced Planck's constant. Considering a graphene layer of thickness D_g, and introducing the concept of volumetric conductivity $\sigma_v = \sigma_{\mathtt{intra}}/D_g$ [9], the equations for the electric field component E_η, ($\eta = x, y, z$), and the associated current density J_η can be written as

$$\varepsilon_0 \frac{\partial E_\eta}{\partial t} = \nabla \times \mathbf{H}|_\eta - \frac{1}{D_g} J_\eta \tag{2}$$

$$\frac{\partial J_\eta}{\partial t} = \sigma_0 E_\eta - v J_\eta \tag{3}$$

Letting the electric field and the associated current density be collocated in time and space, (2) can be discretized as

$$\varepsilon_0 \frac{\delta_t}{\Delta_t} E_\eta\Big|_{r_{Ex}}^{n+\frac{1}{2}} = \widetilde{\nabla} \times \mathbf{H}\Big|_{\eta_{r_{E\eta}}}^{n+\frac{1}{2}} - \frac{1}{D_g} \mu_t\, J_\eta\Big|_{r_{Ex}}^{n+\frac{1}{2}} \tag{4}$$

where $r_{E\eta}$ is the spatial position of E_η, Δ_t is the time step, $\Psi^n = \Psi^n(n\Delta_t)$ ($\Psi = E, H, J$), δ_t is the centered temporal difference operator given by

$$\delta_t u^q_{\alpha,\beta,\gamma} = u^{q+\frac{1}{2}}_{\alpha,\beta,\gamma} - u^{q-\frac{1}{2}}_{\alpha,\beta,\gamma} \tag{5}$$

where $u^{q}_{\alpha,\beta,\gamma} = u\left(q\Delta_t,\ \alpha\Delta_x,\ \beta\Delta_x,\ \gamma\Delta_z\right)$, with $q = \left\{n,\ n+\frac{1}{2}\right\}$, $\alpha = \left\{i,\ i+\frac{1}{2}\right\}$, $\beta = \left\{j,\ j+\frac{1}{2}\right\}$, and $\gamma = \left\{k,\ k+\frac{1}{2}\right\}$, μ_t is the discrete averaging operator defined as

$$\mu_t u^{q}_{\alpha,\beta,\gamma} = \frac{u^{q+\frac{1}{2}}_{\alpha,\beta,\gamma} + u^{q+\frac{1}{2}}_{\alpha,\beta,\gamma}}{2} \tag{6}$$

and, finally, $\widetilde{\nabla}\times$ is the discrete version of $\nabla\times$ given by

$$\widetilde{\nabla}\times = \begin{pmatrix} 0 & -\delta_z & \delta_y \\ \delta_z & 0 & -\delta_x \\ -\delta_y & \delta_x & 0 \end{pmatrix} \tag{7}$$

where $\delta_\eta, \eta \in \{x,\ y,\ z\}$, is the centered spatial difference operator in the η-coordinate, for instance, δ_x is defined at the $(\alpha,\ \beta,\ \gamma)$ grid position as

$$\delta_x u^{q}_{\alpha,\beta,\gamma} = \frac{u^{q}_{\alpha+\frac{1}{2},\beta,\gamma} - u^{q}_{\alpha-\frac{1}{2},\beta,\gamma}}{\Delta_x} \tag{8}$$

with Δ_x being the mesh size along the x-coordinate. Based on (4), the E_η electric field component can be approximated at $n+1$ time step as

$$E^{n+1}_{\eta_{rE\eta}} = E^{n}_{\eta_{rE\eta}} + \frac{\Delta_t}{\varepsilon_0}\widetilde{\nabla}\times \mathbf{H}\Big|^{n+\frac{1}{2}}_{\eta_{rE\eta}} - \frac{\Delta_t}{2\varepsilon_0 D_g}\left(J^{n+1}_{\eta_{rE\eta}} + J^{n}_{\eta_{rE\eta}}\right) \tag{9}$$

By employing the general second-order RK [6] scheme to (3), J^{n+1}_η can be approximated as

$$\begin{cases} \mathcal{K}_1 = \Delta_t f\left(n\Delta_t,\ J^n_\eta\right) \\ \mathcal{K}_2 = \Delta_t f\left(\left(n+\lambda_1\right)\Delta_t,\ J^n_\eta + \lambda_2\mathcal{K}_1\right) \\ J^{n+1}_\eta = J^n_\eta + a_1\mathcal{K}_1 + a_2\mathcal{K}_2 \end{cases} \tag{10}$$

where f is obtained from the right-hand side of (3) as

$$f = \sigma_0 E_\eta - v J_\eta \tag{11}$$

and

$$\begin{cases} a_1 + a_2 = 1 \\ \lambda_1 a_2 = \frac{1}{2} \\ \lambda_2 a_2 = \frac{1}{2} \end{cases} \tag{12}$$

It is important to note that the approximation of (10) is of second-order accuracy, as shown in the Appendix. Employing the midpoint integration rule [6], which is also known as the modified Euler method, these constants can be obtained as

$$a_1 = 0,\ a_2 = 1,\ \text{and}\ \lambda_1 = \lambda_2 = \frac{1}{2} \tag{13}$$

Hence, J_η^{n+1} can be re-arranged from (10) as

$$J_\eta^{n+1} = J_\eta^n + \Delta_t f\left(\left(n + \frac{1}{2}\right)\Delta_t,\ J_\eta^n + \frac{\Delta_t}{2} f\left(n\Delta_t + J_\eta^n\right)\right) \tag{14}$$

which can be written recursively at the $r_{E\eta}$ grid position as

$$J_{\eta_{r_{E\eta}}}^{n+1} = b_1 J_{\eta_{r_{E\eta}}}^n + b_2 E_{\eta_{r_{E\eta}}}^n \tag{15}$$

where $b_1 = 1 - \tilde{v} + \frac{\tilde{v}^2}{2}$ and $b_2 = \sigma_0 \tilde{v}\left(1 - \frac{\tilde{v}}{2}\right)$ with $\tilde{v} = v\Delta_t$. For completeness, the magnetic field component H_η is written in the FDTD form as [2]

$$H_{\eta_{r_{H\eta}}}^{n+\frac{1}{2}} = H_{\eta_{r_{H\eta}}}^{n-\frac{1}{2}} - \frac{\Delta_t}{\mu_0} \tilde{\nabla} \times \mathbf{E}\Big|_{\eta_{r_{H\eta}}}^n \tag{16}$$

In the following section, the stability analysis of the above formulations is investigated by using the root-locus method [7].

3 Root-Locus Stability Analysis

Let the time-harmonic solution of the above field equations with the variables E, H, and J be given by

$$\Psi_{\alpha,\beta,\gamma}^n = \Psi_0 e^{\mathbf{j}\left(\omega n\Delta_t - \tilde{k}_x \alpha\Delta_x - \tilde{k}_y \beta\Delta_y - \tilde{k}_z \gamma\Delta_z\right)} \tag{17}$$

where $\mathbf{j} = \sqrt{-1}$, $\Psi_{\alpha,\beta,\gamma}^n = \Psi\left(n\Delta_t,\ \alpha\Delta_x,\ \beta\Delta_y,\ \gamma\Delta_z\right)$, Ψ_0 is the complex amplitude of the field Ψ, and $\tilde{k}_\eta$ $(\eta = x,\ y,\ z)$ is the wave number in the discrete mode along the η-direction. Upon substituting (17) into (9), (14), and (16), the following system can be obtained:

$$\begin{bmatrix} (\mathcal{Z} - b_1)\mathbf{I}_3 & -b_2\mathbf{I}_3 & \mathbf{0}_3 \\ \frac{\Delta_t(\mathcal{Z}+1)}{2\varepsilon_0 D_g}\mathbf{I}_3 & (\mathcal{Z} - 1)\mathbf{I}_3 & \frac{\Delta_t \mathcal{Z}^{\frac{1}{2}}}{\varepsilon_0}\mathcal{C} \\ \mathbf{0}_3 & \frac{\Delta_t}{\mu_0}\mathcal{C}^T & \left(\mathcal{Z}^{\frac{1}{2}} - \mathcal{Z}^{-\frac{1}{2}}\right)\mathbf{I}_3 \end{bmatrix} \times \begin{bmatrix} \mathbf{J}_0 \\ \mathbf{E}_0 \\ \mathbf{H}_0 \end{bmatrix} = 0 \tag{18}$$

where $\mathcal{Z} = e^{j\omega\Delta_t}$ is the stability factor, $\mathbf{0}_3$ is a 3×3 null matrix, $\mathbf{I}_3$ is a 3×3 identity matrix, and $\mathcal{C}$ contains the eigenvalues of the curl operator of Maxwell's equations given by

$$\mathcal{C} = \begin{pmatrix} 0 & \widehat{\delta}_z & -\widehat{\delta}_y \\ -\widehat{\delta}_z & 0 & \widehat{\delta}_x \\ \widehat{\delta}_y & -\widehat{\delta}_x & 0 \end{pmatrix} \tag{19}$$

where $\widehat{\delta}_\eta = \mathbf{j}2\sin\left(k_\eta\Delta_\eta/2\right)/\Delta_\eta$. Equating the determinant of the coefficient matrix of (18) to zero and taking $\sin\left(\widetilde{k}_\eta\Delta_\eta/2\right) = 1$, to account for the worst possible case, the presented explicit RK-FDTD scheme will have the following stability polynomial:

$$\left(\frac{\mathcal{Z}-1}{\mathcal{Z}^{\frac{1}{2}}}\right)\left(\frac{\mathcal{Z}-b_1}{\mathcal{Z}^{\frac{1}{2}}}\right)^2 \left[(\mathcal{Z}-b_1)(\mathcal{Z}-1) + \frac{b_2\Delta_t}{2\varepsilon_0 D_g}(\mathcal{Z}+1)\right]\left(S^{\mathcal{RK}}(\mathcal{Z})\right)^2 = 0 \tag{20}$$

where $S^{\mathcal{RK}}(\mathcal{Z})$ is given by

$$S^{\mathcal{RK}}(\mathcal{Z}) = 4\mathcal{CN}^2\mathcal{Z} + (\mathcal{Z}-1)^2 + \frac{b_2\Delta_t}{2\varepsilon_0 D_g}\frac{(\mathcal{Z}+1)(\mathcal{Z}-1)}{(\mathcal{Z}-b_1)} \tag{21}$$

and $\mathcal{CN}$ is the Courant number defined as

$$\mathcal{CN} = \frac{\Delta_t}{\Delta_{t_{\max}}^{\mathcal{CFL}}} \tag{22}$$

with $\Delta_{t_{\max}}^{\mathcal{CFL}} = 1/c_0\sqrt{\Delta_x^{-2} + \Delta_y^{-2} + \Delta_z^{-2}}$ being the CFL time-step limit [2], and $c_0 = 1/\sqrt{\varepsilon_0\mu_0}$ being the speed of light in free space. Based on (20) and (21), a reduced stability can be re-arranged from (21) as

$$1 + 4\mathcal{CN}^2\frac{\mathcal{Z}}{(\mathcal{Z}-1)^2}\frac{1}{\widetilde{\varepsilon}_r^{\mathcal{RK}}(\mathcal{Z})} = 0 \tag{23}$$

where $\widetilde{\varepsilon}_r^{\mathcal{RK}}(\mathcal{Z})$ is the numerical permittivity of the presented explicit RK-FDTD scheme, which can be arranged as

$$\widetilde{\varepsilon}_r^{\mathcal{RK}}(\mathcal{Z}) = 1 + \frac{b_2\Delta_t}{2\varepsilon_0 D_g}\frac{(\mathcal{Z}+1)}{(\mathcal{Z}-1)(\mathcal{Z}-b_1)} \tag{24}$$

Based on the root-locus stability analysis [7], the maximum time step of the presented RK-FDTD implementation $\left(\Delta_{t_{\max}}^{\mathcal{RK}}\right)$ can be obtained from the fact that the roots of (23) must lie inside or on the unit circle in the $\mathcal{Z}$-plane, i.e., $|\mathcal{Z}| \leq 1$. To better

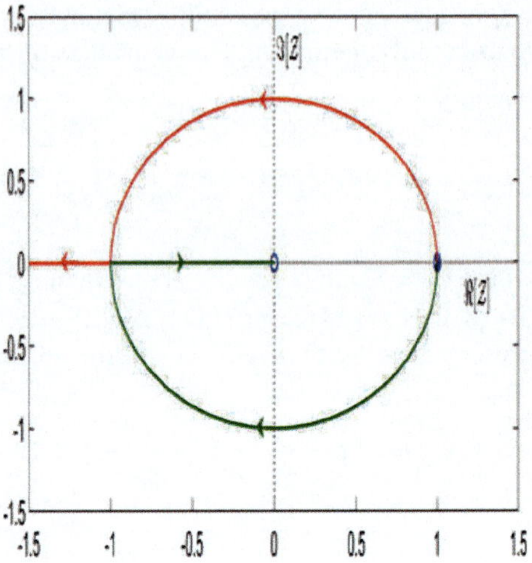

Fig. 1 Root-locus of (23) for the presented explicit RK-FDTD implementation. Graphene parameters are taken as $\upsilon = 2.0$ THz, $\mu_c = 1.0$ eV, $T = 300$ Kelvin, and $d = \Delta = c_0/200 f_{\max}$, with $f_{\max} = 10$ THz

visualize this idea, consider a graphene layer with the following parameters: $\upsilon = 2.0$ THz, $\mu_c = 1.0$ eV, $T = 300$ Kelvin, and let the graphene layer occupy one spatial cell [10], i.e., $D_g = \Delta$, where $\Delta = \Delta_x = \Delta_y = \Delta_z = c_0/200 f_{\max}$ with $f_{\max} = 10$ THz. Figure 1 shows the root-locus of (23), where the initial time step is taken to satisfy $\Delta_{t_{\max}}^{\mathcal{CFL}}$ [11]. As can be seen from Fig. 1, the instability occurs at $\mathcal{Z} = -1$. Hence, by substituting $\mathcal{Z} = -1$ into (23) together with (24) the following can be obtained:

$$1 - \mathcal{CN}^2 = 0 \tag{25}$$

and this implies that $\mathcal{CN}$ is always unity for the presented RK-FDTD scheme and independent from graphene parameters. Noting that $\mathcal{CN} = \Delta_t/\Delta_{t_{\max}}^{\mathcal{CFL}}$, the time-step constraint for the explicit RK-FDTD scheme $\left(\Delta_{t_{\max}}^{\mathcal{RK}}\right)$ can be obtained from (25) as

$$\Delta_{t_{\max}}^{\mathcal{RK}} = \Delta_{t_{\max}}^{\mathcal{CFL}} \tag{26}$$

Hence, the presented RK-FDTD scheme retains $\Delta_t^{\mathcal{CFL}}$.

4 Numerical Stability and Accuracy Verification

In this section, the stability and accuracy of the presented explicit RK-FDTD scheme are verified through a numerical test that investigates electromagnetic wave propagation through an infinite free-standing graphene layer. For this purpose, an electromagnetic wave with E_z and H_y field components propagating in a one-dimensional (1-D) domain along the x-direction is considered. The size of the simulation domain is taken as $8000\Delta_x$, where $\Delta_x = c_0/200\, f_{\max}$ and $f_{\max} = 10$ THz. The convolutional perfectly matched layer (CPML) [12], with a thickness of 10 cells, is used to truncate the computational domain. The graphene layer, with the same parameters used for Fig. 1, occupies one spatial cell at grid point 4000. The simulation domain geometry is shown in Fig. 2. The excitation is a Gaussian pulse with a time dependence of

$$g(t) = e^{-4\pi(t-t_d)^2/t_w^2} \tag{27}$$

where $t_w = 80\Delta_t$ and $t_d = 6t_w$. The excitation source is located at point **S**, and the observation point is located at point **O**, as shown in Fig. 2. The simulation is conducted for the first 100 000 time steps. Figure 3 shows the transmitted E_z field recorded at the observation point **O**, $\left(E_z^{tr}\,(4020\Delta_x)\right)$, computed by the presented explicit RK-FDTD implementation with $\Delta_t = \Delta_{t_{\max}}^{\mathcal{RK}} = \Delta_{t_{\max}}^{\mathcal{CFL}}$. Clearly, E_z remains stable over the complete simulation time and, therefore, the stability of the presented explicit RK-FDTD implementation maintains the conventional time-step CFL constraint. It must be noted that the simulation is also conducted with the presented explicit RK-FDTD implementation for the first 1×10^6 time steps and no instability is observed.

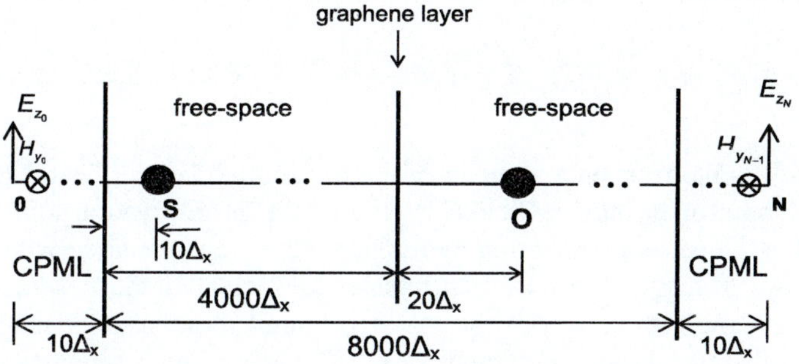

Fig. 2 One-dimensional simulation domain geometry

Fig. 3 Transmitted electric field at node 4020, $E_z^{tr}\,(4020\Delta_x)$ computed by the presented explicit RK-FDTD with $\Delta_t = \Delta_{t_{\max}}^{\mathcal{RK}} = \Delta_{t_{\max}}^{\mathcal{CFL}}$

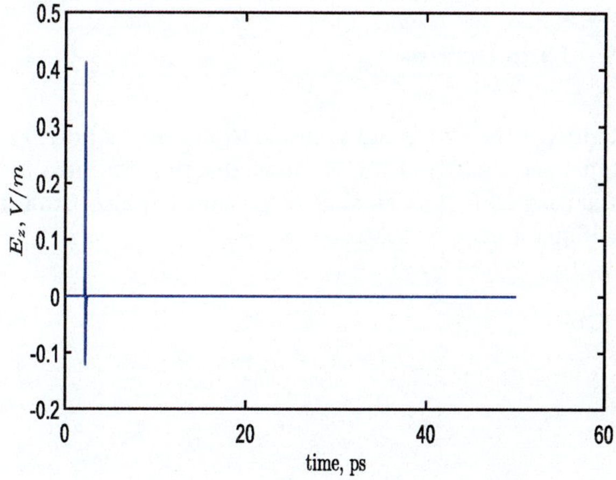

Fig. 4 Transmission coefficient magnitude for a graphene layer computed by the presented explicit RK-FDTD, and the theoretical approaches

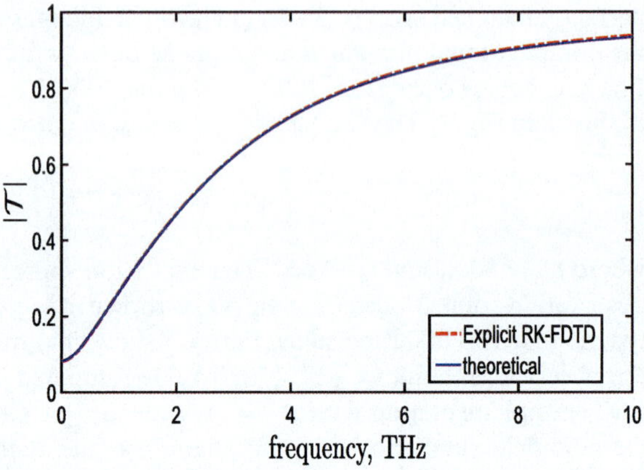

Finally, the accuracy of the presented explicit RK-FDTD implementations is studied. For this purpose, the transmission coefficient ($\mathcal{T}$) of the graphene layer is investigated and computed as

$$\mathcal{T}(\omega) = \frac{E_z^{tr}(\omega)}{E_z^{inc}(\omega)} \tag{28}$$

where $E_z^{tr}(\omega)$ is the frequency domain of the transmitted field E_z^{tr}, and $E_z^{inc}(\omega)$ is frequency domain of the incident field E_z^{inc} recorded at the observation point $\mathbf{O}$, which is obtained in the second simulation by replacing the graphene layer with vacuum. Figure 4 shows the magnitude of the transmission coefficient for the presented explicit RK-FDTD scheme with $\Delta_t = \Delta_{t_{\max}}^{\mathcal{RK}} = \Delta_{t_{\max}}^{\mathcal{CFL}}$. Figure 4 shows also the magnitude of the theoretical transmission coefficient ($\mathcal{T}_{th}$) [8]. Clearly, both schemes give high accuracy as the theoretical results.

5 Conclusions

In this paper, stable and accurate Runge-Kutta FDTD formulations are presented for graphene simulations. It is shown that the presented formulations not only retain the standard CFL time-step stability limit but also exhibit high accuracy as compared with the theoretical results.

6 Appendix: Runge-Kutta Local Truncation Error

Considering $\mathcal{K}_2 = \Delta_t f\left((n+\lambda_1)\,\Delta_t,\, J_\eta^n + \lambda_2\mathcal{K}_1\right)$ given in (10), and applying the multivariate Taylor series expansion, the following can be obtained:

$$\begin{aligned}\mathcal{K}_2 = \Delta_t \Big[& f\left(n\Delta_t + J_\eta^n\right) + \lambda_1\Delta_t f_t'\left(n\Delta, J_\eta^n\right) \\ & + \lambda_2\mathcal{K}_1 f_{J_\eta}'\left(n\Delta, J_\eta^n\right) + O\left(\Delta_t^2, \mathcal{K}_1^2\right)\Big]\end{aligned} \quad (29)$$

where f_t' and f_{J_η}' denote the derivative of f with respect to time and J_η, respectively. Noting that $\mathcal{K}_1 = \Delta_t f\left(n\Delta_t + J_\eta^n\right) + O\left(\Delta_t\right)$, and using (29), J_η^{n+1} in (10) can be arranged as

$$\begin{aligned}J_\eta^{n+1} &= J_\eta^n + (a_1 + a_2)\,\Delta_t f\left(n\Delta_t + J_\eta^n\right) + a_2\Delta_t^2 \\ &\times \left[\lambda_1 f_t'\left(n\Delta, J_\eta^n\right) + \lambda_2 f\left(n\Delta, J_\eta^n\right) f_{J_x}'\left(n\Delta, J_\eta^n\right)\right] \\ &+ O\left(\Delta_t^3\right)\end{aligned} \quad (30)$$

Recalling that J_η^{n+1} can also be written by using the Taylor series expansion as

$$J_\eta^{n+1} = J_\eta^n + \Delta_t J_{\eta_t}' + \frac{\Delta_t^2}{2} J_{\eta_{tt}}'' + O\left(\Delta_t^3\right) \quad (31)$$

and noting that $J_{\eta_t}' = f$ and $J_{\eta_{tt}}'' = f_t' + f_{J_\eta}' f$, (31) can be arranged as

$$\begin{aligned}J_\eta^{n+1} &= J_\eta^n + \Delta_t f\left(n\Delta_t + J_\eta^n\right) + \frac{\Delta_t^2}{2} \\ &\times \left[f_t'\left(n\Delta, J_\eta^n\right) + f\left(n\Delta, J_\eta^n\right) f_{J_\eta}'\left(n\Delta, J_\eta^n\right)\right] \\ &+ O\left(\Delta_t^3\right)\end{aligned} \quad (32)$$

Comparing (30) with (32), it can be easily concluded that the explicit RK-FDTD approximation of (10) is of second-order accuracy if

$$a_1 + a_2 = 1, \text{ and } \lambda_1 a_2 = \lambda_2 a_2 = \frac{1}{2} \quad (33)$$

References

1. Geim, K., Novoselov, K.S.: The rise of graphene. Nat. Mater. **6**(3), 183–191 (2007)
2. Taflove, A., Hangess, S.: Computational Electrodynamics: The Finite-Difference Time-Domain Method, 3rd edn. Artech-House, Norwood (2005)

3. Bouzianas, D., Kantartzis, N., Antonopoulos, S., Tsiboukis, T.: Optimal modeling of infinite graphene sheets via a class of generalized FDTD schemes. IEEE Trans. Magn. **48**(2), 379–382 (2012)
4. Bouzianas, G.D., Kantartzis, N.V., Yioultsis, T.V., Tsiboukis, T.: Consistent study of graphene structures through the direct incorporation of surface conductivity. IEEE Trans. Magn. **50**(2), Article no. 7003804 (2014)
5. Papadimopoulos, A.N., Amanatiadis, S.A., Kantartzis, N.V., Rekanos, I.T., Zygiridis, T.T., Tsiboukis, T.D.: A convolutional PML scheme for the efficient modeling of graphene structures through the ADE-FDTD technique. IEEE Trans. Magn. **53**(6), Article no. 7204504 (2017)
6. Fletcher, S.J.: Numerical solutions to initial value problems. In: Data Assimilation for the Geosciences: From Theory to Application, ch. 8, pp. 273–315. Elsevier, U.K. (2017)
7. Ogata, K.: Discrete-Time Control Systems, 2nd edn. Prentice-Hall, New Jersey (1995)
8. Hanson, G.W.: Dyadic Green's functions and guided surface waves for a surface conductivity model of graphene. J. Appl. Phys. **103**(064302) (2008)
9. Vakil, A., Engheta, N.: Transformation optics using graphene. Science **332**(6035), 1291 (2011)
10. Wang, X.H., Yin, W.Y., Chen, Z.: Matrix exponential FDTD modeling of magnetized graphene sheet. IEEE Antennas Wirel. Propag. Lett. **12**, 1129–1132 (2013)
11. Gutschling, S., Kruger, H., Weiland, T.: Time-domain simulation of dispersive media with the finite integration technique. Int J. Numer. Model. **13**, 329–348 (2000)
12. Roden, J.A., Gedney, S.D.: Convolutional PML (CPML): an efficient FDTD implementation of the CFS-PML for arbitrary media. Microw. Opt. Technol. Lett. **27**(5), 334–339 (2000)

Hydrodynamic Analysis and CFD Modeling of PAWEC Interacted with Regular Waves Using CFX

Ali Shehab, Ahmed M. R. El-Baz, and Abdalla Mostafa Elmarhomy

Abstract The multiplicity of renewable energy sources represents the biggest challenge for environmental scientists and engineers. This research presents a mathematical model and a numerical study using the high-performance ANSYS-CFX software to analyze the dynamic behavior of the point absorber wave energy converter (PAWEC). Two different models were constructed to predict the hydrodynamic response of the wave energy converter in both free and forced oscillations under the action of incident regular waves and external mechanical damping. The differential equations are solved analytically using RKFOM. CFX multiphase model is constructed to solve the 3D Unsteady Reynolds Averaged Navier–Stokes Equation (URANS) using the two-way Fluid–Structure Interaction (FSI) technique. The regular waves were generated in a numerical wave tank, by using a flap-type wavemaker. Mesh densities and solver settings were performed. The numerical results in both models, CFD and RKFOM, are validated against published experimental and numerical data under the same conditions, and the numerical results agreed with both published data. Two additional designs for the body bottom, conical and spherical shapes, were analyzed based on the presented numerical method. The damping coefficient and added mass are obtained for each design in the case of heave motion only.

Keywords CFD · CFX · FSI · Renewable energy · Wave energy · Wave generation · Wave Energy Converter (WEC) · Numerical Wave Tank (NWT)

A. Shehab (✉) · A. M. Elmarhomy
Faculty of Engineering, Ain Shams University in Egypt, Cairo 11517, Egypt
e-mail: Alishehab2@eng.asu.edu.eg

A. M. Elmarhomy
e-mail: abdallah_elmarhoumy@eng.asu.edu.eg

A. M. R. El-Baz
Faculty of Engineering, The British University in Egypt, Cairo 11837, Egypt
e-mail: ahmed.elbaz@bue.edu.eg

D. Zeidan et al. (eds.), *Mathematics and Computation*, Springer Proceedings in Mathematics & Statistics 418, https://doi.org/10.1007/978-981-99-0447-1_14

1 Introduction

The numerical simulation of multiphase applications, coupled with the interaction between the fluids and movable structure, requires a high-performance CFD code. Many CFD platforms are followed for this purpose, especially in ocean engineering. ANSYS-AQWA software is a simple CFD platform accompanied by offshore hydrodynamics in open water which provides an integrated facility for developing primary hydrodynamic parameters required to undertake complex motions and response analysis [1]. OpenFOAM is an open-source CFD software that provides a wide range of features to solve both fluid flows and solid mechanics, it is a C++ toolbox for the development of customized numerical solvers which gave a higher mixing level compared to other software, but it needs a professional programmer [2]. ANSYS-CFX is a high-performance CFD software distinguished for exceptional accuracy and high convergence speed, especially in multiphase applications [3]. CFX can be used to generate a single run to create full operating maps with a simple integration process. Using ANSYS-CFX in the simulation of numerical wave tanks (NWTs) with the corresponding wave energy converters (WECs) provides a broad visualization of both fluid characteristics as well as the hydrodynamic of floating bodies [4].

The numerical simulation of wave energy converters (WECs) is the focus of researchers' efforts in the last decades. The main challenge facing each researcher is to develop the device's efficiency to produce energy and reduce the cost of power [5]. Most of the previous work uses the numerical simulation by ANSYS-AQWA and OpenFOAM; the experimental setup in this field requires special equipment and a great effort, whether in the preparation of movement receptors or the difficulty of extracting results. Jin, Siya [6] construct a 1/50 scale PAWEC in a wave tank to validate the CFD model generated to study the effect of the floating body geometry and the power take-off (PTO) damping on the wave energy absorption. The mathematical model of PAWEC behaviors was performed using two different models, the first is the non-linear state-space model (NSSM) considering a quadratic viscous term, second is the linear state-space model (LSSM). Three different geometries were investigated in this paper using ANSYS/AQWA software: a flat-bottom cylinder, a hemispherical bottom cylinder, and conical bottom with right streamline angle. The experimental data was compared with the two mathematical models and validated with CFD model using ANSYS-AQWA. Jin, Siya concluded that the flat-bottom cylinder has more damping coefficient, and therefore, more added mass than the other designs by 60%. Moreover, the best design is the conical bottom cylinder, which produced max stroke length 100% more than other designs. The selected design was subjected to PTO damping, and therefore, the optimal power is increased by 70% in both regular and irregular waves. Josh et al. [7] used OpenFOAM to investigate the implementation of NWT containing a rigid body solution. The capabilities of NWT were outlined in the case of fluid–structure interaction between cylindrical floating bodies and incident irregular waves generated using the JONSWAP spectrum. Fifty frequencies were used, which were regularly distributed between 0 and 0.5 Hz with 10 s period and different phases. The body motion was analyzed using two ways, one way is the prescribed way to define the hydrodynamic forces concerning the displacement and time, and another way is the numerical simulation to predict the dynamic response of the body against contribution waves and PTO applied force. Shadman et al. [8] used ANSYS-AQWA software to analyze the hydraulic diffraction of cylindrical PAWEC to optimize the geometry based on the maximization of absorbed power and absorption bandwidth in case of natural conditions nearshore region of the Rio de Janeiro coast. The technique of joint probability distribution and resultant wave spectrum was used to perform the optimization method. The two primary advantages of this optimization method are the reduced computational time and the possibility of performing parametric analyses for the WEC geometry. Sjökvist et al. [9] analyzed the hydrodynamic parameters of cylindrical PAWEC using a CFD model built-in Multiphysics simulation software COMSOL and a numerical linear model computed by WAMIT. The linear model of the interaction between the incident waves and a floating structure is solved for the excitation forces, radiation damping, and added mass using green's theorem by integrating the diffraction and radiation velocity potentials in closed surfaces extracted from a 3D panel program WAMIT. The numerical results are validated against an experimental work, the hydrodynamic parameters computed with the COMSOL model show good agreement with the ones computed using WAMIT. Ghasemi et al. [10] presented a numerical computational method to solve the 2D Navier–Stokes equations governing the behavior of flow field interacted with two types of WEC, cylindrical and rectangular cross-sectional shape. The fluid–structure interaction parameters were determined using the VOF method; NWT technique was used to generate the waves by using both Flape-type and piston-type wavemakers. The numerical model was obtained by several degrees of freedom in both floating bodies, heave motion for the cylindrical body against incident waves, and a free-fall test, free rotational pitch motion for the rectangular shape. This paper presented the change in floating system efficiency with respect to the coefficient of damping, the maximum efficiency obtained at single degree of freedom wih value 0.5. Büchner et al. [11] constructed a 3D numerical model using ANSYS-fluent to predict the dynamic response of a single degree of freedom floating cylinder against a regular wave. The horizontal and vertical forces on the float due to drag and inertia loads were presented in case of different values of wave amplitudes and frequencies. The effect of irregular wave on the float dynamics was recommended in the future work.

Devolder et al. [12] construct an experimental measurement to validate the CFD results for an array of up to nine semi-circular WECs under the effect of regular waves. The frictional forces and heave motion are presented in both experimental and numerical studies. Rijnsdorp et al. [13] presents a numerical simulation for the interaction between the incident waves and a fully submerged wave energy converter using the non-hydrostatic framework on large scales. This research demonstrates both linear and non-linear waves with a cylinder shape of WEC connected by a mooring line. The results of the linear waves were validated with an analytical solution in both diffraction, radiation, and dynamic response. Jin et al. [14] investigate the non-linear viscosity effect of a wave energy converter prototype with a scall 1/50 by studying the hydrodynamics loads. The point absorber WEC in this paper was designed with only heave motion in the experimental part. The mathematical model of PAWEC behaviors was performed using two different models, the first is the nonlinear state-space model (NSSM) considering a quadratic viscous term, the second is the linear state-space model (LSSM). The experimental data were compared with the two mathematical models and validated with the CFD model using ANSYS-AQWA. the main objective of this paper is to achieve the optimal power of the device and to maximize the conversion efficiency by indicating the performance of the device at the resonance case. Zhu and Lim [15] constructed a flume experiment for design optimization of a cylindrical wave energy converter which leads to reducing the undesirable motions due to the surrounding environment. A heave plate is recommended to be mounted with the floating body to reduce these motions. Several experimental tests are performed with different heave plates at various gaps in the body. The response amplitude operator RAO for the cylinder with a heave plate was 40% less than that without plates. Weller et al. [16] set up experimental measurements for the 2D motion of WEC under the action of regular and irregular extreme waves. The translation motion, in both directions, and rotation motion (heave, surge, and pitch) were presented using an optical encoder and the analysis of video footage. The relation between the body and the wave is linear in low frequencies, however, in extreme sea-state conditions, the heave motion and wave breaking are presented to predict the hydrodynamic response of WECs in high frequency.

Our need for validated numerical analysis is increasing in the field of wave energy, especially with the presence of published experimental results and with the difficulty of implementing those experiments. Some of these numerical studies will be mentioned in this section, which will contribute to providing the best in this research. The construction of an NWT capable of creating a variety of waves and being able to study the behavior of the buoy under the influence of those waves is the main objective of this research. Shehab et al. [17] constructed an NWT that can simulate both regular and irregular waves in form of wave spectra. Flap-type wavemaker technique was applied for this purpose using CFX software. The generated regular wave was validated against the wavemaker theory WMT, while the irregular wave was validated against experimental data with real sea conditions in the frequency domain. Very useful parameters were discussed in this research that will greatly contribute to the present work. Finnegan and Goggins [18] used CFX to investigate the effect of using a flap-type wave maker to generate regular waves in both deep and shallow water depths. This method is limited to a low normalized wavenumber. The change of hinge location reduced these limitations.

After the previous work, the modeling of the NWTs using CFX is more effective due to the wide range of parameters provided by CFX which leads to a good understanding of the PAWEC hydrodynamics, CFX is a good choice for executing a CFD-based NWT. The main objective of this research is to understand the hydrodynamics of fluid–structure interaction between a partially submerged free-floating body with an incident regular and irregular generated waves inside NWT through a commercial CFD software CFX, which is rarely used in this field. The followed method used in this research has been presented in a detailed manner in addition to studying the extent of its usability compared to the rest of the methods presented above in the literature. The numerical results are validated against previous experimental data in the same conditions and body geometry.

2 Methodology and Problem Setup

Most of the previous research constructs an NWT using simple computational methods that provide a limited range of parameters as mentioned in the literature. Using CFX to construct an NWT requires more effort and additional techniques than other CFD simulation methods, the formation of NWT using CFX is like creating a miniature model of an integrated lab that gives all the useful results for understanding the factors affecting both the fluid and the moving body. The numerical model in this study consists of two phases for the fluid, air, and water, in addition to a solid phase for the floating body. The differential equations governing the fluid flow inside NWT are presented and solved for both velocity and momentum with a proper boundary condition describing the layers of the tank, an additional equation is applied for the solution of Volume Fraction VF of the fluids to determine the vertical position of the free surface at any time. Newton's second low was applied for the solution of solid phase model under the effect of diffraction, radiation, and excitation forces in heave motion only. Two different approaches are used for the solution of the mathematical model, first by using ANSYS-CFX (release 14.5) approach, the second method is to transform the FSI system to a simple harmonic motion followed by a numerical technique to solve the differential equations called RKFOM. A great effort was made to construct the combination between NWT and movable body using CFX. The two approaches are validated, in the case of free-fall test of the float in the still water, with published experimental work carried out in similar conditions by Guo et al. [19]. The validated model is then used to investigate the damping coefficient and added mass term for several designs of the float operating surface, conical and spherical profiles with the same mass are tested in the free-fall conditions to study the effect of bottom design on motion behavior. The design with the best results was tested against regular and irregular waves generated by a Flap-type wavemaker located at the inlet of the tank, the recommendations mentioned in the research submitted by Shehab et al. [17] regarding the wave generation in NWT were taken into consideration.

3 Mathematical Model

The fluid model in CFX consists of two phases separated by a free surface, the flow of water is considered incompressible in the transient scheme. The 3D form of both continuity and momentum differential equations are solved for velocity and pressure, respectively. The governing equations are defined as [20]. Continuity equation

$$\nabla.\vec{v} = 0, \vec{v} = \dot{x}\hat{i} + \dot{y}\hat{j} + \dot{z}\hat{k}, \tag{1}$$

where $\vec{v}$ is the velocity vector, and $\dot{x}$, $\dot{y}$, $\dot{z}$ represent the velocity components along the coordinate axes. Momentum equation

$$\frac{\partial(\rho\vec{v})}{\partial t} + \nabla.(\rho\vec{v}\vec{v}) = -\nabla P + \rho\vec{g} + \nabla.(\vec{\tau}), \tag{2}$$

where P is the static pressure, ρ is the fluid density, $\vec{\tau}$ is the stress tensor, the reference of the coordinate system is located at the centerline of the buoy coincidence with the still water level at the equilibrium position of the buoy. The finite Volume Method FVM is used to solve the previous governing equations in CFX [3]. The simulation of the water-free surface, with respect to time, requires two additional differential equations for the solution of the volume fraction VF for both water q_w and air q_a. The total summation of volume fractions is equal to 1. The instantaneous position of the free surface is estimated using the minimum value of $|q_w - q_a|$ across the domain. The governing equations for this method, Volume of Fraction VOF, are derived by Liang et al. [21] and defined in the following equations.

$$\frac{\partial(q_a)}{\partial t} + \dot{x}\frac{\partial(q_a)}{\partial x} + \dot{y}\frac{\partial(q_a)}{\partial z} + \dot{z}\frac{\partial(q_a)}{\partial z} = 0, \tag{3}$$

$$\frac{\partial(q_w)}{\partial t} + \dot{x}\frac{\partial(q_w)}{\partial x} + \dot{y}\frac{\partial(q_w)}{\partial z} + \dot{z}\frac{\partial(q_w)}{\partial z} = 0. \tag{4}$$

The solid model in CFX is investigated using rigid body dynamics. The float is a cylindrical shape with a curved bottom (S) (see Fig. 1). The mass of the buoy is equal to the displaced fluid in the equilibrium position (at $\Delta z = 0$). A single degree of freedom model is used in only heave motion, therefore, the terms of the rotational moment of inertia are eliminated from the differential equations. The differential equations, governing the dynamics of the float, are derived using Newton's second law (5), and the total forces (gravitational, hydrodynamics, and mechanical) are analyzed in Eq. (6). Sjökvist et al. [9]

$$\sum\vec{F} = \frac{d}{dt}\left(m * \vec{v}\right), \tag{5}$$

$$\vec{F}_g + \vec{F}_e(t) + \vec{F}_{pto}(\dot{z}) + \vec{F}_h(z, s) + \vec{F}_s(\dot{z}) + \vec{F}_a(\ddot{z}) = m\dot{\vec{v}}, \tag{6}$$

$\vec{F}_g = -mg\hat{k}$: Gravitational force, constant and directed along negative Z-Direction. $\vec{F}_e(t) = C_e * \delta(t)\hat{k}$: The excitation force is due to the interaction between the buoy and the externally generated waves. C_e is the coefficient of the radiation response, $\delta(t)$ is the surface elevation function. $\vec{F}_{pto}(\dot{z}) = -C_{pto} * \dot{z}(t)\hat{k}$: The mechanical loads of the power take-off mechanism due to friction and damping of

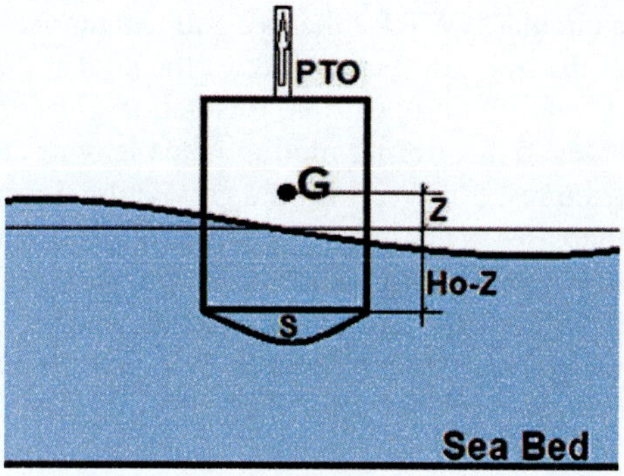

Fig. 1 The floating body at displacement Z above the water-free surface

the mechanical elements. C_{pto} is the damping coefficient of the PTO mechanism, $\dot{z}(t)$ is the vertical component of the velocity as a function of time. $F^{\rightharpoonup}{}_{h}(z, s) = \rho g[A_c(H_0 - Z(t)) + V(S)]\widehat{k}$: The hydrostatic forces due to buoyancy, variable force depending on the position of the float Z(t), and the bottom surface profile of the buoy (S) which is exposed to the fluid. H_0 is the submerged depth of the float in the equilibrium position, A_c is the cross-sectional area of the cylinder, $V(S)$ is the volume of the curved portion below the buoy. $F^{\rightharpoonup}{}_{s}(\dot{z}) = \mu A_s\left(\frac{\partial v^{\rightharpoonup}}{\partial r}\right) = -\frac{\mu A_s}{\Delta r}(\dot{Z})\widehat{k}$: the shear stress forces acting on the longitudinal surface of the float due to fluid viscosity and friction. $\frac{\partial v^{\rightharpoonup}}{\partial r}$ is the flow velocity gradient, μ is the dynamic viscosity of the fluid, A_s is the longitudinal surface area normal to the vertical axis, Δr denotes the radial distance from the body surface to the nearest zero shear stress location. $F^{\rightharpoonup}{}_{a}(\ddot{z}) = -m_a\ddot{z}\widehat{k}$: the equivalent added mass term due to the fluid dissipation around the float as a result of fluid linear inertia. m_a is the added mass. Ghasemi et al. [10].

The numerical model of the float dynamics with a single degree of freedom can be simplified to a spring-damping model with a general Eq. (7) only in heave motion along the vertical axis as

$$(M)\ddot{Z}(t) + (C_d)\dot{Z}(t) + (K)Z(t) = F_{ext}(t), \tag{7}$$

where M denotes the equivalent mass of the system [kg], defined as $M = m + m_a$, C_d represents the equivalent damping coefficient [kg/s], defined as $C_d = C_{pto} + \frac{\mu A_s}{\Delta r}$, K denotes the spring constant of the system [kg/s2], defined as $K = \rho g A_c$, $F_{ext}(t)$ denotes the total external forces of the incident-generated wave excitation [N].

3.1 CFD Model and Boundary Conditions

Two different NWTs are constructed for both the free-fall test model and the interaction between the float and excited wave. Davidson et al. [22] recommended using a circular NWT for the free-fall test model when using OpenFOAM. The main idea of the free-fall test is to adjust the float at a certain vertical displacement (Δz) above its equilibrium position, this action gives the device initial potential energy which leads to a restoring motion when leaving it free to move, then the float is dropped from rest to move with the gravitational force in the downward direction, the speed increases in the first stage of the movement as the forces of gravity overcome the rest of the total hydrodynamic forces acting on the float, the device starts to approach its equilibrium position, the potential energy of the device is converted to kinetic energy which causes increase in its velocity, the float begins to slow down till it comes to rest again in the lowest position as the hydrostatic forces overcome the float weight. The energy was dissipated due to the surrounding fluid damping effects. Accordingly, the device continues to oscillate, up by the buoyancy effects and down by the body weight, around its equilibrium position, the amplitude will be reduced by time depending on the amount of dissipating energy in each stroke and the damping coefficient arising from the shear stresses of the surrounding fluid and the PTO resistance. Finally, the device comes to rest in the equilibrium position. The free-fall test was performed experimentally by Guo et al. [19] in a wide-range tank to absorb the wave at tank walls. Jin et al. [6], Zhu and Lim [15], Angense [23], and Devolder et al. [24] used a rectangular NWT generated by other CFD approaches. A rectangular NWT is used in this research for both the cases as shown in Fig. 2; The free-fall test model is subjected to a simple boundary condition. A symmetric plane was used parallel to XZ coordinate plane, tank wall and seabed were treated as fixed walls with no-slip conditions, the free surface was set as interface subjected to a dynamic mesh related to the change in fluid volume of fraction method, the buoy was defined as a rigid body subjected only to hydrostatic forces in this model without other external effects, the buoy diameter is D with submerged height Ho. The numerical beach is not activated in this model because the wave generated by the free movement of the buoy is rather small compared to the generated wave by the Flap in the second model.

A second model with the sane NWT used for FSI, with incident-generated wave using Flap-type wavemaker, and numerical beach NB is used, as shown (see Fig. 2). The boundary conditions for the three phases are set as followed; inlet plane is a flapper subjected to a rotational motion about fixed hinge located below the free surface by distance ho, the position of the movable plane is defined in Eqs. (8), and (9), Eldeen et al. [17]. The top plane is divided into two regions, the first region in

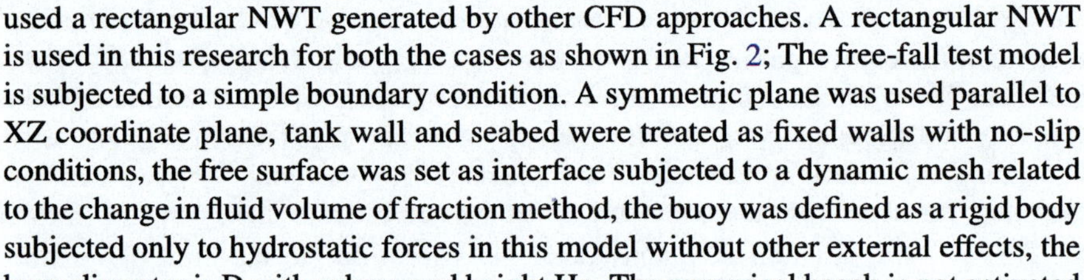

Fig. 2 Rectangular NWT for the free test model

the wave generation region is considered as a wall with a dynamic mesh technique to adjust the motion of the inlet plane, and the second region is considered as a fixed wall with zero-gauge pressure. The outlet plane is defined as the opening domain with a pressure gradient from the seabed to the top plane. The seabed is divided into three sections; the first section inside the wave generation region is defined as a movable surface in XY plane subjected to dynamic mesh technique, the second section inside the wave propagation region is considered as a fixed wall with no-slip conditions, the third section is the numerical beach NB designed with a 1:5 sloped surface to reduce the wave reflections from outlet plane [17]. An asymmetric plane is used to reduce the time of calculations. Lal's et al. [25] recommendations for using a Flap-type wavemaker are taken into consideration. The technical information based on wave simulation and wave absorption in NWT using CFX is explained in detail by the author in a related paper [17].

$$X = S_{max} * \frac{Z + h_o}{h_a + h_o}, \tag{8}$$

$$S_{max} = A_f \sin(\omega t), \tag{9}$$

where S_{max} is the flap position at the upper plane of the model, A_f is the maximum stroke length in the top plane, the air height is h_a, h_o is the fixed hinge depth. The x position of the inlet plane (Flapper) is varying from one point to another depending on the z position and flow time.

3.2 Grid Independence Test

ANSYS design modeler is used to discretize the model into a finite tetrahedron shape compatible with the dynamic mesh in 3D modeling. To detect the small change in buoy position and free surface of the fluid, 15 inflation layers with a total thickness of 0.2 m were used around the water-free surface and the buoy surface. The minimum element size is tested to check the independence of the mesh, seven cases were generated for this purpose. The float is displaced 5 cm above the equilibrium position before it falls from rest, the time step size in this test is selected with a small value of 0.001 s to produce a high sensitivity for buoy oscillations [22]; the float is of 1 m in diameter, 0.5 m immersed in height, and 392.7 kg in total mass (MO-1). The external force on the buoy was eliminated, and the PTO damping coefficient is set to zero in this test. Figure 3 shows the change of the vertical displacement of the buoy against time, in the first 10 s of motion, in each case. The damping coefficient is plotted in each case to monitor the independence of cell size use; Fig. 4.

The zones near the floating body and the free water surface have meshed with fine cells, and the size of elements increased away from the floating body and near the tank walls, top, and bottom planes, this procedure is followed to reduce the total number of elements and therefore reduce the time of calculation. The accepted

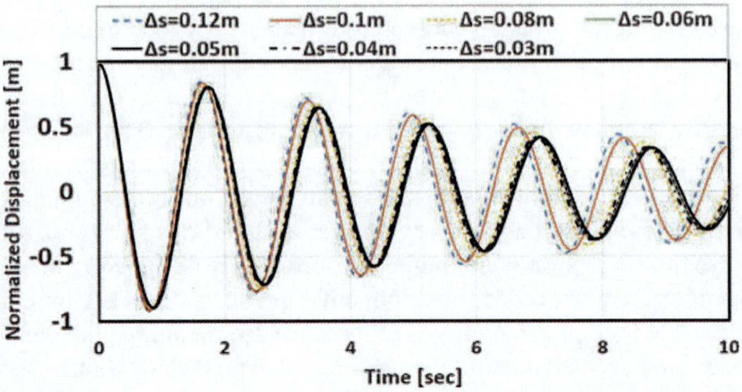

Fig. 3 The normalized vertical displacement of the float at $\Delta z = 0.05$ cm and time step size $\Delta t = 0.001$ s (MO-1)

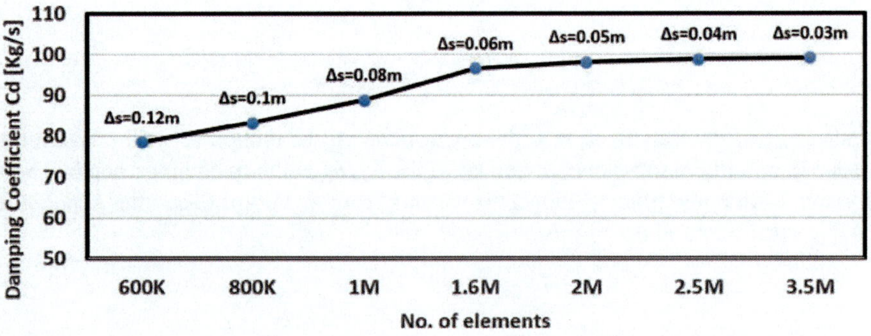

Fig. 4 Damping coefficient variation in each case for grid independence test (MO-1)

minimum element size near the movable surfaces, free water surface, and float, is 0.05 m which is equivalent to 1/20 of the float diameter with a total number of elements is 2 M, average skewness is 0.2, the mesh quality was tested to be larger than 0.75 around the contact surfaces. Figure 5 shows the grid distribution around the float.

3.3 Time Step Size Test

Determining the appropriate time step size value for all the different modeling processes contributes greatly to obtaining more accurate results, as it is a very influential factor in the time of calculations. The values for time step size vary from one simulator system to another, Davidson et al. [22] developed a new methodology to study the hydrodynamic models in the simulation of WECs. The results of this new methodology were validated using boundary-element methods (BEMs) in the linear

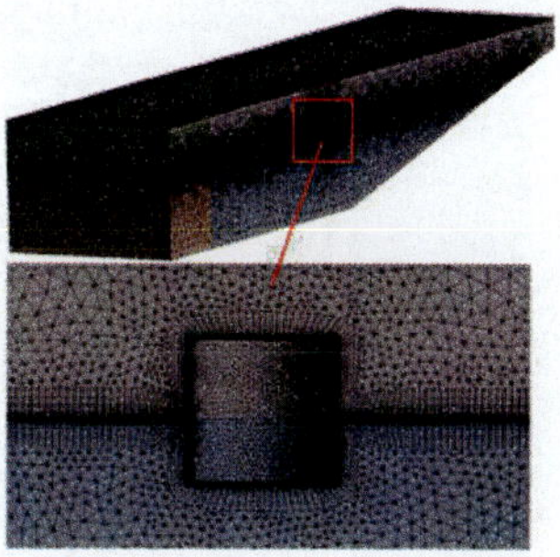

Fig. 5 3D Mesh for the model

case only (heave motion). The recommended value of time step size by Josh in 2015 is 0.001 s when using OpenFOAM software, this time step is selected to produce a maximum courant number of 0.3 which keeps an acceptable accuracy with a high-speed calculation [22]. A numerical case study in similar conditions was constructed using CFX software to detect the difference in numerical methods to predict the small oscillations of the floating body. A cylindrical buoy with a total height of 1 m (0.5 m immersed height) and a diameter of 1 m was used in a free-fall case (MO-2). The initial displacement in this test is 10 cm above the free surface, the total time is 12 s. The transient scheme used in this model is 2nd order backward Euler, and the solution method is second-order for both continuity and momentum equations. The PTO damping is eliminated in this test, five cases were generated with different time step sizes starting with 0.01 s which produces the minimum damping coefficient, a small deviation in dynamic response between the last two cases 0.002 s and 0.001 s. Figure 6 shows the normalized vertical displacement of the float in free-fall test by CFX compared with the published data by OpenFOAM.

The damping coefficient in each case and the independence of time step size are presented in Table 1. The recommended time step size in wave modeling using numerical wave tanks is T/60 according to [17]. In this research, the time step size is recommended to be T/925 in the numerical modeling of wave structure interaction compared with T/1840 as recommended in OpenFOAM.

4 Validation of the Numerical Results

The previous study shows that there is a noticeable difference in the results of both numerical methods, the numerical results in this research are validated against published experimental data by Guo et al. [19] in the same conditions to check the ability of the mathematical model to predict the hydrodynamics and the applied forces of the float. The cylinder used in this test is 0.3 m in diameter, 0.56 m in total height, and the mass is 19.79 kg typically as used in the experiments (MO-3). The numerical results in both models CFX and RKFOM are compared with the experimental results in the free-fall test with an initial displacement of 3 cm; Fig. 7. The

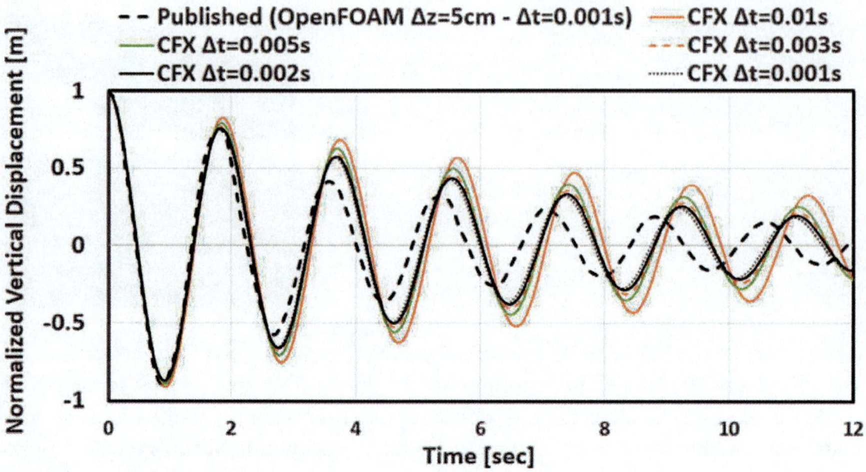

Fig. 6 The dynamic response of the float at different time step sizes by CFX compared with published data ($\Delta z = 10$ cm -MO-2)

Table 1 Independence of time step size using CFX in free-fall test

Model	Case	Normalized Timestep	Periodic time T [s]	Damping coefficient Cd [N.s/m]	Error
CFX-Cylinder Free Fall test Mass = 392 kg $\Delta z = 10$ cm (MO-2)	$\Delta t = 0.01$ s	T/188	1.88	78.4	25%
	$\Delta t = 0.005$ s	T/370	1.85	98	14.8%
	$\Delta t = 0.003$ s	T/615	1.85	112.5	4.8%
	$\Delta t = 0.002$ s	T/925	1.85	117.9	0.93%
	$\Delta t = 0.001$ s	T/1840	1.84	119	–

damping effects of the power take-off mechanism were added to the CFD model by applying variable vertical force on the solid body as a function of its velocity and opposite to the direction of motion, the PTO damping coefficient used is 20 kgs^-1 typically as experiments. The governing differential Eq. (7) is solved by MATLAB software using the numerical method RKFOM, the external force term is ignored in this experiment, and the added mass coefficient term could be neglected in case of lower diameter and small displacements compared to the immersed height of the float ($\Delta z{*}D/ho \leq 0.05$) [15], the term V(s) is zero because the cylinder bottom is flat. Figure 7 shows the agreement of numerical results compared to the published experimental data at time step sizes 0.002 s.

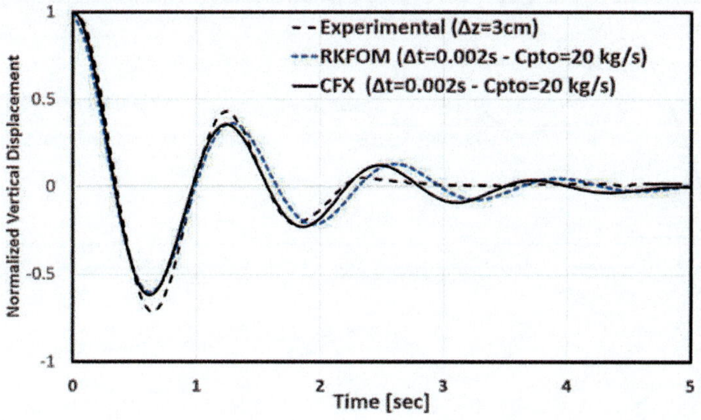

Fig. 7 Normalized vertical displacement for the validation of the numerical results against experimental data (MO-3)

5 Hydrodynamic Analysis

The presented numerical method in this research represents an effective tool that contributes to understanding the factors affecting the dynamic response of the float. A hydrodynamic analysis for both models is performed to understand the effect of each parameter on the dynamic behavior of the floating body. The CFD model, using CFX in the validated case with PTO damping coefficient 20 kgs^-1 and initial displacement 3 cm, and the analytical one using numerical model RKFOM in the same conditions were used to analyze the motion characteristics. Figure 8 shows the time domain comparison of both models to predict the motion kinematics, the extreme lowest position of the float is obtained at the point of maximum positive acceleration and zero velocity at 0.62 s.

The hydrodynamic forces are shown in Fig. 9 with other effects in the time domain for each model. The hydrostatic force at the lowest position is maximum, while the viscous and PTO damping is minimum, this leads to maximizing the total forces on the float and increasing its momentum to begin moving up. The highest position is coincidental with the minimum value of the acceleration and zero velocity, this is understood by monitoring the reduction in hydrostatic forces at the same instant at 1.24 s. The vertical acceleration in RKFOM is slightly declining from its value in the CFX model, and this is due to the elimination of the added mass coefficient [26]. The equilibrium position is reached after 5 s. The perturbed waves around the float in the CFD model are presented in Fig. 10, at the lowest position after 0.62 s and the highest position after 1.24 s.

The CFX approach provides a wide range of useful variables in this study, the pressure distribution on the floating body is variable based on the float position, Fig. 11 shows the change of this distribution in two cases.

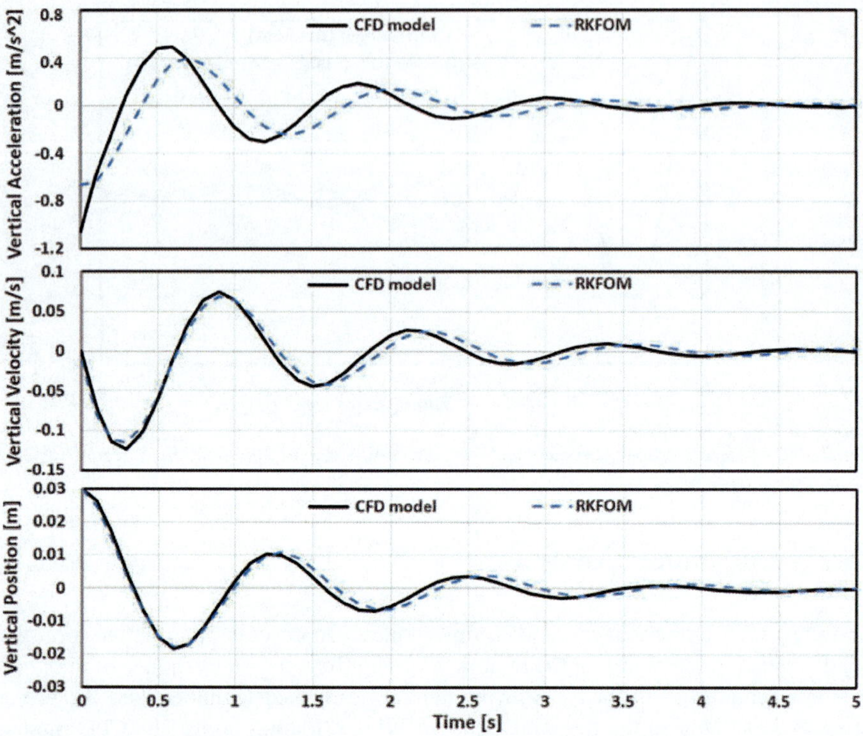

Fig. 8 Time-domain comparison for the kinematic variables of CFD and numerical model at ($\Delta z = 3$ cm–MO-3)

The presented numerical method is used to investigate the added mass term in another case study with different parameters, in which the effect of the added mass coefficient cannot be neglected. A free-fall case is applied on a buoy with a 1 m diameter and 0.5 m immersed height, the total mass of the buoy is 392.7 kg (MO-4). The float was initially located at 5 cm above SWL, this value is rather large compared to the float diameter and immersed height (Δz*D/ho $\geq$ 0.05). The CFX model is initially generated with the same boundary conditions as the free-fall test, the numerical model is used to investigate the hydrodynamic forces and the corresponding value of the damping coefficient and added mass, and the PTO damping coefficient is eliminated in this test to visualize the float dynamics for more time and obtain more accurate results. The time domain of the rigid body kinematics through 12 s is presented in Fig. 12. The agreement between the two models is observed in the first 4 s of motion (2 cycles), and the disparity between the two models begins in the oscillation frequency, despite the convergence in the damping coefficient at $C_d = 154$N.s/m as a result of viscous damping and the absence of a mechanical PTO system. The float reaches its maximum velocity at the equilibrium position, at this instant, the acceleration is zero.

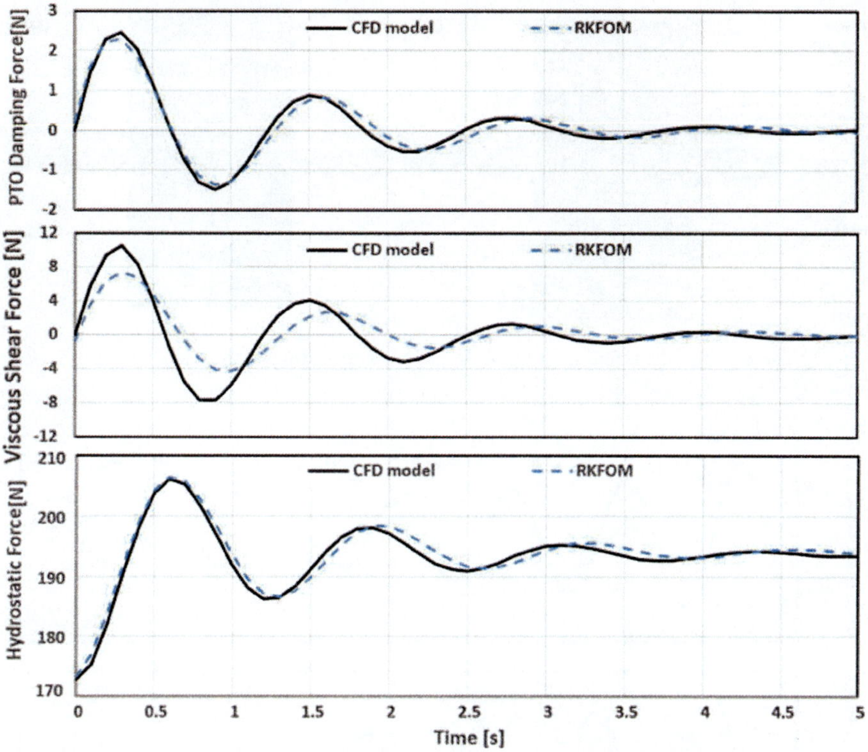

Fig. 9 Time-domain comparison for the hydrodynamic forces in CFD and numerical model at (Δz = 3 cm–MO-3)

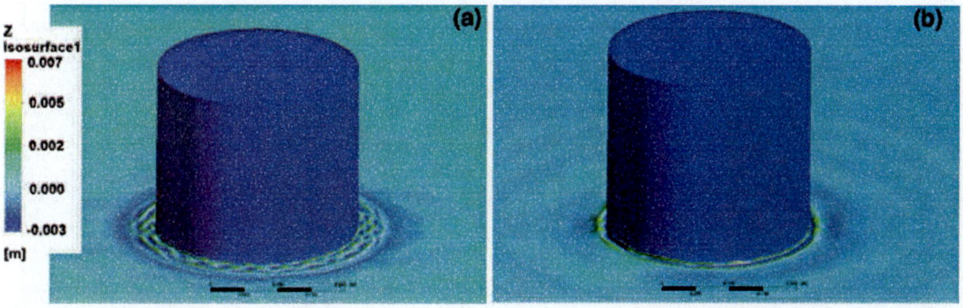

Fig. 10 3D view of the perturbed waves around the float **a** lowest position at t = 0.62 s, **b** highest position at t = 1.24 s–(MO-3)

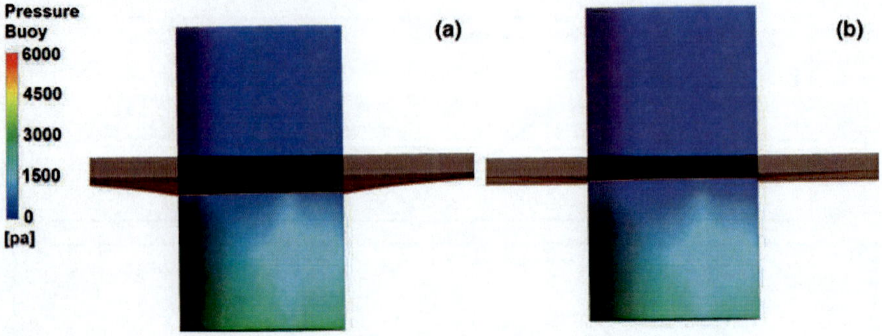

Fig. 11 The pressure distribution on the float surface **a** lowest position at t = 0.62 s, **b** highest position at t = 1.24 s–(MO-3)

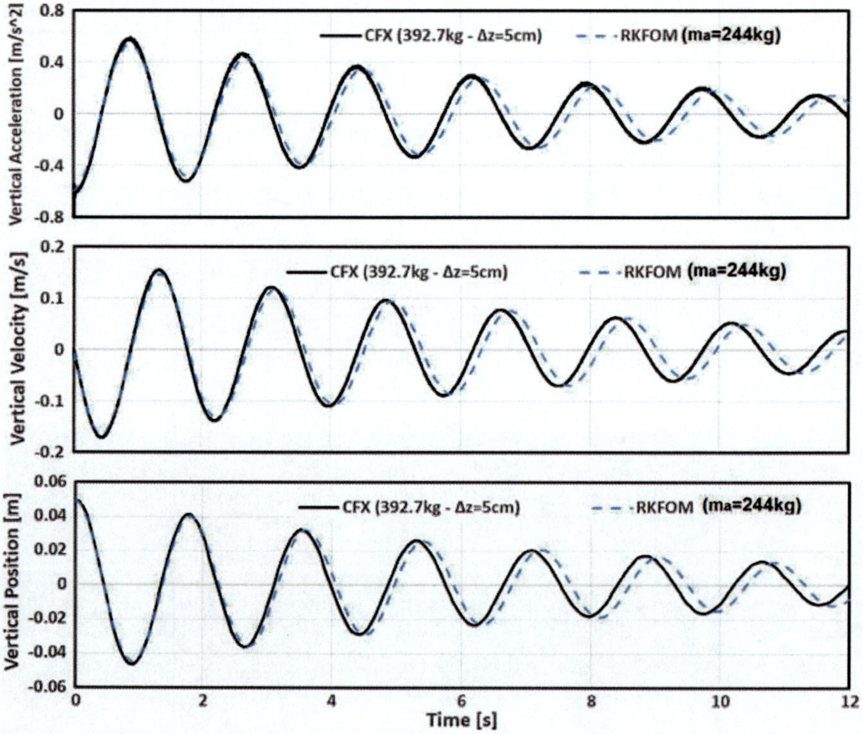

Fig. 12 Time-domain comparison for the kinematic variables of CFD and numerical model at (Δz = 5 cm–MO-4)

The calculation of added mass can be approximated both experimentally and numerically, Zhu and Lim [15] used a separated heave plate added to a circular cylindrical float to calculate the added mass coefficient experimentally. In the present research, the added mass coefficient m_a is calculated using the CFD-CFX model and numerically by RKFOM; Fig. 13 shows the hydrodynamic forces for model MO-4. The added mass term is investigated first in the CFX model in two different ways, first method is to integrate the value of shear stresses on the float surface to get the average viscous damping force $\vec{F}_s(\dot{z})$, the calculated pressure variation on the float bottom determined the hydrostatic forces with respect to time $\vec{F}_h(z, s)$, then the added mass term can be integrated by Eq. 6, $\vec{F}_{pto}(\dot{z})$ the PTO damping term and $\vec{F}_e(t)$ are zero, $\dot{\vec{v}}$ is the calculated vertical acceleration.

The second method is to estimate the value m_a to match the radiation and diffraction forces on the float as the sum of gravitational force $\vec{F}_g$ and hydrostatic force $\vec{F}_s(\dot{z})$, the calculated value of the added mass coefficient is $m_a = 244kg$, the added mass term is investigated using $\vec{F}_a(\ddot{z}) = m_a\ddot{z}$. The results of the two methods are presented in Fig. 13. The differential equations were solved numerically (RKFOM) to validate the CFD results, the analytical solution is shown in Fig. 13, and the

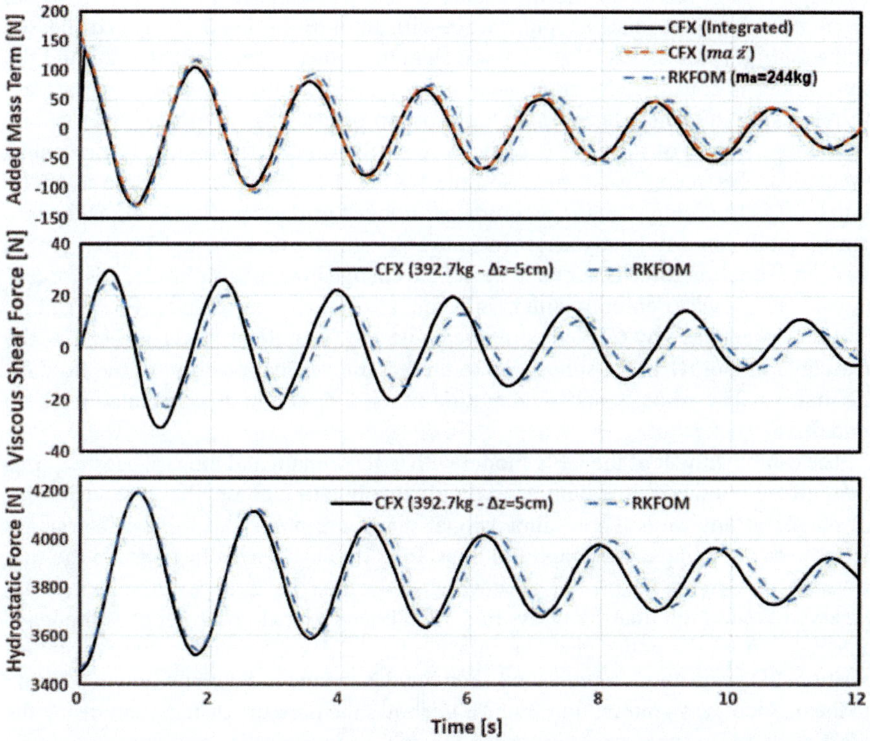

Fig. 13 Time-domain comparison for the hydrodynamic forces in CFD and numerical model at ($\Delta z = 5$ cm–MO-4)

simplified spring-damper system (Eq. 7) is solved based on the present boundary conditions, the effective mass M = 636.7 kg, damping coefficient $C_d = 154$N.s/m, the spring constant $K = 7697 N/m$, the excitation force is zero. The time step size is 0.002 s, and boundary conditions $\dot{z}_o = 0$, $\dot{z}_o = 0.05$ m.

6 Fluid–Structure Interaction

The numerical method in this research provides an effective tool that can be applied to understand the dynamic characteristics in different conditions under the influence of both regular and irregular waves. A case study including regular wave generation is considered to validate the presented numerical model, and to check its ability in different cases. The CFX model is subjected to the second type of boundary conditions presented in the fourth section of this research, NB with a 1/5 sloped surface is activated to reduce wave reflections, and the inlet plane is a movable surface that rotates about a fixed hinge placed at $h_o = 1.5$ m below SWL, the model height is $h_a = 1.5$ m for air region, and 3 m for water. Equations (8) and (9) define the flap position based on the time and vertical position z. The maximum stroke length is 0.15 m to produce a regular wave with amplitude (H = 0.11 m) as discussed in the author's last paper [17]. The wavelength in this experiment is 3 m, and the fluid model is considered a shallow water condition. A floating body with a mass of 392.7 kg (MO-5) is placed in an equilibrium position ($\dot{z_o} = 0$, $\Delta z = 0$) so that the center of mass of the float is coincide with SWL (Z = 0). Figure 14 presents a comparison between CFX results and the RKFOM model, the total time is 13.5 s with 0.002 s time step size, the recommendations of grid generation being taken into consideration, the solid body is free to slide only in heave motion, no PTO damping is used, an excitation force is added to detect the change in wave amplitude, this force is equivalent to a water column with height equal to the wave amplitude $A_f = 0.11$ m. The generated wave by CFX is compared with the theoretical WMT in Fig. 14, the analytical model RKFOM succeeded to predict the vertical position of the float for the first 6 s. The velocity and acceleration in the z-direction are presented for both models.

The establishment of the CFX model with a 3D domain and three phases requires great effort and more time, and accordingly, we obtain a comprehensive vision for all phases at any time. Three-dimensional views are presented for the interaction between the generated wave and the float, Fig. 15, at different instants. In the first lower position of the float (t = 3 s), at which the velocity is zero, perturbed waves are observed around the floating body; Fig. 17. The numerical beach reduces the wave reflections from the outlet plane as shown, after 12.5 s the float reaches its highest position, 3D snapshots are presented showing the change in float position depending on the incident wave interaction. Figure 17 shows the pressure distribution inside the NWT at 13.5 s, the pressure is increasing from the top according to the water column height (Fig. 17).

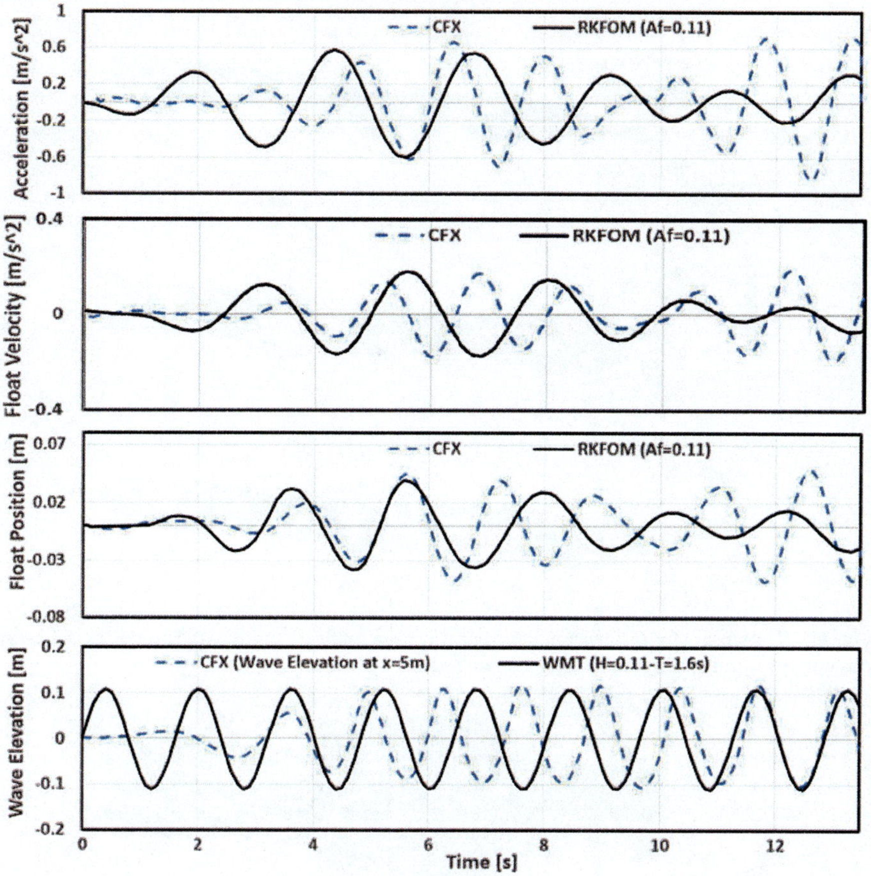

Fig. 14 Time-domain comparison for the kinematic variables of CFD and numerical model at (Regular wave–MO-5)

7 Bottom Shape Optimization

Two additional models were generated in this section with different designs, the flat- bottom shape of the cylindrical float is replaced by another conical shape and spherical shape as shown in Fig. 18. The two models have the same total mass m = 392.7 kg as the cylindrical float in the model (MO-4). The conical bottom shape model (CB: MO-6) has a diameter of 1 m, the height of the conical head is 0.5 m (90° taper angle, 45° base angle), and the immersed cylinder part height is 0.333 m to keep the same mass as the model (MO-4). The spherical bottom shape model (SB: MO-7) has a diameter of 1 m, the height of the hemispherical head is 0.5 m, and the immersed cylinder part height is 0.167 m to keep the same mass as the model (MO-4). Determining the optimal design required a full vision of the kinematics of

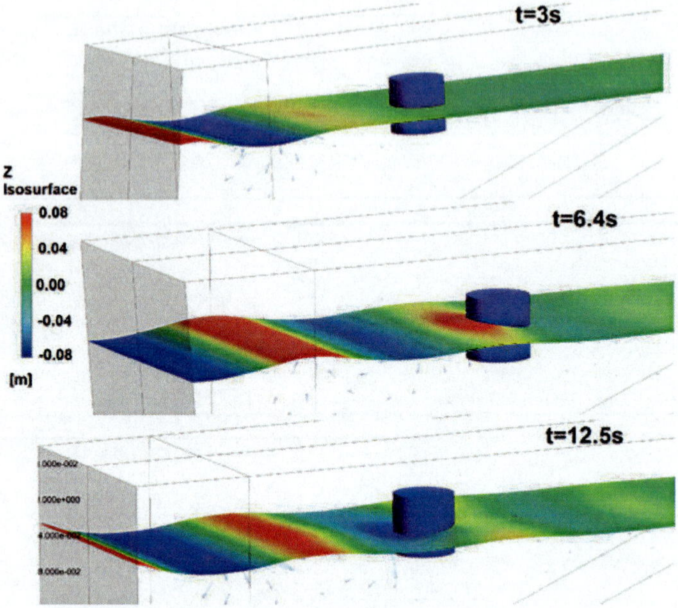

Fig. 15 3D snapshot for the generated regular wave against the float (MO–5) at **a** t = 3 s, **b** lowest position at t = 6.4 s, **c** highest position at t = 12.5 s

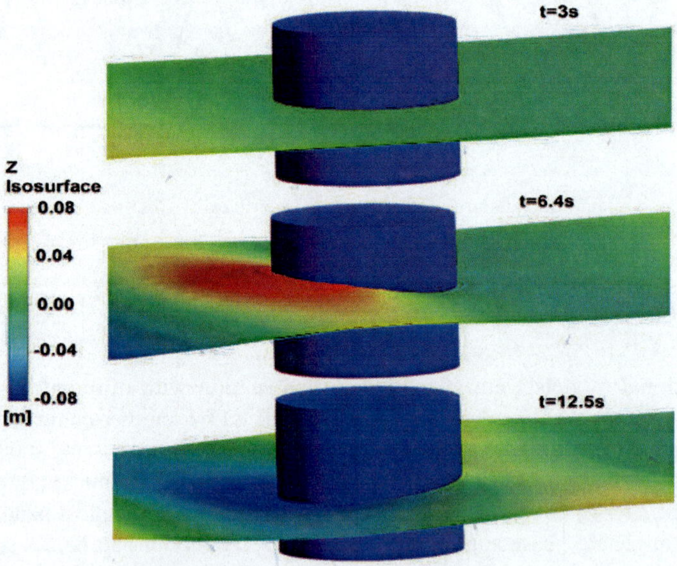

Fig. 16 3D snapshot for the perturbed wave around the float (MO–5) at **a** t = 3 s, **b** lowest position at t = 6.4 s, **c** highest position at t = 12.5 s

Fig. 17 Pressure gradient inside NWT at t = 13.5 s (MO–5)

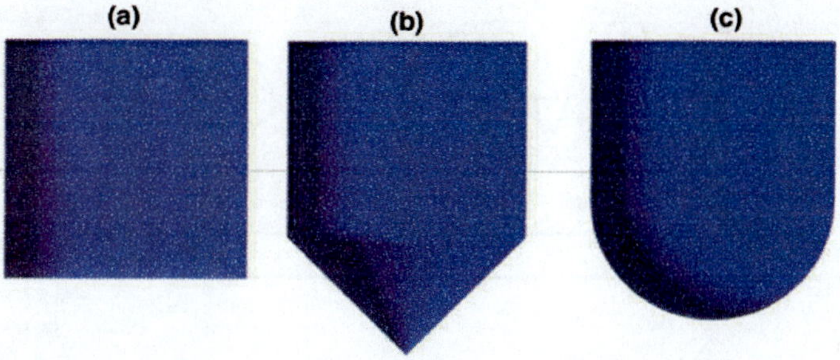

Fig. 18 The three models of floating body PAWEC with different bottom shapes: **a** Flat FB, **b** Conical CB, and **c** Spherical SB

each design. Figure 19 shows the comparison between the modified designs and the original cylindrical design in the time domain kinematics: position, velocity, and acceleration. It is shown that the modified designs started with better performance than the original cylindrical design in the first 6 s, after that time, a dispersion occurred in the SB design due to the increase in the damping coefficient leading to a slowdown of the float. The conical design produces a dynamic response similar to the original design, more results are required to get a clearer vision of the behavior of each design.

The first tough positions of the three designs occur simultaneously after 0.9 s, and the peak position after 1.74 s is investigated for each design. The vertical displacement between the two positions is defined as ΔS. Table 2 shows the hydrodynamic specifications for each design. The volume of the curved portion below the float $V(S)$ is a useful input source in the analytical solution, the second column represents the area of the contact surface with the fluid at the equilibrium position. The damping coefficient and spring constant in each case were investigated in each model from CFX data. The conical design has the highest surface area connected to the fluid; this explains the relative improvement in the conical design. The maximum displacement during the first 2 s is presented in the fourth column in Table 2, a slight relative advantage of the hemispherical design. After the first 6 s, the conical and cylindrical designs have the same response. The damping coefficient in both designs is rather similar to 154 and 156 kg/s, and it is 180 kg/s for the hemispherical design. The spring constant depends on the periodic time of the oscillation and total mass of

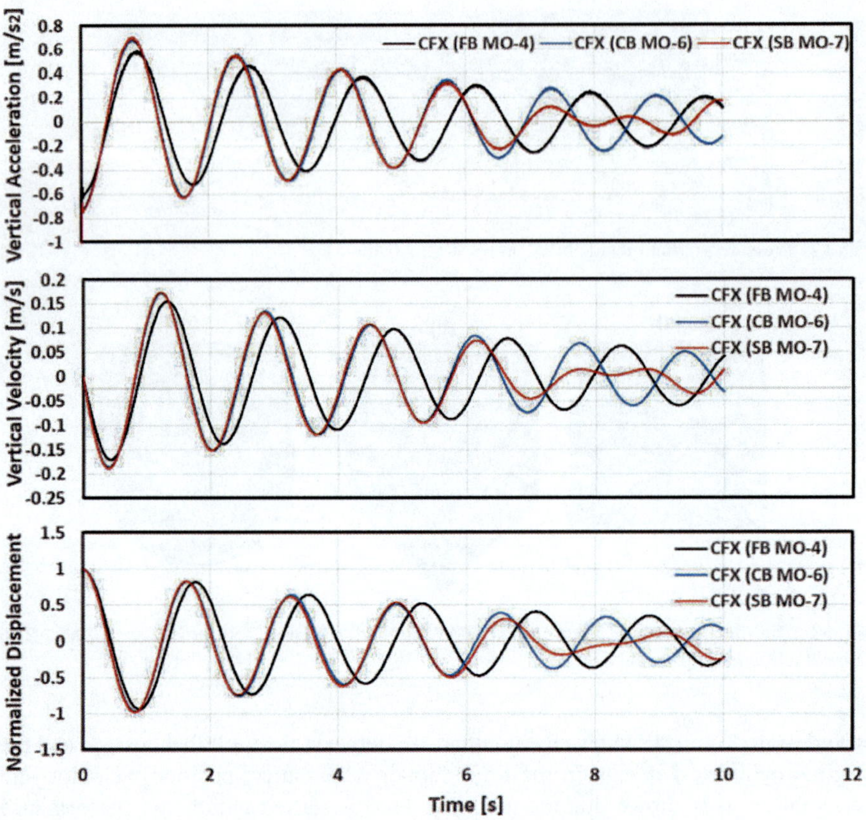

Fig. 19 Time-domain comparison for the kinematic variables in the free-fall test for the two modified models compared with model MO-4 ($\Delta z = 5$ cm–$\Delta t = 0.002$ s)

the system, which tends to increase in the spring constant for the modified designs because of the apparent lack of periodic time.

Table 2 Hydrodynamic specification of each model in the free-fall test (z = 0.05 m)

Model code	Surface area [m^2]	V(S) [m^3]	ΔS max. disp. [m] before 2 s	ΔS max. disp. [m] after 6 s	Damping Coefficient [kg/s]	Spring constant [N/m]
FB (MO-4)	1.5708	0	0.0876	**0.0439**	154	7697
CB (MO-6)	2.1562	0.1309	0.0898	**0.0429**	156	8906
SB (MO-7)	2.0945	0.2618	**0.09**	0.0399	180	9194

8 Conclusion

The numerical method in this research provides an effective tool that can be applied to understand the dynamic characteristics of a floating body in different conditions. Two models were investigated for this purpose, the CFD model using the CFX approach, and the numerical model solved analytically. The 3D form of the differential equations governing the dynamic behavior of the rigid body was analyzed and solved by both methods. The differential equations were simplified to a spring-damper system and solved numerically by RKFOM. The incompressible form of (URANS) is presented for the solution of the fluid model (air and water) in addition to two differential equations for the solution of VF by using the VOF method. The minimum cell size was tested for the grid independence; 0.05 m is recommended near the float and water surface, and the time step size independence test was performed for the minimum value equivalent to T/925 for the periodic time before the stability of the solution. Two case studies were generated for the validation of the numerical model. The numerical results agreed with the published experimental data in both models used. A hydrodynamic analysis was performed for each model to explain the effect of different forces on the solid body. The developed method in this research provided an effective tool to define the added mass coefficient in three ways, the dependency of the float dynamics was displayed in only heave motion. The developed model is used to investigate the hydrodynamics of the float under the action of the incident regular wave by using a flap-type wavemaker. Two additional models were generated: conical and hemispherical bottom shapes for the float. Valuable results are presented for design improvement, the conical design has a better performance based on the damping coefficient and hydrodynamic loads. The proposed numerical and analytical methods proved efficient and accurate in the results, and a practical study was presented to improve the performance of PAWEC. This paper introduced a numerical method to predict the hydrodynamic effects in minor wave amplitudes and body displacement, it is recommended in future work to evaluate the model under significant displacement in the same conditions.

References

1. A. ANSYS: AQWA Theory Manual, ed. Canonsburg, PA 15317, USA (2013)
2. C.F.D. Open: OpenFOAM User Guide, Vol. 2, no. 1, p. 47 (2011)
3. C.F.X. Ansys: Theory Guide. Ansys Inc (2015)
4. Welahettige, P., Vaagsaether, K.: Comparison of OpenFOAM and ANSYS Fluent. In: Proceedings of 9th EUROSIM Congress on Modelling Simulation, EUROSIM 2016, 57th SIMS Conference Simulation Model. SIMS 2016, vol. 142, pp. 1005–1012 (2018)
5. Windt, C., Davidson, J., Ringwood, J.V.: High-fidelity numerical modelling of ocean wave energy systems: a review of computational fluid dynamics-based numerical wave tanks. Renew. Sustain. Energy Rev. **93**, 610–630 (2018). (Elsevier Ltd)
6. Jin, S., Patton, R.J., Guo, B.: Enhancement of wave energy absorption efficiency via geometry and power take-off damping tuning. Energy **169**, 819–832 (2019)

7. Ringwood, J.V.: Implementation of an OpenFOAM numerical wave tank for wave energy experiments. In: Proceedings of 11th European Wave Tidal Energy Conference, pp. 1–10, (2015)
8. Shadman, M., Estefen, S.F., Rodriguez, C.A., Nogueira, I.C.M.: A geometrical optimization method applied to a heaving point absorber wave energy converter. Renew. Energy **115**, 533–546 (2018)
9. Sjökvist, L., et al.: Calculating buoy response for a wave energy converter—A comparison of two computational methods and experimental results. Theor. Appl. Mech. Lett. **7**(3), 164–168 (2017)
10. Ghasemi, A., Anbarsooz, M., Malvandi, A., Ghasemi, A., Hedayati, F.: A nonlinear computational modeling of wave energy converters: a tethered point absorber and a bottom-hinged flap device. Renew. Energy **103**, 774–785 (2017)
11. Büchner, A., Knapp, T., Bednarz, M., Sinn, P., Hildebrandt, A.: Loads and dynamic response of a floating wave energy converter due to regular waves from CFD simulations. In: Proceedings of International Conference Offshore Mechanics Arctic Engineering-OMAE, vol. 2, no. June 2016 (2016)
12. Devolder, B., Stratigaki, V., Troch, P., Rauwoens, P.: CFD simulations of floating point absorber wave energy converter arrays subjected to regular waves. Energies **11**(3), 1–23 (2018)
13. Rijnsdorp, D.P., Hansen, J.E., Lowe, R.J.: Simulating the wave-induced response of a submerged wave-energy converter using a non-hydrostatic wave-flow model. Coast. Eng. **140**(November 2017), 189–204 (2018)
14. Jin, S., Patton, R.J., Guo, B.: Viscosity effect on a point absorber wave energy converter hydrodynamics validated by simulation and experiment. Renew. Energy **129**, 500–512 (2018)
15. Zhu, L., Lim, H.C.: Hydrodynamic characteristics of a separated heave plate mounted at a vertical circular cylinder. Ocean Eng. **131**(January), 213–223 (2017)
16. Weller, S.D., Stallard, T.J., Stansby, P.K.: Experimental measurements of the complex motion of a suspended axisymmetric floating body in regular and near-focused waves. Appl. Ocean Res. **39**, 137–145 (2013)
17. Eldeen, A.S.S., El-Baz, A.M.R., Elmarhomy, A.M.: CFD modeling of regular and irregular waves generated by flap type wave maker. J. Adv. Res. Fluid Mech. Therm. Sci. **85**(2), 128–144 (2021)
18. Finnegan, W., Goggins, J.: Numerical simulation of linear water waves and wavestructure interaction. Ocean Eng. **43**, 23–31 (2012)
19. Guo, B., Patton, R., Jin, S., Gilbert, J., Parsons, D.: Nonlinear modeling and verification of a heaving point absorber for wave energy conversion. IEEE Trans. Sustain. Energy **9**(1), 453–461 (2018)
20. Versteeg, H.K., Malalasekera, W.: Computational fluid dynamics the finite volume method, pp. 1–26 (1995)
21. Liang, X., Yang, J., Li, J., Xiao, L., Li, X.: Numerical simulation of irregular wave-simulating irregular wave train. J. Hydrodyn. **22**(4), 537–545 (2010)
22. Davidson, J., Giorgi, S., Ringwood, J.V.: Linear parametric hydrodynamic models for ocean wave energy converters identified from numerical wave tank experiments. Ocean Eng. **103**, 31–39 (2015)
23. Paci, A., Gaeta, M.G., Antonini, A., Archetti, R.: 3D-numerical analysis of wave-floating structure interaction with OpenFOAM. In: Proceedings of International Offshore Polar Engineering Conference, vol. 2016, pp. 1034–1039 (2016). (-Janua)
24. Devolder, B., Rauwoens, P., Troch, P.: Numerical simulation of an array of heaving floating point absorber wave energy converters using openfoam. In: 7th International Conference Computing Methods, vol. 2017, pp. 777–788 (2017). (Mar. Eng. Mar. 2017, May, no. Oct)
25. Lal, A., Elangovan, M., Setup, A.P.: CFD Simulation and Validation of Flap Type, pp. 76–82 (2008)
26. Westphalen, J.: Extreme Wave Loading on Offshore Wave Energy Devices using CFD. Faculty of Science Technology (2011)

Using Ridge Regression to Estimate Factors Affecting the Number of Births. A Comparative Study

Mowafaq Muhammed Al-Kassab and Salisu Ibrahim

Abstract Ridge regression method is a biased regression analysis method that is used when the data suffer from multicollinearity problem between its explanatory variables, as the use of the least-squares method in analyzing such data will lead to incorrect estimates of the parameters of the regression coefficients and thus lead to incorrect predictions. For this reason, many methods for obtaining the ridge parameters in ridge regression were used previously in dealing with the problem of multicollinearity data, to overcome this problem, we used our formula in previous research named, the performance of the new ridge regression parameter and we applied it to a real data concerning factors affecting the number of births on a group of women who visited the health centers in Babel Governorate, as it was found that these data suffer from collinearity (multicollinearity problem). A comparison was made between this method and the other methods used previously, and results showed the effectiveness of this method as it gave better results than the methods used previously.

Keywords Multiple regression · Collinearity · Biased ridge regression · Unbiased ridge regression · La Soo regression

1 Introduction

Many researchers in the field of regression analysis have used many methods to solve the problem of multicollinearity, whereas regression analysis using the least-squares method does not succeed in finding correct results when performing a regression analysis of a data set that contains a multiplicity of the linear relationship between its explanatory variables. Among the methods that have been used in this field are the usual ridge regression method, which is one of the biased methods, as well as

M. M. Al-Kassab · S. Ibrahim (✉)
Department of Mathematics Education, Tishk International University-Erbil, Kurdistan Region, Iraq
e-mail: salisu.ibrahim@tiu.edu.iq

M. M. Al-Kassab
e-mail: mowafaq.muhammed@tiu.edu.iq

D. Zeidan et al. (eds.), *Mathematics and Computation*, Springer Proceedings in Mathematics & Statistics 418, https://doi.org/10.1007/978-981-99-0447-1_15

the unbiased ridge regression method, and the Lasso method, in addition to other methods suggested by many researchers: [1–27], among others. And recently, the authors in [27–29] presented a new method for the ridge regression suggesting that the ridge parameter k in the matrix $X^T X$ has three cases, the first (constant), the second (vector), and the third (matrix). We applied this method to data presented from a previous study [30] in addition to a comparison with other methods. These lead to the application of serious health conditions that make one vulnerable to Covid 19, see [31].

In this paper, we considered the new ridge regression to estimate factors affecting the number of births of a group of women who visited the health centers in Babel Governorate. As it was found that these data suffer from collinearity (multi-collinearity problem). A comparison was made between this method and the other methods used previously, and results showed the effectiveness of this method as it gave better results than the methods used previously. This paper is scheduled as: Sect. 2 provides the multicollinearity and regression analysis. The biased estimation methods and their applications are presented in Sects. 3 and 4, respectively. The conclusion follows in Sect. 5.

2 Multicollinearity and Regression Analysis

The problem of multicollinearity in the explanatory variables is one of the most important problems that researchers face when they apply regression analysis to a set of data for the purpose of building a good model that has the best characteristics such as estimating the parameters of regression, prediction, etc., as it appears among a number of explanatory variables. Linear relationships affect reaching incorrect results, and therefore it was necessary for researchers to find other ways to solve this problem, and among these methods is the ridge regression method developed by authors in [32]. The aim of this research is to reach the best regression analysis by detecting the problem of collinearity, estimating the best values of the regression function parameters, and arriving at the best regression model that leads to the best predictions.

2.1 Multicollinearity Problem

As we explained previously, what is the problem of multicollinearity, we are now exploring this problem as there are several ways to detect it, [33], the Conditional Index, Variance Inflation Factor (VIF) [34], in addition to other more methods to detect this problem. We will use the VIF method in our study, where it is used as a criterion to detect multicollinearity and determine the explanatory variable responsible for it.

The Variance Inflation Factors used by [35], are mathematically expressed in a mathematical form as follows:

$$VIF_j = \frac{1}{1 - R_{ij}^2} = \frac{1}{Tolerance}, j = 123, \ldots, p, \tag{1}$$

where p represents the number of explanatory variables, and $R_j{}^2$ is the coefficient of determination of the regression model of the independent variable x_j over the rest of the explanatory variables, and $0 \leq R_j^2 \leq 1$.

Authors in [34] suggested that if: $\mathrm{VIF}_j \geq 4$, that means there may be a collinearity problem between the explanatory variables, and we notice through our studies that this number may change according to the type of data. In some types of economic data, multicollinearity appears when $\mathrm{VIF}_j \geq 2$.

2.2 *Methods of Dealing with the Problem of Multicollinearity*

Many methods are presented in order to solve the problem of multicollinearity, and there are several methods to remove multicollinearity, and the authors in [27, 28] studied the application of latent roots regression to multicollinear data. The most important methods are:

1. Adding new data (increasing the sample size) to the original data, increasing the number of observations reducing the standard error, and reducing the effect of multicollinearity on the estimation of the parameters.
2. Excluding one of the variables that have a high correlation, and a problem arises if the excluded variable is important in explaining the change in the dependent variable, which leads in some cases to a process of bias in the estimates [36].
3. Using biased estimation methods such as Latent Roots Regression, Principal Components Regression, and Ridge Regression.

3 Biased Estimation Methods

When it comes to the application perspective, the authors in [37–41] make use of commutativity to study the relation and the sensitivity between systems, the idea can be extended to investigate the commutativity and sensitivity between the independent variables.

3.1 Latent Roots Regression

This method was proposed and developed by authors in [42] and others, and it includes the use of Eigenvalues and Eigenvectors in estimating the parameters of the regression model [43].

3.2 Principal Components Regression

It is proposed by authors in [44, 45] in 1957, and this method aims to convert correlated variables into non-correlated variables.

3.3 Ridge Regression

It was proposed by authors in [32] in (1970), and it involves adding a constant to the diagonal elements of the matrix X' X before taking the inverse of it. The ridge regression method is one of the most popular methods so the ridge regression coefficients will be:

$$\widehat{\beta}_R = (X'X + kI_P)^{-1} X'Y, \tag{2}$$

where k is known as the ridge parameter, this method provides a mean squares error less than the mean squares error obtained from the least-squares method. Many researchers have proposed new proposals and methods.

3.4 Lasso Regression Method

This method was proposed in 1982 in the geophysical literature, and then it was rediscovered and formulated independently by authors in [46] in (1996). The idea of this method is to minimize the total sum of squares of the random errors for the highest limit of the sum of the absolute values of the coefficients of the regression model so that it should be less than a fixed value, and this is a constraint (which is the upper limit). The method is based on solving the following equation:

$$\min\left\{\frac{1}{N}\sum_{i=1}^{N}\left(yi - \beta_0 - x_i^T\beta\right)^2\right\} \text{Subject to} \sum_{i=1}^{p}\left|\beta_j\right| \le t, \tag{3}$$

where t: represents the parameter that specifies the constraint amount and X is the matrix of the explanatory variables.

3.5 Bayesian Ridge Regression Models

In this method, the parameters of the regression model are estimated in the same way as the biased ridge regression with a difference, where the author entered the prior Information in the biased regression formula, so that the formula for the Bayesian ridge regression is [47] as follows:

$$\widehat{\beta}_R = (X'X + kI_P)^{-1}(X'Y + kJ), \tag{4}$$

where, J: represents the initial information vector.

3.6 Al-Kassab and Al-Awjar Method

This is a new method suggested by authors in [48], and it depends on the Eigenvalues and the Eigenvectors of the matrix X'X for finding the ridge parameter k when it is constant or matrix. For the case k constant, the vector of the estimated regression coefficients are:

$$\widehat{\beta}^*_R = (X^T X + \widehat{k}_R I_P)^{-1} X^T Y, \tag{5}$$

where

$$\widehat{k}_R = \frac{1}{\sum_{i=1}^{p} \lambda_i^2}[\lambda^T X^T Y - \lambda^T (X^T X)\lambda].$$

And λ is the vector of the Eigenvalues, For the Case k (K^*) is a diagonal matrix whose elements are either the diagonal elements or the vector–matrix of the matrix X^T X, the estimated regression coefficient

$$\widehat{\beta}^*_R = (X^T X + K^*)^{-1} X^T Y. \tag{6}$$

Based on the mean squares error criterion and on comparing this method with many other methods that were used by several researchers [49] as well as through simulation technique using the Monte Carlo method [50] the researchers concluded that this method for the case K^* diagonal matrix is the best.

4 Application

This study included data from Babil Governorate Health Department records of visits by women to health centers, as well as from the statistical forms for mother and child care in the Public Health Department of Primary Health Care, where a simple random sample consisting of 100 women was taken to study the factors affecting the number of children born, where this was considered as the response variable (dependent Y), while the other explanatory (independent) variables are: the woman's age (X_1), the age at marriage (X_2), the woman's educational attainment (X_3), the husband's academic achievement (X_4), the woman's weight (X_5). Women's use of contraceptives (X_6), smoking women (X_7), the husband's age (X_8), the husband's occupation (X_9), the period of marriage (X_{10}), the number of children who died (X_{11}), the number of hours of sports practice (X_{12}), injury with thyroid diseases (X_{13}), the number of hours a woman sleeps per day (X_{14}), taking medicines by the woman (X_{15}), the duration of breastfeeding (X_{16}), and the profession of the mother (X_{17}).

Table 1 illustrates the descriptive analysis of all variables used in this study. It can be seen that most variables provide similar results of measures of central tendency (mean and median values). It can be said that there is no peculiar observation that might affect these measures. This result is supported by the standard error of the mean. The presence of multicollinearity is investigated using correlation and it is presented in Table 2. *Means that there is a high correlation (dependency) between these pair of variables. This collinearity between the explanatory variables makes the least-squares method of estimation give unreal estimates. Thus, the parameter estimation methods that encounter the multicollinearity problem need to be employed to achieve the aim of the study.

Table 1 shows the descriptive analysis of the study data

Variable	N	Mean	Median	S.E. of the mean
X_1	100	31.030	30.000	0.838
X_2	100	18.640	18.500	0.321
X_3	100	4.440	4.000	0.206
X_4	100	5.080	5.000	0.170
X_5	100	68.520	67.000	0.894
X_6	100	1.1200	1.0000	0.0327
X_7	100	1.3800	1.0000	0.0488
X_8	100	34.170	33.500	0.829
X_9	100	1.4900	1.0000	0.0594
X_{10}	100	12.390	9.500	0.883
X_{11}	100	0.3400	0.0000	0.0655
X_{12}	100	3.850	3.000	0.291
X_{13}	100	1.1300	1.0000	0.0338
X_{14}	100	8.0900	8.0000	0.0740
X_{15}	100	1.7000	2.0000	0.0461
X_{16}	100	23.210	24.000	0.276
X_{17}	100	1.1700	1.0000	0.0428

Table 2 shows the correlation coefficients matrix between the explanatory variables of the study data

	X_1	X_2	X_3	X_4	X_5	X_6	X_7	X_8
X_2	0.047							
	0.640							
X_3	−0.383	0.329						
	0.000[a]	0.001[a]						
X_4	−0.271	0.250	0.706					
	0.006[a]	0.012[a]	0.000[a]					
X_5	0.638	0.024	−0.112	−0.003				
	0.000[a]	0.814	0.268	0.978				
X_6	−0.060	−0.161	0.026	−0.163	0.207			
	0.551	0.110	0.799	0.105	0.039[a]			
X_7	0.084	0.211	0.033	0.182	0.174	0.345		
	0.408	0.035[a]	0.745	0.069	0.083	0.000[a]		
X_8	0.918	0.060	−0.226	−0.184	0.671	−0.064	−0.016	
	0.000[a]	0.556	0.024[a]	0.067	0.000[a]	0.530	0.873	
X_9	−0.094	0.189	0.267	0.421	0.035	−0.098	−0.022	−0.001
	0.351	0.060	0.007[a]	0.000[a]	0.728	0.333	0.831	0.995
X_{10}	0.932	−0.319	−0.483	−0.348	0.597	0.001	0.003	0.850
	0.000[a]	0.001[a]	0.000[a]	0.000[a]	0.000[a]	0.991	0.978	0.000[a]
X_{11}	0.451	0.083	−0.037	0.039	0.354	−0.004	0.129	0.384
	0.000[a]	0.413	0.714	0.701	0.000[a]	0.970	0.201	0.000[a]
X_{12}	−0.198	0.130	−0.002	0.168	−0.008	−0.045	−0.052	−0.199
	0.049[a]	0.196	0.981	0.095	0.935	0.659	0.608	0.048[a]
X_{13}	−0.033	0.155	−0.010	−0.177	0.208	0.498	0.432	−0.058
	0.741	0.123	0.918	0.079	0.038[a]	0.000[a]	0.000[a]	0.563
X_{14}	0.083	0.133	−0.271	−0.159	−0.128	−0.338	−0.040	0.067
	0.414	0.188	0.006[a]	0.115	0.205	0.001[a]	0.695	0.510
X_{15}	−0.317	−0.074	0.087	−0.047	−0.499	−0.497	−0.566	−0.291
	0.001[a]	0.466	0.388	0.646	0.000[a]	0.000[a]	0.000[a]	0.003[a]
X_{16}	−0.047	0.090	0.042	0.009	−0.114	−0.320	0.030	−0.081
	0.642	0.375	0.677	0.927	0.261	0.001[a]	0.766	0.425
X_{17}	−0.066	0.052	0.510	0.426	0.156	0.214	0.026	0.166
	0.513	0.605	0.000[a]	0.000[a]	0.120	0.032[a]	0.796	0.100
	X_9	X_{10}	X_{11}	X_{12}	X_{13}	X_{14}	X_{15}	X_{16}
X_{10}	−0.158							
	0.117							

(continued)

Table 2 (continued)

	X_1	X_2	X_3	X_4	X_5	X_6	X_7	X_8
X_{11}	−0.147	0.398						
	0.145	0.000[a]						
X_{12}	0.084	−0.235	−0.360					
	0.407	0.019[a]	0.000[a]					
X_{13}	−0.019	−0.088	0.072	0.041				
	0.854	0.383	0.476	0.689				
X_{14}	−0.055	0.030	−0.231	0.161	−0.128			
	0.584	0.766	0.021[a]	0.109	0.204			
X_{15}	−0.048	−0.274	−0.328	0.102	−0.331	0.139		
	0.636	0.006[a]	0.001[a]	0.314	0.001[a]	0.167		
X_{16}	0.128	−0.077	−0.118	0.207	−0.062	0.204	−0.037	
	0.206	0.445	0.241	0.039[a]	0.539	0.042[a]	0.712	
X_{17}	0.344	−0.082	−0.208	0.094	−0.085	−0.049	−0.149	−0.005
	0.000[a]	0.418	0.037[a]	0.354	0.403	0.629	0.140	0.962

[a]Means that there is a high correlation (collinearity) between the explanatory variables, P-Value < 0.05

We see from Table 3, according to the mean squares criterion, that **Al-kassab and Al-Awjar** method for the case k vector is the best and the predicted regression model is:

$$y = 1.3689\, x_1 - 0.4294\, x_2 - 0.164\, x_5 - 0.2208\, x_7 - 0.1283\, x_9 - 0.2715\, x_{10} + 0.264\, x_{11} + 0.3177\, x_{13} + 0.1635\, x_{15} + 0.14301\, x_{16} + 0.29326\, x_{17}$$

5 Conclusion

This paper investigates the estimated factors affecting the number of births using ridge regression. Many methods for obtaining the ridge parameters in ridge regression were used previously in dealing with the problem of multicollinearity data. In this work, we consider the new ridge regression parameter and apply it to real data concerning factors that affect the number of births of a group of women who visited the health centers in Babel Governorate. This was because these data suffer from collinearity (multicollinearity problem). A comparison was made between this method and the other methods used previously, and the results showed the effectiveness of this method as it gave better results than the methods used previously. Furthermore, more advanced research techniques can be used to detect collinearity

Table 3 Shows the estimated values of the regression coefficients and the mean square errors for the five methods

Method of estimation	Ridge unbiased	Ridge biased K = 0.09	Lasso	O.L.S	Al-kassab and Al-Awjar for k		
					Constant	Vector	Matrix
X_1	−0.2456[a]	−0.1978[a]	−0.247[a]	0.914	0.1918[a]	1.3689[a]	0.1671
X_2	−0.0854[a]	−0.084[a]	−0.088[a]	−0.333	−0.146	−0.4294[a]	−0.03276
X_3	−0.135[a]	0.105[a]	−0.139[a]	−0.0855	−0.058	−0.06543	−0.09519
X_4	0.015	−0.033	0	−0.032	−0.065	−0.04432	−0.43353
X_5	−0.0751[a]	−0.061[a]	−0.077[a]	−0.167[a]	−0.043	−0.164[a]	1.95601
X_6	−2.719[a]	−2.575[a]	−2.729[a]	−0.312[a]	−0.162[a]	−0.33748	−0.42727
X_7	0.0034	1.1114[a]	0	−0.19[a]	−0.104[a]	−0.2208[a]	0.01158
X_8	−0.1134[a]	−0.081[a]	0.121[a]	−0.258	0.1151[a]	−0.45899	−1.29585
X_9	−0.0559	−0.727	0	−0.128	−0.079[a]	−0.1283[a]	−0.09989
X_{10}	−0.1823[a]	0.2065[a]	−0.187[a]	-[a]	0.2353[a]	−0.2715[a]	0.31395
X_{11}	0.9935[a]	1.2415[a]	0.939*	0.279*	0.1884[a]	0.264[a]	0.25106
X_{12}	0.0336	−0.079	0	−0.0915	−0.078[a]	−0.0976	−0.46485
X_{13}	−2.089[a]	−2.574[a]	−2.083*	0.306*	0.0873[a]	0.3177[a]	−0.10916
X_{14}	−0.194[a]	−0.179[a]	−0.189*	−0.0475	−0.012	−0.04971	0.11606
X_{15}	−1.2245*	1.1746[a]	−1.222*	0.185*	0.1295[a]	0.1635[a]	0.73307
X_{16}	0.0008	0.1687[a]	0	0.157*	0.12[a]	0.14301[a]	0.26925
X_{17}	0.0001	1.6545[a]	0	0.251*	0.0247	0.29326[a]	0.32144
MSE	1.7424	1.8088	1.5923	0.2063	0.0029	0.0022	0.00223
R^2	0.838	0.852	0.82	0.82707	0.82706	0.81743	0.8091

[a]Means that these regression coefficients are significant from zero, P-Value < 0.05.

[51, 52]. The data can be found in [53]. The results are validated with Minitab version 19.

Funding: No funding.

References

1. Hoerl, A.E., Kannard, R.W., Baldwin, K.F.: Ridge regression: some simulations. Commun. Stat.-Theory Methods **4**(2), 105–123 (1975). https://doi.org/10.1080/03610927508827232
2. McDonald, G.C., Galarneau, D.I.: A Monte Carlo evaluation of some ridge-type estimators. J. Am. Stat. Assoc. **70**(350), 407–416 (1975). https://doi.org/10.1080/01621459.1975.10479882
3. Hocking, R.R., Speed, F.M., Lynn, M.J.: A class of biased estimators in linear regression. Technometrics **18**(4), 425–437 (1976). https://doi.org/10.1080/00401706.1976.10489474

4. Lawless, J.F., Wang, P.: A simulation study of the ridge and other regression estimators. Commun. Stat. **A5**, 307–323 (1976)
5. Dempster, A.P., Laird, N.M., Rubin, D.B.: Maximum likelihood from incomplete data via the EM algorithm. J. Roy. Stat. Soc.: Ser. B (Methodol.) **39**(1), 1–22 (1977). https://doi.org/10.1111/j.2517-6161.1977.tb01600.x
6. Gunst, R.F., Mason, R.L.: Biased estimation in regression: an evaluation using mean squared error. J. Am. Stat. Assoc. **72**(359), 616–628 (1977). https://doi.org/10.1080/01621459.1977.10480625
7. Wichern, D.W., Churchill, G.A.: A comparison of ridge estimators. Technometrics **20**(3), 301–311 (1978). https://doi.org/10.1080/00401706.1978.10489675
8. Hemmerle, W.J., Brantle, T.F.: Explicit and constrained generalized ridge estimation. Technometrics **20**(2), 109–120 (1978). https://doi.org/10.1080/00401706.1978.10489634
9. Lawless, J.F.: Ridge and related estimation procedures: theory and practice: ridge and related estimation. Commun. Stat.-Theory Methods **7**(2), 139–164 (1978). https://doi.org/10.1080/03610927808827609
10. Golub, G.H., Heath, M., Wahba, G.: Generalized cross-validation as a method for choosing a good ridge parameter. Technometrics **21**(2), 215–223 (1979). https://doi.org/10.1080/00401706.1979.10489751
11. Gibbons, D.G.: A simulation study of some ridge estimators. J. Am. Stat. Assoc. **76**(373), 131–139 (1981). https://doi.org/10.1080/01621459.1981.10477619
12. Nordberg, L.: A procedure for determination of a good ridge parameter in linear regression. Commun. Stat.-Simul. Comput. **11**(3), 285–309 (1982). https://doi.org/10.1080/03610918208812264
13. Kibria, B.M.G.: On preliminary test ridge regression estimators for linear restriction in a regression model with non—normal disturbances. Commun. Stat. a25, 2349–2369 (1996) https://doi.org/10.1080/03610929608831843
14. Tracy, D.S., Singh, H.P., Singh, R.: Constructing an unbiased estimator of population mean in finite populations using auxiliary information. Stat. Pap. **40**(3), 363–368 (1999). https://doi.org/10.1007/BF02929882
15. Wakgari, N., Wencheko, E.: Risk factors of neonatal mortality in Ethiopia. Ethiopian J. Health Dev. **27**(3), 192–199 (2013)
16. Kim, J.H., Bell, G.A., Bitton, A., Desai, E.V., Hirschhorn, L.R., Makumbi, F., ... Schwarz, D.: Health facility management and primary health care performance in Uganda. BMC Health Serv. Res. **22**(1), 1–11 (2022). https://doi.org/10.1186/s12913-022-07674-3
17. Khalaf, G., Shukur, G.: Choosing ridge parameter for regression problems. Commun. Stat.—Theory Methods **34**(5), 1177–1182 (2005). https://doi.org/10.1081/STA-200056836
18. Zhang, J., Ibrahim, M.: A simulation study on SPSS ridge regression and ordinary least squares regression procedures for multicollinearity data. J. Appl. Stat. **32**(6), 571–588 (2005). https://doi.org/10.1080/02664760500078946
19. Alkhamisi, M., Khalaf, G., Shukur, G.: Some modifications for choosing ridge parameters. Commun. Stat.-Theory Methods **35**(11), 2005–2020 (2006). https://doi.org/10.1080/03610920600762905
20. Alkhamisi, M.A., Shukur, G.: A Monte Carlo study of recent ridge parameters. Commun. Stat.—Simul. Comput.® **36**(3), 535–547 (2007). https://doi.org/10.1080/03610910701208619
21. Mardikyan, S., Cetin, E.: Efficient choice of biasing constant for ridge regression. Int. J. Contemp. Math. Sci. **3**(11), 527–547 (2008)
22. Batah, F.S.M., Ramanathan, T.V., Gore, S.D.: The efficiency of modified jackknife and ridge type regression estimators: a comparison. Surv. Math. Appl. **3**, 111–122 (2008)
23. Muniz, G., Kibria, B.G.: On some ridge regression estimators: an empirical comparisons. Commun. Stat.—Simul. Comput.® **38**(3), 621–630 (2009). https://doi.org/10.1080/03610910802592838
24. Al-Hassan, Y.M., Al-, M.M.: A Monte Carlo Comparison between Ridge and. Appl. Math. Sci. **3**(42), 2085–2098 (2009)

25. Månsson, K., Shukur, G., Golam Kibria, B.M.: A simulation study of some ridge regression estimators under different distributional assumptions. Commun. Stat.-Simul. Comput. **39**(8), 1639–1670 (2010). https://doi.org/10.1080/03610918.2010.508862
26. Dorugade, A.V., Kashid, D.N.: Alternative method for choosing ridge parameter for regression. Appl. Math. Sci. **4**(9), 447–456 (2010)
27. Al-kassab, M.M., Dilnas S.Y.: Application of latent roots regression to multicollinear data. J. Adv. Res. Comput. Sci. Eng. **4**(12), 1–11 (2017). https://doi.org/10.53555/nncse.v4i12.393
28. Al-Kassab, M.M., Adnan, M.A., Dilnas, S.Y.: Studying the Effect of Some Variables on the economic Growth Using Latent Roots Method (11), 1–10, (2019)
29. Al-Kassab, M.M., Al-Awjar, M.Q.: A Monte Carlo comparison between least squares and the new ridge regression parameters. J. Adv. Appl. Stat. **62**(1), 97–105 (2020). https://doi.org/10.17654/AS062010097
30. Dormann, C.F., Elith, J., Bacher, S., Buchmann, C., Carl, G., Carré, G., Lautenbach, S.: Collinearity: a review of methods to deal with it and a simulation study evaluating their performance. Ecography **36**(1), 27–46 (2013). https://doi.org/10.1111/j.1600-0587.2012.07348.x
31. Ibrahim, S., Al-Kassab, M.M.: Using Linear Regression Analysis to Study the Recovery Cases of COVID 19 in Erbil, Kurdistan Region. Drugs Cell Therap. Hematol. **10**(1), 1226–1239 (2021)
32. Horel, A.E., Kannard, R.W.: Ridge regression: biased estimation for non-orthogonal problems. Technometrics **12**, 55–68 (1970). https://doi.org/10.1080/00401706.1970.10488634
33. Farrar, D.E., Glauber, R.: Multicollineanty in regression analysis: the problem revisited. Rev. Econ. Stat. **49**, 92–107 (1967). https://doi.org/10.2307/1937887
34. Gunst, R.F., Mason, R.L.: Regression Analysis and its Application. Marcel Dekker Inc. New York, U.S.A., vol. 114 (1980). https://doi.org/10.1201/9780203741054
35. Goldstein, M., Smith, A.F.: Ridge-type estimators for regression analysis. J. Roy. Stat. Soc.: Ser. B (Methodol.) **36**(2), 284–291 (1974). https://doi.org/10.1111/j.2517-6161.1974.tb01006.x
36. Draper, D.C.: Rank-based robust analysis of linear models. University of California, Berkeley (1981)
37. Ibrahim, S., Koksal, M.E.: Commutativity of sixth-order time-varying linear systems. Circuits Syst. Signal Process. **40**(10), 4799–4832 (2021). https://doi.org/10.1007/s00034-021-01709-6
38. Ibrahim, S., Koksal, M.E.: Realization of a fourth-order linear time-varying differential system with nonzero initial conditions by cascaded two second-order commutative pairs. Circuits Syst. Signal Process. **40**(6), 3107–3123 (2021). https://doi.org/10.1007/s00034-020-01617-1
39. Ibrahim, S., Rababah, A.: Decomposition of fourth-order euler-type linear time-varying differential system into cascaded two second-order euler commutativepairs. Complexity **1–9**, 2022 (2022)
40. Ibrahim, S.: Numerical approximation method for solving differential equations. Eurasian J. Sci. Eng. **6**(2), 157–168 (2020). https://doi.org/10.23918/eajse.v6i2p157
41. Rababah, A., Ibrahim, S.: Weighted G^1-Multi-Degree reduction of Bézier curves. Int. J. Adv. Comput. Sci. Appl. **7**(2), 540–545 (2016)
42. Hawkins, D.M.: On the investigation of alternative regressions by principal component analysis. Appl. Statist. **22**, 275–286 (1973). https://doi.org/10.2307/2346776
43. Gunst, R.F., Webster, J.T., Mason, R.L.: A comparison of least squares and latent root regression estimators. Technometrics **18**(1), 75–83 (1976). https://doi.org/10.1080/00401706.1976.10489403
44. Kendall, M. G.: Multivariate analysis: London. Charles Griffin & Co. 2, (1975).
45. Hotelling, H.: The relations of the newer multivariate statistical methods to factor analysis. Brit. J. Stat. Psychol. **10**, 69–79 (1957). https://doi.org/10.1111/j.2044-8317.1957.tb00179.x
46. Tibshirani, R.: Regression shrinkage and selection via the lasso. J. R. Stat. Soc. Ser. B (Methodol.) 267–288 (1996)
47. Swindel, B.F.: Good ridge estimators based on prior information. Commun. Stat.-Theory Methods **5**(11), 1065–1075 (1976). https://doi.org/10.1080/03610927608827423

48. Al-Kassab, M.M., Al-Awjar, M.Q.: Performance of the new ridge regression parameters. J. Adv. Math. Comput. Sci. **34**(5), 1–9 (2019). https://doi.org/10.9734/JAMCS/2019/v34i530225
49. Al-Hassan, Y.: Performance of New Ridge regression estimators. J. Assoc. Arab Univ. Basic Appl. Sci. **9**, 23–26 (2010). https://doi.org/10.1016/j.jaubas.2010.12.006
50. Al-Hassan, Y.M.M., Al-Kassab, M.M.: A comparison between ridge and principal components regression methods using simulation technique. Al Al-Bayt University (2000)
51. Ibrahim, S., Al-Kassab, M. M., Al-Awjar, M. Q.: Investigating multicollinearity in factors affecting number of born children in Iraq. J. Alg. Stat. **13**(2), 967–974 (2022). https://doi.org/10.52783/jas.v13i2.250
52. Ibrahim, S., Al-Kassab, M. M., Al-Awjar, M.Q.: Factors affecting number of born children in Iraq. J. Alg. Stat. **13**(2), 955–966 (2022). https://doi.org/10.52783/jas.v13i2.248
53. Majid, S., Alsabah, S.: Parameters Estimation of the Multiple Linear Regression (2015)

Discrete Maximum Principle and Positivity Certificates for the Bernstein Dual Petrov–Galerkin Method

Tareq Hamadneh, Jochen Merker, and Gregor Schuldt

Abstract In this article, we discuss the validity of the discrete maximum principle for the spectral method called Bernstein-Dual-Petrov-Galerkin method [4] in case of a uniformly elliptic second-order linear partial differential equation (PDE) in divergence form and corresponding Dirichlet boundary values problems on simply connected domains, which have no holes and are therefore diffeomorphic to a cube.

Keywords Discrete maximum principle · Positivity certificates · Bernstein dual Petrov–Galerkin method · Numerical analysis

1 Introduction

Consider Poisson's equation for homogeneous Dirichlet boundary conditions

$$-\Delta u = f \text{ in } \Omega\,, \quad u = 0 \text{ on } \partial\Omega \tag{1}$$

on a bounded domain $\Omega \subset \mathbb{R}^N$ with boundary $\partial\Omega$ piecewise sufficiently smooth. By the weak maximum principle, $f \geq 0$ implies $u \geq 0$ in Ω for the solution u of (1). [1] proved an analogous discrete maximum principle (DMP) for a finite difference (FD) discretization of (1), and [2] presented a DMP suitable for both finite element (FE) and FD discretizations by providing a practically convenient set of sufficient conditions on matrix blocks implying validity of a DMP. While these conditions imply for a piecewise linear triangular FE discretization of (1) that the inverse of the stiffness matrix is positive under the interior edge condition (the sum of the two

T. Hamadneh
Al Zaytoonah University of Jordan, Airport Rd., Amman, Jordan

J. Merker (✉) · G. Schuldt
Leipzig University of Applied Sciences, PF 30 11 66, 04251 Leipzig, Germany
e-mail: jochen.merker@htwk-leipzig.de

G. Schuldt
e-mail: gregor.schuldt@htwk-leipzig.de

D. Zeidan et al. (eds.), *Mathematics and Computation*, Springer Proceedings in Mathematics & Statistics 418, https://doi.org/10.1007/978-981-99-0447-1_16

angles opposite to every interior edge is $\leq \pi$), the DMP may fail for certain meshes [3].

In this article, we discuss the validity of the discrete maximum principle for the spectral method called Bernstein-Dual-Petrov-Galerkin method [4] in case of a uniformly elliptic second-order linear partial differential equation (PDE) in divergence form and corresponding Dirichlet boundary values problems on simply connected domains $\Omega \subset \mathbb{R}^N$, which have no holes and are therefore diffeomorphic to a cube. This numerical method combines two advantages, the exponential fast convergence of a spectral method in the interior of Ω for analytic data, and the good approximation properties of Bernstein polynomials [5]. Particularly, the latter will allow us to certify the positivity of numerical solutions. For Helmholtz equation subject to homogeneous Neumann boundary conditions and Bernstein Bubnov-Galerkin method, see [6].

1.1 Outline

In Sect. 2, we provide basic information about linear elliptic PDEs in divergence form; about positivity, the maximum principle and the comparison principle for classical and weak solutions; about Bernstein polynomials and the induced dual polynomials resp. the modal basis functions; and about different certificates of non-negativity resp. positivity. In Sect. 3, the Bernstein dual Petrov–Galerkin method is formulated for general linear elliptic PDEs in divergence form on a domain diffeomorphic to a cube. In the main Sect. 4 of this paper, we discuss algebraic and functional discrete maximum principles for this method as well as Bernstein certificates of non-negativity resp. positivity for the approximate solution in a way, which easily generalizes to dual Petrov–Galerkin methods with arbitrary non-negative basis functions. Hereby, we provide numerical examples and a summary that concludes the article.

2 Preliminaries

2.1 Linear Elliptic PDEs

A linear second order differential operator $L = \sum_{ij} a_{ij}\partial_{x_i}\partial_{x_j} + \sum_i b_i\partial_{x_i} + c$ with possibly spatially varying measurable coefficients a_{ij} (w.l.o.g. symmetric), b_i, c on a bounded domain $\Omega \subset \mathbb{R}^N$ is said to be strictly elliptic if there exists a constant $\lambda > 0$, such that

$$\sum_{ij} a_{ij}(x)\xi_i\xi_j \geq \lambda|\xi|^2 \text{ for every } \xi \in \mathbb{R}^N \text{ and a.e. } x \in \Omega. \tag{2}$$

In this article, we consider linear second-order differential operators

$$Lu := \operatorname{div}(a\nabla u) - cu \tag{3}$$

in divergence form, and require uniform ellipticity in the sense that the coefficients $a \in L^\infty(\Omega, \operatorname{Sym}(n \times n))$, $c \in L^\infty(\Omega, \mathbb{R})$ are at least bounded and the symmetric matrices $a = (a_{ij})$ are positive definite with smallest eigenvalue bounded away from zero on Ω by a constant $\lambda > 0$. Note that (3) can be rewritten under the additional assumption $a \in C^1(\Omega, \operatorname{Sym}(n \times n))$ in the above general form, and uniform ellipticity for bounded coefficients just means validity of (2). The corresponding Dirichlet boundary value problem reads as

$$-Lu = -\operatorname{div}(a\nabla u) + cu = f \text{ in } \Omega\,, \quad u = g \text{ on } \partial\Omega\,, \tag{4}$$

where we assume that the right-hand side (r.h.s., or inhomogeneity) satisfies at least $f \in (H_0^1(\Omega))^*$ and the boundary data satisfies $g \in H^{1/2}(\partial\Omega)$. In the case $c \geq 0$, the bilinear form

$$B(u, v) := \int_\Omega (a\nabla u) \cdot \nabla v + cuv\,dx \tag{5}$$

induced by $-L$ on the Sobolev space $H_0^1(\Omega)$ is coercive in the sense that $B(u, u) \geq \lambda\|\nabla u\|_2^2$ and bounded due to $|B(u, v)| \leq (\|a\|_\infty + C^2\|c\|_\infty)\|\nabla u\|_2\|\nabla v\|_2$ with the constant C in Sobolev's inequality $\|u\|_2 \leq C\|\nabla u\|_2$ for $u \in H_0^1(\Omega)$. Hence, by Lax–Milgram, (4) has a unique solution $u \in H^1(\Omega)$ satisfying the Dirichlet boundary condition in the sense that $u - g \in H_0^1(\Omega)$ for an extension of g from trace space $H^{1/2}(\partial\Omega)$ to $H^1(\Omega)$.

2.2 *Positivity, Maximum and Comparison Principle*

Definition 1 We say for a linear operator L on a space of functions on Ω that

- weak positivity holds if the validity of $-Lu \geq 0$ in Ω and $u \geq 0$ on $\partial\Omega$ implies the non-negativity $u \geq 0$ in Ω (resp. we say that strong positivity holds, if either $u \equiv 0$ or $u > 0$ in Ω is implied),
- the weak maximum principle holds if the validity of $Lu \geq 0$ in Ω implies that a non-negative maximum is attained by u on the boundary $\partial\Omega$ (resp. we say that the strong maximum principle holds, if either u is constant equal to its maximum or a non-negative maximum is attained by u only on the boundary $\partial\Omega$ and not inside Ω), or equivalently the weak minimum principle holds if the validity of $-Lu \geq 0$ in Ω implies that a non-positive minimum is attained by u on the boundary $\partial\Omega$ (resp. we say that the strong minimum principle holds, if either u is constant equal to its minimum or a non-positive minimum is attained by u only on the boundary $\partial\Omega$ and not inside Ω),

- the weak comparison principle holds if the validity of $-Lu_i = f_i$ in Ω, $u_i = g_i$ on $\partial\Omega$, $i = 1, 2$, implies for data $f_1 \geq f_2$, $g_1 \geq g_2$, that $u_1 \geq u_2$ holds in Ω (resp. the strong comparison principle holds, if either $u_1 \equiv u_2$ or $u_1 > u_2$ in Ω is implied).

To obtain the equivalence of minimum principle and maximum principle claimed in this definition, just substitute $-u$ for u. Note that Definition 1 is not fully precise, because no function space for u is provided. If $u \in C^2(\Omega) \cap C(\overline{\Omega})$ is assumed, then Definition 1 is called positivity, maximum principle or comparison principle for classical solutions. Positivity, maximum principle or comparison principle for weak solutions $u \in H^1(\Omega)$ requires a more precise definition indicated at the end of this subsection.

The weak minimum principle (and thus also the weak maximum principle) implies weak positivity (for classical solutions): If $-Lu \geq 0$ in Ω and $u \geq 0$ on $\partial\Omega$, then u cannot be negative in Ω, because by the weak minimum principle, a non-positive minimum would be attained on $\partial\Omega$ in contradiction to $u \geq 0$ on $\partial\Omega$, and hence $u \geq 0$ in Ω. Similarly, the strong minimum (or maximum) principle implies strong positivity.

Denote by $u = u_+ - u_-$ the decomposition of a function u into its positive part $u_+ := \max(u, 0) \geq 0$, and its negative part $u_- := \max(-u, 0) \geq 0$. For the convenience of the reader, we provide here proof of two well-known facts.

Lemma 1 *Every uniformly elliptic linear second-order differential operator L in divergence form* (3) *with $c \geq 0$ satisfies weak positivity (of weak solutions).*

Proof Assume that $-Lu = f \geq 0$ in Ω and $u \geq 0$ on $\partial\Omega$. Test $-Lu = f$ by u_- (note that $u \geq 0$ on $\partial\Omega$ implies $u_- = 0$ on $\partial\Omega$, thus u_- can act as a test function) and use $c \geq 0$ to obtain $\lambda\|u_-\|_{L^2}^2 \leq \int_\Omega (a\nabla u_-) \cdot \nabla u_- + c\,|u_-|^2\,dx = -\int_\Omega (a\nabla u) \cdot \nabla u_-\,dx - \int_\Omega c\,uu_-\,dx = \langle Lu, u_-\rangle = -\langle f, u_-\rangle \leq 0$ (because $f, u_- \geq 0$), i.e. $u_- \equiv 0$ and thus $u = u_+ \geq 0$ in Ω.

Remark 1 Weak positivity still holds for slightly negative c, as long as c is larger than the negative $-\lambda_1$ of the smallest Dirichlet eigenvalue λ_1 of $-L$.

Lemma 2 *For the differential operator L given by* (3) *with $c \geq 0$, weak positivity implies the weak maximum principle (for classical solutions).*

Proof Let L satisfy weak positivity, and let u be such that $Lu \geq 0$ and the maximum M of u on $\partial\Omega$ is non-negative, i.e. $M \geq 0$. Then $-L(M - u) \geq Lu \geq 0$ in Ω (as $c \geq 0$) and $M - u \geq 0$ on $\partial\Omega$. Thus, by weak positivity of L we have $M - u \geq 0$ in Ω and hence $u \leq M$ in Ω, i.e. a non-negative maximum of u is attained on the boundary $\partial\Omega$.

Similarly, for L given by (3) with $c \geq 0$, strong positivity implies the strong maximum principle (for classical solutions), and the weak (strong) comparison principle is equivalent to weak (strong) positivity, just put $u := u_1 - u_2$.

Yet, to show the strong maximum principle (or strong positivity) for weak solutions is more demanding. In its precise form, inequalities $f \geq 0$ for functionals $f \in (H_0^1(\Omega))^*$ have to be interpreted in the functional sense that $\langle f, v \rangle \geq 0$ for every $v \in H_0^1(\Omega)$ with $v \geq 0$ a.e. in Ω, and maximum / minimum have to be replaced by essential supremum/infimum. The strong maximum principle then states for a uniformly elliptic linear second order differential operator L in divergence form (3) with $c \geq 0$ (or even $c > -\lambda_1$) that $Lu \geq 0$ and $\sup_B u = \sup_\Omega u \geq 0$ for some closed ball B with positive radius in Ω imply $u \equiv \sup_\Omega u$ a.e. constant in Ω. For a proof of this strong maximum principle for weak solutions, the weak Harnack inequality can be applied to show that if $\sup_B u = \sup_\Omega u =: M \geq 0$ for some closed ball B with positive radius r in Ω, then $u \equiv M$ is constant on an even larger ball in Ω with radius greater than r, and a covering argument then allows to conclude $u \equiv M$ a.e. in Ω, see, e.g. [7, Theorem 8.19].

2.3 Bernstein Polynomials and Their Duals

As we aim to discuss in this article the Bernstein dual Petrov–Galerkin method for the approximation of a solution of (4), we need to discuss Bernstein polynomials and their dual polynomials over the N-dimensional unit cube $(0, 1)^N$. To formulate the Bernstein expansion of a real polynomial N-variate function, we use component-wise comparisons and arithmetic operations on *multiindices* $i = (i_1, \ldots, i_n) \in \mathbb{N}_0^N$. For $x \in \mathbb{R}^N$ and a multiindex $i \in \mathbb{N}_0^N$, its monomial is $x^i := x_1^{i_1} \ldots x_N^{i_N}$. Using compact notation $D = (D_1, \ldots, D_N) \in \mathbb{N}_0^N$, we put $\sum_{i=0}^{D} := \sum_{i_1=0}^{D_1} \cdots \sum_{i_N=0}^{D_N}$ and $\binom{D}{i} := \prod_{\mu=1}^{N} \binom{D_\mu}{i_\mu}$. An N-variate polynomial function u is expressed in *monomial form* as

$$u(x) = \sum_{i=0}^{d} a_i x^i, \tag{6}$$

where $d = (d_1, \ldots, d_n)$, and can be represented in *Bernstein form* by

$$u(x) = \sum_{j=0}^{D} u_j^{(D)} \, S_j^{(D)}(x), \quad x \in (0, 1)^N. \tag{7}$$

In (7), the jth Bernstein polynomial of degree $D \geq d$ is

$$S_j^{(D)}(x) = \binom{D}{j} x^j (1-x)^{D-j}, \ x \in (0, 1)^N \tag{8}$$

and can be considered as tensor product of univariate Bernstein polynomials, i.e. $S_j^{(D)}(x) = S_{j_1}^{D_1}(x_1) \cdot \ldots \cdot S_{j_N}^{D_N}(x_N)$. Moreover, the *Bernstein coefficients* $u_j^{(D)}$ of degree D are given analytically in terms of the coefficients a_i in (6) by the formula

$$u_j^{(D)} = \sum_{i=0}^{j} \frac{\binom{j}{i}}{\binom{D}{i}} a_i, \; 0 \leq j \leq D. \tag{9}$$

Conversely, the following theorem from the literature provides a way of converting a polynomial from the Bernstein form to the monomial form.

Theorem 1 ([8, Theorem 3.3]) *Let $u(x)$ be a polynomial in Bernstein form of any degree D. Then its monomial form is*

$$u(x) = \sum_{i=0}^{D} a_i x^i,$$

where

$$a_i = \sum_{j=0}^{i} (-1)^{i-j} \binom{D}{i} \binom{i}{j} u_j^{(D)}, \;\; 0 \leq i \leq D.$$

We highlight two important properties of Bernstein polynomials, namely, the *end-point interpolation property*

$$u_j^{(D)} = u\left(\frac{j}{D}\right),$$

for $0 \leq j \leq D$ with $j_k \in \{0, D_k\}$, $k = 1, \ldots, N$, and the *enclosing property* [9]

$$\min_{0 \leq j \leq D} u_j^{(D)} \leq u(x) \leq \max_{0 \leq j \leq D} u_j^{(D)},$$

for all $x \in (0, 1)^N$. The parameters $D = (D_1, \ldots, D_N) \in \mathbb{N}_0^N$ determine in the mesh-free Bernstein dual Petrov–Galerkin method the resolution of the approximation in each coordinate direction, in analogy as the number of subdivisions of each interval in $(0, 1)^N$ determines how fine a rectangular mesh is in a FE method. In the following, for the convenience of the reader, we usually suppress the upper index containing the fixed degree D and mention it only, where it is helpful for understanding.

The dual polynomials to (one-dimensional) Bernstein polynomials in $L^2(0, 1)$ have been introduced by [10], who found a recurrence relation involving Legendre polynomials. We denote by $\tilde{\Psi}_i^{(D)}$ the N-variate dual Bernstein polynomials of degree $D \in \mathbb{N}_0^N$ determined by biorthogonality

$$\int_{(0,1)^N} S_j^{(D)} \tilde{\Psi}_i^{(D)} \, d\vec{x} = \delta_{ij}\,. \tag{10}$$

The coefficients c_{ij} in the decomposition $\tilde{\Psi}_i = \sum_{j=0}^{D} c_{ij} S_j$ are explicitly known (in one dimension) due to [11], see also [4, (2.4), (2.5)]. Here, we focus on linear combinations Ψ_i, $1 \leq i \leq D-1$, of the dual Bernstein polynomials called modal basis functions, which vanish on the boundary of the unit cube $(0, 1)^N$. In contrast to [4, Proposition 1], we use an index shifted by one and a scaling so that relation [4, (2.8)] becomes

$$\Psi_i(x) := \tilde{a}_i \tilde{\Psi}_{i-1}(x) + \tilde{\Psi}_i(x) + \tilde{b}_i \tilde{\Psi}_{i+1}(x) \tag{11}$$

for $1 \leq i \leq D-1$ with $\tilde{a}_i = \frac{D-i+2}{2(i+1)}$, $\tilde{b}_i = \frac{i+2}{2(D-i+1)}$. Particularly, the space $\Pi_0^{(D)}$ of polynomial functions of maximal degree $D \in \mathbb{N}_0^N$ in each coordinate vanishing on the boundary of $(0, 1)^N$ is not only spanned by S_j, $1 \leq j \leq D-1$, but also by Ψ_i, $1 \leq i \leq D-1$.

2.4 Certificates of Non-negativity Resp. Positivity

In a broad sense, every identity that gives immediate proof of non-negativity resp. positivity for a (multivariate) real function u is considered to be a certificate of non-negativity resp. positivity, and thus there are many different certificates of non-negativity resp. positivity. For example, a sum of squares (SOS) certificate of non-negativity for a real polynomial function u on $\mathbb{R}^N$ is a representation $u = \sum_{k=1}^{m} p_k^2$ of u by sums of squares of polynomials $p_1, \ldots, p_m$ on $\mathbb{R}^N$. However, not every real polynomial function $u \geq 0$ can be decomposed into a sum of squares of polynomials, e.g. the Motzkin polynomial $u(x, y) := x^4y^2 + x^2y^4 + 1 - 3x^2y^2$ on $\mathbb{R}^2$. On $\mathbb{R}_+^N = (0, \infty)^N$, a certificate of positivity for a real polynomial function u of degree $d \in \mathbb{N}_0^N$ in monomial form $u(x) = \sum_{i=0}^{d} a_i x^i$ is the validity of $a_i > 0$ for all $0 \leq i \leq d$. However, again not every real polynomial function $u > 0$ on $\mathbb{R}_+^N$ has positive monomial coefficients.

In this article, we consider Bernstein certificates: As the Bernstein basis polynomials $S_j^{(D)}$ in (8) are by construction non-negative on the closed unit cube $[0, 1]^N$ and positive in its interior $(0, 1)^N$, i.e. $S_j^{(D)}(x) > 0$ for all $x \in (0, 1)^N$ and $0 \leq j \leq D$, where 0 denotes the multiindex with all components equal to zero, for a real polynomial function u on $\mathbb{R}^N$ the validity of $u_j^{(D)} \geq 0$ for all $0 \leq j \leq D$ implies non-negativity $u \geq 0$ on $[0, 1]^N$. Thus, the non-negativity of Bernstein coefficients is a certificate of non-negativity on $[0, 1]^N$. Further, if additionally $u_j^{(D)} > 0$ for one $0 \leq j \leq D$, then $u > 0$ on $(0, 1)^N$, hence $u_j^{(D)} \geq 0$ for all $0 \leq j \leq D$ and $u_j^{(D)} \neq 0$ is a certificate of positivity on $(0, 1)^N$. However, again there exist positive polynomials over a box which have non-positive Bernstein coefficients, as shown in the following example.

Example 1 Consider the polynomial $u(x) = 7x^2 - 3x + 5$. It is immediate to check that u is positive on $[-1, 1]$, but the list of Bernstein coefficients $(u_j^{(2)}) = (15, -2, 9)$ has a negative value. However, the polynomial u has a certificate of positivity at degree 3 on $[-1, 1]$, since $(u_j^{(3)})) = (15, 3.6, 1.6, 9)$.

3 Bernstein Dual Petrov–Galerkin Method

In using the Bernstein dual Petrov–Galerkin method [4] for solving (4) on a simply connected bounded domain $\Omega \subset \mathbb{R}^N$, the first step is to fix the degree $D = (D_1, \ldots, D_N) \in \mathbb{N}_0^N$ as a parameter which determines the resolution of the approximation in each coordinate, and a diffeomorphism $T : \Omega \to (0, 1)^N$ which extends continuously to a homeomorphism $T : \overline{\Omega} \to [0, 1]^N$. As basis functions, then the transformed Bernstein polynomials $S_j^{(D)}(T(x))$ are used (where in the following we usually suppress the upper index containing the fixed degree D), and particularly we search for an approximation of the solution of the form

$$u(x) = g(x) + \sum_{j=1}^{D-1} u_j S_j(T(x)) \,. \tag{12}$$

with an extension g of the boundary data to $H^1(\Omega)$. Note that the sum vanishes on the boundary $\partial\Omega$ due to $1 \le j \le D - 1$, where 1 denotes the multiindex with all components equal to one. Putting the ansatz (12) into the weak formulation $B(u, v) = \langle f, v\rangle$ with the bilinear form B from (5) and using as test functions v the transformed modal basis functions $\Psi_i \circ T$ vanishing on the boundary of the unit cube $(0, 1)^N$, which are induced via (11) by the dual Bernstein polynomials $\tilde{\Psi}_i$, we obtain a linear system

$$\mathbf{A}\vec{u} = \vec{b} \tag{13}$$

with stiffness matrix $\mathbf{A} = \left(\int_\Omega (a\nabla(S_j \circ T)) \cdot \nabla(\Psi_i \circ T) + c(S_j \circ T)(\Psi_i \circ T)\, d\vec{x}\right)$ and r.h.s. $\vec{b} = (\langle f, \Psi_i \circ T\rangle - B(g, \Psi_i \circ T))$.

Example 2 If $\Omega := (\underline{x}_1, \overline{x}_1) \times \cdots \times (\underline{x}_N, \overline{x}_N)$ is a general open n-dimensional cube with

$$\underline{x}_\mu < \overline{x}_\mu, \ \mu = 1, \ldots, N$$

identified by the usual affine linear transform $T : \Omega \to (0, 1)^N$ with the unit cube, then the transformed jth Bernstein polynomial of degree $D \in \mathbb{N}_0^N$ is

$$S_j^{(D)}(T(x)) = \binom{D}{i}(x - \underline{x})^i(\overline{x} - x)^{D-i} w(\Omega)^{-D}, \tag{14}$$

where $w(\Omega) = (\overline{x}_1 - \underline{x}_1, \ldots, \overline{x}_N - \underline{x}_N)$ denotes the width of intervals.

Due to the chain rule and transformation formula, system (13) is identical with the system obtained from (4) on the unit cube $(0, 1)^N$ with coefficients replaced by $\left(\frac{1}{|\det DT|}(DT)a(DT)^*\right) \circ T^{-1}$ resp. $\left(\frac{1}{|\det DT|}c\right) \circ T^{-1}$. Note that this transformation does not change the uniform ellipticity and boundedness of the coefficients. Therefore, we can restrict our discussion to the case of the unit cube.

In the special case, where after the transformation the coefficients of (4) on the unit cube $(0, 1)^N$ are given by constants $a = \mathbf{I}$ equal to the identity matrix $\mathbf{I}$ and $c = 0$, and the boundary value $g = 0$ vanishes, we are in case of Poisson's equation (1) on the unit cube $(0, 1)^N$ subject to homogeneous Dirichlet boundary conditions. For this case, in dimension $N = 2$ it can be seen from [4, 3.2.1] that the stiffness matrix $\mathbf{A}$ in (13) can be written as tensor product $\mathbf{A} = A \otimes B + B \otimes A$ with sparse matrices $A = \left(\int_0^1 S_j'(x)\Psi_i'(x)\,dx\right)$, $B = \left(\int_0^1 S_j(x)\Psi_i(x)\,dx\right)$, containing one-dimensional integrals of univariate (dual) Bernstein polynomials and their first derivatives. Hereby, sparsity follows from the validity of the three-term recurrence relation [12], [4, (2.3)],

$$S_j'(x) = (D - j + 1)S_{j-1}(x) - (D - 2j)S_j(x) - (j + 1)S_{j+1}(x) \qquad (15)$$

for the one-dimensional derivative of Bernstein polynomials.

Similarly, [4, Corollary 1] offers a five-term recurrence relation for the derivative of one-dimensional dual Bernstein polynomials, and together with the three-term recurrence relation for the derivative of one-dimensional Bernstein polynomials and biorthogonality (10) sparsity of the one-dimensional matrices A, B and thus of the stiffness matrix $\mathbf{A} = A \otimes B + B \otimes A$ follows. Yet, even in case of Poisson's equation (1) on the unit cube $(0, 1)^N$ subject to homogeneous Dirichlet boundary conditions, this stiffness matrix does not have a non-negative inverse.

Example 3 In the case $N = 2$, $D = (6, 6)$, the one-dimensional matrices $A, B \in \mathbb{R}^{5,5}$ are given by

$$A = \begin{pmatrix} 58 + \frac{2}{7} & -10 - \frac{2}{7} & -10 - \frac{2}{7} & -1 - \frac{5}{7} & 0 \\ -9 & 28.5 & -1.5 & -9 & -1.5 \\ -8.64 & -1.44 & 21.76 & -1.44 & -8.64 \\ -1.5 & -9 & -1.5 & 28.5 & -9 \\ 0 & -1 - \frac{5}{7} & -10 - \frac{2}{7} & -10 - \frac{2}{7} & 58 + \frac{2}{7} \end{pmatrix}$$
$$B = \begin{pmatrix} 1 & 1 & 0 & 0 & 0 \\ \frac{5}{8} & 1 & \frac{5}{8} & 0 & 0 \\ 0 & 1 & 1 & 1 & 0 \\ 0 & 0 & \frac{7}{4} & 1 & \frac{1}{4} \\ 0 & 0 & 0 & 4 & 1 \end{pmatrix} \qquad (16)$$

Note that in contrast to [4] due to our scaling (11) here A has the symmetry $a_{6-i,6-j} = a_{i,j}$ for all $i, j \in \{1, 2, 3, 4, 5\}$ and B has values 1 on the diagonal. Further, while A^{-1} is positive, the stiffness matrix $\mathbf{A} = A \otimes B + B \otimes A \in \mathbb{R}^{25,25}$ has negative entries.

Remark 2 Moreover, in the general situation of an arbitrary domain and arbitrary coefficients, the stiffness matrix **A** is neither sparse nor has a non-negative inverse $\mathbf{A}^{-1}$.

While we will see in the next section that a non-negative inverse of the stiffness matrix **A** is related to algebraic positivity, for the formulation of the algebraic discrete maximum principle let us fix an approximation by Bernstein polynomials of the right- hand side $f(x) = \sum_{j=0}^{D} f_j S_j(T(x))$ and of the (extended) boundary data $g(x) = \sum_{j=0,D} g_j S_j(T(x))$, where $j = 0, D$ means that at least one index satisfies $j_k \in \{0, D_k\}$ for $k = 1, \ldots, N$, and let us extend the system (13) to

$$\begin{pmatrix} \mathbf{A} & \mathbf{A}^{\partial} \\ \mathbf{0} & \mathbf{I} \end{pmatrix} \begin{pmatrix} \vec{u} \\ \vec{g} \end{pmatrix} = \begin{pmatrix} \mathbf{M}\vec{f} \\ \vec{g} \end{pmatrix} \tag{17}$$

with $\mathbf{A}^{\partial} = \left(\int_{\Omega}(a\nabla(S_j \circ T)) \cdot \nabla(\Psi_i \circ T) + c\,(S_j \circ T)(\Psi_i \circ T)\, d\vec{x}\right)_{1\le i\le D-1,\, j=0,D}$, and $\mathbf{M} := \left(\int_{\Omega}(S_j \circ T)(\Psi_i \circ T)\, d\vec{x}\right)_{1\le i\le D-1,\, 0\le j\le D}$. In contrast to (13), where the right hand side f and the boundary data g are hidden within the vector $\vec{b} = (\langle f, \Psi_i \circ T\rangle - B(g, \Psi_i \circ T))$, in (17) the dependence of the algebraic solution on the coefficients of the data is made explicit.

Remark 3 Note that the discussion of a discrete maximum principle by [2] is based on the extended linear system (17). Not only in case of FEM or the Bernstein dual Petrov–Galerkin method, but in an arbitrary Galerkin method also, such an extension can always be obtained by projecting boundary data on a finite dimensional space spanned by boundary basis functions.

4 Discrete Maximum Principle and Positivity Certificates

Discrete positivity and the discrete maximum/minimum principle can be valid in two different ways, algebraically or functionally. Let us begin our discussion by defining algebraic discrete positivity and the algebraic discrete maximum/minimum principle.

Definition 2 We say that the extended linear system (17) satisfies

- algebraic weak discrete positivity, if data $\vec{f}, \vec{g} \ge 0$ implies a solution $\vec{u} \ge 0$ (resp. algebraic strong discrete positivity, if either $(\vec{u}, \vec{g}) = 0$ or $\vec{u} > 0$ is implied),
- the algebraic weak discrete maximum principle, if data $\vec{f} \le 0$ implies that a non-negative maximal component of $(\vec{u}, \vec{g})$ already occurs in $\vec{g}$ (resp. the algebraic strong discrete maximum principle, if it is implied that either all components of $(\vec{u}, \vec{g})$ are identical or a non-negative maximal component of $(\vec{u}, \vec{g})$ occurs only in $\vec{g}$),

or equivalently the algebraic weak discrete minimum principle, i.e. data $\vec{f} \geq 0$ implies that a non-positive minimal value of $(\vec{u}, \vec{g})$ already occurs in $\vec{g}$ (resp. the algebraic strong discrete minimum, if it is implied that either all components of $(\vec{u}, \vec{g})$ are identical or a non-positive minimal component of $(\vec{u}, \vec{g})$ occurs only in $\vec{g}$)

The following discrete analogue of Lemma 2 allows to restrict our attention to algebraic discrete positivity.

Lemma 3 *If the matrix in* (17) *has non-negative row sums and* $\mathbf{M} \geq 0$*, then algebraic weak (resp. strong) discrete positivity implies the algebraic weak (resp. strong) discrete maximum principle.*

Proof Let (17) satisfy algebraic weak positivity, let $\vec{f} \leq 0$, and assume that the maximal value M in a component of $\vec{g}$ is non-negative. Then

$$\mathbf{A}(M\vec{1} - \vec{u}) + \mathbf{A}^{\partial}(M\vec{1} - \vec{g}) \geq -\mathbf{A}\vec{u} - \mathbf{A}^{\partial}\vec{g} = -\mathbf{M}\vec{f} \geq 0 \tag{18}$$

due to non-negative row sums $\mathbf{A}\vec{1} + \mathbf{A}^{\partial}\vec{1} \geq 0$ and non-negativity $\mathbf{M} \geq 0$, and $M\vec{1} - \vec{g} \geq 0$. Thus, by algebraic weak positivity of (17) we have $M\vec{1} - \vec{u} \geq 0$ and hence $\vec{u} \leq M\vec{1}$, i.e. a non-negative maximal component of $(\vec{u}, \vec{g})$ already occurs in $\vec{g}$. In case of algebraic strong positivity, we obtain in the last step of the former proof either $(M\vec{1} - \vec{u}, M\vec{1} - \vec{g}) = 0$ or $M\vec{1} - \vec{u} > 0$, i.e. either all components of $(\vec{u}, \vec{g})$ are identical with M or $\vec{u} < M\vec{1}$ so that a non-negative maximal component of $(\vec{u}, \vec{g})$ occurs only in $\vec{g}$.

The following Lemma allows to prove algebraic weak (resp. strong) positivity and hence the algebraic weak (resp. strong) discrete maximum principle for several Galerkin methods, e.g. piecewise linear FEM on simplices under the interior edge condition.

Lemma 4 *If the matrix in* (17) *has non-negative row sums, if* $\mathbf{A}$ *is a non-singular* M*-matrix,* $\mathbf{A}^{\partial} \leq 0$ *and* $\mathbf{M} \geq 0$*, then algebraic weak discrete positivity holds. If moreover, at least one-row sum is positive,* $\mathbf{A}$ *is irreducible and* $\mathbf{M}$ *is surjective, then even algebraic strong discrete positivity holds.*

Proof By [13], $\mathbf{A}$ is a non-singular M-matrix iff $\mathbf{A}^{-1}$ exists and is non-negative. Therefore, the matrix in (17) hat the inverse $\begin{pmatrix} \mathbf{A}^{-1} & -\mathbf{A}^{-1}\mathbf{A}^{\partial} \\ 0 & \mathbf{I} \end{pmatrix}$, which is non-negative due to $\mathbf{A}^{\partial} \leq 0$. Moreover, under the additional assumptions, $\mathbf{A}$ is an irreducibly diagonally dominant real square matrix with strictly positive diagonal and non-positive off-diagonal entries, and thus $\mathbf{A}^{-1} > 0$ by [14].

If the stiffness matrix $\mathbf{A}$ of the linear system (13) does not have a non-negative inverse $\mathbf{A}^{-1}$, then algebraic (weak) positivity is not valid. In fact, if $\mathbf{A}^{-1}$ has an element $a_{ij}^{-1} < 0$, then for $\vec{g} := \vec{0}$ the jth component of $\mathbf{M}\vec{f} \geq 0$ can be chosen so large that $\vec{u} = \mathbf{A}^{-1}\mathbf{M}\vec{f}$ has a negative ith component. This is the case for Bernstein dual Petrov–Galerkin method by Example 3 and Remark 2. Yet, on the one hand, for some data, it is still possible to obtain a Bernstein certificate of non-negativity (resp. positivity) for the approximate solution.

Definition 3 For data $\vec{f}$ and $\vec{g} \geq 0$, we say that the extended linear system (17) allows a Bernstein certificate of non-negativity for the approximate solution u given by (12), if $\vec{u} \geq 0$ holds (resp. a Bernstein certificate of positivity, if additionally $\vec{u} \neq 0$ holds).

On the other hand, instead of algebraic weak discrete positivity $\vec{u} \geq 0$ it may be possible to conclude merely functional weak discrete positivity, where the weaker conclusion $u \geq 0$ for the approximate solution (12) with Bernstein coefficients $\vec{u}$ is drawn.

Definition 4 We say that the extended linear system (17) satisfies

- algebraic-functional weak discrete positivity, if data $\vec{f}, \vec{g} \geq 0$ implies $u \geq 0$ for the approximate solution (12) (resp. algebraic-functional strong discrete positivity, if either $u = 0$ or $u > 0$ is implied).
- functional-functional weak discrete positivity, if data $f, g \geq 0$ implies $u \geq 0$ for the approximate solution (12) (resp. functional-functional strong discrete positivity, if either $u = 0$ or $u > 0$ is implied).

Note that in this double notation, the first point in Definition 2 would be algebraic-algebraic discrete positivity (while for functional-algebraic weak discrete positivity the weakest conditions $f, g \geq 0$ would need to imply the strongest condition $\vec{u} \geq 0$). It is not astonishing that many spectral methods and particularly the Bernstein dual Petrov–Galerkin method do not satisfy algebraic discrete positivity, because signs of the coefficients $\langle f, \Psi_i \circ T \rangle$ do not say much about non-negativity resp. positivity of the function f, as the test functions Ψ_i itself are sign-changing. Therefore, functional discrete positivity is more important, however, also more difficult to verify, because the convex cone of non-negative functions is not finitely generated. To clarify, whether functional weak (resp. strong) positivity is valid, or for which data we can provide a Bernstein certificate of non-negativity (resp. positivity), let us apply the general theory of convex cones.

Definition 5 A subset $K \subset X$ of a real vector space X is called

- a cone, if $x \in K$ implies $rx \in K$ for every $r \geq 0$,
- convex, if $x, y \in K$ imply $\lambda x + (1 - \lambda)y \in K$ for every $\lambda \in [0, 1]$.

Thus, a convex cone $K \subset X$ is a subset such that $ax + by \in K$ for every linear combination with non-negative coefficients $a, b \geq 0$. If X is a Banach space, then a convex cone K is called closed if it is closed w.r.t. norm topology. For example, $\{\vec{u} \in \mathbb{R}^n \mid \vec{u} \geq 0\}$ is a closed convex cone in $\mathbb{R}^n$, and the subset $\{u \in H^1(\Omega) \mid u \geq 0 \text{ a.e.}\}$ is a closed convex cone in $H^1(\Omega)$.

Definition 6 The polar cone of a convex cone $K \subset X$ in a real Banach space is the subset $K^o := \{f \in X^* \mid \forall x \in K : \langle f, x \rangle \leq 0\}$ of the dual space X^*.

The polar of a convex cone is automatically closed in X^*, and for a closed convex cone $K \subset X$ in a reflexive Banach space $X \cong X^{**}$ the bipolar theorem $(K^o)^o = K$ holds. This fact allows the following characterization of data b such that the solution u of $Au = b$ lies in K.

Theorem 2 *Let $K \subset X$ be a closed convex cone in a reflexive Banach space X, and let $A : X \to X^*$ a bijective continuous linear map. Then the image $A(K)$ is a closed convex cone in X^*, and with the polar cone $P := (A(K))^o \subset X^{**} = X$ of $A(K)$ the following characterization holds:*

The unique solution $u \in X$ of $Au = b$ satisfies $u \in K$ iff $b \in P^o$.

Proof Eventually, $A(K)$ is a convex cone, and by the open mapping / closed graph theorem the image $A(K)$ is closed. By the bipolar theorem, $b \in P^o$ is equivalent to $b \in ((A(K))^o)^o = A(K)$, and thus $P^o \ni b = Au$ is equivalent to $u \in K$ by the uniqueness of solutions.

While the former using the bipolar theorem is more formal, the Theorem shows that those data b, for which $u \in K$ can be concluded, form a closed convex cone (namely P^o). As a consequence, the following results about positivity certificates and functional discrete positivity are valid.

Corollary 1 *Precisely for data $\vec{f}$ and $\vec{g} \geq 0$ satisfying $\mathbf{A}^{-1}\mathbf{M}\vec{f} - \mathbf{A}^{-1}\mathbf{A}^{\partial}\vec{g} \geq 0$ (resp. additionally a strict inequality > 0 in at least one component) a Bernstein certificate of non-negativity (resp. positivity) for the approximate solution can be provided.*

Proof Let $K := \{\begin{pmatrix} \vec{u} \\ \vec{g} \end{pmatrix} \mid \vec{u} \geq 0,\ \vec{g} \geq 0\}$, then for $\vec{g} \geq 0$ we have $\begin{pmatrix} \mathbf{M}\vec{f} \\ \vec{g} \end{pmatrix} \in P^o = \begin{pmatrix} \mathbf{A} & \mathbf{A}^{\partial} \\ \mathbf{0} & \mathbf{I} \end{pmatrix}(K)$ iff $\mathbf{A}^{-1}\mathbf{M}\vec{f} - \mathbf{A}^{-1}\mathbf{A}^{\partial}\vec{g} \geq 0$ (and if additionally a strict inequality > 0 holds in at least one component, then $\vec{u} \neq 0$ and thus the approximate solution u is positive).

Although this precise characterization involves the inverse, the inequalities in Corollary 1 define a convex cone, which can be practically used to verify positivity of an approximate solution without solving the linear system (17).

Example 4 In the special case of Poisson's problem (1) on the unit cube $(0, 1)^N$ with homogeneous Dirichlet boundary conditions, $N = 2$, $D = (6, 6)$, and with the matrices $A, B \in \mathbb{R}^{5,5}$ provided in Example 3, the stiffness matrix is given by $\mathbf{A} = A \otimes B + B \otimes A \in \mathbb{R}^{25,25}$. Its inverse (for better readability scaled and rounded) $100\mathbf{A}^{-1}$ reads as

$$
\begin{pmatrix}
0.9 & -0.2 & -0.5 & 0.5 & -0.1 & -0.2 & 0.2 & -0 & -0.2 & 0.1 & -0.5 & -0 & 1.1 & -0.5 & 0 & 0.5 & -0.2 & -0.5 & 0.4 & -0 & -0.1 & 0.1 & 0 & -0 & -0 \\
-0.1 & 0.5 & 1 & -0.8 & 0.2 & 0.2 & -0 & -0.7 & 0.7 & -0.1 & -0 & 0.8 & -0.8 & 0.5 & -0.1 & -0.1 & -0.1 & 0.4 & -0.3 & 0.1 & 0.1 & -0.1 & 0.1 & -0 & 0 \\
-0.4 & 1.5 & -1.3 & 1.1 & -0.1 & -0.2 & -1 & 2 & -0.9 & 0.1 & 1.2 & -1.2 & 1 & -0.6 & 0.2 & -0.6 & 0.6 & -0.3 & 0.3 & -0.1 & 0 & 0.2 & -0.2 & 0.1 & -0.0 \\
1.6 & -3.6 & 2.9 & -0.5 & 0.2 & -1.3 & 2.9 & -2 & 0.6 & -0.1 & -0.7 & 1.9 & -2.2 & 1.2 & -0.1 & 0.7 & -1.4 & 1.3 & -0.4 & 0 & -0 & 0 & 0.2 & -0.1 & 0.0 \\
-5.5 & 10.1 & -6.9 & 1.3 & 0.4 & 6.9 & -7.9 & 2.3 & 0.1 & -0 & -1.8 & -4.8 & 8.4 & -2.7 & 0.1 & -0.4 & 4.6 & -5 & 1 & 0.2 & 0 & -0.4 & 0.3 & 0.2 & -0.1 \\
-0.1 & 0.2 & -0 & -0.1 & 0.1 & 0.5 & -0 & 0.8 & -0.1 & -0.1 & 1 & -0.7 & -0.8 & 0.4 & 0.1 & -0.8 & 0.7 & 0.5 & -0.3 & -0 & 0.2 & -0.1 & -0.1 & 0.1 & 0.0 \\
0.1 & -0 & -0.4 & 0.4 & -0 & -0 & 1.7 & 0.2 & -0.2 & 0.1 & -0.4 & 0.2 & -0.6 & 0.4 & 0 & 0.4 & -0.2 & 0.4 & -0.1 & -0 & -0 & 0.1 & 0 & -0 & 0.0 \\
-0.1 & -0.5 & 1.2 & -0.5 & 0.1 & 1 & 0.3 & -0.2 & 0.5 & 0.1 & -1.2 & -0.8 & 2.4 & -0.9 & -0 & 0.7 & 0.5 & -1.2 & 0.5 & 0.1 & -0.1 & 0 & 0.1 & 0 & -0.0 \\
-0.8 & 1.7 & -1.1 & 0.4 & -0 & 0.9 & -1.7 & 0.6 & 1.1 & 0.2 & 0 & 0.8 & -0.8 & 0.4 & -0.2 & -0.4 & 0.3 & 0.3 & -0.2 & 0.2 & 0.1 & -0.2 & 0.2 & 0 & -0.0 \\
4.2 & -4.8 & 1.3 & 0.2 & -0 & -8.1 & 5 & 2.7 & -1.6 & 0.6 & 6.1 & 0.9 & -6.1 & 0.6 & 0.6 & -1.8 & -2.9 & 4.3 & 0.1 & -0.4 & 0.1 & 0.9 & -1 & 0.1 & 0.1 \\
-0.4 & -0.2 & 1.2 & -0.6 & 0 & 1.5 & -1 & -1.2 & 0.6 & 0.2 & -1.3 & 2 & 0.9 & -0.3 & -0.2 & 1.1 & -0.9 & -0.6 & 0.3 & 0.1 & -0.1 & 0.1 & 0.2 & -0.1 & -0.0 \\
-0.1 & 1 & -1.2 & 0.7 & -0.1 & -0.5 & 0.3 & -0.8 & 0.5 & 0 & 1.2 & -0.2 & 2.4 & -1.2 & 0.1 & -0.5 & 0.5 & -0.9 & 0.5 & 0 & 0.1 & 0.1 & -0 & 0.1 & -0.0 \\
1.2 & -1.7 & 1.6 & -1 & 0.3 & -1.7 & -1.2 & 3.5 & -1.3 & -0 & 1.6 & 3.5 & -5.4 & 2.9 & -0 & -1 & -1.3 & 2.9 & -1.1 & 0 & 0.3 & -0 & -0 & 0 & 0.1 \\
-0.9 & 2.5 & -3 & 1.6 & -0.2 & 0 & 1.1 & -1.2 & 0.6 & -0.3 & 0.7 & -3.8 & 4.5 & -1.3 & 0.8 & 0 & 1.1 & -1.4 & 0.8 & -0.3 & -0.1 & 0.2 & -0.2 & 0.2 & 0.0 \\
-1 & -5.7 & 9.2 & -3.4 & 0.2 & 8.7 & 1.4 & -8.8 & 0.8 & 0.8 & -12 & 3 & 4.8 & 2.2 & -0.9 & 5.3 & 0.9 & -4.7 & -0.4 & 0.7 & -0.6 & -1 & 1.6 & -0.3 & -0.0 \\
1.6 & -1.3 & -0.7 & 0.7 & -0.1 & -3.7 & 2.9 & 1.9 & -1.4 & 0 & 2.9 & -2 & -2.2 & 1.3 & 0.2 & -0.5 & 0.6 & 1.2 & -0.4 & -0.1 & 0.2 & -0.1 & -0.1 & 0 & 0.0 \\
-0.8 & 0.9 & 0 & -0.4 & 0.1 & 1.7 & -1.7 & 0.8 & 0.3 & -0.2 & -1.1 & 0.6 & -0.8 & 0.3 & 0.2 & 0.4 & 1.1 & 0.4 & -0.2 & 0 & -0 & 0.2 & -0.2 & 0.2 & -0.0 \\
-0.9 & 0 & 0.7 & 0 & -0.1 & 2.5 & 1.1 & -3.8 & 1.1 & 0.2 & -3 & -1.2 & 4.5 & -1.4 & -0.2 & 1.6 & 0.6 & -1.3 & 0.8 & 0.2 & -0.2 & -0.3 & 0.7 & -0.3 & 0.0 \\
1.8 & -2.8 & 1.3 & 0.2 & -0.1 & -2.8 & 3.7 & -0.9 & -0.8 & 0.5 & 1.3 & -0.9 & -0.1 & 0.4 & -0.4 & 0.2 & -0.8 & 0.4 & 0.7 & 0.3 & -0.1 & 0.5 & -0.4 & 0.3 & -0.0 \\
-3.3 & 12 & -9 & -0.3 & 1 & -0.6 & -20 & 20.2 & 0.4 & -2.2 & 9.4 & 10.2 & -18.8 & 1.6 & 1.6 & -7.9 & -0 & 7.4 & -1.4 & 0 & 1.6 & -0.7 & -0.6 & 0 & 0.2 \\
-5.5 & 6.9 & -1.8 & -0.4 & 0 & 10.1 & -7.9 & -4.8 & 4.6 & -0.4 & -6.9 & 2.3 & 8.4 & -5 & 0.3 & 1.3 & 0.1 & -2.7 & 1 & 0.2 & 0.4 & -0 & 0.1 & 0.2 & -0.1 \\
4.2 & -8.1 & 6.1 & -1.8 & 0.1 & -4.8 & 5 & 0.9 & -2.9 & 0.9 & 1.3 & 2.7 & -6.1 & 4.3 & -1 & 0.2 & -1.6 & 0.6 & 0.1 & 0.1 & -0 & 0.6 & 0.6 & -0.4 & 0.1 \\
-1 & 8.7 & -12 & 5.3 & -0.6 & -5.7 & 1.4 & 3 & 0.9 & -1 & 9.2 & -8.8 & 4.8 & -4.7 & 1.6 & -3.3 & 0.8 & 2.2 & -0.4 & -0.3 & 0.2 & 0.8 & -0.9 & 0.7 & -0.0 \\
-3.3 & -0.6 & 9.4 & -7.9 & 1.6 & 12 & -19.9 & 10.2 & -0 & -0.7 & -9 & 20.2 & -18.8 & 7.4 & -0.6 & -0.3 & 0.4 & 1.6 & -1.4 & 0 & 1 & -2.2 & 1.6 & 0 & 0.2 \\
13.1 & -22.7 & -0.8 & 16.6 & -5.2 & -22.7 & 79.8 & -60.8 & -0.5 & 5.9 & -0.8 & -60.8 & 75.6 & -17.9 & -1.9 & 16.6 & -0.5 & -17.9 & 6.5 & 0.1 & -5.2 & 5.9 & -1.9 & 0.1 & 0.3
\end{pmatrix}
$$

and obviously is not non-negative. The mass matrix $\mathbf{M} \geq 0$ is given by the extension of $B \otimes B$ to a (25×49)-matrix containing additionally the values of $\int_0^1 S_j(x)\Psi_i(x)\,dx$ for $1 \leq i \leq D-1$, $j = 0, D$, and again $\mathbf{A}^{-1}\mathbf{M} \in \mathbb{R}^{25\times 49}$ is not non-negative. Thus, by Corollary 1 for Poisson's problem (1) on the unit cube $(0, 1)^2$ with boundary data $g = 0$, a Bernstein certificate of non-negativity for the approximate solution u can be provided for r.h.s. f with Bernstein coefficients $\vec{f}$ in the cone C given by inequalities $\mathbf{A}^{-1}\mathbf{M}\vec{f} \geq 0$. Due to missing non-negativity of $\mathbf{A}^{-1}\mathbf{M}$, the cone $\{\vec{f} \mid \vec{f} \geq 0\}$ is not a subset of C, and while e.g. $\mathbb{R}^{25+24} \ni \vec{f} = (1, 0, \ldots, 0)^T \notin C$, the Bernstein coefficients $\mathbb{R}^{25+24} \ni \vec{f} = (1, \ldots, 1, 0, \ldots, 0)^T$ satisfy $\mathbf{A}^{-1}\mathbf{M}\vec{f} \geq 0$ and thus lie in C. Hence, $\vec{f} = (1, \ldots, 1, 0, \ldots, 0)^T$ (and $\vec{g} = 0$) allow a Bernstein certificate of non-negativity for the approximate solution u, precisely the (scaled) Bernstein coefficients of u are given by

$$10\vec{u} = (0.41, 0.67, 0.76, 0.71, 0.57, 0.67, 1.15, 1.29, 1.23, 0.87, 0.76, 1.29, 1.45, 1.41, 1.04, 0.71, 1.23, 1.41, 1.32, 0.91, 0.57, 0.87, 1.04, 0.91, 0.97)^T.$$

Of course, it would be nice to have more simple inequalities for $\hat{f}$ (and $\hat{g} \geq 0$) or equivalently more simple closed convex cones, which guarantee a certificate of non-negativity $\vec{u} \geq 0$. Obtaining such simplification strongly depends on the chosen method and will be a task for a forthcoming paper about the Bernstein dual Petrov–Galerkin method.

Note that the cone $K = \{\begin{pmatrix}\vec{u}\\ \vec{g}\end{pmatrix} \mid \vec{u} \geq 0,\ \vec{g} \geq 0\} = \operatorname{cone}(\{\vec{e}_j \mid 0 \leq j \leq D\})$ in the former proof is finitely generated by the unit vectors. This is not the case in the next Corollary characterizing functional weak positivity, what makes it more difficult to apply the Corollary.

Corollary 2 *Denote by* $K := \{\begin{pmatrix}\vec{u}\\ \vec{g}\end{pmatrix} \mid \forall x \in \Omega : \sum_{j=0,D} g_j S_j(T(x)) + \sum_{j=1}^{D-1} u_j S_j(T(x)) \geq 0\}$ *the convex cone of Bernstein coefficients of non-negative Bernstein polynomials of degree D, then algebraic-functional weak discrete positivity holds iff the* *matrix* $\begin{pmatrix} \mathbf{A}^{-1}\mathbf{M} & -\mathbf{A}^{-1}\mathbf{A}^{\partial} \\ 0 & \mathbf{I} \end{pmatrix}$ *maps the convex cone* $C_1 := \{\begin{pmatrix} \vec{f} \\ \vec{g} \end{pmatrix} \mid \vec{f}, \vec{g} \geq 0\}$ *into* K, *and functional-functional weak discrete positivity holds iff the convex cone*

$$C_2 := \{\begin{pmatrix} \vec{f} \\ \vec{g} \end{pmatrix} \mid \forall x \in \Omega : \sum_{j=0,D} g_j S_j(T(x)) \geq 0, \sum_{j=0}^{D} f_j S_j(T(x)) \geq 0\}$$

is mapped into K.

Similarly, algebraic-functional or functional-functional strong discrete positivity can be characterized, using cones K without zero $K \setminus \{0\}$ or the pointed interior cone $\overset{o}{K} \cup \{0\}$. Hereby, in the infinite-dimensional case it is important to work in $H^1(\Omega)$ or a stronger space, because else even the standard cone does not have a non-empty interior.

Remark 4 While the convex cone $\{u \in L^2(\Omega) \mid u \geq 0 \text{ a.e.}\}$ is closed in $L^2(\Omega)$, it has no interior points. In fact, a function $u \in L^2(\Omega)$ satisfying $u > 0$ a.e. and $u < M$ a.e. on $B_{\epsilon_0}(x_0)$ can be perturbed by subtracting $M1_{B_\epsilon(x_0)}$ for sufficiently small $\epsilon < \epsilon_0$. The resulting function $u - M1_{B_\epsilon(x_0)}$ is negative on $B_\epsilon(x_0)$, although the norm $\|M1_{B_\epsilon(x_0)}\|_{L^2} = M\,\mathrm{Vol}(B_\epsilon(x_0))$ is arbitrarily small as $\epsilon \searrow 0$.

Let us conclude this section with a discussion of algebraic-functional weak discrete positivity in our main example.

Example 5 In the special case of Poisson's problem (1) on the unit cube $(0, 1)^N$ with homogeneous Dirichlet boundary conditions, $N = 2$, $D = (6, 6)$, already discussed

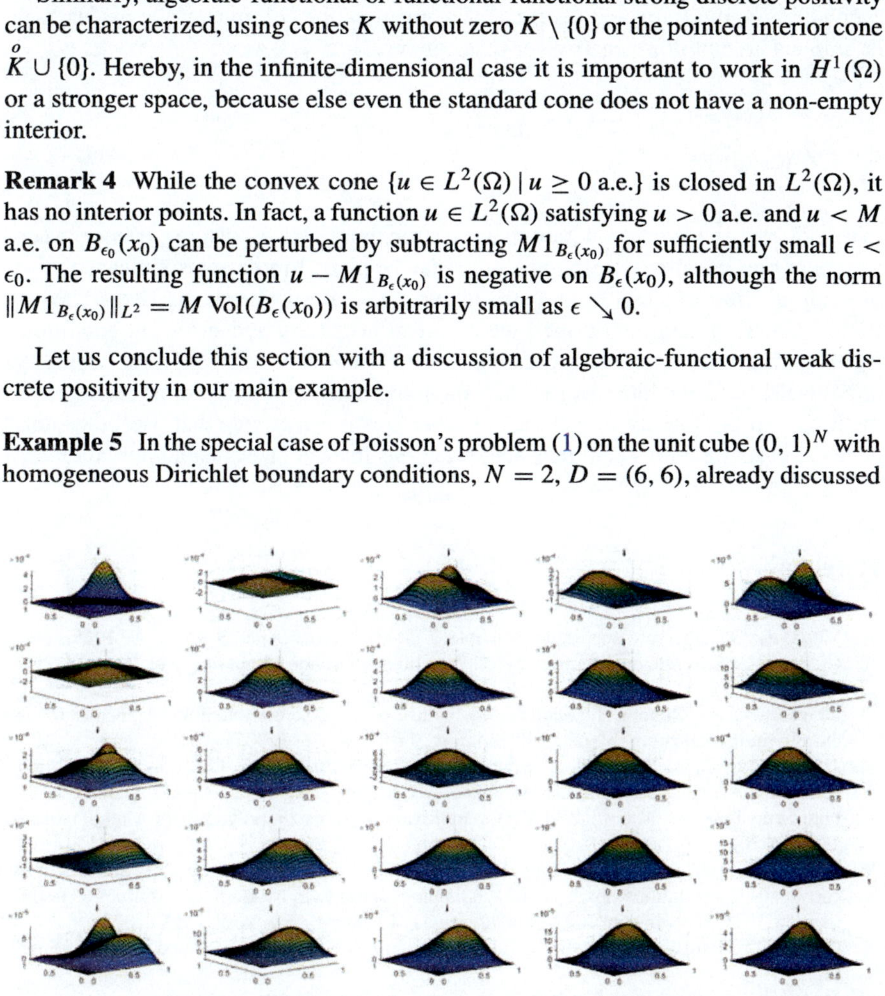

Fig. 1 The 25 approximate solutions of Poisson's problem (1) on the unit cube $(0, 1)^2$ obtained by Bernstein dual Petrov–Galerkin method for $D = (6, 6)$ and Bernstein polynomials $f = S_i$, $1 \leq i \leq D - 1$, vanishing on $\partial(0, 1)^2$ as r.h.s

in Examples 3 and 4, Fig. 1 shows the 25 approximate solutions u_i to those Bernstein polynomials $f = S_i$, $1 \leq i \leq D - 1$ as r.h.s. which vanish on $\partial(0, 1)^N$, i.e. those corresponding to $\vec{g} = 0$ and unit vectors $\vec{f} = \vec{e}_i$. As not all of these approximate solutions are non-negative, the cone C_1 from Corollary (2) is not mapped into K, i.e. algebraic-functional weak discrete positivity (and thus also the stronger functional-functional weak discrete positivity) does not hold for the Bernstein dual Petrov–Galerkin method. However, as merely the functions $u_{(1,2)}$, $u_{(2,1)}$, $u_{(1,3)}$, $u_{(3,1)}$, $u_{(1,4)}$, $u_{(4,1)}$, $u_{(2,5)}$, $u_{(5,2)}$ and $u_{(3,3)}$ become negative, we can guarantee non-negativity of approximate solutions u to r.h.s. having Bernstein coefficients $\vec{f}$ with vanishing components at indices $(1, 2)$, $(2, 1)$, $(1, 3)$, $(3, 1)$, $(1, 4)$, $(4, 1)$, $(2, 5)$, $(5, 2)$, $(3, 3)$, even although the Bernstein coefficients $\vec{u}$ given by the columns of the matrix $\mathbf{A}$ in Example 4 are not non-negative for other indices, too.

5 Conclusion

In this article, we have completely characterized those data for which a Bernstein certificate of non-negativity (resp. positivity) can be given for the approximate solution of an elliptic linear second-order PDEs in divergence form when using Bernstein dual Petrov–Galerkin method. Further, we provided necessary and sufficient conditions for the validity of algebraic-functional or functional-functional discrete positivity, or equivalently for validity of the corresponding discrete maximum principles. Our methods can be directly transferred to other spectral methods that use other non-negative basis functions and their dual functions instead of Bernstein polynomials.

References

1. Varga, R.: On a discrete maximum principle. J. SIAM Numer. Anal. **3**, 355–359 (1966)
2. Ciarlet, P.G.: Discrete maximum principle for finite-difference operators. Aequ. Math. **4**, 266–268 (1970)
3. Drăgănescu, A., Dupont, T., Scott, L.R.: Failure of the discrete maximum principle for an elliptic finite element problem. Math. Comput. **74**(249), 1–23 (2005)
4. Jani, M., Javadi, S., Babolian, E., Bhatta, D.: Bernstein dual-Petrov-Galerkin method: application to 2D time fractional diffusion equation. Comput. Appl. Math. **37**, 2335–2353 (2018)
5. Foupouagnigni, M., Wouodjié, M.M.: On multivariate Bernstein polynomials. Mathematics **8**, 1397 (2020)
6. Hamadneh, T., Merker, J., Schimmel, W., Schuldt, G.: Simplicial Bernstein form and positivity certificates for solutions obtained in a stationary digital twin by Bernstein Bubnov-Galerkin method. In: Proceedings of ICoMS 2022, pp. 41–46. ACM, New York, USA (2022)
7. Gilbarg, D., Trudinger, N.: Elliptic Partial Differential Equations of Second Order. Springer (2001)
8. Narkawicz, A., Garloff, J., Smith, A.P., Munoz, C.A.: Bounding the range of a rational function over a box. Reliab. Comput. **17**, 34–39 (2012)

9. Garloff, J.: Convergent bounds for the range of multivariate polynomials. In: Proceedings of the International Symposium on Interval Mathematics 1985, LNCS, vol. 212, pp. 37–56. Springer (1986)
10. Ciesielski, Z.: The basis of B-splines in the space of algebraic polynomials. Ukr. Math. J. **38**, 311–315 (1986)
11. Jüttler, B.: The dual basis functions for the Bernstein polynomials. Adv. Comput. Math. **8**, 345–352 (1998)
12. Jani, M., Babolian, E., Javadi, S., Bhatta, D.: Banded operational matrices for Bernstein polynomials and application to the fractional advection-dispersion equation. Numer. Algorithms (2017)
13. Plemmons, R.J.: M-matrix characterizations. I - nonsingular M-matrices. Linear Algebr. Appl. **18**, 175–188 (1977)
14. Varga, R.: Matrix Iterative Analysis. Prentice Hall (1962)

On the Dynamic Geometry of Kasner Triangles with Complex Parameter

Dorin Andrica and Ovidiu Bagdasar

Abstract We explore the dynamics of the sequence of Kasner triangles $(A_n B_n C_n)_{n \geq 0}$ when α is a complex number, and we find the values α for which the iterations are convergent. We also investigate the parameter values for which the resulting patterns are periodic or divergent. The results further extend previous research concerning Kasner triangles with a fixed real parameter, where it was found that iterations were convergent if and only if $0 < \alpha < 1$, that is the triangles in the sequence are nested.

Keywords Kasner triangles · Dynamical systems · Convergence · Orbits · Characteristic polynomial · Nested triangles

1 Introduction

For a real number α and an initial triangle $A_0 B_0 C_0$, one can construct the triangle $A_1 B_1 C_1$ such that A_1, B_1 and C_1 divide the segments $[B_0 C_0]$, $[C_0 A_0]$ and $[A_0 B_0]$, respectively, in the ratio $1 - \alpha : \alpha$. Continuing this process one obtains a sequence of triangles $A_n B_n C_n$, $n \geq 0$ whose terms are called Kasner triangles (after E. Kasner (1878–1955)), or nested triangles in other references.

Related examples of iterative processes inspired by simple geometrical configurations are reviewed in the expository article [5]: the dynamic geometry generated by the incircle and the circumcircle of a triangle, the pedal triangle [14], the orthic triangle, and the incentral triangle. Other such recursive systems describing dynamic geometries are considered by S. Abbot [1], G. Z. Chang and P. J. Davis [7],

D. Andrica
Babeş-Bolyai University, 400084 Cluj-Napoca, Romania
e-mail: dandrica@math.ubbcluj.ro

O. Bagdasar (✉)
University of Derby, Kedleston Road, DE22 1GB, Derby, United Kingdom
e-mail: o.bagdasar@derby.ac.uk

Department of Mathematics, Faculty of Exact Sciences, 1 Decembrie 1918 University of Alba Iulia, 510009 Alba Iulia, Romania

D. Zeidan et al. (eds.), *Mathematics and Computation*, Springer Proceedings in Mathematics & Statistics 418, https://doi.org/10.1007/978-981-99-0447-1_17

R. J. Clarke [8], J. Ding, L. R. Hitt and X-M. Zhang [9], L. R. Hitt and X-M. Zhang [12], or D. Ismailescu and J. Jacobs [13].

A natural problem is to determine all real numbers α for which the sequence $(A_nB_nC_n)_{n\geq 0}$ is convergent. In the paper [4], we proved that the sequence is convergent if and only if $\alpha \in (0, 1)$, also providing the order of convergence. To this end we used the complex coordinates of the vertices $A_n(a_n)$, $B_n(b_n)$, $C_n(c_n)$, $n \geq 0$, which can be defined recursively for $n \geq 0$ as:

$$\begin{cases} a_{n+1} = \alpha b_n + (1-\alpha)c_n \\ b_{n+1} = \alpha c_n + (1-\alpha)a_n \\ c_{n+1} = \alpha a_n + (1-\alpha)b_n. \end{cases} \tag{1}$$

The main purpose of this paper is to investigate the geometry of the sequence $(A_nB_nC_n)_{n\geq 0}$ when α is a complex number, and to find the values α for which the resulting sequence of Kasner triangles is convergent (Theorem 1). Clearly, when α is complex, the triangles $A_nB_nC_n$ may not always be nested.

The following result is useful in what follows.

Lemma 1 (Propositions 2 and 3, [2]) *Consider the distinct points A, B, C in the complex plane, with coordinates a, b, c. ABC is a positively oriented equilateral triangle if and only if*

$$a + b\omega + c\omega^2 = 0,$$

where $\omega = e^{\frac{2\pi}{3}i}$. Furthermore, the triangle ABC is equilateral and negatively oriented if and only if

$$a + b\omega^2 + c\omega = 0.$$

Proof By $1 + \omega + \omega^2 = 0$, the first relation yields

$$a + b\omega + c\omega^2 = (b-a)\omega + (c-a)\omega^2 = 0,$$

which can be written as

$$(c-a) = -(b-a)\omega^2 = (b-a)e^{\frac{4\pi i}{3}}e^{-\pi i} = (b-a)e^{\frac{\pi i}{3}}.$$

This is equivalent to C being obtained from B by a rotation about A through an angle of $\frac{\pi}{3}$, i.e., the triangle ABC is equilateral and positively oriented.

In the other case one obtains similarly that

$$0 = a + b\omega^2 + c\omega = (b-a)\omega^2 + (c-a)\omega = 0,$$

from where we deduce that

$$(c-a) = -(b-a)\omega = (b-a)e^{\frac{2\pi i}{3}}e^{-\pi i} = (b-a)e^{\frac{-\pi i}{3}}.$$

This is equivalent to C being obtained from B by a rotation about A through an angle of $-\frac{\pi}{3}$, i.e., the triangle ABC is equilateral and negatively oriented. □

Using the factorization

$$a^2 + b^2 + c^2 - ab - bc - ca = \left(a + b\omega + c\omega^2\right)\left(a + b\omega^2 + c\omega\right),$$

we obtain the following consequence.

Corollary 1 *The triangle ABC is equilateral if and only if*

$$a^2 + b^2 + c^2 - ab - bc - ca = 0.$$

2 Kasner Triangles with a Complex Parameter

The system (1) can be written in matrix form as

$$X_{n+1} = \begin{pmatrix} a_{n+1} \\ b_{n+1} \\ c_{n+1} \end{pmatrix} = \begin{pmatrix} 0 & \alpha & 1-\alpha \\ 1-\alpha & 0 & \alpha \\ \alpha & 1-\alpha & 0 \end{pmatrix} \begin{pmatrix} a_n \\ b_n \\ c_n \end{pmatrix} = TX_n, \tag{2}$$

where $X_n = (a_n, b_n, c_n)^T$, $n \geq 0$. In this notation one can write

$$X_n = T^n X_0. \tag{3}$$

The matrix T has the characteristic polynomial

$$p_T(u) = (u-1)\left(u^2 + u + 3\alpha^2 - 3\alpha + 1\right),$$

whose roots are $u_0 = 1$ and denoting $\omega = \exp\left(\frac{2\pi i}{3}\right)$ we have

$$u_1 = -\frac{1}{2} - \frac{\sqrt{3}}{2}i + \alpha\sqrt{3}i = \omega^2 + \alpha\sqrt{3}i, \tag{4}$$

$$u_2 = -\frac{1}{2} + \frac{\sqrt{3}}{2}i - \alpha\sqrt{3}i = \omega - \alpha\sqrt{3}i. \tag{5}$$

If follows that

$$T = F^{-1}\begin{pmatrix} 1 & 0 & 0 \\ 0 & u_1 & 0 \\ 0 & 0 & u_2 \end{pmatrix} F,$$

where the matrices F and F^{-1} are given by

$$F = \begin{pmatrix} 1 & 1 & 1 \\ 1 & \omega^2 & \omega \\ 1 & \omega & \omega^2 \end{pmatrix}, \quad F^{-1} = \frac{1}{3}\overline{F} = \frac{1}{3}\begin{pmatrix} 1 & 1 & 1 \\ 1 & \omega & \omega^2 \\ 1 & \omega^2 & \omega \end{pmatrix}. \tag{6}$$

For every positive integer n we have the following relations

$$T^n = F^{-1}\begin{pmatrix} 1 & 0 & 0 \\ 0 & u_1^n & 0 \\ 0 & 0 & u_2^n \end{pmatrix} F \tag{7}$$

$$= \frac{1}{3}\begin{pmatrix} 1 & 1 & 1 \\ 1 & \omega & \omega^2 \\ 1 & \omega^2 & \omega \end{pmatrix}\begin{pmatrix} 1 & 0 & 0 \\ 0 & u_1^n & 0 \\ 0 & 0 & u_2^n \end{pmatrix}\begin{pmatrix} 1 & 1 & 1 \\ 1 & \omega^2 & \omega \\ 1 & \omega & \omega^2 \end{pmatrix} \tag{8}$$

$$= \frac{1}{3}\begin{pmatrix} 1+u_1^n+u_2^n & 1+\omega^2u_1^n+\omega u_2^n & 1+\omega u_1^n+\omega^2u_2^n \\ 1+\omega u_1^n+\omega^2u_2^n & 1+u_1^n+u_2^n & 1+\omega^2u_1^n+\omega u_2^n \\ 1+\omega^2u_1^n+\omega u_2^n & 1+\omega u_1^n+\omega^2u_2^n & 1+u_1^n+u_2^n \end{pmatrix}. \tag{9}$$

Finally, this is multiplied by $(a_0, b_0, c_0)^T$ to give

$$X_n = T^n\begin{pmatrix} a_0 \\ b_0 \\ c_0 \end{pmatrix} = \frac{1}{3}\begin{pmatrix} S+\left(a_0+\omega^2b_0+\omega c_0\right)u_1^n+\left(a_0+\omega b_0+\omega^2c_0\right)u_2^n \\ S+\left(a_0\omega+b_0+c_0\omega^2\right)u_1^n+\left(a_0\omega^2+b_0+c_0\omega\right)u_2^n \\ S+\left(a_0\omega^2+b_0\omega+c_0\right)u_1^n+\left(a_0\omega+b_0\omega^2+c_0\right)u_2^n \end{pmatrix}, \tag{10}$$

where we denote $S = a_0 + b_0 + c_0 = 3g_0$ (constant).

The latest relation can be written explicitly as

$$\begin{aligned} a_n &= \frac{a_0+b_0+c_0}{3} + \frac{a_0+b_0\omega^2+c_0\omega}{3}u_1^n + \frac{a_0+b_0\omega+c_0\omega^2}{3}u_2^n \\ b_n &= \frac{a_0+b_0+c_0}{3} + \frac{a_0\omega+b_0+c_0\omega^2}{3}u_1^n + \frac{a_0\omega^2+b_0+c_0\omega}{3}u_2^n \\ c_n &= \frac{a_0+b_0+c_0}{3} + \frac{a_0\omega^2+b_0\omega+c_0}{3}u_1^n + \frac{a_0\omega+b_0\omega^2+c_0}{3}u_2^n. \end{aligned} \tag{11}$$

By Corollary 1, if $a_0^2 + b_0^2 + c_0^2 - a_0b_0 - b_0c_0 - c_0a_0 \neq 0$ (i.e., the initial triangle is not equilateral), then none of the coefficients of u_1^n or u_2^n vanishes.

3 Dynamical Properties

We now discuss orbits obtained for various values of α. By (4) and (5) we get

$$u_1 = \sqrt{3}i\left[\alpha - \left(\frac{1}{2} - \frac{\sqrt{3}}{6}i\right)\right], \tag{12}$$

$$u_2 = -\sqrt{3}i\left[\alpha - \left(\frac{1}{2} + \frac{\sqrt{3}}{6}i\right)\right], \tag{13}$$

which can also be written as

$$\begin{aligned} u_1 &= r_1 e^{2\pi i\theta_1} = \sqrt{3}i\,(\alpha - z_1) = \sqrt{3}\,(\alpha - z_1)\,e^{\frac{2\pi i}{4}}, \\ u_2 &= r_2 e^{2\pi i\theta_2} = -\sqrt{3}i\,(\alpha - z_2) = \sqrt{3}\,(\alpha - z_2)\,e^{\frac{6\pi i}{4}}, \end{aligned} \tag{14}$$

where $r_1, r_2, \theta_1, \theta_2$ are real numbers, where we denote

$$z_1 = \frac{1}{2} - \frac{\sqrt{3}}{6}i, \quad z_2 = \frac{1}{2} + \frac{\sqrt{3}}{6}i. \tag{15}$$

The following sets also play an important role

$$D_1 = \left\{z \in \mathbb{C} : |z - z_1| < \frac{\sqrt{3}}{3}\right\}, \qquad D_2 = \left\{z \in \mathbb{C} : |z - z_2| < \frac{\sqrt{3}}{3}\right\}$$

$$C_1 = \left\{z \in \mathbb{C} : |z - z_1| = \frac{\sqrt{3}}{3}\right\}, \qquad C_2 = \left\{z \in \mathbb{C} : |z - z_2| = \frac{\sqrt{3}}{3}\right\}.$$

Notice that $z_2 - z_1 = \frac{\sqrt{3}}{3}i$, while $z_1 \in C_2$ and $z_2 \in C_1$. The circles C_1, C_2, the discs D_1, D_2 and the points z_1, z_2 are depicted in Fig. 1. By (14) we get

1. If $\alpha \in D_1 \cap D_2$, then $0 < r_1, r_2 < 1$;
2. If α is in the interior of the complement of $D_1 \cap D_2$, then $\max\{r_1, r_2\} > 1$;
3. If $\alpha \in C_1$ or $\alpha \in C_2$ then $r_1 = 1$ or $r_2 = 1$, respectively, with $C_1 \cap C_2 = \{0, 1\}$;
4. If $\alpha = z_1$, then $r_1 = 0$ and $r_2 = 1$;
5. If $\alpha = z_2$, then $r_1 = 1$ and $r_2 = 0$.

The boundary of the shaded region in Fig. 1 consists of two arcs

$$U_1 = C_1 \cap D_2, \quad U_2 = C_2 \cap D_1,$$

which can be parametrized as

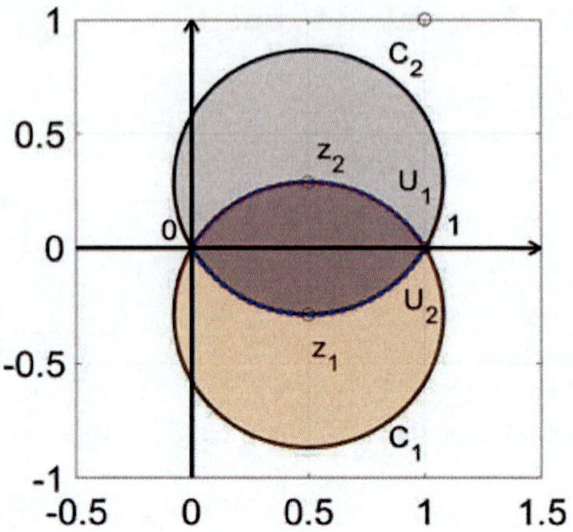

Fig. 1 Circles C_1 and C_2 determining the possible configurations

$$\alpha(t) = \begin{cases} z_1 + \frac{\sqrt{3}}{3}(\cos t + i \sin t), & t \in \left[\frac{\pi}{6}, \frac{5\pi}{6}\right] \\ z_2 + \frac{\sqrt{3}}{3}(\cos t + i \sin t), & t \in \left[\frac{7\pi}{6}, \frac{11\pi}{6}\right]. \end{cases} \tag{16}$$

To describe the orbits of the sequences $(a_n)_{n\geq 0}$, $(b_n)_{n\geq 0}$ and $(c_n)_{n\geq 0}$, one needs to understand the behaviour of $(z^n)_{n\geq 0}$, where $z \in \mathbb{C}$ (see, for example, Lemma 2.1 in [6], or Lemma 5.2 in [3]). These configurations are depicted in Fig. 2.

Lemma 2 *Let $z = re^{2\pi i\theta}$, where $r \geq 0$, $\theta \in \mathbb{R}$. The orbit of $(z^n)_{n\geq 0}$ is*
(a) a spiral convergent to 0 for $r < 1$;
(b) a divergent spiral for $r > 1$;
(c) a regular k-gon if z is a primitive kth root of unity, $k \geq 3$;
(d) a dense subset of the unit circle if $r = 1$ and $\theta \in \mathbb{R} \setminus \mathbb{Q}$.
When $\theta = j/k \in \mathbb{Q}$ is irreducible, then the terms of the spirals obtained in (a) and (b) align along k rays.

To prove part (*d*) we have $z^n = e^{2\pi i\, n\theta} = e^{2\pi i\,(n\theta+m)}$ for $n \geq 0$ and m integers. By Kronecker's Lemma [11, Theorem 442], the set $\{n\theta + m : m, n \in \mathbb{Z}, n \geq 0\}$ is dense in $\mathbb{R}$, hence z^n is dense within the unit circle.

As linear combinations of $(u_1^n)_{n\geq 0}$ and $(u_2^n)_{n\geq 0}$, given by the explicit formula (11) in the complex plane, and we have the following possibilities, assuming that the initial triangle is not equilateral.

Lemma 3 *The patterns produced by formula* (11) *are summarized below:*

1. *Convergent if* $0 < r_1, r_2 < 1$;
2. *Divergent if* $\max\{r_1, r_2\} > 1$;
3. *Periodic if $r_1 = r_2 = 1$ (when $\alpha = 0$ or $\alpha = 1$), or when* $\min\{r_1, r_2\} = 0$ *and* $\max\{r_1, r_2\} = 1$ *(when $\alpha = z_1$ or $\alpha = z_2$);*
4. *There are two distinct patterns when* $0 < \min\{r_1, r_2\} < \max\{r_1, r_2\} = 1$. *Denoting $\theta = \theta_1$ if $r_1 = 1$ or $\theta = \theta_2$ if $r_2 = 1$, then the orbit*

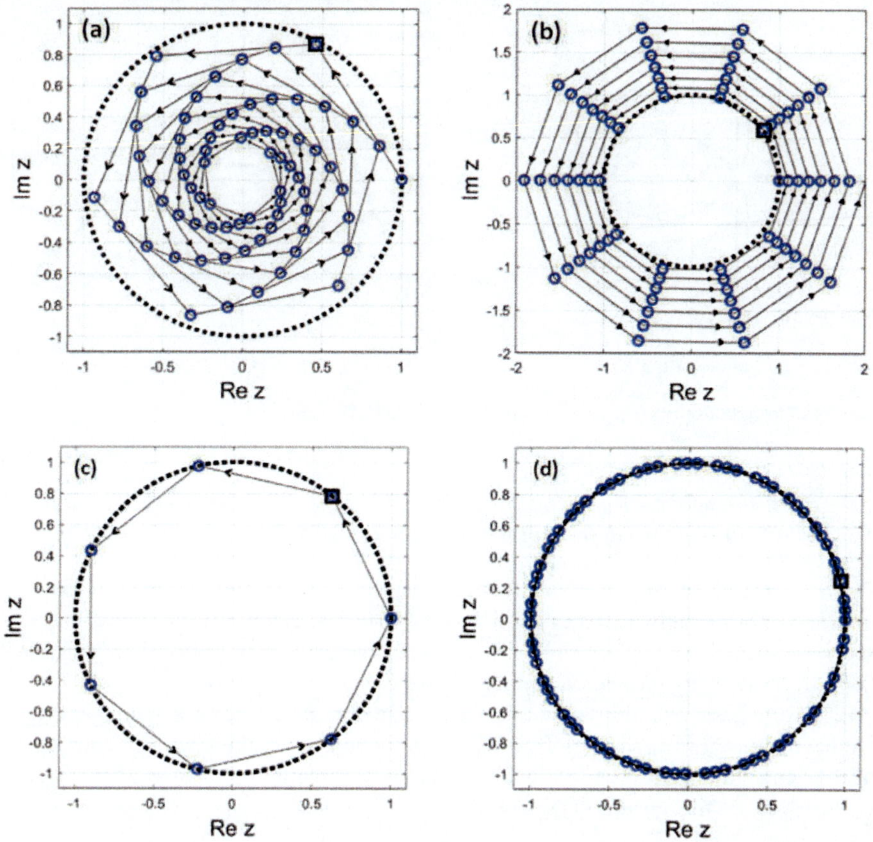

Fig. 2 The terms z^n, $n = 0, \ldots, 70$ obtained for **a** $r = 0.98$ and $x = \sqrt{3}/10$; **b** $r = 1.01$ and $x = 1/10$; **c** $r = 1$ and $x = 1/10$; **d** $r = 1$ and $x = \sqrt{2}/35$. Arrows indicate the direction of the orbit, and the dotted line represents the unit circle. The point $z = r\exp(2\pi i x)$ is shown as a square

(a) has k convergent subsequences if $\theta = \frac{j}{k}$ *is an irreducible fraction;*
(b) is dense within a circle when θ *is irrational.*

The details of geometric patterns obtained in each case are presented below.

3.1 Convergent Orbits

If $0 < r_1, r_2 < 1$, then by (14) u_1^n and u_2^n are convergent and $\alpha \in D_1 \cap D_2$. Therefore, by (11) the sequences $(a_n)_{n\geq 0}$, $(b_n)_{n\geq 0}$ and $(c_n)_{n\geq 0}$ converge to g_0.

Theorem 1 *1° The sequence of triangles* $(A_n B_n C_n)_{n\geq 0}$ *is convergent if and only if* $\alpha \in D_1 \cap D_2$.

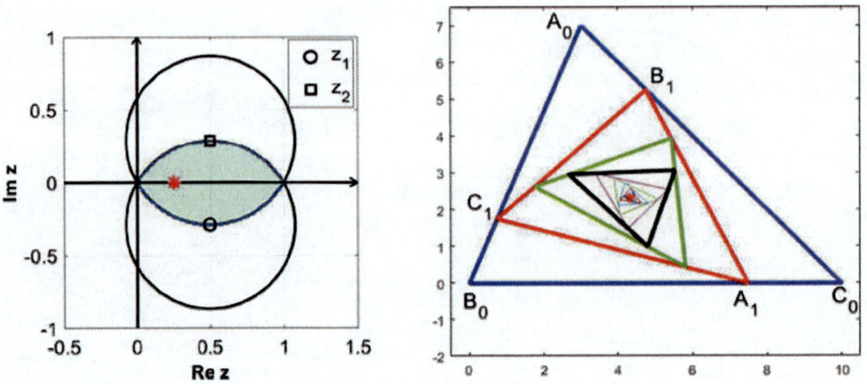

Fig. 3 Convergent orbits (right) obtained for $\alpha = 0.25$ (left)

2° *When the sequence $(A_n B_n C_n)_{n\geq 0}$ is convergent, its limit is the degenerated triangle at G_0, the centroid of the initial triangle $A_0 B_0 C_0$.*

Proof The intersection $D_1 \cap D_2$ represents the shaded area in Fig. 1. Clearly, $\alpha \in D_1 \cap D_2$ is equivalent to $r_1 < 1$ and $r_2 < 1$. The relation (11) shows that the sequences $(a_n)_{n\geq 0}$, $(b_n)_{n\geq 0}$ and $(c_n)_{n\geq 0}$ are convergent if and only if $(u_1^n)_{n\geq 0}$ and $(u_2^n)_{n\geq 0}$ are convergent, which happens when $u_1^n \to 0$ and $u_2^n \to 0$.

2°. Adding the equation in the system (1) one obtains that for every integer $n \geq 0$ we have $a_n + b_n + c_n = a_0 + b_0 + c_0 = 3g_0$, where g_0 is the complex coordinate of the centroid G_0 of the initial triangle $A_0 B_0 C_0$. Assume that $a_n \to a^*$, $b_n \to b^*$, $c_n \to c^*$. From the system (1) we obtain

$$\begin{cases} a^* = \alpha b^* + (1-\alpha)c^* \\ b^* = \alpha c^* + (1-\alpha)a^* \\ c^* = \alpha a^* + (1-\alpha)b^*. \end{cases} \tag{17}$$

Using $c^* = \alpha a^* + (1-\alpha)b^*$ in the first relation we get $a^* = \alpha b^* + \alpha(1-\alpha)a^* + (1-\alpha)^2 b^*$, hence $(\alpha^2 - \alpha + 1)a^* = (\alpha^2 - \alpha + 1)b^*$. However, if $\alpha^2 - \alpha + 1 = 0$ we have $\alpha^3 = -1$, hence $|\alpha| = 1$, which is not possible for $\alpha \in D_1 \cap D_2$. From here we deduce that $a^* = b^*$. From the first equation of the system (17) we get $(1-\alpha)a^* = (1-\alpha)c^*$, hence $a^* = c^*$, since $\alpha \neq 1$.

Therefore, from $a_n + b_n + c_n = 3g_0$ it follows that $a^* = b^* = c^* = g_0$. □

For $0 < \alpha < 1$ one has $\alpha \in D_1 \cap D_2$, and moreover, in this case the vertices $A_{n+1}, B_{n+1}, C_{n+1}$ are interior points of the segments $[B_n, C_n]$, $[A_n, C_n]$ and $[A_n, B_n]$, respectively. Such an example is depicted in Fig. 3.

On the other hand, when the parameter $\alpha \in D_1 \cap D_2$ is not real, the orbit is convergent, but the points are not aligned any more, as illustrated in Fig. 4.

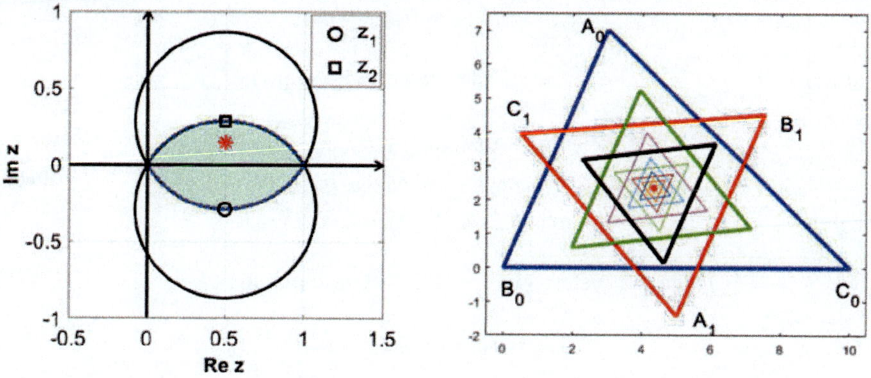

Fig. 4 Convergent orbits (right) obtained for $\alpha = \frac{1}{2} + \frac{\sqrt{3}}{12} i$ (left)

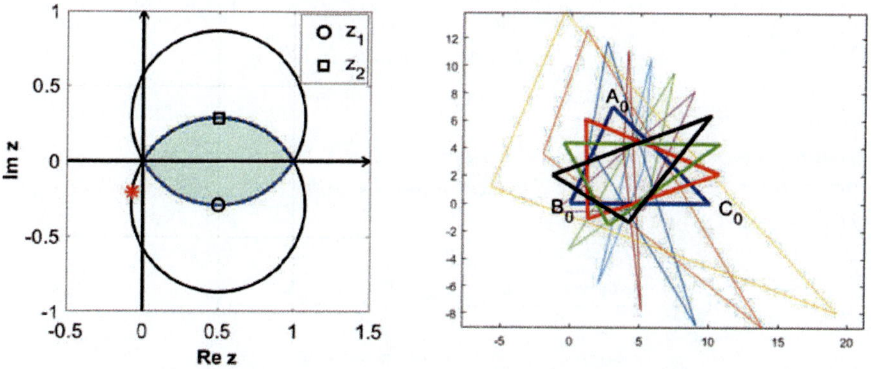

Fig. 5 Divergent orbits (right) obtained for $\alpha = z_1 + \frac{\sqrt{3}}{3} (\cos 3 + i \sin 3)$ (left)

3.2 *Divergent Orbits*

If $\max\{r_1, r_2\} > 1$, then $\alpha \in \operatorname{int}(D_1 \cap D_2)^c$ by (14) either u_1^n or u_2^n are divergent. Therefore, by (11) the sequences $(a_n)_{n\geq 0}$, $(b_n)_{n\geq 0}$ and $(c_n)_{n\geq 0}$ are divergent (as long as the corresponding coefficient is not vanishing, which is the case when the starting triangle a_0, b_0, c_0 are not the coordinates of an equilateral triangle.

Figure 5 depicts a divergent iteration.

3.3 *Periodic Orbits*

If $r_1 = r_2 = 1$, then $|\alpha - z_1| = |\alpha - z_2| = \frac{\sqrt{3}}{3}$, hence $\alpha \in C_1 \cap C_2 = \{0, 1\}$.

Case 1. $\alpha = 0$. From the system (1), for all $n \geq 0$ one obtains

$$a_{n+3} = c_{n+2} = b_{n+1} = a_n.$$

Similarly, $b_{n+3} = b_n$ and $c_{n+3} = c_n$, and the periodic sequences are given by

$$\begin{cases} a_n: & a_0, c_0, b_0, a_0, c_0, b_0, a_0, \ldots \\ b_n: & b_0, a_0, c_0, b_0, a_0, c_0, b_0, \ldots \\ c_n: & c_0, b_0, a_0, c_0, b_0, a_0, c_0, \ldots. \end{cases} \tag{18}$$

Case 2. $\alpha = 1$. From the system (1), for all $n \geq 0$ one obtains

$$a_{n+3} = c_{n+2} = b_{n+1} = a_n.$$

In the same way, $b_{n+3} = b_n$ and $c_{n+3} = c_n$, and explicitly we can write

$$\begin{cases} a_n: & a_0, b_0, c_0, a_0, b_0, c_0, a_0, \ldots \\ b_n: & b_0, c_0, a_0, b_0, c_0, a_0, b_0, \ldots \\ c_n: & c_0, a_0, b_0, c_0, a_0, b_0, c_0, \ldots. \end{cases} \tag{19}$$

Other stable orbits are obtained for $r_1 = 0$ ($\alpha = z_1$), or $r_2 = 0$ ($\alpha = z_2$).

Case 3. $\alpha = z_1$. By (14), here we have $u_1 = 0$ and $u_2 = -\sqrt{3}i(z_1 - z_2) = \sqrt{3}i \cdot \frac{\sqrt{3}}{3}i = -1$, hence by (11) we get

$$\begin{aligned} a_n &= \frac{a_0 + b_0 + c_0}{3} + \frac{a_0 + b_0\omega + c_0\omega^2}{3}(-1)^n \\ b_n &= \frac{a_0 + b_0 + c_0}{3} + \frac{a_0\omega^2 + b_0 + c_0\omega}{3}(-1)^n \\ c_n &= \frac{a_0 + b_0 + c_0}{3} + \frac{a_0\omega + b_0\omega^2 + c_0}{3}(-1)^n, \end{aligned} \tag{20}$$

which is clearly periodic with period 2, i.e., $a_{n+2} = a_n$, $b_{n+2} = b_n$ and $c_{n+2} = c_n$.

For all $k \geq 1$, one obtains the following explicit formulae:

$$\begin{aligned} a_{2k-1} &= \frac{1-\omega}{3}b_0 + \frac{1-\omega^2}{3}c_0, & a_{2k} &= \frac{2}{3}a_0 + \frac{1+\omega}{3}b_0 + \frac{1+\omega^2}{3}c_0 \\ b_{2k-1} &= \frac{1-\omega^2}{3}a_0 + \frac{1-\omega}{3}c_0, & b_{2k} &= \frac{1+\omega^2}{3}a_0 + \frac{2}{3}b_0 + \frac{1+\omega}{3}c_0 \\ c_{2k-1} &= \frac{1-\omega}{3}a_0 + \frac{1-\omega^2}{3}b_0, & c_{2k} &= \frac{1+\omega}{3}a_0 + \frac{1+\omega^2}{3}b_0 + \frac{2}{3}c_0. \end{aligned} \tag{21}$$

These configurations are depicted in Fig. 6(left).

Case 4. $\alpha = z_2$. By (14) we have $u_1 = \sqrt{3}i(z_2 - z_1) = \sqrt{3}i \cdot \frac{\sqrt{3}}{3}i = -1$ and $u_2 = 0$, hence by (11) it follows that

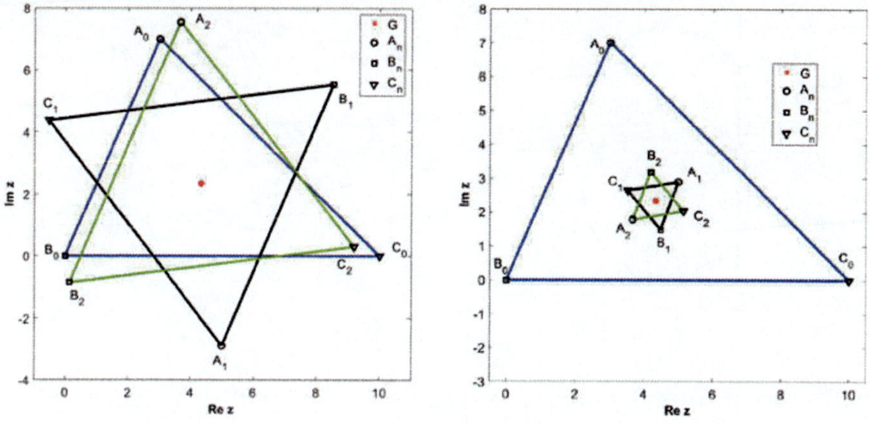

Fig. 6 Periodic orbits obtained for: $\alpha = z_1 = \frac{1}{2} - \frac{\sqrt{3}}{6}i$ (left) and $\alpha = z_2 = \frac{1}{2} + \frac{\sqrt{3}}{6}i$ (right)

$$
\begin{aligned}
a_n &= \frac{a_0 + b_0 + c_0}{3} + \frac{a_0 + b_0\omega^2 + c_0\omega}{3}(-1)^n \\
b_n &= \frac{a_0 + b_0 + c_0}{3} + \frac{a_0\omega + b_0 + c_0\omega^2}{3}(-1)^n \\
c_n &= \frac{a_0 + b_0 + c_0}{3} + \frac{a_0\omega^2 + b_0\omega + c_0}{3}(-1)^n,
\end{aligned} \tag{22}
$$

which is periodic with period 2 for $n \geq 1$, i.e., $a_{n+2} = a_n$, $b_{n+2} = b_n$ and $c_{n+2} = c_n$. The orbit obtained for $\alpha = z_2$ is shown in Fig. 6(right). For $k \geq 1$ we get

$$
\begin{aligned}
a_{2k-1} &= \frac{1-\omega^2}{3}b_0 + \frac{1-\omega}{3}c_0, & a_{2k} &= \frac{2}{3}a_0 + \frac{1+\omega^2}{3}b_0 + \frac{1+\omega}{3}c_0 \\
b_{2k-1} &= \frac{1-\omega}{3}a_0 + \frac{1-\omega^2}{3}c_0, & b_{2k} &= \frac{1+\omega}{3}a_0 + \frac{2}{3}b_0 + \frac{1+\omega^2}{3}c_0 \\
c_{2k-1} &= \frac{1-\omega^2}{3}a_0 + \frac{1-\omega}{3}b_0, & c_{2k} &= \frac{1+\omega^2}{3}a_0 + \frac{1+\omega}{3}b_0 + \frac{2}{3}c_0.
\end{aligned} \tag{23}
$$

3.4 Orbits with Convergent Subsequences

If $0 < \min\{r_1, r_2\} < \max\{r_1, r_2\} = 1$ then one either has $\alpha \in C_1 \cap D_2$ for $r_1 = 1$, or $\alpha \in C_2 \cap D_1$ for $r_2 = 1$. The orbit has a finite number of limit points if the complex argument θ of u_1 if $r_1 = 1$, or the argument of u_2 if $r_2 = 1$ is rational. First, assume that $r_1 = \max\{r_1, r_2\} = 1$, i.e., α is on the upper arc $C_1 \cap D_2$.

As $\alpha \in C_1$, there is $t_1 \in \left[\frac{\pi}{6}, \frac{5\pi}{6}\right]$ with $\alpha = z_1 + \frac{\sqrt{3}}{3}e^{2\pi i t}$, so by (14) we get

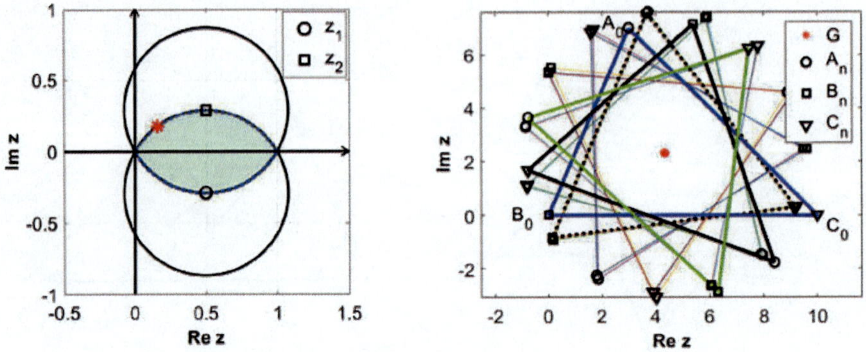

Fig. 7 Orbits for $\theta_1 = p/k = 3/5$ where $\alpha = z_1 + \frac{\sqrt{3}}{3}e^{2\pi i\left(\frac{1}{4}+\frac{3}{5}\right)}$ (left)

$$u_1 = e^{2\pi i\theta_1} = \sqrt{3}i\,(\alpha - z_1) = e^{2\pi i\left(t-\frac{1}{4}\right)}.$$

When $\theta_1 = \frac{p}{k}$ is an irreducible fraction, the orbit has a finite number of convergent subsequences. Therefore, we have the following result (Fig. 7).

Theorem 2 *If for the integers* $0 < p < k$ *we have* $\theta_1 = \frac{p}{k} \in \left[\frac{5\pi}{12}, \frac{13\pi}{12}\right]$ *is an irreducible fraction, then* $u_1 = e^{2\pi i\frac{p}{k}}$ *and by formula (11) the sequences* $(a_n)_{n\geq 0}$, $(b_n)_{n\geq 0}$ *and* $(c_n)_{n\geq 0}$ *have k subsequences convergent to the vertices of three regular k-gons centred in* g_0. *Explicitly, for each* $j = 0, \ldots, k-1$ *one has*

$$\lim_{n\to\infty} a_{nk+j} = g_0 + \frac{a_0 + b_0\omega^2 + c_0\omega}{3}u_1^j$$
$$\lim_{n\to\infty} b_{nk+j} = g_0 + \frac{a_0\omega + b_0 + c_0\omega^2}{3}u_1^j$$
$$\lim_{n\to\infty} c_{nk+j} = g_0 + \frac{a_0\omega^2 + b_0\omega + c_0}{3}u_1^j. \tag{24}$$

Proof By (11), one can write $a_n = g_0 + Au_1^n + Bu_2^n$, where $A = \frac{a+b\omega^2+c\omega}{3}$ and $B = \frac{a+b\omega+c\omega^2}{3}$. Since $u_2^n \to 0$, the long term behaviour of sequence $(a_n)_{n\geq 0}$ is that of sequence $(g_0 + Au_1^n)_{n\geq 0}$, which is a regular polygon centred in g_0 having radius $|A|$. This behaviour is illustrated in Fig. 8. □

The sequences $(a_n)_{n\geq 0}$, $(b_n)_{n\geq 0}$ and $(c_n)_{n\geq 0}$ are plotted in Fig. 8.

Similarly, if $r_2 = \max\{r_1, r_2\} = 1$, so α is on the arc $C_2 \cap D_1$ defined by (16). Therefore, there is $t \in \left[\frac{7\pi}{6}, \frac{11\pi}{6}\right]$ with $\alpha = z_2 + \frac{\sqrt{3}}{3}e^{2\pi i t}$, and we get

$$u_2 = e^{2\pi i\theta_2} = -\sqrt{3}i\,(\alpha - z_2) = e^{2\pi i\left(t-\frac{3}{4}\right)}.$$

When θ_2 is rational, the orbit has a finite number of convergent subsequences.

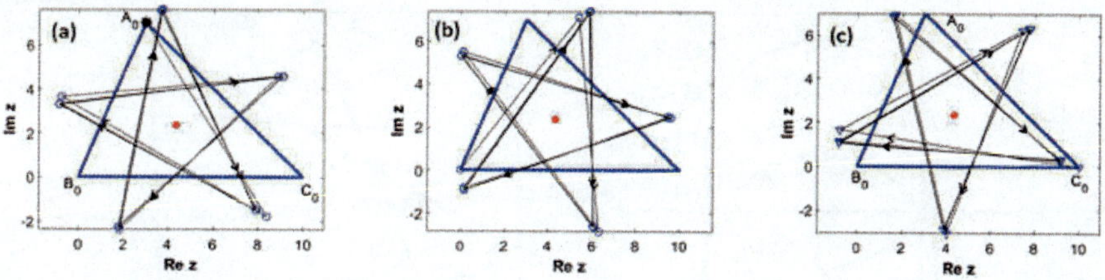

Fig. 8 Orbits obtained for $\theta_1 = p/k = 3/5$. **a** $(a_n)_{n\geq 0}$; **b** $(b_n)_{n\geq 0}$; **c** $(c_n)_{n\geq 0}$

3.5 Dense Orbits

As in the previous section, when $0 < \min\{r_1, r_2\} < \max\{r_1, r_2\} = 1$ but with θ_1 irrational, the orbits of $(a_n)_{n\geq 0}$, $(b_n)_{n\geq 0}$ and $(c_n)_{n\geq 0}$ are dense within circles. First, assume that $r_1 = \max\{r_1, r_2\} = 1$, i.e., α is on the upper arc $C_1 \cap D_2$.

The following result follows by Lemma 2 (d).

Theorem 3 *If* $\theta_1 \in \left[\frac{5\pi}{12}, \frac{13\pi}{12}\right]$ *is irrational, then the sequences* $(a_n)_{n\geq 0}$, $(b_n)_{n\geq 0}$ *and* $(c_n)_{n\geq 0}$ *are each dense within a circle centred at* g_0.

Proof By (11), one can write $a_n = g_0 + Au_1^n + Bu_2^n$, where $A = \frac{a_0+b_0\omega^2+c_0\omega}{3}$ and $B = \frac{a_0+b_0\omega+c_0\omega^2}{3}$. Since $u_2^n \to 0$, one can deduce that

$$\lim_{n\to\infty} |a_n - g_0| = \lim_{n\to\infty} |b_n - g_0| = \lim_{n\to\infty} |c_n - g_0| = |A|, \tag{25}$$

where we used $a_0\omega^2 + b_0\omega + c_0 = \omega(a_0\omega + b_0 + c_0\omega^2) = \omega^2(a_0 + b_0\omega^2 + c_0\omega)$. Furthermore, the complex arguments of the sequences $(a_n)_{n\geq 0}$, $(b_n)_{n\geq 0}$ and $(c_n)_{n\geq 0}$ are dense in $[0, 2\pi]$, therefore their orbits are dense within the (same) circle centred in g_0 and having radius $|A|$. □

Figure 9 illustrates the sides of the triangles obtained for $n = 10$ iterations respectively, when $\alpha \in C_1 \cap D_2$.

On the other hand, Fig. 10 depicts all the vertices of the original triangle together in part (a), but also the individual sequences $(a_n)_{n\geq 0}$, $(b_n)_{n\geq 0}$ and $(c_n)_{n\geq 0}$, taken separately.

A similar behaviour is encountered for $r_2 = \max\{r_1, r_2\} = 1$, so when α is on the arc $C_2 \cap D_1$ defined by (16). Therefore, there is $t \in \left[\frac{7\pi}{6}, \frac{11\pi}{6}\right]$ with $\alpha = z_2 + \frac{\sqrt{3}}{3}e^{2\pi i t}$, and we get

$$u_2 = e^{2\pi i\theta_2} = -\sqrt{3}i\,(\alpha - z_2) = e^{2\pi i\left(t-\frac{3}{4}\right)}.$$

When θ_2 is irrational, the orbit is again dense, as seen in Fig. 11.

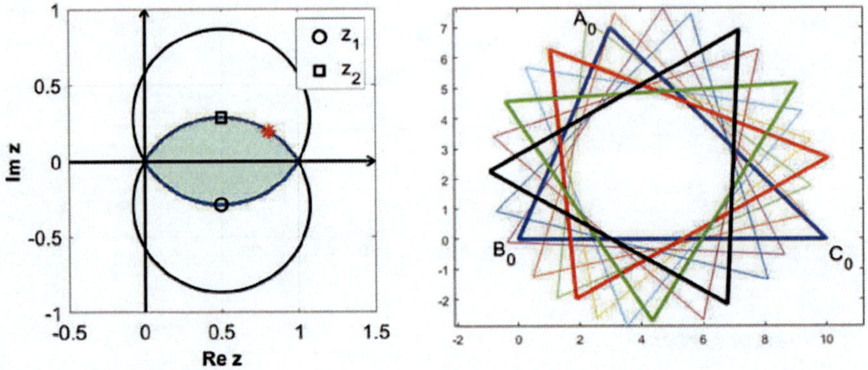

Fig. 9 Orbits obtained for $n = 10$ iterations (right), for the parameter value $\alpha = \frac{1}{2} - \frac{\sqrt{3}}{6}i + \frac{\sqrt{3}}{3}(\cos 1 + i \sin 1)$ (left figure)

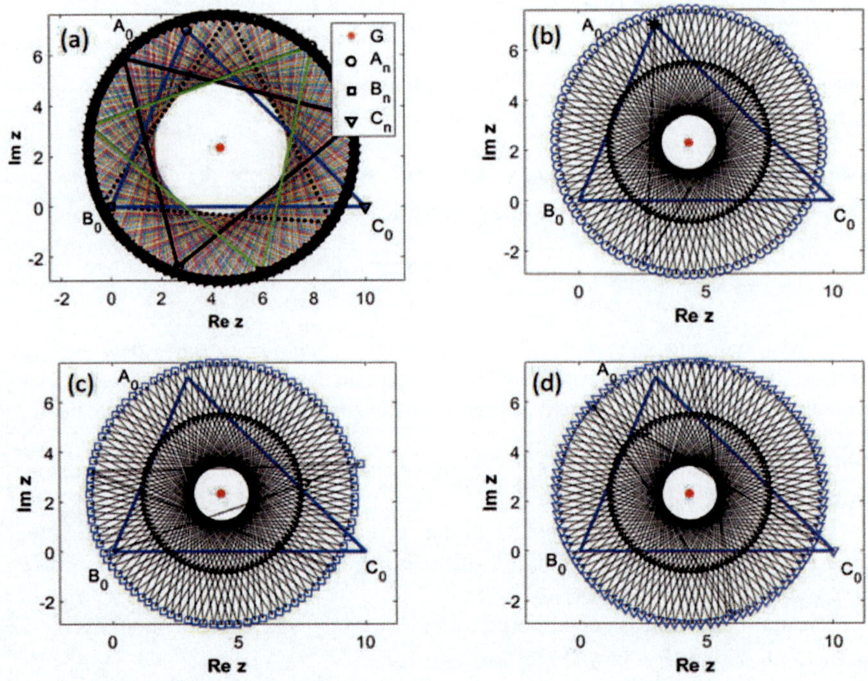

Fig. 10 Orbits given for $\theta_1 = \frac{\sqrt{3}}{4}$. **a** Sequences $(a_n)_{n\geq 0}$, $(b_n)_{n\geq 0}$ and $(c_n)_{n\geq 0}$; **b** $(a_n)_{n\geq 0}$; **c** $(b_n)_{n\geq 0}$; **d** $(c_n)_{n\geq 0}$

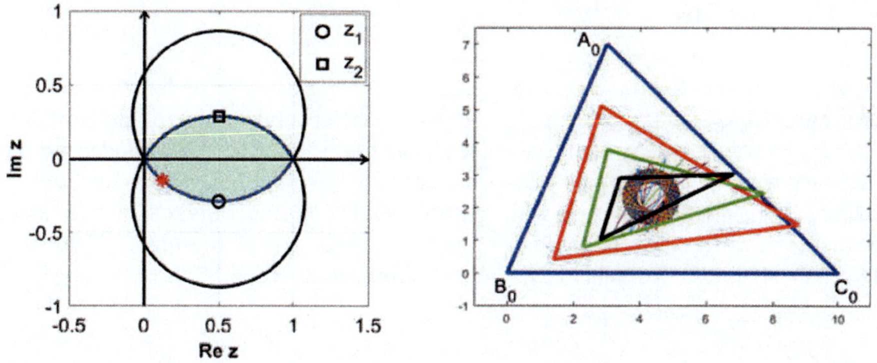

Fig. 11 Dense orbits obtained after $n = 10$ iterations (right), generated for $\alpha = z_2 + \frac{\sqrt{3}}{3}(\cos 4 + i \sin 4)$ (left), when $u_2 = e^{2\pi i \theta_2}$, with $\theta_2 = \frac{4}{2\pi} + \frac{3}{4}$

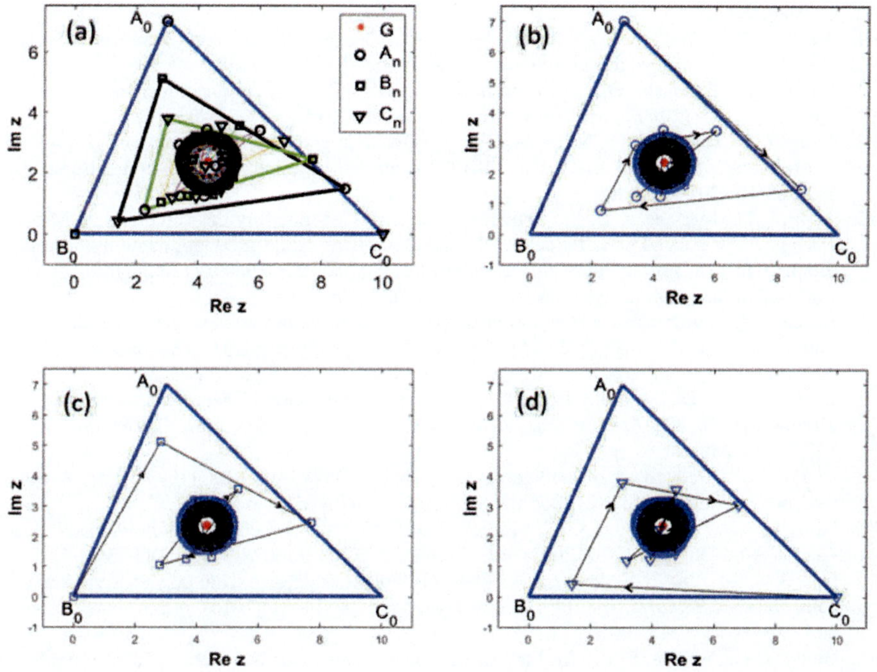

Fig. 12 Orbits given by $\theta_2 = \frac{4}{2\pi} + \frac{3}{4}$. **a** Sequences $(a_n)_{n\geq 0}$, $(b_n)_{n\geq 0}$ and $(c_n)_{n\geq 0}$; **b** $(a_n)_{n\geq 0}$; **c** $(b_n)_{n\geq 0}$; **d** $(c_n)_{n\geq 0}$

Figure 12 depicts all the vertices of the original triangle together in part (a), but also the individual sequences $(a_n)_{n\geq 0}$, $(b_n)_{n\geq 0}$ and $(c_n)_{n\geq 0}$, taken separately are illustrated at the points (b), (c) and (d).

4 Conclusions

In this paper we have investigated the behaviour of Kasner triangles with a complex parameter, identifying the values for which the generated patterns are periodic, convergent, or divergent. Some of these iterative processes can be extended for an arbitrary m-polygon to obtain classical Kasner polygons, Kasner polygons with a fixed weight, Kasner polygons with m fixed weights, Kasner polygons with a fixed sequence of weights, as discussed in the papers by S. Donisi, H. Martini, G. Vincenzi, G. Vitale [10] or O. Roeschel [15], and the references therein.

References

1. Abbot, S.: Average sequences and triangles. Math. Gaz. **80**, 222–224 (1996). https://www.jstor.org/stable/3620354
2. Andreescu, T., Andrica, D.: Complex Numbers from A to ... Z. 2nd edn, Birkhäuser, Boston (2014). https://link.springer.com/book/10.1007/978-0-8176-8415-0
3. Andrica, D., Bagdasar, O.: Recurrent Sequences: Key Results, Applications, and Problems. Springer Nature (2020). https://link.springer.com/book/10.1007/978-3-030-51502-7
4. Andrica, D., Bagdasar, O., Marinescu, D.-Şt.: Dynamic geometry of Kasner triangles with a fixed weight. Int. J. Geom. **11**(2), 101–110 (2022). https://ijgeometry.com/wp-content/uploads/2022/03/10.-101-110.pdf
5. Andrica, D., Marinescu, D.-Şt.: Dynamic Geometry Generated by the Circumcircle Midarc Triangle. In: Rassias, Th.M., Pardalos, P.M. (eds.) Analysis. Nonlinear Optimization and Applications. World Scientific Publishing Company Ltd, Singapore, Geometry (2023). https://www.worldscientific.com/worldscibooks/10.1142/13002#t=aboutBook
6. Bagdasar, O., Larcombe, P.J.: On the characterization of periodic complex Horadam sequences. Fibonacci Q. **51**(1), 28–37 (2013). https://www.fq.math.ca/Papers1/51-1/BagdasarLarcombe.pdf
7. Chang, G.Z., Davis, P.J.: Iterative processes in elementary geometry. Amer. Math. Monthly **90**(7), 421–431 (1983). https://www.tandfonline.com/doi/abs/10.1080/00029890.1983.11971250
8. Clarke, R.J.: Sequences of polygons. Math. Mag. **90**(2), 102–105 (1979). https://www.tandfonline.com/doi/abs/10.1080/0025570X.1979.11976761
9. Ding, J., Hitt, L.R., Zhang, X-M.: Markov chains and dynamic geometry of polygons. Linear Algebr. Appl. **367**, 255–270 (2003). https://doi.org/10.1016/S0024-3795(02)00634-1
10. Donisi, S., Martini, H., Vincenzi, G., Vitale, G.: Polygons derived from polygons via iterated constructions. Electron. J. Differ. Geom. Dyn. Syst. **18**, 14–31 (2016). http://www.mathem.pub.ro/dgds/v18/D18-do-b77.pdf
11. Hardy, G.H., Wright, E.M.: An Introduction to the Theory of Numbers. Oxford University Press, Oxford, Fifth Edition (1979)
12. Hitt, L.R., Zhang, X-M.: Dynamic geometry of polygons. Elem. Math. **56**(1), 21–37 (2001). https://doi.org/10.1007/s000170050086
13. Ismailescu, D., Jacobs, J.: On sequences of nested triangles. Period. Math. Hung. **53**(1-2), 169–184 (2006). https://doi.org/10.1007/s10998-006-0030-3
14. Kingston, J.G., Synge, J.L.: The sequence of pedal triangles. Amer. Math. Monthly **95**(7), 609–620 (1988). https://doi.org/10.1080/00029890.1988.11972056
15. Roeschel, O.: Polygons and iteratively regularizing affine transformations. Beitr. Algebra Geom. **58**, 69–79 (2017). https://doi.org/10.1007/s13366-016-0313-7

Application of Conformable Fractional Nakagami Distribution

Dana Amr and Ma'mon Abu Hammad

Abstract The paper introduces conformable fractional analogs of some basic concepts related to probability distributions of random variables, namely density, cumulative distribution, survival, and hazard functions. Moreover, it introduces conformable fractional analogs to expected values, rth moments, rth central moments, mean, variance, skewness, and kurtosis. In addition, it introduces conformable fractional analogs to some entropy measures, namely, Shannon, Renyi, and Tsallis entropy. All these concepts had been applied to the conformable fractional Nakagami probability distribution (Abu Hammad et al (2020) J Math Comput Sci 1239–1250; Gaeddert and Annamalai (2005) IEEE 9:22–24).

Keywords Conformable · Conformable entropy · Distributions

1 Introduction

The Nakagami distribution is flexible and provides a fit for failure in data sets. Nakagami appeared in (1960) Distribution for Radio-Signal Fading Modeling. Several parametric models are used in the analysis of age data and issues related to the Failure process[2, 5, 6].

Ultrasound modeling has applications in medical imaging studies especially in Photographing different forms of tumors, including breast tumors. It is also useful for modeling high-frequency seismogram envelopes. Reliability engineering makes extensive use of the Nakagami distribution [9, 10, 14].

The first to use this distribution was Hoffmann in modeling the attenuation of radio signals Crossing multiple paths. Lin and Yang research and derive a statistical model for the chromatic spatial distribution of images. By comprehensive evaluation

D. Amr · M. A. Hammad (✉)
Al-Zaytoonah University of Jordan, Queen Alia Airport St. 594 11942, Amman, Jordan
e-mail: m.abuhammad@zuj.edu.jo

D. Amr
e-mail: Danaamr1998@icloud.com

D. Zeidan et al. (eds.), *Mathematics and Computation*, Springer Proceedings in Mathematics & Statistics 418, https://doi.org/10.1007/978-981-99-0447-1_18

of large image databases. Shanker, Tsui, and others used Nakagami Distribution for modeling ultrasound data in medical imaging studies. Use Kim and Latch man Nakagami distribution in their multimedia analysis. It was obtained by Azzam Zakka and Ahmed Saeed Akhtar Bayes estimator for Nakagami distribution [10, 14].

In 2014 Khalil et al. [10] created a new definition of fractional derivative.

This research proposal investigates conformable fractional analogs of some basic concepts related to the probability distribution of random variables, namely density, cumulative distribution, survival, and hazard functions. Moreover, it introduces conformable fractional analogs to expected values, rth moments, mean, variance, skewness, and kurtosis. In addition, it introduces conformable fractional analogs to some entropy measures such as Shannon, Renyi, and Tsallis measures.

After using the Nakagami probability distribution equation that was solved by Abu Hammad et al. [4, 8] using the new definition of Khalil and finding a new probability distribution, we will find all the properties of this distribution and generalize them to Nakagami's special case.

2 Application on Conformable is α–Nakagami Distribution

Abu Hammad et al. [4, 13], solved the conformable Nakagami differential equation.

They obtained that

$$f(\mathrm{x}) = Ax^{2\beta-1}e^{\left(\frac{-\beta x^{2\alpha}}{(\theta\alpha)}\right)}.$$

2.1 *The Conformable Fractional Probability Density Function is (CFPDF)*

Now to get the conformable fractional probability density function Is (CFPDF) of a random variable X. It's conformable integral on the interval $(0, \infty)$ [1, 3, 7, 11, 12] $\int_0^\infty f_\alpha(x)d^\alpha x = 1$. To evaluate this integral use the substitution $z = \frac{\beta x^{2\alpha}}{(\theta\alpha)}$ to get

$$f_\alpha(x) = \frac{2\alpha\beta^{\frac{2\beta+\alpha-1}{2\alpha}}}{(\theta\alpha)^{\frac{2\beta+\alpha-1}{2\alpha}}\Gamma\left(\frac{2\beta+\alpha-1}{2\alpha}\right)}x^{2\beta-1}e^{-\frac{\beta x^{2\alpha}}{\theta\alpha}} \tag{1}$$

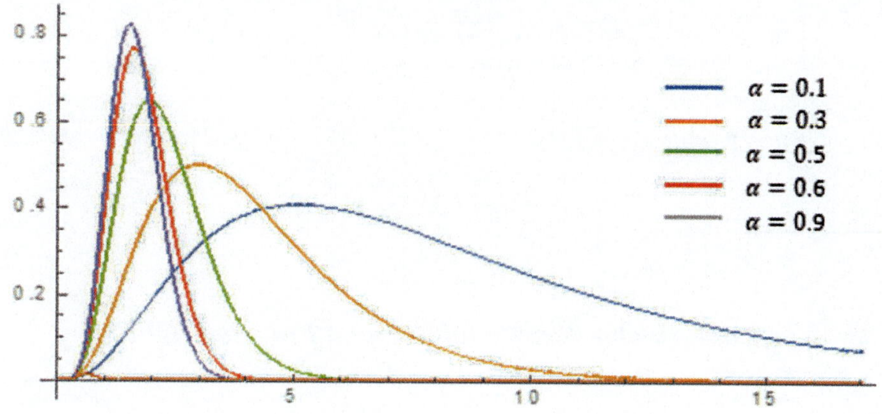

Fig. 1 The (CFPDF) graph when X: NAK (3,2.3)

2.2 *The Conformable Fractional Cumulative Distribution Function (CFCDF)*

The CFCDF of X is

$$F_{\alpha}(\mathrm{x}) = \int_{0}^{\mathrm{x}} f_{\alpha}(\mathrm{y}) d^{\alpha}\mathrm{y}$$

Thus,

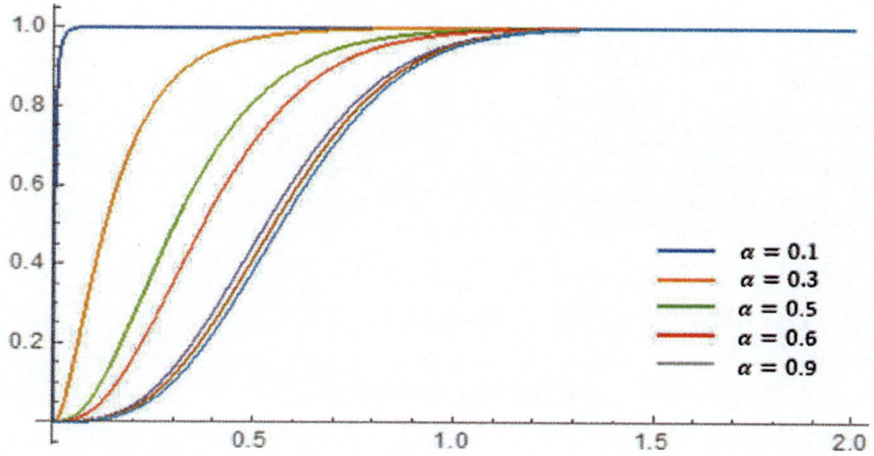

Fig. 2 The (CFCDF) graph of Nakagami distribution

$$F_\alpha(x) = 1 - \frac{\psi\left(\frac{-1+2\beta+\alpha}{2\alpha}, \frac{\beta x^{2\alpha}}{\theta\alpha}\right)}{\Gamma\left(\frac{-1+2\beta+\alpha}{2\alpha}\right)}, 2\beta + \alpha > 1 \tag{2}$$

where ψ is polygama function.

$$F_\alpha(0) = 0\, and \lim_{x\to\infty} F_\alpha(x) = 1,$$

2.3 *The Conformable Fractional Survival Function (CFSF) Sα*

Is defined as

S_α (x) = 1 − F_α (x). Using Eq. (2).

Hence,

$$S_\alpha(x) = \frac{\psi\left(\frac{-1+2\beta+\alpha}{2\alpha}, \frac{\beta x^{2\alpha}}{\theta\alpha}\right)}{\Gamma\left(\frac{-1+2\beta+\alpha}{2\alpha}\right)} \tag{3}$$

2.4 *The Conformable Fractional Hazard Function (CFHF) hα*

Is defined as $H_\alpha(x) = \frac{f_\alpha(x)}{S_\alpha(x)}$. Using Eqs. (1) and (3).

Hence,

$$h_\alpha(x) = \frac{2(\theta\alpha)^{-\frac{-1+2\beta+\alpha}{2\alpha}}\ \beta^{\frac{-1+2\beta+\alpha}{2\alpha}}\ \alpha}{\psi\left(\frac{-1+2\beta+\alpha}{2\alpha}, \frac{\beta x^{2\alpha}}{\theta\alpha}\right)}\ \mathrm{e}^{-\frac{\beta x^{2\alpha}}{\theta\alpha}}\quad x^{-1+2\beta}$$

2.5 *The rth Non-central α–Moment ($E_\alpha(X^r)$)*

$$E_\alpha\left(X^r\right) = \int_0^\infty x^r f_\alpha(x) d_\alpha x$$

Let $y = \frac{\beta x^{2\alpha}}{\theta\alpha}$. We get.

$$E_\alpha(X^r) = \frac{\theta\alpha^{\frac{r}{2\alpha}}\,\beta^{-\frac{r}{2\alpha}}\,\Gamma\left(\frac{2\beta+r+\alpha-1}{2\alpha}\right)}{\Gamma\left(\frac{2\beta+\alpha-1}{2\alpha}\right)}$$

Remark 1

(1) When $r = 1$. We get

$$E_\alpha(X) = \frac{\left(\frac{\theta\alpha}{\beta}\right)^{-\frac{-1+2\beta+\alpha}{2\alpha}}\left(\frac{\beta}{\theta\alpha}\right)^{-\frac{1}{2}-\frac{\beta}{\alpha}}\Gamma\left(\frac{1}{2}+\frac{\beta}{\alpha}\right)}{\Gamma\left(\frac{2\beta+\alpha-1}{2\alpha}\right)} \tag{4}$$

(2) When $r = 2$. We get

$$E_\alpha X^2 = \frac{\left(\frac{\theta\alpha}{\beta}\right)^{\frac{1}{\alpha}}\Gamma\left(\frac{1+2\beta+\alpha}{2\alpha}\right)}{\Gamma\left(\frac{-1+2\beta+\alpha}{2\alpha}\right)} \tag{5}$$

(3) The conformable fractional variance (σ_α^2).

$$\sigma_\alpha^2 = E_\alpha X^2 - (E_\alpha X)^2$$

By using Eqs. (4) and (5).
Then,

$$\sigma_\alpha^2 = \frac{\theta\alpha^{\frac{1}{\alpha}}\beta^{-1/\alpha}\left(-\Gamma\left(\frac{1}{2}+\frac{\beta}{\alpha}\right)^2 + \Gamma\left(\frac{2\beta+\alpha-1}{2\alpha}\right)\Gamma\left(\frac{2\beta+\alpha+1}{2\alpha}\right)\right)}{\Gamma\left(\frac{2\beta+\alpha-1}{2\alpha}\right)^2}$$

(4) The Conformable Standard Deviation (σ_α).

Is defined by

$$\sigma_\alpha = \sqrt{\sigma_\alpha^2}$$

$$\sigma_\alpha = \frac{\theta\alpha^{\frac{1}{2}/\alpha}\beta^{-\frac{1}{2}/\alpha}\sqrt{-\Gamma\left(\frac{1}{2}+\frac{\beta}{\alpha}\right)^2 + \Gamma\left(\frac{2\beta+\alpha-1}{2\alpha}\right)\Gamma\left(\frac{2\beta+\alpha+1}{2\alpha}\right)}}{\Gamma\left(\frac{2\beta+\alpha-1}{2\alpha}\right)}$$

(5) When $r = 3$. We get

$$E_\alpha X^3 = \frac{(\frac{\theta\alpha}{\beta})^{\frac{3}{2}/\alpha}\Gamma(\frac{2+2\beta+\alpha}{2\alpha})}{\Gamma(\frac{-1+2\beta+\alpha}{2\alpha})} \tag{6}$$

The Conformable Skewness (αskw) defined by

$$\alpha\text{skw} = \frac{E_\alpha(X-\mu)^3}{(\sigma_\alpha{}^2)^{\frac{3}{2}}}$$

By Eqs. (4), (5) and (6). We get the α skw is

$$\alpha skw = \frac{2\Gamma\left(\frac{1}{2}+\frac{\beta}{\alpha}\right)^3 - 3\Gamma\left(\frac{1}{2}+\frac{\beta}{\alpha}\right)\Gamma\left(\frac{2\beta+\alpha-1}{2\alpha}\right)\Gamma\left(\frac{1+2\beta+\alpha}{2\alpha}\right) + \Gamma\left(\frac{2\beta+\alpha-1}{2\alpha}\right)^2\Gamma\left(\frac{2+2\beta+\alpha}{2\alpha}\right)}{\left(-\Gamma\left(\frac{1}{2}+\frac{\beta}{\alpha}\right)^2 + \Gamma\left(\frac{2\beta+\alpha-1}{2\alpha}\right)\Gamma\left(\frac{1+2\beta+\alpha}{2\alpha}\right)\right)^{3/2}}$$

(6) When $r = 4$. We get

$$E_\alpha X^4 = \frac{(\frac{\theta\alpha}{\beta})^{2/\alpha}\Gamma(\frac{3+2\beta+\alpha}{2\alpha})}{\Gamma(\frac{-1+2\beta+\alpha}{2\alpha})}$$

The Conformable Kurtosis (αkur) defined by

$$\alpha kur = \frac{E_\alpha(x-\mu)^4}{\sigma_\alpha^4} \tag{7}$$

By Eqs. (4), (5), (6), and (7). We get the αkur is

$$\alpha kur = \frac{-3\Gamma\left(\frac{1}{2}+\frac{\beta}{\alpha}\right)^4 + 6\Gamma\left(\frac{1}{2}+\frac{\beta}{\alpha}\right)^2\Gamma\left(\frac{-1+2\beta+\alpha}{2\alpha}\right)\Gamma\left(\frac{1+2\beta+\alpha}{2\alpha}\right)}{\left(\Gamma\left(\frac{1}{2}+\frac{\beta}{\alpha}\right)^2 - \Gamma\left(\frac{-1+2\beta+\alpha}{2\alpha}\right)\Gamma\left(\frac{1+2\beta+\alpha}{2\alpha}\right)\right)^2} + \frac{-4\Gamma\left(\frac{1}{2}+\frac{\beta}{\alpha}\right)\Gamma\left(\frac{2\beta+\alpha-1}{2\alpha}\right)^2\Gamma\left(\frac{2+2\beta+\alpha}{2\alpha}\right) + \Gamma\left(\frac{2\beta+\alpha-1}{2\alpha}\right)^3\Gamma\left(\frac{3+2\beta+\alpha}{2\alpha}\right)}{\left(\Gamma\left(\frac{1}{2}+\frac{\beta}{\alpha}\right)^2 - \Gamma\left(\frac{-1+2\beta+\alpha}{2\alpha}\right)\Gamma\left(\frac{1+2\beta+\alpha}{2\alpha}\right)\right)^2}.$$

2.6 Conformable Fractional Shannon Entropy αH

Defined by

$$\alpha H = -\int_{0}^{\infty} f_\alpha(x)(\log f_\alpha(x))d^\alpha x$$

$$\alpha H = \frac{1}{2\alpha\Gamma\left(\frac{2\beta+\alpha-1}{2\alpha}\right)}\theta\alpha^{1+\frac{1-2\beta-3\alpha}{2\alpha}}\beta^{\frac{-1+2\beta+\alpha}{2\alpha}}\left(\frac{\beta}{\theta\alpha}\right)^{-\frac{-1+2\beta+\alpha}{2\alpha}}$$

$$\left(\begin{array}{l} 2\alpha\Gamma\left(\frac{2\beta+3\alpha-1}{2\alpha}\right)-\Gamma\left(\frac{2\beta+\alpha-1}{2\alpha}\right) \\ \left(\begin{array}{l}(1-2\beta)\log\left(\frac{\beta}{\theta\alpha}\right)+\alpha\left(\log(4)+2\log(\alpha)-2\log\left(\Gamma\left(\frac{2\beta+\alpha-1}{2\alpha}\right)\right)\right) \\ +(-1+2\beta)\psi\left(\frac{2\beta+\alpha-1}{2\alpha}\right)\end{array}\right)\end{array}\right)$$

2.7 Conformable Fractional Tsallis Entropy $\alpha H_{T,c}$

Defined by

$\alpha H_{T,c} = \frac{1}{1-c}(\log(\int_0^\infty \alpha f^{c-1}(x)dx) - 1)$

Then,

$\alpha H_{T,c} = \frac{2-2^c\theta^{\frac{1}{2}-\frac{c}{2}}\beta^{\frac{1}{2}(-1+c)}\alpha^{\frac{1}{2}(-1+c)}c^{-\frac{\alpha-c+2\beta c}{2\alpha}}\Gamma\left(\frac{-1+2\beta+\alpha}{2\alpha}\right)^{-c}\Gamma\left(\frac{\alpha-c+2\beta c}{2\alpha}\right)}{2(-1+c)}.$

The $\lim\limits_{c\to 1}\alpha H_{T,c} = \alpha H$. Hence,

$$\lim_{c\to 1}\alpha H_{T,c} = \frac{1}{2\alpha}(-1+2\beta+\alpha-\alpha Log(4)+\alpha Log(\theta)-\alpha Log(\beta)) + \frac{1}{2\alpha}\left(\begin{array}{c}-\alpha Log(\alpha)+2\alpha Log\left(\Gamma\left(\frac{-1+2\beta+\alpha}{2\alpha}\right)\right) \\ +(1-2\beta)\psi\left(\frac{-1+2\beta+\alpha}{2\alpha}\right)\end{array}\right),$$

which is the αH.

Table 1 Appendix (1): The percentiles of the distribution are: $\beta = 1$ and $= \theta$

α / P	0.1	0.2	0.3	0.4	0.5	0.6	0.7	0.8	0.9
0.1	0.013	0.063	0.12	0.172	0.22	0.262	0.301	0.336	0.369
0.2	0.04	0.144	0.238	0.314	0.377	0.43	0.475	0.515	0.549
0.3	0.086	0.25	0.374	0.465	0.534	0.59	0.635	0.673	0.705
0.4	0.16	0.39	0.535	0.632	0.701	0.754	0.794	0.827	0.854
0.5	0.281	0.577	0.733	0.826	0.888	0.931	0.962	0.986	1.005
0.6	0.484	0.838	0.987	1.062	1.105	1.132	1.149	1.16	1.168
0.7	0.848	1.226	1.332	1.366	1.375	1.374	1.368	1.361	1.353
0.8	1.592	1.868	1.849	1.795	1.741	1.693	1.651	1.614	1.582
0.9	3.651	3.222	2.816	2.54	2.345	2.199	2.086	1.996	1.921

2.8 Conformable Fractional Renyi Entropy $\alpha H_{R,c}$

Defined by

$\alpha H_{R,c} = \frac{1}{1-c}(\log \int_0^\infty f^{c-1}(x)dx)$.

Then,

$$\alpha H_{R,c} = \frac{1}{2\alpha(-1+c)}\left(\begin{array}{c}\alpha(-1+c)\mathrm{Log}(\theta) + (\alpha - \alpha c)\mathrm{Log}(4\beta) + \alpha\mathrm{Log}(\alpha)\\ -\alpha c\mathrm{Log}(\alpha) + \alpha\mathrm{Log}(c) - c\mathrm{Log}(c) + 2\beta c\mathrm{Log}(c)\\ +2\alpha c\mathrm{Log}\left(\Gamma\left(\frac{-1+2\beta+\alpha}{2\alpha}\right)\right) - 2\alpha\mathrm{Log}\left(\Gamma\left(\frac{\alpha-\beta+2\beta c}{2\alpha}\right)\right)\end{array}\right).$$

The $\lim_{c\to 1}\alpha H_{R,c} = \alpha H$. Hence,

$$\lim_{c\to 1}\alpha H_{R,c} = \frac{\begin{array}{c}-1+2\beta+\alpha-\alpha Log(4)+\alpha Log(\theta)-\alpha Log(\beta)-\alpha Log(\alpha)\\ +2\alpha Log\left(\Gamma\left(\frac{-1+2\beta+\alpha}{2\alpha}\right)\right)+(1-2\beta)\psi\left(\frac{-1+2\beta+\alpha}{2\alpha}\right)\end{array}}{2\alpha},$$

which is the αH

Table 2 Appendix (2): The percentiles of the distribution are: $\beta = 0.9$ and $\theta = 0.9$

α / p	0.1	0.2	0.3	0.4	0.5	0.6	0.7	0.8	0.9
0.1	0.001	0.011	0.036	0.069	0.105	0.142	0.179	0.214	0.247
0.2	0.002	0.027	0.077	0.135	0.193	0.247	0.296	0.341	0.382
0.3	0.004	0.05	0.128	0.208	0.283	0.349	0.407	0.457	0.502
0.4	0.007	0.081	0.19	0.293	0.382	0.456	0.519	0.572	0.617
0.5	0.013	0.125	0.269	0.393	0.493	0.574	0.639	0.692	0.736
0.6	0.024	0.189	0.372	0.517	0.626	0.709	0.774	0.824	0.864
0.7	0.043	0.287	0.516	0.679	0.793	0.874	0.933	0.977	1.011
0.8	0.086	0.454	0.736	0.912	1.022	1.092	1.14	1.172	1.194
0.9	0.214	0.82	1.158	1.323	1.403	1.443	1.46	1.465	1.464

References

1. Aalen, Borgan, Gjessing: Survival and Event History Analysis: A Process Point of View (2008)
2. Abu Hammad, M., Awad, A.: Distribution of Shannon Statistic from Normal Sample, Metron, pp. 259–275 (2007)
3. Abu Hammad, M., Awad, A., Khalil, R.: Properties of conformable fractional chi-square probability distribution. J. Math. Comput. Sci, 1239–1250 (2020)
4. Abu Hammad, M.A., Khalil, A., Aldabbas, R.E.: Fractional distributions and probability density functions of random variables generated using FDE. J. Math. Comput. Sci. 522–534 (2020)
5. Awad, A.: A statistical information measures. Dirasat (science) 7–20 (1987)
6. Awad, A.: Statistical view of information theory. In: International Encyclopedia of Statistical Science, pp. 1473–14758. Springer (2011)
7. Batiha, I.M., Albadarneh, R.B., Momani, S., Jebril, I.H.: Dynamics analysis of fractional-order Hopfield neural networks. Int. J. Biomath. **13**(8), Art. no. 2050083. Cited 15 times (2020)
8. Bezziou, M., Jebril, I., Dahmani, Z.: A new nonlinear duffing system with sequential fractional derivatives Chaos. Solitons Fract. **151**, art. no. 111247, Cited 1 time (2021)
9. Cheng, J., Beaulieu, N.C.: Generalized moment estimators for the Nakagami m fading parameter, "Communications Letters." IEEE **6**(4), 144–146 (2002)
10. Gaeddert, J., Annamalai, A.: Further results on Nakagami m parameter estimation, "Communications Letters." IEEE **9**(1), 22–24 (2005)
11. Khalil, R.A., Abu Hammad, M.M.: Geometric meaning of conformable derivative via fractional cords. J. Math. Comput. Sci. 241–245
12. Khalil, R., AlHorani, M., Yousef, A., Sababheh, M.: A new definition of fractional derivative. J. Comput. Appl. Math. **264**, 65–70 (2014)
13. Ouannas, A., Batiha, I.M., Khennaoui, A.-A., Jebril, I.H.: On the 0–1 test for chaos applied to the generalized fractional-order Arnold Map 2021. In: Proceedings of the International Conference on Information Technology, ICIT 2021, Art. no. 9491633, pp. 242–245 (2021)
14. Parsons, J.D.: The Mobile Radio Propagation Channel. Wiley, New York (1992)

Self-Consistent Single-Particle Spectra with Delta Excitations

Mohammed Hassen Eid Abu-Sei'leek

Abstract Single-particle energies of spherical double magic rich-neutron ^{208}Pb nucleus are investigated by using a realistic effective baryon-baryon Hamiltonian. The results showed that the computed spectrum followed the expected arrangement of the shell model in the dominant nucleon orbitals. In this spectrum, visible gaps between the shells are clearly shown. By compressed nucleus, the arrangement of single-particle orbitals and their gaps is maintained. When the nucleus is compressed, the general trend of single-particle energies shifts to higher energies. When the orbitals approach the surface, their curvature rises more and more. For Δ^0 orbitals, in some root mean square radii (r_{rms}), some orbitals close together but do not intersect. There is no clear evidence for the gaps in the nuclear shells. However, a gap of about 251.2 MeV was observed between the last dominant neutron orbital and the first predominant Δ^0 orbitals, in this work. This is attributed to the difference in the rest mass of baryons that are neutron (ns and Δ^0) particles.

Keywords Nuclear structure · Single-particle energy · Compressed finite nuclei · Δ-resonance · Shell model · Heavy spherical doubly magic ^{208}Pb nucleus

PACS 21.10.Dr · 21.10.Pc · 21.60.Cs · 27.80.+w

1 Introduction

The nuclear shell model, developed by Mayer and Jensen in 1952, is now a very successful and highly developed microscopic theory for the structure of finite nuclei [1]. It would be hoped to reproduce a basic systematic feature of the set of atomic nuclei: the "magic numbers". Empirically, it is found that there are large deviations from the smooth Bethe-Weizäcker formula for nuclear binding energies near certain "magic" values of N and Z [2]. The nuclei that have the same values of Z, N = 2, 8, 20, 28, 50, 82, 126, 184, and 258 have an excess of binding energy [3]. The

M. H. Eid Abu-Sei'leek (✉)
Department of Physics, Faculty of Science, Zarqa University, Al-Zarqa, Jordan
e-mail: mseileek@zu.edu.jo; moh2hassen@yahoo.com

D. Zeidan et al. (eds.), *Mathematics and Computation*, Springer Proceedings in Mathematics & Statistics 418, https://doi.org/10.1007/978-981-99-0447-1_19

Woods-Saxon potential [4] with a spin-orbital force is the mean field which is the basis for the nuclear shell model. In addition, one can add residual interactions and correlations to produce more of the details of the structure of nuclei.

Studies of the compression of stable nuclei found on Earth have already taught scientists a great deal of invaluable information about the mechanical properties of the dense matter of which they are formed: nuclear matter. For several decades, physicists have been able to compress these stable nuclei through collisions with light nuclei. Until now, however, it has been impossible to compress unstable nuclei, since they are found in the form of beams produced by an accelerator [5].

In the dense phase of relativistic heavy ion collisions, 30% of the baryon population is presented as delta-resonances [6–9]. It leads to a great interest in the investigation of delta matter formation at the deep interior of compact stars [10].

The primary purpose of this paper is to investigate the self-consistent spectra of a single particle of ^{208}Pb with delta excitations. The detailed calculations demonstrated the effective Hamiltonian [11–14], H_{eff}, and the calculation procedures [15–17]. Based on these presented studies, the two body matrix elements in the N-N sector are scaled to the optimum value of $\hbar\omega'$, the oscillator energy for the ^{208}Pb nucleus in the 9 nuclear shells with the 10 Δ orbitals [15–24]. Using the modification parameters, λ_1, λ_2, and $\hbar\omega'$, it is possible to obtain such a fit to the balance of the binding energy and r_{rms} radius [25]. In this study, the modification parameters, λ_1, λ_2, and $\hbar\omega'$ are 0.997, 1.001, and 7.345, respectively.

This paper is written as follows: Sect. 2 includes results and discussions. Section 3 specifies the conclusion and outlook.

2 Results and Discussion

In our previous studies, the nuclei ^{40}Ca, ^{90}Zr, ^{100}Sn, $1^{32}Sn$, and ^{208}Pb were concentrated to define methods and demonstrate sensitivity for choosing a nuclear model space size ?, [11–24]. One major conclusion from these investigations was that the zero temperature compressibility was decreased between 20% and 40% when the Δ degree of freedom was activated.

Now, properties of heavier nucleus ^{208}Pb are examined in the largest of the N-Δ model space. A section of results for ^{208}Pb concentrates on the self-consistent behavior of single-particle spectra for nucleons and Δs in the lowest state as a function of the compression on the nucleus.

Figure 1 displays the 37 consistent self-occupied orbitals of zero-charge baryon. These 37 orbitals are occupied by 258 baryons which are mixtures of the neutron and the Δ^0. Recall that this single-particle spectrum is generated from the underlying microscopic Hamiltonian [26]. Furthermore, the gaps in the conventional shell are clearly visible. When ^{208}Pb is also compressed, the arrangement of single-particle levels and the shell gaps are conserved. The orbitals closest to zero single-particle

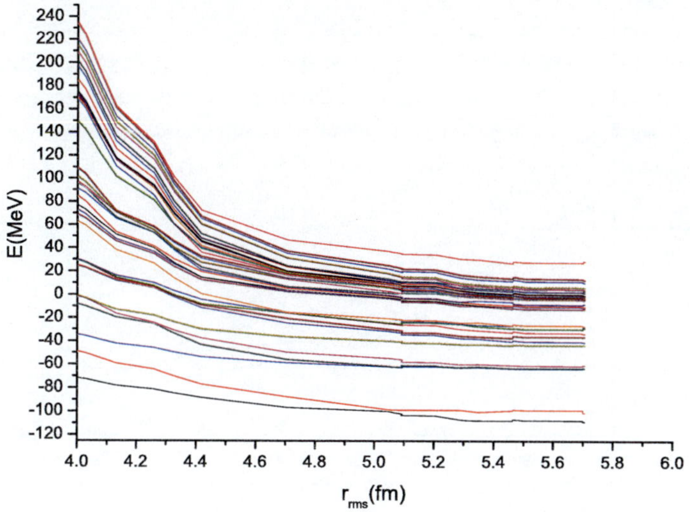

Fig. 1 Lowest 37 constrained spherical Hartree-Fock single-particle orbitals for ^{208}Pb as a function of the root mean square (rms) radius. These orbitals absorb 258 charge-zero baryons in the system

energy are the most compression-sensitive. This means that the surface of the nucleus is more responsive to the compression load than the inner region of the nucleus.

Ten self-consistent zero-charge levels of a single particle that are dominant Δ^0 characters are shown in Fig. 2 versus the rms radius. As for the occupied orbitals, the expected trend toward higher energy is indicated by the compressed system. It is interesting to note that the ranking of the level, which progresses from lowest to highest, is $0s_{3/2}$, $0p_{3/2}$, $0p_{1/2}$, $0d_{5/2}$, $0d_{3/2}$, $0d_{1/2}$, $1s_{3/2}$, $0f_{7/2}$, $0f_{5/2}$, and $0f_{3/2}$. Although some orbitals come close to each other in some rms radii, they do not intersect. There is no clear evidence of the nuclear shell gaps in Fig. 2.

Figure 3 sees the three uninhabited orbitals of zero charge (the lower curves in the figure) and the 10 orbitals, which are Δ^0 (overriding the upper curves in the figure). A gap of about 251.2 MeV is observed between the last dominant orbital of neutrons and the first predominantly Δ^0 orbitals due to the difference in the rest mass of baryons (neutrons and Δ^0s). These results show the gap between nucleons and delta levels.

Finally, the behavior of the positively charged baryon orbitals is not shown separately here because it exhibits properties similar to those of the chargeless baryons.

Perhaps the most prominent feature of these ^{208}Pb results, in contradiction to all our previous results, is the large presence of Δs in the ground state of equilibrium. This strongly stimulates more efforts to move forward with larger model spaces and heavier nuclei.

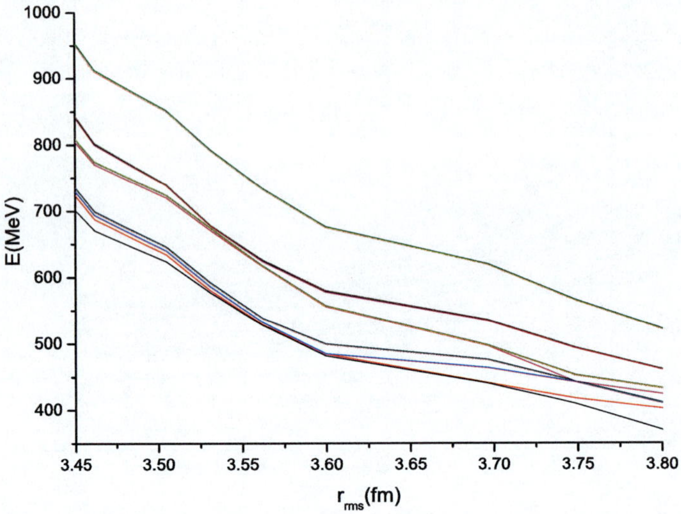

Fig. 2 The 10 self-consistent zero-charge levels of single-particle orbitals for the ^{208}Pb nucleus which are dominant Δ^0 in character versus the rms radius

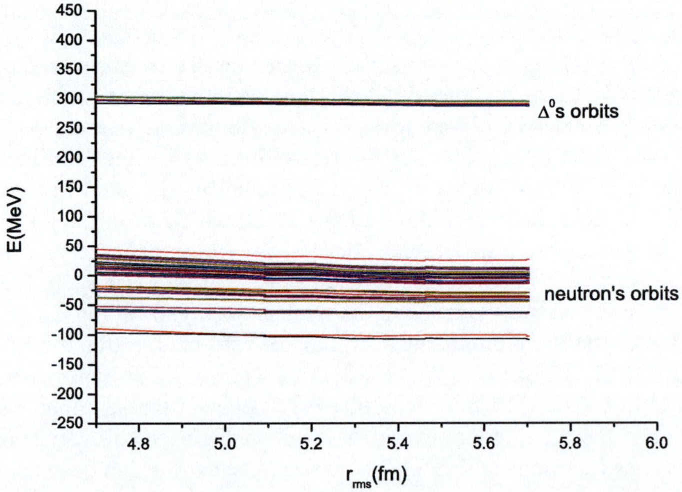

Fig. 3 Orbitals of single-particle energy as a function of r_{rms} for the zero-charge unoccupied orbitals (low curves) and the 10 orbitals, which are dominant Δ^0 for the ^{208}Pb nucleus

3 Conclusion and Outlook

By using an effective baryon-baryon realist Hamiltonian, self-consistent single-particle spectra with delta excitations of the spherical nucleus ^{208}Pb have been investigated in a bound Hartree-Fock approximation. In this approach, the nuclear shell model is derived with single-particle energy levels occupied by baryons, which are a mixture of nucleons and Δs. As shown in ^{208}Pb, the results compare favorably with those of the success phenomena of the shell model. In the near future, we are excited to extend this study to finite temperature, into larger nuclear model space sizes and into heavier nuclei.

Remember that the spectrum of a single particle is constructed entirely from the basic microscopic Hamiltonian. Thus, it is a remarkable result of these calculations that the computed spectrum follows the expected order of shell model phenomena in the orbits of dominant nucleons, and the spectrum gaps appear clearly between the shells. When the nucleus is compressed, the arrangement of single-particle orbitals and gaps is maintained. It should also be noted that orbitals closer to a single-particle zero energy are more compression-sensitive; this means that there is more compression response from inside the nucleus. The general trend for exhibited single-particle energies is to shift to higher energies when the nucleus is compressed.

The behavior of single-particle energy orbitals matches well with the orbital arrangement of the nuclear standard shell model. The gap between the shells is very clear. The splitting of the orbitals in each shell is an indication that the L-S coupling is sufficiently strong in Reid soft core potential, i.e., L-S coupling becomes stronger when the static load on the nucleus is increased. When the nucleus is compressed, the cleavage of the orbitals becomes more pronounced, especially in delta orbitals.

Acknowledgements The author acknowledges that this research was supported by the Deanship of Scientific Research at Zarqa University/Jordan.

References

1. Cottle, P.: Nature **465**, 430 (2010)
2. Krane, K.: Introductory Nuclear Physics, 2nd edn. Wiley (1988)
3. Bohr, A., Mottelson, B.: Nuclear Structure, vol. 1. W. A. Beujamin, New York (1969)
4. Arda, A., Sever, R.: Int. J. Mod. Phys. C **22**, 651 (2009)
5. Reisman, D., Toor, A., Cauble, R., Hall, C., Asay, J., Knudson, M., Furnish, M.: J. Appl. Phys. **89**, 1625 (2001)
6. Hjort, E., et al.: Phys. Rev. Lett. **79**, 4345 (1997)
7. Hofmann, M., Mattiello, R., Sorge, H., Stöcher, H., Greiner, W.: Phys. Rev. C **51**, 2095 (1995)
8. Hong, B., et al.: Phys. Lett. B **407**, 115 (1997)
9. Gaitanos, T., Colonna, M., Toro, M., Wolter, H., Ferini, G.: Phys. Rev. Lett. **97**, 202301 (2006)
10. Silva, A., Oliveira, J., Rodrigues, H., Duarte, S., Chiapparini, M.: AIP Conf. Proc. **1351**, 83 (2011)
11. Abu-Sei'leek, M.H.: Nucl. Phys. Rev. **27**, 399 (2010)
12. Abu-Sei'leek, M.H.: Commun. Theor. Phys. **55**, 115 (2011)

13. Abu-Sei'leek, M.H.: Int. J. Pure Appl. Phys. **7**, 73 (2011)
14. Abu-Sei'leek, M.H.: Pramana-J. Phys. **76**, 573 (2011)
15. Abu-Sei'leek, M.H.: Turk. J. Phys. **55**, 115 (2011)
16. Abu-Sei'leek, M.H.: Nucl. Phys. Rev. **28**, 416 (2011)
17. Abu-Sei'leek, M.H.: J. Phys. Soc. Jpn. **80**, 104201 (2011)
18. Abu-Sei'leek, M.H.: Turk. J. Phys. **38**, 253 (2014)
19. Abu-Sei'leek, M.H., Farrag, E., Masharfe, R.: IJISM **4**, 37 (2016)
20. Abu-Sei'leek, M.H.: J. Appl. Math. Phys. **4**, 586 (2016)
21. Abu-Sei'leek, M.H.: J. Appl. Math. Phys. **6**, 458 (2018)
22. Abu-Sei'leek, M.H.: Iran. J. Sci. Technol. Trans. Sci. **43**, 1365 (2019)
23. Abu-Sei'leek, M.H.: Nucl. Phys. A **1027**, 122522 (2022)
24. Abu-Sei'leek, M.H.: Pramana-J. Phys. **98** (2022)
25. Sharma, M., Lalazissis, G., Köning, J., Ring, P.: Phys. Rev. Lett. **74**, 3744 (1995)
26. Klopp, F., Zenk, H.: Adv. Math. Phys. **1**, 1 (2009)

Fractional-Order SEIR Covid-19 Model: Discretization and Stability Analysis

Iqbal M. Batiha, Noureddine Djenina, Adel Ouannas, and Taki-Eddine Oussaeif

Abstract From the perspective of the fact that confirms all statistics on epidemics can be classified as discrete, we aim in this paper to provide a new discrete-time version of a recent SEIR mathematical model. In other words, a new nabla fractional-order discrete-time system associated with the SEIR model is investigated in terms of its stability analysis including its positively invariant region, fixed points, and basic reproductive number. Several numerical simulations are illustrated to verify our findings.

Keywords Discrete fractional calculus · Fractional-order discrete-time system · Positively invariant region · Fixed points · Basic reproductive number

1 Introduction

Epidemics have been a source of danger to humanity since ancient times, and they still pose a great threat to life, and also cause great disruptions from an economic point of view and cause many problems. Therefore, understanding the behavior of epidemics is deemed a very important issue to try to control. It is common knowledge that the mathematical modeling can provide us with great efficiency and reliability for understanding the behavior of diseases and epidemics. More recently, several models have been studied in this field trying to understand the behavior of Covid-19

I. M. Batiha (✉)
Department of Mathematics, Al Zaytoonah,
University of Jordan, Amman 11733, Jordan
e-mail: i.batiha@zuj.edu.jo

Nonlinear Dynamics Research Center (NDRC), Ajman University, Ajman, UAE

N. Djenina · A. Ouannas · T.-E. Oussaeif
Department of Mathematics and Computer Science, University of Larbi Ben M'hidi,
Oum El Bouaghi, Algeria
e-mail: Ouannas.adel@univ-oeb.dz

T.-E. Oussaeif
e-mail: taki_maths@live.fr

D. Zeidan et al. (eds.), *Mathematics and Computation*, Springer Proceedings
in Mathematics & Statistics 418, https://doi.org/10.1007/978-981-99-0447-1_20

Table 1 Initial values of the SEIR model

Variable	Description
S	Susceptible compartment
E	Exposed compartment
I	Infected compartment
R	Recovered compartment
μ'	Corona death rate
d_0'	Natural death
β'	Reducing infection rate
b'	Recruitment rate
a_1'	Contact rate
γ'	The saturation constant
a_2'	Interaction of infected and exposed
w'	Recovery rate
τ'	Individuals goes to Exposed class

diseases and other diseases, see [5–11]. In light of this fact, we intend to be interest with a recent Covid-19 model presented in [12] that can be described as follow:

$$\begin{cases} \frac{dS(t)}{dt} = b' - \frac{a_1' S(t) I(t)}{1+\gamma' I(t)} - (d_0' + \tau')S(t), \\ \frac{dE(t)}{dt} = \tau' S(t) - d_0' E(t) - a_2' \beta' E(t) I(t), \\ \frac{dI(t)}{dt} = a_2' \beta' E(t) I(t) + \frac{a_1' S(t) I(t)}{1+\gamma' I(t)} - (d_0' + \mu' + w') I(t), \\ \frac{dR(t)}{dt} = w' I(t) - d_0' R(t), \end{cases} \tag{1}$$

subject to the following initial conditions:

$$S(0), E(0), I(0), R(0) \geq 0. \tag{3}$$

where $t \in \mathbb{R}^+$, and where the parameters as well as the states of the above system are described in Table 1.

Adding up the equations in system (1) yields the following assertion:

$$N = S + E + I + R.$$

In fact, due to the statistics associated with this epidemic being discrete, then the system that we aim to propose it will be proper to be modeled in its discrete-time case. In light of this view, we will use the following approximation:

$$\frac{dX(t)}{dt} \simeq \frac{X(t) - X(t-h)}{h}. \tag{4}$$

This would make system (1) to be expressed as follows:

$$\begin{cases} \frac{S(t)-S(t-h)}{h} = b' - \frac{a_1' S(t) I(t)}{1+\gamma' I(t)} - (d_0' + \tau')S(t), \\ \frac{E(t)-E(t-h)}{h} = \tau' S(t) - d_0' E(t) - a_2' \beta' E(t) I(t), \\ \frac{I(t)-I(t-h)}{h} = a_2' \beta' E(t) I(t) + \frac{a_1' S(t) I(t)}{1+\gamma' I(t)} - (d_0' + \mu' + w') I(t), \\ \frac{R(t)-R(t-h)}{h} = w' I(t) - d_0' R(t). \end{cases} \quad t \in \mathbb{R}^+.$$

By multiplying both sides of the above system by h, using the notations $X(t) = X(n)$ as well as $X(t-h) = X(n-1)$, where $n \in \mathbb{N}$, $X(t) = (S(t), E(t), I(t), R(t))$, and then by setting:

$$\mu = h\mu'; \; d_0 = hd_0'; \; \beta = h\beta'; \\ b = hb'; \; a_1 = ha_1'; \; \gamma = \gamma' \\ a_2 = ha_2'; \; w = hw'; \; \tau = h\tau',$$

then we will gain the following system:

$$\begin{cases} \nabla S(n) = b - \frac{a_1 S(n) I(n)}{1+\gamma I(n)} - (d_0 + \tau) S(n), \\ \nabla E(n) = \tau S(n) - d_0 E(n) - a_2 \beta E(n) I(n), \\ \nabla I(n) = a_2 \beta E(n) I(n) + \frac{a_1 S(n) I(n)}{1+\gamma I(n)} - (d_0 + \mu + w) I(n), \\ \nabla R(n) = w I(n) - d_0 R(n). \end{cases} \quad n \in \mathbb{N}, \tag{5}$$

where ∇ is the backward difference operator (i.e., $\nabla X(n) = X(n) - X(n-1)$).

In more recent time, several modeling chemical and physical phenomena have broadly been carried out using the theory of fractional-order discrete-time systems, see [1, 2]. To obtain a full overview of the topic handled here, the reader may refer to [13] to get many definitions of the fractional-order discrete-time operators, while the reader may refer to [14] to get a sufficient knowledge about the stability analysis of the delta commensurate fractional-order operators. In the same regard, the incommensurate fractional-order case was addressed in [3, 4], whereas the stability analysis of the nabla operator was just studied in [15] in its commensurate fractional-order case.

2 Preliminaries

In order to propose a new discrete-time version of the SEIR model given in the system (5), this section introduces briefly some basic definitions and preliminaries associated with discrete fractional calculus. In all of the definitions below, the function f is defined on $\mathbb{N}_a = \{a, a+1, a+2, \ldots\}$, for $a \in \mathbb{R}$.

Definition 1 ([13]) For a function $f : N_a \to \mathbb{R}$, the nabla fractional sum of order $\alpha > 0$ is defined by

$$\nabla_a^{-\alpha} f(t) := \frac{1}{\Gamma(\alpha)} \sum_{s=a+1}^{t} (t-s+1)^{\overline{\alpha-1}} f(s), \text{ for } t \in \mathbb{N}_a, \tag{10}$$

where $\Gamma(\cdot)$ is the Euler's gamma function and $t^{\overline{\alpha}} = \frac{\Gamma(t+\alpha)}{\Gamma(t)}$.

Definition 2 ([13]) The nabla Riemann-Liouville fractional-order difference operator of order $0 < \alpha \leq 1$ is defined by

$$\nabla_a^{\alpha} f(t) := \left(\nabla\, \nabla_a^{-(1-\alpha)} f\right)(t) = \frac{1}{\Gamma(1-\alpha)} \nabla \sum_{s=a+1}^{t} (t-s+1)^{\overline{-\alpha}} f(s), \text{ for } t \in \mathbb{N}_{a+1}, \tag{11}$$

where a is the starting point and $\nabla f(t) = f(t) - f(t-1)$.

Definition 3 ([13]) Assume that $0 < \alpha \leq 1$, $a \in \mathbb{R}$, and f is defined on N_a. Then the nabla Caputo fractional-order difference operator of order α is defined by

$$^{C}\nabla_a^{\alpha} f(t) := \left(\nabla_a^{-(1-\alpha)} \nabla f\right)(t) = \frac{1}{\Gamma(1-\alpha)} \sum_{s=a+1}^{t} (t-s+1)^{\overline{-\alpha}} (\nabla f)(s), \tag{12}$$

where a is the starting point and $t \in \mathbb{N}_{a+1}$.

In view of the previous arguments, we intend to propose what we are interested in this paper; the fractional-order discrete-time system associated with system (5). In particular, we propose the following system:

$$\begin{cases} ^{C}\nabla_a^{\alpha} S(t) = b - \frac{a_1 S(t) I(t)}{1+\gamma I(t)} - (d_0 + \tau) S(t), \\ ^{C}\nabla_a^{\alpha} E(t) = \tau S(t) - d_0 E(t) - a_2 \beta E(t) I(t), \\ ^{C}\nabla_a^{\alpha} I(t) = a_2 \beta E(t) I(t) + \frac{a_1 S(t) I(t)}{1+\gamma I(t)} - (d_0 + \mu + w) I(t), \\ ^{C}\nabla_a^{\alpha} R(t) = w I(t) - d_0 R(t). \end{cases} \quad t \in \mathbb{N}_1. \tag{13}$$

Clearly, the above system is more general than the system (5), and all the stability results that we will obtain soon remain true for system (5) when $\alpha = 1$. As a matter of fact, we can see that the last equation of system (13) can be abandoned, and so we aim to take care of the following system:

$$\begin{cases} ^{C}\nabla_a^{\alpha} S(t) = b - \frac{a_1 S(t) I(t)}{1+\gamma I(t)} - (d_0 + \tau) S(t), \\ ^{C}\nabla_a^{\alpha} E(t) = \tau S(t) - d_0 E(t) - a_2 \beta E(t) I(t), \\ ^{C}\nabla_a^{\alpha} I(t) = a_2 \beta E(t) I(t) + \frac{a_1 S(t) I(t)}{1+\gamma I(t)} - (d_0 + \mu + w) I(t), \end{cases} \quad t \in \mathbb{N}_1. \tag{14}$$

3 Stability Analysis

In this part, we will investigate the stability analysis of a fractional-order discrete-time system (14) in light of its positively invariant region, fixed points, and the basic reproductive number.

3.1 Positively Invariant Region

In order to address the invariant region of the fractional-order discrete-time system (13), we state and prove the next theoretical result.

Theorem 1 *The set:*

$$\Psi = \left\{ (S, E, I, R) \in \mathbb{R}_+^4 \text{ and } S + E + I + R \leq \frac{b}{d_0} \right\}, \tag{6}$$

of system (14) is invariant region, where

$$\mathbb{R}_+^4 = \left\{ (x_1, x_2, x_3, x_4) \in \mathbb{R}^4 \text{ and } x_i \geq 0 \text{ for } i = 1, 2, 3, 4 \right\}.$$

Proof Without loss of generality, we assume that $\alpha = 1$ in system (13). This means that system (13) and system (5) are equivalent. Now, to prove the positivity of the solution of the system (5), we assume by contrary the opposite. In particular, we assume that the first component of S is negative at $n_0 \in \mathbb{N}$. This would imply

$$S(n_0) - S(n_0 - 1) = b - \frac{a_1 S(n_0) I(n_0)}{1 + \gamma I(n_0)} - (d_0 + \tau) S(n_0) \geq 0,$$

or

$$S(n_0) \geq S(n_0 - 1) \geq 0,$$

which is a contradiction. Therefore, we have $S(n) \geq 0$, for $n > n_0$. In a similar manner, we can prove the following assertions:

$$E(n) \geq 0, \ I(n) \geq 0, \ R(n) \geq 0.$$

Consequently, by adding the equations of system (5) to each other, we get

$$\nabla N(n) = b - d_0 N(n) - \mu I(n).$$

Due to the class I is positive, we obtain

$$\nabla N(n) \leq b - d_0 N(n),$$

where

$$N(0) = S(0) + E(0) + I(0) + R(0).$$

Applying comparison theorem yields to the following inequality:

$$N(n) \le \frac{b + N(n-1)}{1 + d_0}.$$

Let $N(0) \le \frac{b}{d_0}$, and suppose that $N(n-1) \le \frac{b}{d_0}$ for some natural number n, then we get

$$N(n) \le \frac{b + N(n-1)}{1 + d_0} \le \frac{b + \frac{b}{d_0}}{1 + d_0} = \frac{b}{d_0}.$$

By induction, we can find that the solution of system (5) exists when:

$$0 \le N(n) \le \frac{b}{d_0},$$

for all n. Hence, this solution belongs to the following invariant region:

$$\Psi = \left\{ (S, E, I, R) \in \mathbb{R}_+^4 \text{ and } S + E + I + R \le \frac{b}{d_0} \right\}.$$

3.2 Fixed Points and Basic Reproduction Number

In order to compute the basic reproduction number, which is deemed one of the most important epidemical concepts, it is necessary to study the dynamics of the proposed system in terms of its fixed points. Such points can be found by solving the following system of equations:

$$\begin{cases} b - \frac{a_1 S^* I^*}{1+\gamma I^*} - (d_0 + \tau) S^* = 0, \\ \tau S^* - d_0 E^* - a_2 \beta E^* I^* = 0, \\ a_2 \beta E^* I^* + \frac{a_1 S^* I^*}{1+\gamma I^*} - (d_0 + \mu + w) I^* = 0, \\ w I^* - d_0 R^* = 0. \end{cases} \tag{7}$$

As a matter of fact, there are two kinds of fixed points; the first one is called the disease-free fixed point which can be yielded by considering $I^* = 0$, whereas the second one is called the endemic fixed point which can be obtained by considering $I^* \neq 0$. In view of this point, the previous equations can yield $\varepsilon_0 = \left(\frac{b}{\tau + d_0}, \frac{\tau b}{d_0(\tau + d_0)}, 0, 0 \right)$ as a disease-free fixed point. On the other hand, if one supposes that $I^* \neq 0$, then we get

$$\begin{aligned} S^* &= \frac{b(1+\gamma I^*)}{d_0+\tau+a_1 I^*}, \\ E^* &= \frac{\tau b(1+\gamma I^*)}{(d_0+\tau+a_1 I^*)(a_2\beta I^*-d_0)}, \\ R^* &= \frac{w}{d_0} I^*. \end{aligned} \tag{8}$$

This means that the endemic fixed point is $\varepsilon^* = (S^*, E^*, I^*, R^*)$. It was reported in Ref. [12] that the basic reproductive number R_0 has the form:

$$R_0 = \frac{b(a_1 d_0 + a_2 \beta \tau)}{d_0\,(d_0+\tau)\,(d_0+\mu+w)}. \tag{9}$$

In fact, the basic reproduction number has increasingly been a principal quantity used for determining the force of required interferences for controlling the epidemics. It is common knowledge that if $R_0 < 1$, then there is an absence of epidemics in natural populations. On the contrary, if $R_0 > 1$, then the disease will increasingly spread in the susceptible population. From this point of view together with the next result, we can then give some other theoretical results connected with the stability analysis of the system (14).

Theorem 2 ([15]) *If all eigenvalues of the jacobian matrix J at the fixed point x^* of system (14) are located in the set $\{z \in \mathbb{C} : |\arg z| > \frac{\alpha\pi}{2}$ or $|z| > \left(2\cos\frac{\arg z}{\alpha}\right)^{\alpha}\}$, then system (14) has a unique solution for all initial vectors close enough to x^* and moreover x^* is asymptotically stable.*

Theorem 3 *Suppose that $R_0 < 1$. Then the disease-free fixed point ε_0 of system (14) is locally asymptotically stable.*

Proof To prove this result, we find the Jacobian matrix at ε_0 as follows:

$$J_0 = \begin{pmatrix} -(\tau+d_0) & 0 & \frac{a_1 b}{\tau+d_0} \\ \tau & -d_0 & \frac{a_2\beta\tau b}{d_0(\tau+d_0)} \\ 0 & 0 & R_0-1 \end{pmatrix}.$$

The characteristic polynomial of J_0 can be then given by

$$\lambda^3 + A\lambda^2 + B\lambda + C = 0, \tag{15}$$

where

$$\begin{aligned} A &= d_0(\tau+d_0) + (\mu+d_0+w)(1-R_0) > 0 \\ B &= d_0(\tau+d_0)\,(1+(\mu+d_0+w)(1-R_0)) > 0 \\ C &= d_0(\tau+d_0)(\mu+d_0+w)(1-R_0) > 0. \end{aligned}$$

Consequently, we have

$$AB - C = d_0(\tau+d_0)\left((\mu+d_0+w)^2 + d_0(\tau+d_0)\,((\mu+d_0+w)+1)\right)(1-R_0).$$

Then by Routh-Hurtwriz criteria the roots of polynomial (15) have a negative real part if $R_0 < 1$. This shows, according to Theorem 2, that system (14) is locally asymptotically stable at ε_0.

Theorem 4 *Under the condition* $R_0 > 1$, *system (14) is locally asymptotically stable at* ε^*.

Proof The jacobian matrix of system (14) at the endemic fixed point ε^* can be given as

$$J_* = \begin{pmatrix} -(\tau + d_0) - \frac{a_1 I^*}{1+\gamma I^*} & 0 & -\frac{a_1 S^*}{(1+\gamma I^*)^2} \\ \tau & -d_0 - a_2\beta I^* & -a_2\beta E^* \\ \frac{a_1 I^*}{1+\gamma I^*} & a_2\beta I^* & \frac{a_1 S^*}{(1+\gamma I^*)^2} + a_2\beta E^* - (\mu + d_0 + w) \end{pmatrix}.$$

According to [12], all eigenvalues of J_* have negative real parts when $R_0 > 1$, Thus, according to Theorem 2, system (14) is asymptotically stable at ε^*.

4 Numerical Simulations

In this section, we will provide several graphical simulations to explain the results obtained in the previous section. Therefore, we choose a divided population (in millions) as follows [16]:

$$\begin{aligned} S(0) &= 219.87904, \\ E(0) &= 0, \\ I(0) &= 0.521211, \\ R(0) &= 0.486225. \end{aligned} \tag{16}$$

In the same regard, we take the parameters of system (14) as follows:

$$\begin{array}{lll} \mu = 1.9 \times 10^{-5}; & d_0 = 1.9 \times 10^{-5}; & \beta = 9 \times 10^{-6}; \\ b = 3 \times 10^{-4}; & a_1 = 10^{-5}; & \gamma = 9.8601 \times 10^{-6}; \\ a_2 = 2 \times 10^{-5}; & w = 5.8 \times 10^{-6}; & \tau = 7.28 \times 10^{-5}. \end{array} \tag{17}$$

To apply Theorem 3, one must first calculate R_0, which would be as follows:

$$R_0 = \frac{b(a_1 d_0 + a_2\beta\tau)}{d_0\,(d_0 + \tau)\,(d_0 + \mu + w)},$$

or

$$R_0 = \frac{0.0003(0.00001 \times 0.000019 + 0.00002 \times 0.000009 \times 0.0000728)}{0.000019 \times (0.000019 + 0.0000728)\,(0.000019 + 0.000019 + 0.0000058)}.$$

This implies that $R_0 = 0.74616 < 1$. Therefore, according to Theorem 3, the disease-free fixed point ε_0 of system (14) is locally asymptotically stable. However, Figs. 1,

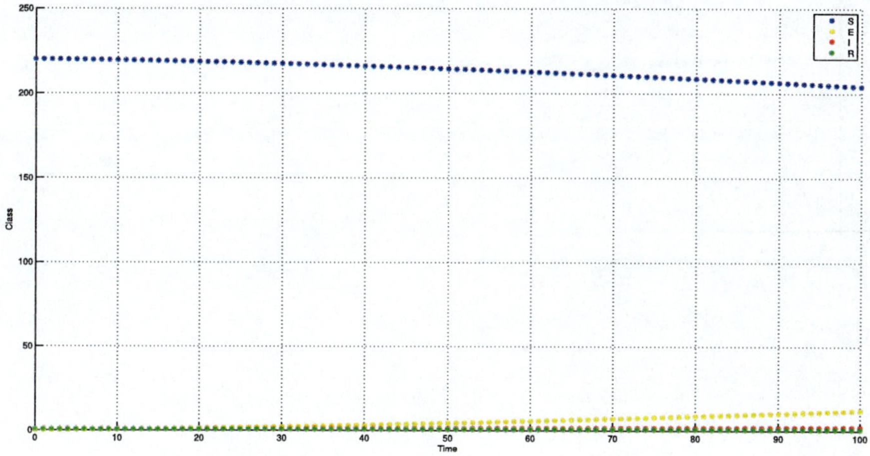

Fig. 1 Numerical simulation when $a_1 = 10^{-5}$ ($R_0 < 1$) and $\alpha = 0.5$

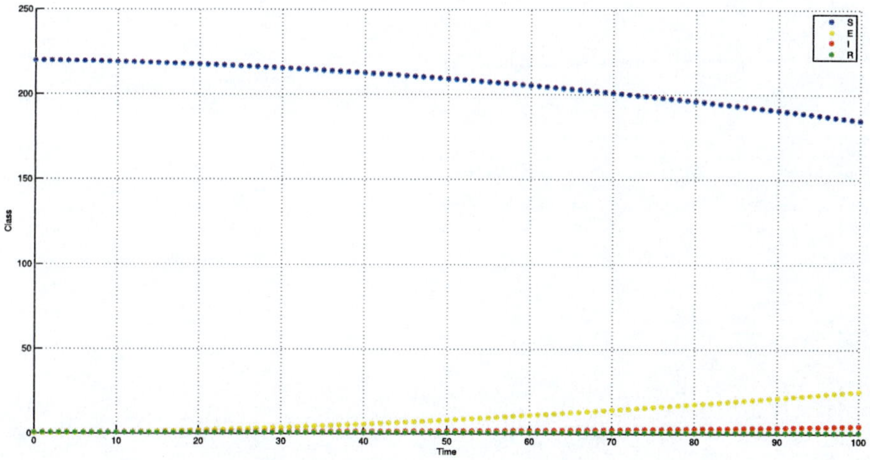

Fig. 2 Numerical simulation when $a_1 = 10^{-5}$ ($R_0 < 1$) and $\alpha = 0.7$

2 and 3 illustrate the dynamics of system (14) in the case of $R_0 < 1$. It can be seen from these graphs that a slight change in the fractional-order values allows one to control the curvature of the system's curves.

In the same context, if one takes the value $a_1 = 0.00003$, we get $R_0 = 2.2384 > 1$. Then the endemic fixed point ε^* of system (14) is locally asymptotically stable. Thus, it is expected that the Covid-19 diseases will be increased over time, see Figs. 4, 5, and 6.

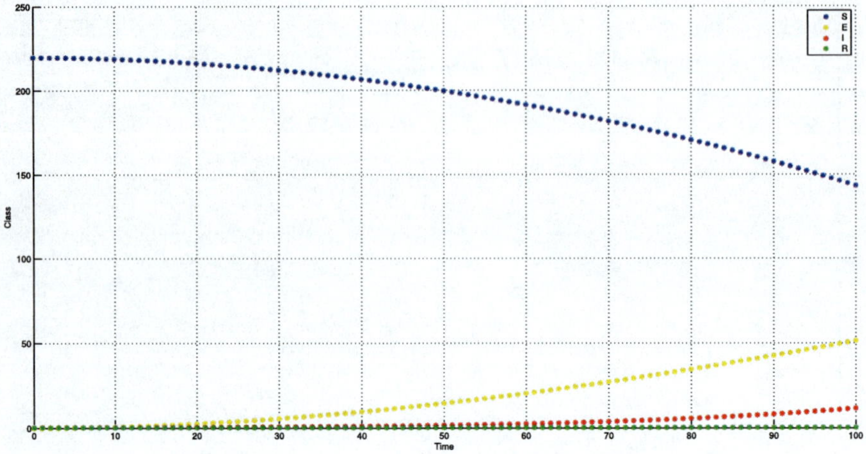

Fig. 3 Numerical simulation when $a_1 = 10^{-5}$ ($R_0 < 1$) and $\alpha = 0.9$

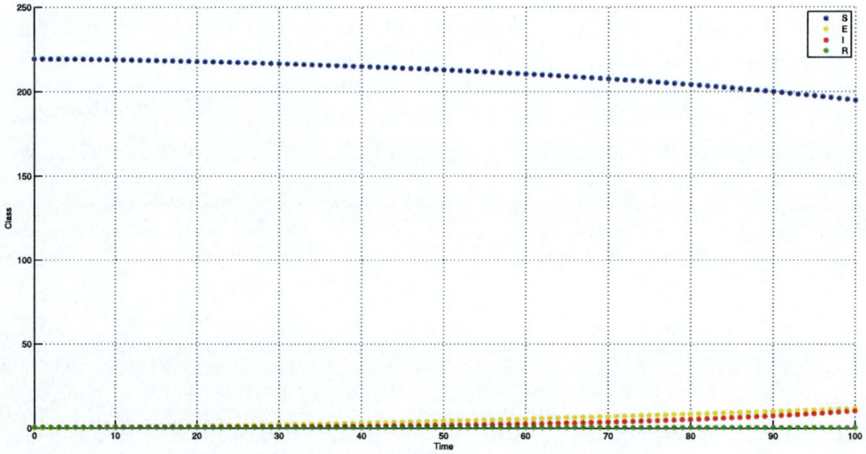

Fig. 4 Numerical simulation when $a_1 = 3 \times 10^{-5}$ ($R_0 > 1$) and $\alpha = 0.5$

5 Conclusions

In this work, we have provided a new discrete-time version of a recent SEIR mathematical model. This model has been investigated in terms of its stability analysis including its positively invariant region, fixed points, and the basic reproductive number.

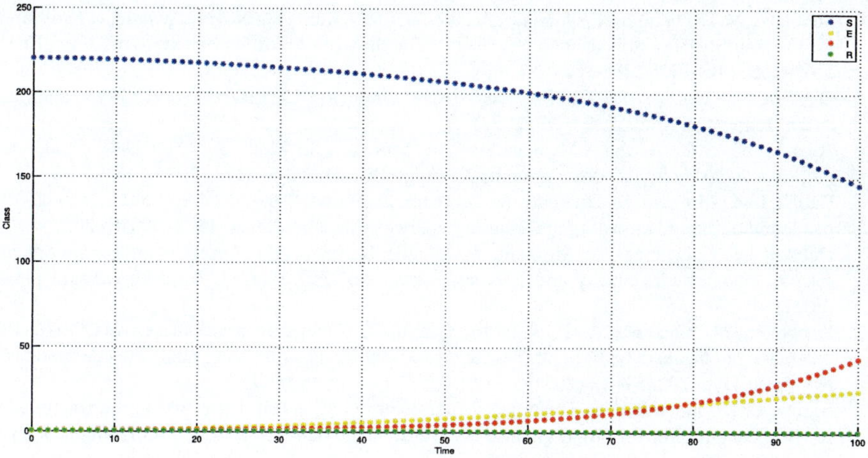

Fig. 5 Numerical simulation when $a_1 = 3 \times 10^{-5}$ $(R_0 > 1)$ and $\alpha = 0.7$

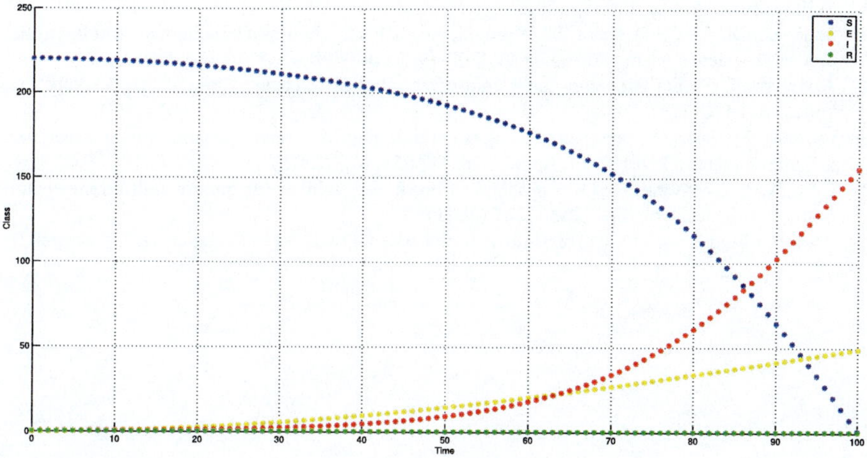

Fig. 6 Numerical simulation when $a_1 = 3 \times 10^{-5}$ $(R_0 > 1)$ and $\alpha = 0.9$

References

1. Wang, B., Ouannas, A., Xia, W.F., Jahanshahi, H., Alotaibi, N.D.: Synchronizing between two reaction-diffusion systems of integer-and fractional-order applied on certain chemical models. Fractals, preprint (2022)
2. Gasri, A., Ouannas, A., Khennaoui, A.A., Grassi, G., Oussaeif, T.E., Pham, V.T.: Chaotic fractional discrete neural networks based on the Caputo h-difference operator: stabilization and linear control laws for synchronization. Eur. Phys. J. Special Top. **2022**, 1–15 (2022)
3. Djenina, N., Ouannas, A., Batiha, I.M., Grassi, G., Pham, V.T.: On the stability of linear incommensurate fractional-order difference systems. Mathematics (10), 1754 (2020)

4. Shatnawi, M.T., Djenina, N., Ouannas, A., Batiha, I.M., Grassi, G.: Novel convenient conditions for the stability of nonlinear incommensurate fractional-order difference systems. Alex. Eng. J. **61**, 1655–1663 (2022)
5. Djenina, N., Ouannas, A.: The fractional discrete model of COVID-19: solvability and simulation. Innovat. J. Math. **1**, 23–33 (2022)
6. Albadarneh, R.B., Batiha, I.M., Ouannas, A., Momani, S.: Modeling COVID-19 pandemic outbreak using fractional-order systems. Comput. Sci. **16**, 1405–1421 (2021)
7. Batiha, I.M., Momani, S., Ouannas, A., Momani, Z., Hadid, S.B.: Fractional-order COVID-19 pandemic outbreak: modeling and stability analysis. Int. J. Biomath. **15**, 2150090 (2022)
8. Debbouche, N., Ouannas, A., Momani, S., Cafagna, D., Pham, V.T.: Fractional-order biological system: chaos, multistability and coexisting attractors. Eur. Phys. J. Spec. Top. **2021**, 1–10 (2021)
9. Debbouche, N., Ouannas, A., Batiha, I.M., Grassi, G.: Chaotic dynamics in a novel COVID-19 pandemic model described by commensurate and incommensurate fractional-order derivatives. Nonlinear Dyn. **2021**, 1–13 (2021)
10. Batiha, I.M., Oudetallah, J., Ouannas, A., Al-Nana, A.A., Jebril, I.H.: Tuning the fractional-order PID-controller for blood glucose level of diabetic patients. J. Adv. Soft Comput. Appl. **13**, 1–10 (2021)
11. Debbouche, N., Almatroud, A.O., Ouannas, A., Batiha, I.M.: Chaos and coexisting attractors in glucose-insulin regulatory system with incommensurate fractional-order derivatives. Chaos, Solitons Fractals **143**, 110575 (2021)
12. Shah, K., Din, R.U., Deebani, W., Kumam, P., Shah, Z.: On nonlinear classical and fractional order dynamical system addressing COVID-19. Results Phys. **24**, 104069 (2021)
13. Abdeljawad, T.: On Riemann and Caputo fractional differences. Comput. Math. Appl. **62**, 1602–1611 (2011)
14. Čermák, J., Györi, I., Nechvátal, L.: On explicit stability conditions for a linear fractional difference system. Fract. Calc. Appl. Anal. **18**, 651–672 (2015)
15. Čermák, J., Nechvátal, L.: On a problem of linearized stability for fractional difference equations. Nonlinear Dyn. **104**, 1253–1267 (2021)
16. www.worldometers.info, Currentinformation about COVID-19 in Pakistan, 18 January, 2021

A q-Starlike Class of Harmonic Meromorphic Functions Defined by q-Derivative Operator

Abdullah Alsoboh and Maslina Darus

Abstract In this present work, we introduces a subclass of harmonic meromorphic functions associated with q-calculus operator. With that, we study various interesting properties of this class, like, the coefficients bounds, distortion theorems, extreme points, convolution, convex combinations. Further, q-integral operator is also defined and we show that the new class aforementioned is closed under this q-derivative operator.

Keywords Harmonic function · Meromorphic function · q-Starlike function · q-Calculus · q-Integral operator

1 Introduction and Preliminaries

Quantum calculus (or q-calculus) has attracted the interest of many researchers due to its several applications in different branches of mathematics and physics, especially geometric function theory. The structure of q-calculus enhances the method of conventional complements for various modules of orthogonal polynomials and functions. One of the most useful and well-designed tools for analysing the characteristics of special functions in mathematical analysis and mathematical physics is the connection between equilibriums of differential formulae (equations, operators, and inequalities) and their solutions. The q-calculus was initiated by Euler and Jacobi in 18th century. The application of q-calculus was initiated and developed in a systematic way by [17, 18]. Aral and Gupta [10, 11] proposed q-analogue of Baskakov and Durrmeyer operator depends on quantum calculus. Some other applications of

A. Alsoboh (✉)
Department of Mathematics, Al-Leith University College, Umm Al-Qura University, Mecca 24231, Saudi Arabia
e-mail: amsoboh@uqu.edu.sa

M. Darus
Department of Mathematical Sciences, Faculty of Science and Technology, Universiti Kebangsaan Malaysia, 43600 Bangi, Selangor, Malaysia
e-mail: maslina@ukm.edu.my

D. Zeidan et al. (eds.), *Mathematics and Computation*, Springer Proceedings in Mathematics & Statistics 418, https://doi.org/10.1007/978-981-99-0447-1_21

q-operator are studied by Aral et. al [12] and Elhaddad et. al [15]. The harmonic type of q-analogues calculus are found in ([1–3, 13, 23]). Many different problems related to q-calculus can be seen recently in [21] and ([4–8]). In the future, we believe that deriving operators upon q-analogues in the classes of harmonic functions will gain significant importance.

Now, we present some definitions and concepts of q-calculus used throughout this paper by assuming that q satisfies the condition $q \in (0, 1)$, (see for more details [16]).

Definition 1 Let $s \in \mathbb{N}$ and $q \in (0, 1)$. The q-number, denoted by $[s]_q$, is defined by

$$[s]_q = \frac{1 - q^s}{1 - q}.$$

When $s \in \mathbb{N}$, we obtain $[s]_q = 1 + q + \cdots + q^{s-1}$, and when $q \to 1^-$, then $[s]_q = s$.

Definition 2 The q-derivative (or q-difference operator) of a function f, defined on

$$\partial_q f(z) = \begin{cases} \frac{f(z)-f(qz)}{z-qz} & z \in \mathbb{C}\backslash\{0\} \\ 1 & z = 0, \end{cases}$$

We note that $\lim\limits_{q \to 1^-} \partial_q f(z) = f'(z)$ if f is differentiable at $z \in \mathbb{C}$.

Jackson [17] also introduced the q-integral of any function f by

$$\int_0^z f(t) d_q t = (1 - q)z \sum_{n=0}^{\infty} q^n f(q^n z) \tag{1}$$

provided that the series on right-hand side converges. For $z \in \Delta^* = \Delta \setminus \{0\}$, let $\mathcal{M}_{\mathcal{H}}$ be the class of functions:

$$f(z) = h(z) + \overline{g(z)} = \frac{1}{z} + \sum_{s=1}^{\infty} a_s z^s + \sum_{s=1}^{\infty} \overline{b_s z^s}, \tag{2}$$

which are harmonic in the punctured unit disc Δ^*, where $h(z)$ and $g(z)$ are analytic in Δ^* and Δ, respectively, and $h(z)$ has a simple pole at the origin with residue 1, this class was investigated studied by Jahangiri and Silverman [20] and then studied by [2, 9, 14].

We now define the class $\mathcal{M}_{\mathcal{H}_q}$ consisting of q-harmonic meromorphic functions in Δ^*.

Definition 3 A harmonic function $f = h + \overline{g}$ defined by (2) is said to be q-harmonic meromorphic, locally univalent and sense-preserving in Δ^* denoted by $\mathcal{M}_{\mathcal{H}_q}$, if and only if

$$\left|\frac{\partial_q g(z)}{\partial_q h(z)}\right| < 1, \qquad (q \in (0,1),\ z \in \Delta^*). \tag{3}$$

Note that for $\lim\limits_{q\to 1^-} \mathcal{M}_{\mathcal{H}_q} = \mathcal{M}_{\mathcal{H}}$, the harmonic meromorphic was studied by Jahangiri and Silverman [20] and Jahangiri [19].

Let $\kappa \geq 0$, $0 < q < 1$ and $f \in \mathcal{M}_{\mathcal{H}_q}$, we now define the operator $\mathcal{D}_q^\kappa f(z)$: $\mathcal{M}_{\mathcal{H}_q} \to \mathcal{M}_{\mathcal{H}_q}$ as

$$D_q^\kappa f(z) = D_q^\kappa h(z) + (-1)^\kappa \overline{D_q^\kappa g(z)}, \tag{4}$$

where

$$D_q^\kappa h(z) = qz\partial_q (D_q^{\kappa-1} h(z)) = \frac{(-1)^\kappa}{z} + \sum_{s=1}^{\infty} q^\kappa [s]_q^\kappa a_s z^s,$$

$$D_q^\kappa g(z) = qz\partial_q (D_q^{\kappa-1} g(z)) = \sum_{s=1}^{\infty} q^\kappa [s]_q^\kappa b_s z^s.$$

Using the operator D_q^κ, we define a generalised harmonic meromorphically starlike functions in Δ^*, with the definition below.

Definition 4 For $0 < q < 1$ and $0 \leq \alpha < 1$, $\mathcal{M}_{\mathcal{H}_q^*}(\alpha, \kappa)$ denotes the class of harmonic meromorphic functions f as in (2) if it satisfies the condition

$$\Re e \left\{ -\frac{qz\partial_q (D_q^\kappa h(z)) - \overline{qz\partial_q (D_q^\kappa g(z))}}{D_q^\kappa h(z) + \overline{D_q^\kappa g(z)}} \right\} > \alpha, \quad (z \in \Delta^*). \tag{5}$$

When $\kappa = 0$, then $\mathcal{M}_{\mathcal{H}_q^*}(\alpha, \kappa)$ is reduced to $\mathcal{M}_{\mathcal{H}_q^*}(\alpha)$ introduced by Aldweby and Darus [2], and $\lim_{q\to 1^-} \mathcal{M}_{\mathcal{H}_q^*}(0, 0)$ is the harmonic starlike class introduced by Jahangiri and Silverman [20].

Also, $\mathcal{M}_{\overline{\mathcal{H}_q^*}}(\alpha, \kappa) \subset \mathcal{M}_{\mathcal{H}_q^*}(\alpha, \kappa)$ consists of meromorphically harmonic functions of the form $f_\kappa(z) = h_\kappa + \overline{g_\kappa}$ such that h_κ and g_κ are of the form

$$h_\kappa(z) = \frac{(-1)^\kappa}{z} + \sum_{s=1}^{\infty} |a_s| z^s, \qquad g_\kappa(z) = (-1)^\kappa \sum_{s=1}^{\infty} |b_s| z^s, \qquad z \in \Delta^*. \tag{6}$$

2 Coefficient Condition

In our first theorem, we determine the sufficient coefficient bounds for the functions f in the class $\mathcal{M}_{\mathcal{H}_q^*}(\alpha,\kappa)$.

Theorem 1 *For $0 \le \alpha < 1$ and $f = h + \overline{g}$ defined by (2) which satisfies the condition*

$$\sum_{s=1}^{\infty} q^{\kappa}[s]_q^{\kappa}\Big((q[s]_q+\alpha)|a_s| + (q[s]_q-\alpha)|b_s|\Big) \le 1-\alpha, \tag{7}$$

where $\kappa \in \{0,1,2,\dots\}$ and $q \in (0,1)$. Then

(a) f is harmonic univalent and sense-preserving in Δ^*.
(b) $f \in \mathcal{M}_{\mathcal{H}_q^*}(\alpha,\kappa)$.

□

Proof Consider $f = h + \overline{g}$ as in (2), satisfying the inequality (7). For $0 < |z_1| \le |z_2| < 1$, we have

$$\begin{aligned}
|f(z_1) - f(z_2)| &\ge \frac{|z_1 - z_2|}{|z_1 z_2|}\left(1 - |z_2|^2 \sum_{s=1}^{\infty}(|a_s| + |b_s|)\frac{|{z_1}^s - {z_2}^s|}{|z_1 - z_2|}\right)\\
&\ge \frac{|z_1 - z_2|}{|z_1 z_2|}\left(1 - |z_2|^2 \sum_{s=1}^{\infty}(|a_s| + |b_s|)\left|z_1^{s-1} + \cdots + z_2^{s-1}\right|\right)\\
&\ge \frac{|z_1 - z_2|}{|z_1 z_2|}\left(1 - |z_2|^2 \sum_{s=1}^{\infty}(|a_s| + |b_s|)q^{\kappa}[s]_q^{\kappa}\right)\\
&> |z_1 - z_2|\left(1 - \sum_{s=1}^{\infty}\left\{\frac{q^{\kappa}[s]_q^{\kappa}([s]_q+\alpha)}{1-\alpha}|a_s| + \frac{([s]_q-\alpha)q^{\kappa}[s]_q^{\kappa}}{1-\alpha}|b_s|\right\}\right).
\end{aligned}$$

By condition (7) the last expression is non-negative. Therefore, f is univalent in Δ^*. To prove that f is sense-preserving, it is enough to show that $|\partial_q h(z)| > |\partial_q g(z)|$, as follows.

$$\begin{aligned}
\left|q\partial_q h(z)\right| &= \left|\frac{-1}{z^2} + \sum_{s=1}^{\infty} a_s q[s]_q z^{s-1}\right| \ge \left|\frac{-1}{z^2}\right| - \sum_{s=1}^{\infty}|a_s| q[s]_q |z|^{s-1}\\
&\ge \frac{1}{r^2} - \sum_{s=1}^{\infty}|a_s| q[s]_q r^{s-1} \ge 1 - \sum_{s=1}^{\infty} q[s]_q|a_s| \ge 1 - \sum_{s=1}^{\infty}|a_s|\left(\frac{q[s]_q+\alpha}{1-\alpha}\right)\\
&\ge \sum_{s=1}^{\infty}\left(\frac{q[s]_q-\alpha}{1-\alpha}\right)|b_s| > \sum_{s=1}^{\infty}|b_s| q[s]_q r^{s-1} > |q\partial_q g(z)|.
\end{aligned}$$

In order to prove that $f \in \mathcal{M}_{\mathcal{H}_q^*}(\alpha,\kappa)$, it suffices to show that

$$\Re e\left\{-\frac{qz\partial_q(D_q^\kappa h(z)) - \overline{qz\partial_q(D_q^\kappa g(z))}}{D_q^\kappa h(z) + \overline{D_q^\kappa g(z)}} - \alpha\right\} > 0, \quad (z \in \Delta^*).$$

Since, $\Re e(\rho(z)) > 0$ if and only if $\left|\frac{\rho(z)-1}{\rho(z)+1}\right| < 1$ for an analytic function $\rho(z) = 1 + d_1 z + d_2 z^2 + \ldots$. Let

$$\Gamma(z) = -qz\partial_q(D_q^\kappa h(z)) + \overline{qz\partial_q(D_q^\kappa g(z))} - \alpha D_q^\kappa h(z) - \alpha\overline{D_q^\kappa g(z)}, \tag{8}$$

and

$$\Lambda(z) = D_q^\kappa h(z) + \overline{D_q^\kappa g(z)}. \tag{9}$$

Then, we have to show that

$$\aleph(z) = |\Gamma(z) + \Lambda(z)| - |\Gamma(z) - \Lambda(z)| > 0 \tag{10}$$

Now, by substituting (4) in the left-hand side of the inequality (10) yields

$$\begin{aligned}
\aleph(z) &\geq \frac{2-2\alpha}{r} - 2\sum_{s=1}^{\infty}(q^{\kappa+1}[s]_q^{\kappa+1} + \alpha q^\kappa [s]_q^\kappa)|a_s|r^s - 2\sum_{s=1}^{\infty}(q^{\kappa+1}[s]_q^{\kappa+1} - \alpha q^\kappa [s]_q^\kappa)|b_s|r^s \\
&\geq 2\left((1-\alpha) - \sum_{s=1}^{\infty} q^\kappa [s]_q^\kappa (q[s]_q + \alpha)|a_s| - \sum_{s=1}^{\infty} q^\kappa [s]_q^\kappa (q[s]_q - \alpha)|b_s|\right) \\
&\geq 2(1-\alpha)\left(1 - \sum_{s=1}^{\infty} \frac{q^\kappa [s]_q^\kappa (q[s]_q + \alpha)}{1-\alpha}|a_s| - \sum_{s=1}^{\infty} \frac{q^\kappa [s]_q^\kappa (q[s]_q - \alpha)}{1-\alpha}|b_s|\right).
\end{aligned}$$

This expression is positive by condition (7), which completes the proof.

In the following theorem, it is shown that the condition (7) is also necessary for $f \in \mathcal{M}_{\overline{\mathcal{H}}_q^*}(\alpha, \kappa)$. □

Theorem 2 *For* $f_\kappa = h_\kappa + \overline{g_\kappa} \in \mathcal{M}_{\overline{\mathcal{H}}_q}$ *of the form* (6), *then* $f_\kappa \in \mathcal{M}_{\overline{\mathcal{H}}_q^*}(\alpha, \kappa)$ *if and only if the inequality*

$$\sum_{s=1}^{\infty} q^\kappa [s]_q^\kappa \left((q[s]_q + \alpha)|a_s| + (q[s]_q - \alpha)|b_s|\right) \leq 1 - \alpha, \quad (0 \leq \alpha < 1), \tag{11}$$

is satisfied. □

Proof In view of Theorem 1, it suffices to show that the "if" part holds true. Suppose that $f_\kappa \in \mathcal{M}_{\overline{\mathcal{H}}_q^*}(\alpha, \kappa)$, then

$$\Re e\left\{-\frac{qz\partial_q(D_q^\kappa h_\kappa(z)) - (-1)^\kappa \overline{qz\partial_q(D_q^\kappa g_\kappa(z))} + \alpha D_q^\kappa h_\kappa(z) + (-1)^\kappa \alpha\overline{(D_q^\kappa g_\kappa(z))}}{D_q^\kappa h_\kappa(z) + (-1)^\kappa \overline{D_q^\kappa g_\kappa(z)}}\right\} > 0$$

$$= \Re e\left\{\frac{\frac{1-\alpha}{z} - \sum\limits_{s=1}^{\infty} q^\kappa [s]_q^\kappa |a_s|([s]_q + \alpha) z^s - \sum\limits_{s=1}^{\infty} q^\kappa [s]_q^\kappa |b_s|([s]_q - \alpha) z^s}{\frac{1}{z} + \sum\limits_{s=1}^{\infty} q^\kappa [s]_q^\kappa |a_s| z^s - \sum\limits_{s=1}^{\infty} q^\kappa [s]_q^\kappa |b_s| z^s}\right\} > 0. \tag{12}$$

The expression (12) must hold for all $z \in \Delta^*$. Upon choosing the value of z on the positive real axis, where $0 < z = r < 1$, we have

$$\frac{1 - \alpha - \sum_{s=1}^{\infty} q^\kappa [s]_q^\kappa |a_s|([s]_q + \alpha) r^{s+1} - \sum_{s=1}^{\infty} q^\kappa [s]_q^\kappa |b_s|([s]_q - \alpha) r^{s+1}}{1 + \sum_{s=1}^{\infty} q^\kappa [s]_q^\kappa |a_s| r^{s+1} - \sum_{s=1}^{\infty} q^\kappa [s]_q^\kappa |b_s| r^{s+1}} > \alpha. \tag{13}$$

If the condition (11) does not hold, then the numerator of (13) is negative for $r \to 1^-$. Hence, there exists $z_0 = r_0$ in the interval $(0, 1)$ for which the left-hand side of the inequality (13) is negative. This contradicts condition (11), which completes the proof.

□

Corollary 1 *For $n = 0$, Theorems 1 and 2 yields the results obtained by the Aldweby and Darus* [2] (Theorems 1 and 2). □

3 Distortion Bounds and Extreme Points

A growth property for the class $\mathcal{M}_{\overline{\mathcal{H}}_q^*}(\alpha, \kappa)$ are obtained in the following theorem:

Theorem 3 *Let $f_\kappa(z) = h_\kappa(z) + \overline{g_\kappa(z)}$ be defined by (6) in the class $\mathcal{M}_{\overline{\mathcal{H}}_q^*}(\alpha, \kappa)$, then we have for $|z| = r < 1$*

$$\frac{1}{r} - \frac{(1-\alpha)r^2}{q[2]_q^\kappa (q[2]_q - \alpha)} \leq |f_\kappa(z)| \leq \frac{1}{r} + \frac{(1-\alpha)r^2}{q[2]_q^\kappa (q[2]_q - \alpha)}.$$

□

Proof Taking the absolute value for $f_\kappa(z)$ given by (6), we have

$$
\begin{aligned}
|f_\kappa(z)| &= \left|\frac{(-1)^\kappa}{z} + \sum_{s=1}^{\infty} a_s z^s + (-1)^\kappa \sum_{s=1}^{\infty} \overline{b_s z^s}\right| \\
&\leq \frac{1}{r} + \sum_{s=1}^{\infty} (|a_s| + |b_\kappa|) r^s \\
&\leq \frac{1}{r} + \sum_{s=1}^{\infty} (|a_s| + |b_\kappa|) r \\
&\leq \frac{1}{r} + \frac{1-\alpha}{q[2]_q^\kappa (q[2]_q - \alpha)} \sum_{s=1}^{\infty} \Big(\frac{q^\kappa [s]_q^\kappa (q[s]_q + \alpha)}{1-\alpha} |a_s| + \frac{q^\kappa [s]_q^\kappa (q[s]_q - \alpha)}{1-\alpha} |b_\kappa|\Big) r \\
&\leq \frac{1}{r} + \frac{(1-\alpha) r}{q[2]_q^\kappa (q[2]_q - \alpha)}.
\end{aligned}
$$

For the left-hand side of the inequality, we have

$$
\begin{aligned}
|f_\kappa(z)| &\geq \frac{1}{r} - \sum_{s=1}^{\infty} (|a_s| + |b_\kappa|) r^s \\
&\geq \frac{1}{r} - \sum_{s=1}^{\infty} (|a_s| + |b_\kappa|) r \\
&\geq \frac{1}{r} - \frac{1-\alpha}{q[2]_q^\kappa (q[2]_q - \alpha)} \sum_{s=1}^{\infty} \Big(\frac{q^\kappa [s]_q^\kappa (q[s]_q + \alpha)}{1-\alpha} |a_s| + \frac{q^\kappa [s]_q^\kappa (q[s]_q - \alpha)}{1-\alpha} |b_\kappa|\Big) r \\
&\geq \frac{1}{r} - \frac{(1-\alpha) r}{q[2]_q^\kappa (q[2]_q - \alpha)}.
\end{aligned}
$$

This proves the required result. □

Corollary 2 *If $f_\kappa \in \mathcal{M}_{\overline{\mathcal{H}_q^*}}(\alpha, \kappa)$, then*

$$
f(\Delta^*) \subseteq \left\{\omega : |\omega| < \frac{q[2]_q^\kappa (q[2]_q - \alpha) - (1-\alpha) r}{q[2]_q^\kappa (q[2]_q - \alpha) r}\right\}.
$$

□

Next, we determine the extreme points of the closed convex halls of $\mathcal{M}_{\overline{\mathcal{H}_q^*}}(\alpha, \kappa)$, denoted by $clco\mathcal{M}_{\overline{\mathcal{H}_q^*}}(\alpha, \kappa)$.

Theorem 4 *Let $f_\kappa = h_\kappa + \overline{g_\kappa}$ of the form (6), then $f \in clco\mathcal{M}_{\overline{\mathcal{H}_q^*}}(\alpha, \kappa)$ if and only if $f_{\kappa,s}(z)$ can be expressed as*

$$
f_\kappa(z) = \sum_{s=1}^{\infty} \varrho_s h_{\kappa,s}(z) + \Psi_s g_{\kappa,s}(z),
$$

where

$$h_{\kappa,0}(z)=\frac{(-1)^\kappa}{z},\quad h_{\kappa,s}(z)=\frac{(-1)^\kappa}{z}+\frac{1-\alpha}{q^\kappa[s]_q^\kappa(q[s]_q+\alpha)}z^s,\qquad s=1,2,\ldots,$$

$$g_{\kappa,0}(z)=\frac{(-1)^\kappa}{z},\quad g_{\kappa,s}(z)=\frac{(-1)^\kappa}{z}+(-1)^\kappa\frac{1-\alpha}{q^\kappa[s]_q^\kappa(q[s]_q-\alpha)}\overline{z}^s,\qquad s=1,2,\ldots,$$

where $\varrho_s\geq 0$, $\Psi_s\geq 0$ *and* $\sum_{s=0}^{\infty}(\varrho_s+\Psi_s)=1$. *The extreme points of* $\mathcal{M}_{\overline{\mathcal{H}}_q^*}(\alpha,\kappa)$ *are* $\{h_{\kappa,s}\}$ *and* $\{g_{\kappa,s}\}$. □

Proof For $f(z)=\sum_{s=0}^{\infty}(\varrho_s h_{\kappa,s}+\Psi_s g_{\kappa,s})$ where $\sum_{s=0}^{\infty}(\varrho_s+\Psi_s)=1$, we have

$$\begin{aligned}f_\kappa(z)&=\varrho_0h_{0,s}+\Psi_0g_{0,s}+\sum_{s=1}^{\infty}(\varrho_sh_{\kappa,s}+\Psi_sg_{\kappa,s})\\&=\sum_{s=0}^{\infty}\frac{(-1)^\kappa(\varrho_s+\Psi_s)}{z}+\sum_{s=1}^{\infty}\varrho_s\left(\frac{1-\alpha}{q^\kappa[s]_q^\kappa([s]_q+\alpha)}\right)z^s+(-1)^\kappa\sum_{s=1}^{\infty}\Psi_s\left(\frac{1-\alpha}{q^\kappa[s]_q^\kappa([s]_q-\alpha)}\right)\overline{z}^s\\&=\frac{(-1)^\kappa}{z}+\sum_{s=1}^{\infty}\left(\frac{1-\alpha}{q^\kappa[s]_q^\kappa([s]_q+\alpha)}\right)\varrho_sz^s+(-1)^\kappa\sum_{s=1}^{\infty}\left(\frac{1-\alpha}{q^\kappa[s]_q^\kappa([s]_q-\alpha)}\right)\Psi_s\overline{z}^s.\end{aligned}$$

This belongs to $\mathcal{M}_{\overline{\mathcal{H}}_q^*}(\alpha,\kappa)$ because

$$\begin{aligned}&\sum_{s=1}^{\infty}(q^\kappa[s]_q^\kappa([s]_q+\alpha))\left(\frac{1-\alpha}{q^\kappa[s]_q^\kappa([s]_q+\alpha)}\right)\varrho_s+(q^\kappa[s]_q^\kappa([s]_q-\alpha))\left(\frac{1-\alpha}{q^\kappa[s]_q^\kappa([s]_q-\alpha)}\right)\Psi_s\\&=\sum_{s=1}^{\infty}(1-\alpha)\varrho_s+(1-\alpha)\Psi_s=(1-\alpha)\sum_{s=1}^{\infty}\varrho_s+\Psi_s=(1-\alpha)(1-\varrho_0-\Psi_0)\leq 1-\alpha.\end{aligned}$$

Conversely, suppose that $f\in clco\mathcal{M}_{\overline{\mathcal{H}}_q^*}(\alpha,\kappa)$. For $s=1,2,3,\ldots,$ set

$$\begin{aligned}\varrho_s&=\frac{q^\kappa[s]_q^\kappa([s]_q+\alpha)}{1-\alpha}|a_s|,\qquad 0\leq\varrho_s\leq 1\\\Psi_s&=\frac{q^\kappa[s]_q^\kappa([s]_q-\alpha)}{1-\alpha}|b_s|,\qquad 0\leq\Psi_s\leq 1\\\varrho_0+\Psi_0&=1-\sum_{s=1}^{\infty}\varrho_s-\sum_{s=1}^{\infty}\Psi_s.\end{aligned}$$

Therefore, f_κ can be written as

$$
\begin{aligned}
f_\kappa(z) &= \frac{(-1)^\kappa}{z} + \sum_{s=1}^{\infty} |a_s| z^s + (-1)^\kappa \sum_{s=1}^{\infty} |b_s| \overline{z^s} \\
&= \frac{(-1)^\kappa}{z} + \sum_{s=1}^{\infty} \left(\frac{1-\alpha}{q^\kappa [s]_q^\kappa ([s]_q + \alpha)} \right) \varrho_s z^s + (-1)^\kappa \sum_{s=1}^{\infty} \left(\frac{1-\alpha}{q^\kappa [s]_q^\kappa ([s]_q - \alpha)} \right) \Psi_s \overline{z}^s \\
&= \frac{\varrho_0 + \Psi_0}{z} + \sum_{s=1}^{\infty} \left(h_{\kappa,s}(z) - \frac{(-1)^\kappa}{z} \right) \varrho_s + \sum_{s=1}^{\infty} \left(g_{\kappa,s}(z) - \frac{(-1)^\kappa}{z} \right) \Psi_s \\
&= \sum_{s=0}^{\infty} (\varrho_s h_{\kappa,s} + \Psi_s g_{\kappa,s}), \text{ as required.}
\end{aligned}
$$

□

4 Convex Combination and Convolution

Next, we show that the class $\mathcal{M}_{\overline{\mathcal{H}}_q^*}(\alpha, \kappa)$ is closed under convolution and convex combination.

Theorem 5 *For $0 \leq \delta \leq \alpha < 1$, let $f_\kappa(z) \in \mathcal{M}_{\overline{\mathcal{H}}_q^*}(\alpha, \kappa)$ and $\beta_\kappa(z) \in \mathcal{M}_{\overline{\mathcal{H}}_q^*}(\delta, n)$, then $(f_\kappa * \beta_\kappa)(z) \in \mathcal{M}_{\overline{\mathcal{H}}_q^*}(\alpha, \kappa) \subseteq \mathcal{M}_{\overline{\mathcal{H}}_q^*}(\delta, \kappa)$.* □

Proof The convolution of $f_\kappa(z)$ and $\beta_\kappa(z)$ is given by

$$
(f_\kappa * \beta_\kappa)(z) = \frac{(-1)^\kappa}{z} + \sum_{s=1}^{\infty} |a_s||c_s| z^s + (-1)^\kappa \sum_{s=1}^{\infty} |b_s||d_s| z^s.
$$

We want to show that the coefficients of $f_\kappa * \beta_\kappa$ satisfy condition (11). For $\beta_\kappa(z) \in \mathcal{M}_{\overline{\mathcal{H}}_q^*}(\delta, \kappa)$, we note that $|c_s| \leq 1$ and $|d_s| \leq 1$,

$$
\begin{aligned}
&\sum_{s=1}^{\infty} \frac{q^\kappa [s]_q^\kappa (q[s]_q + \delta)}{1-\delta} |a_s||c_s| + \sum_{s=1}^{\infty} \frac{q^\kappa [s]_q^\kappa (q[s]_q - \delta)}{1-\delta} |b_s||d_s| \\
&\leq \sum_{s=1}^{\infty} \frac{q^\kappa [s]_q^\kappa (q[s]_q + \delta)}{1-\delta} |a_s| + \sum_{s=1}^{\infty} \frac{q^\kappa [s]_q^\kappa (q[s]_q - \delta)}{1-\delta} |b_s| \\
&\leq \sum_{s=1}^{\infty} \frac{q^\kappa [s]_q^\kappa (q[s]_q + \alpha)}{1-\alpha} |a_s| + \sum_{s=1}^{\infty} \frac{q^\kappa [s]_q^\kappa (q[s]_q - \alpha)}{1-\alpha} |b_s| \leq 1,
\end{aligned}
$$

since $f_\kappa(z) \in \mathcal{M}_{\overline{\mathcal{H}}_q^*}(\alpha, \kappa)$ and $0 \leq \delta \leq \alpha < 1$. Therefore, $(f * \beta)(z) \in \mathcal{M}_{\overline{\mathcal{H}}_q^*}(\alpha, \kappa)$ $\subseteq \mathcal{M}_{\overline{\mathcal{H}}_q^*}(\delta, \kappa)$. □

Theorem 6 *Let $f_{m,\kappa}$ defined as*

$$f_{m,\kappa} = \frac{(-1)^\kappa}{z} + \sum_{s=1}^{\infty} |a_{s,m}| z^s + (-1)^\kappa \sum_{s=1}^{\infty} |b_{s,m}| \overline{z}^s$$

be in class $\mathcal{M}_{\overline{\mathcal{H}_q^*}}(\alpha, \kappa)$ for every $m = 1, 2, \ldots, l$, then the function

$$\Im_m(z) = \sum_{m=1}^{l} c_m f_{m,\kappa}(z), \qquad (0 \le c_m \le 1), \tag{14}$$

are also in the class $\mathcal{M}_{\overline{\mathcal{H}_q^*}}(\alpha, \kappa)$, where $\sum\limits_{m=1}^{l} c_m = 1$. □

Proof According to the definition of $\Im_m(z)$ given by (14), we can write

$$\Im_m(z) = \frac{(-1)^\kappa}{z} + \sum_{s=1}^{\infty} \left(\sum_{m=1}^{l} c_m |a_{s,m}| \right) z^s + (-1)^\kappa \sum_{s=1}^{\infty} \left(\sum_{m=1}^{l} c_m |b_{s,m}| \right) \overline{z}^s.$$

Furthermore, for every $m = 1, 2, \ldots, l$, we have $f_{m,\kappa} \in \mathcal{M}_{\overline{\mathcal{H}_q^*}}(\alpha, \kappa)$. Then, by (??), we have

$$\begin{aligned}
&\sum_{s=1}^{\infty} q^\kappa [s]_q^\kappa ([s]_q + \alpha) \left\{ \sum_{m=1}^{l} c_s |a_{s,m}| \right\} + \sum_{s=1}^{\infty} q^\kappa [s]_q^\kappa ([s]_q - \alpha) \left\{ \sum_{m=1}^{l} c_m |b_{s,m}| \right\} \\
&= \sum_{m=1}^{l} c_s \left(\sum_{s=1}^{\infty} q^\kappa [s]_q^\kappa ([s]_q + \alpha) |a_{s,m}| + \sum_{s=1}^{\infty} q^\kappa [s]_q^\kappa ([s]_q - \alpha) |b_{s,m}| \right) \\
&\le \sum_{m=1}^{l} c_m (1 - \alpha) \le 1 - \alpha.
\end{aligned}$$

Therefore, $\Im_m(z) \in \mathcal{M}_{\overline{\mathcal{H}_q^*}}(\alpha, \kappa)$. □

Corollary 3 *The class $\mathcal{M}_{\overline{\mathcal{H}_q^*}}(\alpha, \kappa)$ is closed under convex combination.* □

5 Generalised q-Integral Operator

In the following definition, we define q-integral operator on a function f_κ defined by (6). We also prove that this operator belongs to $\mathcal{M}_{\overline{\mathcal{H}_q^*}}(\alpha, \kappa)$.

Definition 5 Let $f_\kappa = h_\kappa + \overline{g_\kappa}$ be defined by (6). Then, the q-integral operator $F_{\kappa,q}$: $\mathcal{M}_{\overline{\mathcal{H}_q^*}}(\alpha, \kappa) \longrightarrow \mathcal{M}_{\overline{\mathcal{H}_q^*}}(\alpha, \kappa)$ is defined by the relation

$$F_{\kappa,q}(z) = \frac{[\varsigma]_q}{z^{\varsigma+1}} \int\limits_0^z \tau^{\varsigma} h_\kappa(\tau) \partial_q \tau + \overline{\frac{[\varsigma]_q}{z^{\varsigma+1}} \int\limits_0^z \tau^{\varsigma} g_\kappa(\tau) \partial_q \tau}, \qquad (\varsigma > 0, z \in \Delta^*). \tag{15}$$

Theorem 7 *Let f_κ be defined by (6) and belongs to the class $\mathcal{M}_{\overline{\mathcal{H}}_q^*}(\alpha, \kappa)$, then $F_{\kappa,q}(z)$ defined by (15) also belongs to $\mathcal{M}_{\overline{\mathcal{H}}_q^*}(\alpha, \kappa)$.* □

Proof From Definition 5, we conclude that

$$\begin{aligned}
F_{\kappa,q}(z) &= \frac{[c]_q}{z^{\varsigma+1}} \int\limits_0^z \left\{ (-1)^\kappa \tau^{\varsigma-1} + \sum_{s=1}^{\infty} |a_s| \tau^{\varsigma+s} \right\} \partial_q \tau + \overline{(-1)^\kappa \int\limits_0^z \left\{ \sum_{s=1}^{\infty} |b_s| \tau^{\varsigma+s} \right\} \partial_q \tau} \\
&= \frac{[\varsigma]_q}{z^{\varsigma+1}} \left\{ (-1)^\kappa (1-q) z \sum_{m=0}^{\infty} (zq^m)^{\varsigma-1} q^m + \sum_{s=1}^{\infty} |a_s| \left((1-q) z \sum_{m=0}^{\infty} (zq^m)^{\varsigma+s} q^m \right) \right. \\
&\quad \left. + \overline{(-1)^\kappa \sum_{s=1}^{\infty} |b_s| \left((1-q) z \sum_{m=0}^{\infty} (zq^m)^{c+s} q^m \right)} \right\} \\
&= \frac{[\varsigma]_q (1-q)}{z^{\varsigma+1}} \left\{ (-1)^\kappa \left(\sum_{m=0}^{\infty} q^{m\varsigma} \right) z^{\varsigma} + \sum_{s=1}^{\infty} |a_s| z^{\varsigma+s+1} \left(\sum_{m=0}^{\infty} q^{m(\varsigma+s+1)} \right) \right. \\
&\quad \left. + \overline{(-1)^\kappa \sum_{s=1}^{\infty} |b_s| z^{\varsigma+s+1} \left(\sum_{m=0}^{\infty} q^{m(\varsigma+s+1)} \right)} \right\} \\
&= \frac{(-1)^\kappa}{z} + \sum_{s=1}^{\infty} \frac{[\varsigma]_q}{[s+\varsigma+1]_q} |a_s| z^s + (-1)^n \sum_{s=1}^{\infty} \frac{[\varsigma]_q}{[s+\varsigma+1]_q} |b_s| \overline{z^s}.
\end{aligned}$$

We want to show that the coefficients of $F_{\kappa,q}(z)$ satisfy the condition (11). Therefore, we have

$$\begin{aligned}
&\sum_{s=1}^{\infty} q^\kappa [s]_q^\kappa \left((q[s]_q + \alpha) \left(\frac{[c]_q}{[s+c+1]_q} |a_s| \right) + (q[s]_q - \alpha) \left(\frac{[c]_q}{[s+c+1]_q} |b_s| \right) \right) \\
&\quad \leq \sum_{s=1}^{\infty} q^\kappa [s]_q^\kappa \Big((q[s]_q + \alpha) |a_s| + (q[s]_q - \alpha) |b_s| \Big) \leq 1 - \alpha.
\end{aligned}$$

Hence, $F_{\kappa,q}(z) \in \mathcal{M}_{\overline{\mathcal{H}}_q^*}(\alpha, \kappa)$. □

6 Concluding Remarks and Observations

We introduced and studied systematically a new subclass of the family of harmonic meromorphic q-starlike functions associated with the q-calculus. This has led us to a study of the coefficient estimates, distortion theorems, extreme points, convolution, and convex combinations. Further, we defined the q-integral operator and showed that the new class aforementioned is closed under this q-derivative operator for this harmonic meromorphic function class. Deriving new classes of the family of harmonic meromorphic functions using post-quantum calculus and q-fractional derivative and the integral operator will be important in the future, we believe.

Acknowledgements

Conflict of interest The authors declare that there is no conflict of interests regarding the publication of this paper.

References

1. Ahuja, O.P., Çetinkaya, A.: Connecting quantum calculus and harmonic starlike functions. Filomat. **34**(5), 1431–1441 (2020)
2. Aldweby, H., Darus, M.: A new subclass of harmonic meromorphic functions involving quantum calculus. J. Class. Anal **6**(2), 153–162 (2015)
3. Aldweby, H., Darus, M.: A note on q-integral operators. Electron. Notes Discret. Math. **67**, 25–30 (2018)
4. Alsoboh, A., Darus, M.: Certain subclass of meromorphic functions involving q-ruscheweyh operator. Transylv. J. Math. Mech. **6**, 01–08 (2019)
5. Alsoboh, A., Darus, M.: On fekete-szegö problems for certain subclasses of analytic functions defined by differential operator involving q-ruscheweyh operator. J. Funct. Space. Article ID 8459405 (2019)
6. Alsoboh, A., Darus, M.: On Fekete-Szego problem associated with q-derivative operator. J. Phys.: Conf. Ser. **1212**(1), 012003 (2019)
7. Amourah, A., Alsoboh, A., Ogilat, O., Gharib, G.M., Saadeh, R., Al Soudi, M.: A generalization of Gegenbauer polynomials and bi-univalent functions. Axioms **12**, 128 (2023). https://doi.org/10.3390/axioms12020128
8. Alsoboh, A., Amourah, A., Darus, M., Sharefeen, R.I.A.: Applications of neutrosophic q-Poisson distribution series for subclass of analytic functions and bi-univalent functions. Mathematics **11**, 868 (2023). https://doi.org/10.3390/math11040868
9. Al-Shaqsi, K., Darus, M.: On meromorphic harmonic functions with respect to k-symmetric points. J. Inequalities Appl. Article ID 259205, 11 pages (2008)
10. Aral, A., Gupta, V.: Generalized q-Baskakov operators. Math. Slovaca. **61**(4), 619–634 (2011)
11. Aral, A., Gupta, V.: On the durrmeyer type modification of the q-baskakov type operators. Nonlinear Anal. Theory Methods Appl. **72**(3–4), 1171–1180 (2010)
12. Aral, A., Gupta, V., Agarwal, R.: Applications of q-Calculus in Operator Theory. Springer, NewYork (2013)
13. Arif, M., Barkub, O., Srivastava, H.M., Abdullah, S., Khan, S.A.: Some Janowski type harmonic q-starlike functions associated with symmetrical points. Math. **8**(4), Article ID: 629, 16 pages (2020)
14. Bostancı, H., Oztürk, M.: New classes of salagean type meromorphic harmonic functions. Int. J. Math. Math. Sci. **2**, 52–57 (2008)
15. Elhaddad, S., Aldweby, H., Darus, M.: Some properties on a class of harmonic univalent functions defined by q-analogue of ruscheweyh operator. J. Math. Anal. **9**(4), 28–35 (2018)
16. Gasper, G., Rahman, M.: Basic Hypergeometric Series. Cambridge University Press, Cambridge (2004)

17. Jackson, F.H.: On q-definite integrals. Q. J. Pure Appl. Math. **41**, 193–203 (1910)
18. Jackson, F.H.: On q-functions and a certain difference operator. Trans. R. Soc. Edinb. **46**(2), 253–281 (1909)
19. Jahangiri, J.: Harmonic meromorphic starlike functions. Bull. Korean Math. Soc. **37**(2), 291–301 (2000)
20. Jahangiri, J., Silverman, H.: Meromorphic univalent harmonic functions with negative coefficients. Bull. Korean Math. Soc. **36**(4), 763–770 (1999)
21. Mohammed, A., Darus, M.: A generalized operator involving the q-hypergeometric function. Mat. Vesn. **65**(4), 454–465 (2014)
22. Srivastava, H.: Operators of basic (or q-)calculus and fractional q-calculus and their applications in geometric function theory of complex analysis. Iranian Journal of Science and Technology-Transactions of Mechanical Engineering **44**(1), 454–45 (2020)
23. Khan, N., Srivastava, H.M., Rafiq, A., Arif, M., Arjika, S.: Some applications of q-difference operator involving a family of meromorphic harmonic functions. Adv. Differ. Equ. **2021**(1), 01–18 (2021)

Theoretical Study of Explosion Phenomena for a Semi-parabolic Problem

Jamal Oudetallah, Zainouba Chebana, Taki-Eddine Oussaeif, Adel Ouannas, and Iqbal M. Batiha

Abstract This paper aims to present the explosion phenomena for a special semi-parabolic problem with a classical Neumann condition where we are interested in the finite time to blow up by using the energy method. A new theoretical result is provided with its proof.

Keywords Parabolic equation · Nonlinear equations · Finite-time blow-up of solution

1 Introduction

The topic of nonlinear partial differential equations can exhibit a number of nonlinear properties that are often related to several important features of real-world phenomena. These equations have become the most active area of many mathematical research in the last century and the current era. This is actually due to the successful methods of analysis that can enable mathematicians to provide a lot of rigorous answers to different important questions like existence and uniqueness, stability and also other domains [1–3, 6, 9, 12–16].

J. Oudetallah (✉) · I. M. Batiha
Department of Mathematics, Faculty of Science and Technology, Irbid National University, Irbid 2600, Jordan
e-mail: jamalayasrah12@gmail.com

I. M. Batiha
e-mail: ibatiha@inu.edu.jo

Z. Chebana · T.-E. Oussaeif · A. Ouannas
Department of Mathematics and Computer Science, University of Larbi Ben M'hidi, Oum El Bouaghi, Algeria
e-mail: taki_maths@live.fr

A. Ouannas
e-mail: Ouannas.adel@univ-oeb.dz

I. M. Batiha
Nonlinear Dynamics Research Center (NDRC), Ajman University, Ajman, UAE

D. Zeidan et al. (eds.), *Mathematics and Computation*, Springer Proceedings in Mathematics & Statistics 418, https://doi.org/10.1007/978-981-99-0447-1_22

The blowing-up phenomenon is the most striking question in the recent research world. In fact, this subject was started in a Russian school in the 40s and 50s. It was raised in the context of Semenov's chain reaction theory until the 60s of the last century, mainly after general approaches to the blow-up problems that were addressed by Fujita, Kaplan, Friedman and some others [17]. Motivated by these research works, we are interested in this manuscript to answer one of the most important questions when we granted that the explosion phenomenon occurs. We will definitely move on to answering exactly the question when does the blow-up occur? or what time does the explosion occur? Anyhow, the remaining of this paper is organized as follows. In the next section, we will formulate the main problem of this work, whereas Sect. 3 will exhibit the primary results associated with our formulation, followed by the final section that will summarize the conclusions of the paper.

2 Problem's Formulation

In this section, we will let $Q = \{(x, t) \in [\Omega] \times [0, T]\}$ where Ω is an open bounded domain and $T < \infty$. To this aim, we intend to consider a certain class of semilinear parabolic equations defined on a bounded domain. Such class of equations can be outlined by the following nonlinear problem:

$$\begin{cases} u_t - (a(x,t)u_x)_x + bu = u^3 & \forall (x,t) \in Q \\ u(x,0) = \varphi(x) & \forall x \in [0,1] \\ u_x(0,t) = 0 & \forall t \in [0,T] \\ u_x(1,t) = 0 & \forall t \in [0,T] \end{cases} . \tag{P}$$

In other words, the parabolic equation regarding the above nonlinear problem can be given as follows:

$$u_t - (a(x,t)u_x)_x + bu = u^3, \tag{1}$$

subject to the following initial condition:

$$u(x,0) = \varphi(x), \quad x \in [0,1], \tag{2}$$

and to the following boundary conditions of the Neumann type:

$$u_x(0,t) = u_x(1,t) = 0, \tag{3}$$

where a and φ are two known functions and b is a positive constant. It should be noted that the function φ satisfies the following compatibility condition:

$$\varphi_x(1) = 0, \ \varphi_x(0) = 0. \tag{4}$$

3 The Main Results

From the perspective of several current literature, there are more than six methods that can be considered to deal with the blow-up phenomenon. The most famous of them are the concavity method, Kaplan's first eigenvalue method, comparison method, upper-lower method, energy method and others. Herein, for the purpose of analyzing the blow-up phenomenon of the reaction-diffusion equation of problem (P), we choose the energy method which is deemed one of the most important methods that can be employed to address the finite-time blow-up of semilinear parabolic equations. In order to accomplish this objective, a new theoretical result is provided next with its proof.

Theorem 1 *Under the condition* $a_t(x,t) < 0, \forall (x,t) \in Q$. *The solution to problem* (P) *blows up in the finite-time* T^* *where*

$$T^* = \frac{1}{\Pi(0)},$$

with

$$\Pi(0) = \int_{\Omega} u^2(x,t)dx.$$

Proof To prove this result, we first multiply Eq. (1) by u and then integrate the result over Ω to obtain the following assertion:

$$\int_{\Omega} u_t(x,t) \cdot u(x,t)dx - \int_{\Omega} (a(x,t)u_x)_x \cdot u(x,t)dx + b\int_{\Omega} u(x,t) \cdot u(x,t) = \int_{\Omega} u^4(x,t)dxdt.$$

By using integration by part and using Neumann conditions (3), we get

$$\frac{1}{2}\frac{d}{dt}\int_{\Omega} u^2(x,t)dx + \int_{\Omega} a(x,t)u_x^2 dx + b\int_{\Omega} u^2(x,t)dx = \int_{\Omega} u^4(x,t)dx. \quad (5)$$

In this regard and to continue this proof, we should find a formal expression for the energy function $E(\cdot)$. In order to achieve this goal, we multiply the above equation by u_t, and then integrate the result over Ω to obtain

$$\int_{\Omega} u_t^2 - \int_{\Omega} (a(x,t)u_x)_x u_t + b\int_{\Omega} uu_t = \int_{\Omega} u^3 u_t. \quad (6)$$

It follows

$$\int_{\Omega} u_t^2 - \int_{\Omega} (a(x,t)u_x)_x u_t + \frac{b}{2}\frac{d}{dt}\int_{\Omega} u^2 = \frac{1}{4}\frac{d}{dt}\int_{\Omega} u^4. \quad (7)$$

Again, using integration by part yields immediately the following equality:

$$\int_{\Omega} u_t^2 - \frac{1}{2} \int_{\Omega} a(x,t) \frac{d}{dt} u_x^2 + \frac{b}{2} \frac{d}{dt} \int_{\Omega} u^2 = \frac{1}{4} \frac{d}{dt} \int_{\Omega} u^4. \tag{8}$$

In order to simplify the last expression, we use the law of derivation to find two functions as follows:

$$\frac{d}{dt}\left(a(x,t)u_x^2\right) = a(x,t)\frac{d}{dt}u_x^2 + a_t(x,t)u_x^2, \tag{9}$$

under the condition $a_t(x,t) < 0, \forall (x,t) \in Q$. Consequently, based on Eq. (9), we can obtain

$$\frac{d}{dt}\left(a(x,t)u_x^2\right) \leq a(x,t)\frac{d}{dt}u_x^2. \tag{10}$$

By combining (6) and (10), we get

$$\int_{\Omega} u_t^2 + \frac{1}{2} \frac{d}{dt} \int_{\Omega} \left(a(x,t)u_x^2\right) + \frac{b}{2} \frac{d}{dt} \int_{\Omega} u^2 \leq \frac{1}{4} \frac{d}{dt} \int_{\Omega} u^4. \tag{11}$$

Multiplying (11) by 2 yields the following inequality:

$$\frac{d}{dt}\left(\int_{\Omega} \left(a(x,t)u_x^2\right) + b \int_{\Omega} u^2 - \frac{1}{4} \frac{d}{dt} \int_{\Omega} u^4 \right) \leq 0.$$

This observation seems to be known. Thus, the energy function can be given by

$$E(t) = \int_{\Omega} a(x,t)u_x^2 dx + b \int_{\Omega} u^2(x,t)dx - \frac{1}{2} \int_{\Omega} u^4(x,t)dx,$$

where E is a decreasing function over $[0, T]$. From this point of view, we can write Eq. (5) by using the energy function E as follows:

$$\begin{aligned} \frac{1}{2} \frac{d}{dt} \int_{\Omega} u^2(x,t)dx + E(t) = \frac{1}{2} \int_{\Omega} u^4(x,t)dx \frac{1}{2} \frac{d}{dt} \int_{\Omega} u^2(x,t)dx - \frac{1}{2} \int_{\Omega} u^4(x,t)dx \\ = -E(t). \end{aligned} \tag{12}$$

It is important to know that we have prepared our study on the basis of the assumption that the initial value of energy is negative as we mentioned earlier, and therefore the energy function is negative over time t. This would imply the following inequality:

$$\frac{1}{2}\frac{d}{dt}\int\limits_{\Omega} u^2(x,t)dx - \frac{1}{2}\int\limits_{\Omega} u^4(x,t)dx \geq 0.$$

In addition, based on the fact that asserts the problem at hand is already defined in a bounded domain, then by using Holder inequality we can have

$$\left(\int\limits_{\Omega} u^4(x,t)dx\right)^{\frac{1}{2}} > \int\limits_{\Omega} u^2(x,t)dx,$$

or

$$\frac{d}{dt}\int\limits_{\Omega} u^2(x,t)dx > \left(\int\limits_{\Omega} u^2(x,t)dx\right)^2.$$

By putting $\Pi(t) = \int\limits_{\Omega} u^2(x,t)dx$, we can get

$$\frac{d\Pi(t)}{dt} > \Pi(t)^2, \tag{13}$$

i.e.,

$$\frac{d\Pi(t)}{(\Pi(t))^2} > dt. \tag{14}$$

After integrating the above inequality over $[0, t]$, we can obtain

$$\left[-\frac{1}{\Pi(t)} + \frac{1}{\Pi(0)}\right] > t, \tag{15}$$

or

$$\frac{1}{-t + \frac{1}{(\Pi(0))}} < \Pi(t). \tag{16}$$

As $t \longrightarrow T^*$, we can obtain

$$T^* = \frac{1}{\Pi(0)},$$

which confirms that the solution u of the problem at hand must blow up.

4 Conclusion

In this work, we have analyzed a class of semilinear parabolic equations defined on a bounded domain. In particular, we have selected a finite-time blow-up for a class of solutions with a negative initial energy of the considered problem.

References

1. Bouziani, A., Oussaeif, T.E., Ben Aoua, L.: A mixed problem with an integral two-space-variables condition for parabolic equation with the bessel operator. J. Math. **2013**, 457631 (2013)
2. Dhelis, S., Bouziani, A., Oussaeif, T.E.: Study of solution for a parabolic integro-differential equation with the second kind integral condition. Int. J. Anal. Appl. **16**, 569–593 (2018)
3. Oussaeif, T.E., Bouziani, A.: A priori estimates for weak solution for a time-fractional nonlinear reaction-diffusion equations with an integral condition. Chaos, Solitons & Fractals **103**, 79–89 (2017)
4. Oussaeif, T.E., Bouziani, A.: Inverse problem of a hyperbolic equation with an integral overdetermination condition. Electron. J. Differ. Equ. **2016**, 1–7 (2016)
5. Oussaeif, T.E., Bouziani, A.: Existence and uniqueness of solutions to parabolic fractional differential equations with integral conditions. Electron. J. Differ. Equ. **2014**, 1–10 (2014)
6. Oussaeif, T.E., Bouziani, A.: Mixed problem with an integral two-space variables condition for a parabolic equation. Int. J. Evol. Equ. **9**, 181–198 (2016)
7. Oussaeif, T.E., Bouziani, A.: Mixed problem with an integral two-space-variables condition for a class of hyperbolic equations. Int. J. Anal. **2013**, 957163 (2013)
8. Oussaeif, T.E., Bouziani, A.: Solvability of nonlinear viscosity equation with a boundary integral condition. J. Nonl. Evol. Equ. Appl. **3**, 31–45 (2015)
9. Oussaeif, T.E., Bouziani, A.: Mixed problem with an integral two-space-variables condition for a third order parabolic equation. Int. J. Anal. Appl. **12**, 98–117 (2016)
10. Oussaeif, T.E., Bouziani, A.: Nonlocal problem for a third order partial differential equation of mixed type with an integral two-space variables condition. Commun. Optim. Theory. **2017**, 16 (2017)
11. Oussaeif, T.E., Bouziani, A.: On a class of hyberbolic equation with an integral two-space-variables condition with the bessel operator. Adv. Stud. Contemp. Math. **28**, 83–92 (2018)
12. Bahia, G., Ouannas, A., Batiha, I.M., Odibat, Z.: The optimal homotopy analysis method applied on nonlinear time-fractional hyperbolic partial differential equations. Numer. Methods Partial Differ. Equ. **37**, 2008–2022 (2021)
13. Shatnawi, M.T., Ouannas, A., Bahia, G., Batiha, I.M., Grassi, G.: The optimal homotopy asymptotic method for solving two strongly fractional-order nonlinear benchmark oscillatory problems. Math. **9**, 2218 (2021)
14. Djenina, N., Ouannas, A., Batiha, I.M., Grassi, G., Pham, V.T.: On the stability of linear incommensurate fractional-order difference systems. Math. **8**, 1754 (2020)
15. Chebana, Z., Oussaeif, T.E., Ouannas, A.: Solvability of dirichlet problem for a fractional partial differential equation by using energy inequality and faedo-galerkin method. Innov. J. Math. **1**, 34–44 (2022)
16. Shatnawi, M.T., Djenina, N., Ouannas. A., Batiha, I.M., Grassi, G.: Novel convenient conditions for the stability of nonlinear incommensurate fractional-order difference systems. Alex. Eng. J. **61**, 1655–1663 (2022)
17. Galakyionov, A., Vazquez, J.L.: The problem of blow up in nonlinear parablic equations. Discret. Contin. Dyn. Syst. **8**, 399–433 (2002)

Explicit Formulae of Linear Recurrences

László Szalay

Abstract One important and widely studied problem in the theory of linear recurrences is to find explicit formulae for the general term of the sequences. Having an explicit formula facilitates the research of the properties of the sequence we investigate. The main tool is to apply the fundamental theorem of homogeneous linear recurrences, but other approaches may work as well. In the present paper, we concentrate on a specific case when the characteristic polynomial of the sequence has a double zero, and on a general formula.

Keywords Recurrence sequence · Explicit formula · Simple zero · Double zero

1 Introduction

Assume that the complex numbers $G_0, G_1, \ldots, G_{k-1}$ are the initial values of a homogeneous linear recurrence $\{G\}_{n=0}^{\infty}$ of order k given by

$$G_n = A_1 G_{n-1} + A_2 G_{n-2} + \cdots + A_k G_{n-k}, \qquad n \geq k, \tag{1}$$

where the coefficients $A_1, A_2, \ldots, A_k$ are also complex numbers.

The fundamental theorem of linear recurrences, roughly speaking says that there exists a suitable positive integer s with the condition $s \leq k$, and there exist complex numbers $\alpha_1, \alpha_2, \ldots, \alpha_s$ and polynomials $c_1(x), c_2(x), \ldots, c_s(x)$ with complex coefficients such that

$$G_n = \sum_{i=1}^{s} c_i(n) \alpha_i^n$$

holds for any non-negative integer n. (See, for instance, Theorem C.1 in [3].)

L. Szalay (✉)
University of Sopron, Sopron, Hungary
e-mail: szalay.laszlo@uni-sopron.hu

J. Selye University, Komárno, Slovakia

D. Zeidan et al. (eds.), *Mathematics and Computation*, Springer Proceedings in Mathematics & Statistics 418, https://doi.org/10.1007/978-981-99-0447-1_23

In order to find the values α_i we need to determine the zeros of the characteristic polynomial

$$p(x) = x^k - A_1 x^{k-1} - \cdots - A_{k-1}x - A_k$$

of the recurrence $\{G\}_{n=0}^{\infty}$. Then the multiplicities of the zeros and the initial values together fix the polynomials $c_i(x)$.

For example, if $\{G\}_{n=0}^{\infty}$ is a binary recurrence (i.e. when $k = 2$),

$$p(x) = x^2 - A_1 x - A_2 = (x - \alpha_1)(x - \alpha_2),$$

(α_1 and α_2 are not necessarily distinct) then the general term is given as follows.

Theorem 1 *The general term of a binary recurrence* $\{G\}_{n=0}^{\infty}$ *satisfies*

$$G_n = \begin{cases} \dfrac{(G_1 - \alpha_2 G_0)\alpha_1^n - (G_1 - \alpha_1 G_0)\alpha_2^n}{\alpha_1 - \alpha_2}, & \text{if } \alpha_1 \neq \alpha_2, \\ n\alpha_1^{n-1}G_1 - (n-1)\alpha_1^n G_0, & \text{if } \alpha_1 = \alpha_2 \end{cases}. \tag{2}$$

2 Explicit Formulae

Many specific cases, for example, the binary recurrences (see Theorem 1), or when $p(x)$ has only single zeros (Theorem 2) are widely studied.

In the main part of the section, as a new result, we describe the situation of one double zero (Theorem 3), and we also deal with a general formula (Theorem 4).

2.1 *Case of One Double Zero*

This subsection assumes that $G_0 = \cdots = G_{k-2} = 0$, $G_{k-1} = 1$. The first statement here (Theorem 2) is a known result, but it plays the role of a springboard for the proof of Theorem 3 (the case of a double zero).

Theorem 2 *Suppose that the characteristic polynomial* $p(x)$ *has distinct roots* $\alpha_1, \alpha_2, \ldots, \alpha_k$. *Then*

$$G_n = \sum_{j=1}^{k} \frac{\alpha_j^n}{p'(\alpha_j)}, \qquad n \geq 0. \tag{3}$$

Theorem 3 *If the characteristic polynomial* $p(x)$ *has two equal roots* $\alpha_1 = \alpha_2$, *and distinct roots* $\alpha_i \neq \alpha_1$ $(i = 3, \ldots, k)$, *then*

$$G_n = \left(n - \frac{\tilde{p}'(\alpha_1)}{\tilde{p}(\alpha_1)}\alpha_1\right)\frac{\alpha_1^{n-1}}{\tilde{p}(\alpha_1)} + \sum_{j=3}^{k}\frac{\alpha_j^n}{p'(\alpha_j)}, \qquad n \geq 0, \tag{4}$$

where $p(x) = (x - \alpha_1)^2 \cdot \tilde{p}(x)$.

Observe that both formulae (3) and (4) are harmonized with (2) if we assume $k = 2$ in Theorems 2 and 3, and suppose $G_0 = 0$, $G_1 = 1$ in (2).

Proof We apply Theorem 2 with the limit $\alpha_2 \to \alpha_1$. Split (3) into two parts as follows. Let

$$G_n = \underbrace{\sum_{j=1}^{2}\frac{\alpha_j^n}{p'(\alpha_j)}}_{G_n^\star} + \sum_{j=3}^{k}\frac{\alpha_j^n}{p'(\alpha_j)},$$

and consider $G_n^\star$ as $\alpha_2 \to \alpha_1$. First only the limit calculus is prepared. Let $\tilde{p}(x)$ be defined by $p(x) = (x - \alpha_1)(x - \alpha_2) \cdot \tilde{p}(x)$. For the derivative

$$p'(x) = (x - \alpha_2)\tilde{p}(x) + (x - \alpha_1)\tilde{p}(x) + (x - \alpha_1)(x - \alpha_2)\tilde{p}'(x)$$

we see that

$$p'(\alpha_1) = (\alpha_1 - \alpha_2)\tilde{p}(\alpha_1), \qquad p'(\alpha_2) = (\alpha_2 - \alpha_1)\tilde{p}(\alpha_2).$$

Now

$$G_n^\star = \frac{\alpha_1^n}{p'(\alpha_1)} + \frac{\alpha_2^n}{p'(\alpha_2)} = \frac{1}{(\alpha_1 - \alpha_2)\tilde{p}(\alpha_1)}\alpha_1^n - \frac{1}{(\alpha_1 - \alpha_2)\tilde{p}(\alpha_2)}\alpha_2^n$$

$$= \frac{1}{\tilde{p}(\alpha_1)}\left[\underbrace{\left(\frac{\alpha_1^n}{\alpha_1 - \alpha_2} - \frac{\alpha_2^n}{\alpha_1 - \alpha_2}\right)}_{m_1} - \underbrace{\left(\frac{\tilde{p}(\alpha_1)}{\tilde{p}(\alpha_2)} - 1\right)\frac{\alpha_2^n}{\alpha_1 - \alpha_2}}_{m_2}\right].$$

Obviously,

$$m_1 = \frac{\alpha_1^n - \alpha_2^n}{\alpha_1 - \alpha_2} = \sum_{j=0}^{n-1}\alpha_1^{n-1-j}\alpha_2^j$$

tends to $n\alpha_1^{n-1}$ as α_2 tends to α_1. On the other hand,

$$\lim_{\alpha_2 \to \alpha_1} m_2 = \lim_{\alpha_2 \to \alpha_1}\frac{\tilde{p}(\alpha_2) - \tilde{p}(\alpha_1)}{\alpha_2 - \alpha_1} \cdot \frac{\alpha_2^n}{\tilde{p}(\alpha_2)} = \tilde{p}'(\alpha_1) \cdot \frac{\alpha_1^n}{\tilde{p}(\alpha_1)}.$$

Putting together what we have, we obtain (4).

Example 1 Let $G_0 = G_1 = 0$, $G_2 = 1$, and $G_n = 5G_{n-1} - 8G_{n-2} + 4G_{n-3}$. The characteristic polynomial has the form

$$p(x) = x^3 - 5x^2 + 8x - 4 = (x-2)^2(x-1).$$

Then $p'(x) = 3x^2 - 10x + 8$, $p'(1) = 1$, $\tilde{p}(x) = x - 1$, $\tilde{p}(2) = 1$, $\tilde{p}'(x) = 1$. Thus

$$G_n = (n-2)2^{n-1} + 1.$$

2.2 *Arbitrary Initial Values but Single Zeros*

Now we assume that the initial values $G_0, G_1, \ldots, G_{k-1}$ are fixed arbitrarily, the recursive rule (1) holds, furthermore the characteristic polynomial $p(x)$ has only single zeros, like in Theorem 2. Supposing $A_k \neq 0$ it ensures $\alpha_i \neq 0$. During surveying the antecedents of the theme it turned out that Wolfram [4] sketched the method we use here.

The coefficients and zeros of $p(x)$ satisfy

$$A_t = (-1)^{t+1} \sum \alpha_{i_1}\alpha_{i_2}\cdots\alpha_{i_t}, \qquad t = 1, 2, \ldots, k.$$

Put

$$\sigma_0^{(i)} = -1, \qquad i = 1, 2, \ldots, k, \tag{5}$$

and for each $i \in \{1, 2\ldots, k\}$ define

$$\sigma_t^{(i)} = A_t + \sigma_{t-1}^{(i)}\alpha_i, \qquad t = 1, 2, \ldots, k-1. \tag{6}$$

Note that if $t = k-1$, then $\sigma_{k-1}^{(i)} = -A_k/\alpha_i$. We also introduce the notation

$$g(\alpha_i^\star) = G_{k-1} - \sigma_1^{(i)}G_{k-2} - \cdots - \sigma_{k-1}^{(i)}G_0 = -\sum_{t=0}^{k-1} \sigma_t^{(i)} G_{k-1-t}.$$

based on the initial values of the sequence and (6). The main result of this subsection is

Theorem 4 *The general term of the sequence* $\{G\}_{n=0}^{\infty}$ *can be given by*

$$G_n = \frac{\sum_{i=1}^{k} c_i g(\alpha_i^\star)\alpha_i^n}{A_k \sum_{i=1}^{k} \frac{c_i}{\alpha_i}}, \qquad n \geq 0, \tag{7}$$

where $c_k = -1$, *and for* $i = 1, 2, \ldots k-1$ *we have*

$$c_i = (-1)^{k-1-i}\left(\frac{\alpha_k}{\alpha_i}\right)^{k-2}\frac{\prod_{j=1,j\neq i}^{k-1}\left(\frac{1}{\alpha_k}-\frac{1}{\alpha_j}\right)}{\prod_{j=1}^{i-1}\left(\frac{1}{\alpha_i}-\frac{1}{\alpha_j}\right)\cdot\prod_{j=i+1}^{k-1}\left(\frac{1}{\alpha_j}-\frac{1}{\alpha_i}\right)}. \tag{8}$$

Proof For any $1 \le i \le k$ a straightforward computation shows for $n \ge 0$ that

$$\begin{aligned} &G_{n+k-1} - \sigma_1^{(i)}G_{n+k-2} - \cdots - \sigma_{k-1}^{(i)}G_n \\ &\quad = \alpha_i^n\left(G_{k-1} - \sigma_1^{(i)}G_{k-2} - \cdots - \sigma_{k-1}^{(i)}G_0\right) = g(\alpha_i^\star)\alpha_i^n. \end{aligned} \tag{9}$$

Let the matrix $\mathbf{V}$ consist of the column vectors $\overline{v}_i = [1, -\sigma_1^{(i)}, \ldots, -\sigma_{k-2}^{(i)}]^\top \in \mathbb{C}^{k-1}$ for $i = 1, 2, \ldots, k-1$. Define $\overline{v}_k$ analogously. We can see that

$$D = \det(\mathbf{V}) = (-1)^{k_1}\frac{A_k^{k-2}}{\alpha_k^{k-2}}\cdot\mathcal{V}\left(\frac{1}{\alpha_1}, \frac{1}{\alpha_2}, \ldots, \frac{1}{\alpha_{k-1}}\right) \neq 0, \tag{10}$$

where $k_1 = \lfloor k/2 \rfloor$, and $\mathcal{V}(\,)$ denotes the Vandermonde determinant. Hence the vectors $\overline{v}_1, \overline{v}_2, \ldots, \overline{v}_{k-1}$ form a basis in $\mathbb{C}^{k-1}$. Consequently, there uniquely exist complex numbers $c_1, c_2, \ldots, c_{k-1}$ such that

$$\overline{v}_k = c_1\overline{v}_1 + c_2\overline{v}_2 + \cdots + c_{k-1}\overline{v}_{k-1}.$$

The coordinates c_i can be determined by the Cramer's rule in the form

$$c_i = \frac{D_i}{D}, \qquad i = 1, 2 \ldots, k-1, \tag{11}$$

where D_i turns up

$$D_i = (-1)^{k_1+k-1-i}\frac{A_k^{k-2}}{\alpha_i^{k-2}}\cdot\mathcal{V}\left(\frac{1}{\alpha_1}, \ldots, \frac{1}{\alpha_{i-1}}, \frac{1}{\alpha_{i+1}}, \ldots, \frac{1}{\alpha_{k-1}}, \frac{1}{\alpha_k}\right). \tag{12}$$

Thus (8) follows from (10), (12), and (11), but we must explain the role of c_i in (7). In order to do that multiply (9) by c_i for $i = 1, 2, \ldots, k-1$, and subtract the linear combination from the case $i = k$ of (9). This manipulation results

$$\left(-\sigma_{k-1}^{(k)} + \sum_{i=1}^{k-1} c_i\sigma_{k-1}^{(i)}\right)G_n = g(\alpha_k^\star)\alpha_k^n - \sum_{i=1}^{k-1} c_i g(\alpha_i^\star)\alpha_i^n.$$

Since $\sigma_{k-1}^{(i)} = -A_k/\alpha_i$, we obtain

$$A_k\left(\frac{1}{\alpha_k}-\sum_{i=1}^{k-1}\frac{c_i}{\alpha_i}\right)G_n=g(\alpha_k^\star)\alpha_k^n-\sum_{i=1}^{k-1}c_ig(\alpha_i^\star)\alpha_i^n.$$

Then eliminate G_n from the equality above, and after some simplifications, it leads to (7). The proof is complete.

Example 2 Let $k=3$, $G_n=A_1G_{n-1}+A_2G_{n-2}+A_3G_{n-3}$. Assume that G_0, G_1 and G_2 are arbitrary. Put

$$p(x)=x^3-A_1x^2-A_2x-A_3=(x-\alpha_1)(x-\alpha_2)(x-\alpha_3),$$

where the linear forms are distinct. Thus

$$A_1=\alpha_1+\alpha_2+\alpha_3,\quad A_2=-(\alpha_1\alpha_2+\alpha_1\alpha_3+\alpha_2\alpha_3),\quad A_3=\alpha_1\alpha_2\alpha_3.$$

By (5) and (6), we have

$$\begin{array}{lll}\sigma_0^{(1)}=-1 & \sigma_0^{(2)}=-1 & \sigma_0^{(3)}=-1,\\ \sigma_1^{(1)}=\alpha_2+\alpha_3 & \sigma_1^{(2)}=\alpha_1+\alpha_3 & \sigma_1^{(3)}=\alpha_1+\alpha_2,\\ \sigma_2^{(1)}=-\alpha_2\alpha_3 & \sigma_2^{(2)}=-\alpha_1\alpha_3 & \sigma_2^{(3)}=-\alpha_1\alpha_2.\end{array}$$

For $i=1,2,3$, we see that

$$G_{n+2}-\sigma_1^{(i)}G_{n+1}-\sigma_2^{(i)}G_n=\alpha_i^n(G_2-\sigma_1^{(i)}G_1-\sigma_2^{(i)}G_0)=\alpha_i^ng(\alpha_i^\star)$$

follows. Then

$$D=(-1)\frac{A_3}{\alpha_3}\cdot\mathcal{V}\left(\frac{1}{\alpha_1},\frac{1}{\alpha_2}\right)=\alpha_2-\alpha_1.$$

Similarly, we can determine $D_1=\alpha_2-\alpha_3$, and $D_2=\alpha_3-\alpha_1$, and then subsequently c_1 and c_2. Hence

$$(-\sigma_2^{(3)}+c_1\sigma_2^{(1)}+c_2\sigma_2^{(2)})G_n=g(\alpha_3^\star)\alpha_3^n-c_1g(\alpha_1^\star)\alpha_1^n-c_2g(\alpha_2^\star)\alpha_2^n,$$

and then G_n is expressible. Scrutiny reveals that, using the equivalent form

$$g(\alpha_i^\star)=G_2+(\alpha_i-A_1)G_1+\frac{A_3}{\alpha_i}G_0$$

we finally obtain

$$G_n=\frac{(\alpha_3-\alpha_2)g(\alpha_1^\star)\alpha_1^n-(\alpha_3-\alpha_1)g(\alpha_2^\star)\alpha_2^n+(\alpha_2-\alpha_1)g(\alpha_3^\star)\alpha_3^n}{(\alpha_2-\alpha_1)(\alpha_3-\alpha_1)(\alpha_3-\alpha_2)}.$$

Note that this formula, assuming $G_0 = G_1 = 0$ and $G_2 = 1$ leads to

$$G_n = \frac{\alpha_1^n}{(\alpha_1 - \alpha_2)(\alpha_1 - \alpha_3)} + \frac{\alpha_2^n}{(\alpha_2 - \alpha_1)(\alpha_2 - \alpha_3)} + \frac{\alpha_3^n}{(\alpha_3 - \alpha_1)(\alpha_3 - \alpha_2)},$$

which is equivalent to the corresponding result in [2] for the Tribonacci sequence.

3 Outline

There are other types of formulae derived not from the fundamental theorem of linear recurrences but from other suitable, often combinatorial properties. For example, it is known that the terms of the Fibonacci sequence (given by $F_0 = 0$, $F_1 = 1$, $F_n = F_{n-1} + F_{n-2}$ for $n \geq 2$) satisfy

$$F_{n+1} = \sum_{j=0}^{\lfloor n/2 \rfloor} \binom{n-j}{j}.$$

Recently, Açikel et al. [1] found a new identity for the terms of the so-called k-generalized Lucas numbers $\{L_n^{(k)}\}_{n\geq 0}$. This sequence starts with the positive integer initial values $L_0^{(k)} = k$, $L_1^{(k)} = 1$, $L_2^{(k)} = 3$, ..., $L_{k-1}^{(k)} = 2^{k-1} - 1$, and each term afterward is the sum of the k consecutive preceding elements. Let $n \in \mathbb{N}$ and $r = \lfloor n/(k+1) \rfloor$.

Theorem 5 ([1]) *Using the notation above, we have*

$$L_n^{(k)} = -1 + \sum_{j=0}^{r} (-1)^j \left(\binom{n-kj}{j} + k\binom{n-kj-1}{j-1} \right) 2^{n-(k+1)j}.$$

Note that the explicit formula

$$L_n^{(k)} = \alpha_1^n + \alpha_2^n + \cdots + \alpha_k^n$$

based on the fundamental theorem provides also the terms $L_n^{(k)}$, where $\alpha_1, \alpha_2, \ldots, \alpha_k$ are the roots, all are simple, of the characteristic polynomial

$$p_k(x) = x^k - x^{k-1} - \cdots - x - 1$$

of the sequence.

This area is fast developing and very challenging because it has had a lot of applications, for instance, in the theory of Diophantine equations.

Acknowledgements The research was supported in part by National Research, Development and Innovation Office Grant 2019-2.1.11-TÉT-2020-00165, by Hungarian National Foundation for Scientific Research Grant No. 128088, and No. 130909, and by the Slovak Scientific Grant Agency VEGA 1/0776/21.

References

1. Açikel, A., Irmak, N., Szalay, L.: The *k*-generalized lucas numbers close to a power of 2. Math. Slovaca.
2. Spickerman, W.R.: Binet's formula for thew tribonacci sequence. Fibonacci Q. **20**, 118–120 (1982)
3. Shorey, T., Tijdeman, R.: Exponential Diophantine Equations (Cambridge Tracts in Mathematics). Cambridge University Press, Cambridge (1986). https://doi.org/10.1017/CBO9780511566042
4. Wolfram, D.A.: Solving generalized fibonacci recurrences. Fibonacci Q. **36**, 129–145 (1998)

The Influence of S-quasinormal Subgroups on the Structure of Finite Groups

Jehad Al Jaraden and Rashad Abu Sallik

Abstract A subgroup H of a group G is called S-quasinormal in G if it permutes with every Sylow subgroup of G. The purpose of this paper is to study the structure of a finite group under the assumption that some subgroups are S-quasinormal in G and Give some examples of groups with these conditions.

Keywords Finite group · S-quasinormal subgroup · Sylow subgroup

1 Introduction

All groups considered in this paper will be finite. Two subgroups H and K of a group G are said to permute if HK = KH. A subgroup of a group G is said to be quasinormal in a group G if it is permuted with all subgroups of the group G, this concept was introduced by Øystein Ore [1], and he proved that such a subgroup is subnormal in a finite group. A subgroup of a group G is said to be S-quasinormal in G if it permutes with every Sylow subgroup of G. This concept was introduced by Kegel [2]. Several authors have investigated the structure of a finite group when some subgroups of the prime power order of the group are well-situated in the group. Buckley [3] proved that if all minimal subgroups of an odd order group are normal, then the group is supersolvable. It turns out that the group which has many S-quasinormal subgroups have well-described structure.

J. Al Jaraden (✉) · R. A. Sallik
Al-Zaytoonah University of Jordan, Amman, Jordan
e-mail: jjaradeen@ahu.edu.jo

J. Al Jaraden
Al-Hussein Bin Talal University, Ma'an, Jordan

D. Zeidan et al. (eds.), *Mathematics and Computation*, Springer Proceedings in Mathematics & Statistics 418, https://doi.org/10.1007/978-981-99-0447-1_24

2 Preliminaries

Definition 2.1 (Ore 1937) A subgroup A of group G is called quasinormal or permutable in G if it permutes with all subgroups of G.

Example 2.2 In symmetric group S_3, the alternating subgroup A_3 is a quasinormal subgroup. Since A_3 is permuted with all subgroups of S_3.

Example 2.3 In quaternion group Q_8, the subgroups $\{\pm 1, \pm i\}, \{\pm 1, \pm j\}$ *and* $\{\pm 1, \pm k\}$, are normal subgroups in Q_8, and therefore it is quasinormal subgroup in Q_8.

Remark 2.4 ([5]) Every normal subgroup is quasinormal, the converse is not true.

Example 2.5 let $G = \{a, b | a^8 = b^2 = 1, b^{-1}ab = a^5\}$, then.

$G = \{1, a, a^2, a^3, a^4, a^5, a^6, a^7, b, ab, a^2b, a^3b, a^4b, a^5b, a^6b, a^7b\}, and$ the subgroups of G is

$$K_1 = \{1\}, K_2 =< b >= \{1, b\}, K_3 =< a >= \{1, a, a^2, a^3, a^4, a^5, a^6, a^7\}$$

$$K_4 =< a^6 >=< a^{-2} >= \{1, a^2, a^4, a^6\}, K_5 =< a^4 >= \{1, a^4\}$$

$$K_6 = \{1, a^4\text{b}\}, K_7 = \{1, \text{b}, a^4\text{b}, a^4\},$$

$$K_8 = \{1, a^2, , a^4, a^6, , ba, ba^3, ba^5, ba^7\},$$

$$K_9 = \{1, a^2b, a^4b, a^6b\}, K_{10} = \{1, a^2, a^4, a^6, b, a^2b, a^4b, a^6b\}, K_{11} = G$$

Now $K_2 =< b >= \{1, b\}$ is not normal in G, since.
$aK_2 = \{a, ab\}, K_2a = \{a, ba\} = \{a, a^5b\}\ aK_2 \neq K_2a$ and $K_2 \ntrianglelefteq G$
but K_2 is quasinormal in G, as $K_2K_1 = K_2K_1, K_2K_6 = K_6K_2, Since\ K_2 \subset K_6$

$$K_2K_7 = K_7K_2, Since\ K_2 \subset K_7, K_2K_{10} = K_{10}K_2, Since K_2 \subset K_{10},$$

$$K_2K_{11} = K_{11}K_2, Since\ K_2 \subset K_{11}$$

$$K_2K_3 = K_3K_2 = \{b, ab, a^2b, a^3b, a^4b, a^5b, a^6b, a^7b\}$$

$$K_2K_4 = K_4K_2 = \{b, a^2b, a^4b, a^6b\}$$

$$K_2K_5 = K_5K_2 = \{b, a^4b\}$$

$$K_2K_8 = K_8K_2 = \{b, a^2b, a^4b, a^6b, a, a^3, a^5, a^7\}$$

$$K_2K_9 = K_9K_2 = \{b, a^2, a^4, a^6\}$$

Lemma 2.6 ([4]) *If H is a subgroup of G, then the following conditions are equivalent:*

(i) H is quasinormal in G.

(ii) For every $g \in Gandh \in H$, *there exist* $r \in \mathbb{Z}andh' \in HSuchthatgh = g^r h'$.

Definition 2.7 ([2]) A subgroup H of a group *G* is called *S*-quasinormal in *G* if HP = PH, for every Sylow subgroup P of G.

Example 2.8 Let G = S_3, $S_3| = 6 = 3 \cdot 2$,

Then ∃ a Sylow 3-subgroups of order 3 is $A_3 = \{\rho_0, \rho_1, \rho_2\}$, and Sylow 2-subgroups of order 2, are $H_1 = \{\rho_0, \mu_1\}$, $H_2 = \{\rho_0, \mu_2\}$, $H_3 = \{\rho_0, \mu_3\}$, A_3 it permutes with each Sylow subgroup of S_3, since: $A_3H_i = H_iA_3 = S_3$, where i = 1, 2 3.

Then A_3 is the S-quasinormal of a group S_3.

Example 2.9 Let G = A_4, $|A_4| = 12 = 3 \cdot 2^2$,

then ∃ a Sylow 3-subgroup of order 3 are.

$$H_1 = \{\rho_0, \delta_3, \mu_2\}, H_2 = \{\rho_0, \delta_4, \mu_3\}, H_3 = \{\rho_0, \mu_4, \delta_2\}, H_4 = \{\rho_0, \mu_1, \delta_1\},$$

and then ∃ a 2-Sylow subgroup of order 4 is $H_5 = \{\rho_0, \rho_1, \rho_2, \rho_3\}$

H_5 it permutes with each Sylow subgroup of A_4, since: now,

$H_5H_i = H_iH_5 = \{\rho_0, \rho_1, \rho_2, \rho_3, \mu_1, \mu_2, \mu_3, \mu_4, \delta_1, \delta_2, \delta_3, \delta_4\} = A_4$, where i = 1, 2, 3, 4

Then H_5 is *S*-quasinormal group in A_4.

Lemma 2.10 Let G be a group and H a normal subgroup in G. Then G/H has prime order if and only if H is maximal.

Proof Assume that H is maximal and suppose that the order of G/H is not prime, that is, there exists a proper, nontrivial normal subgroup K of G/H. Let π: G → G/H be the projection g → gH. Then since π is surjective, π^{-1}(K) is a normal subgroup of G containing H. Since K is nontrivial, it contains a non-identity element kH where k ∈ H. Then k ∈ π^{-1}(K) but k ∈ H, so H is a proper subset of π^{-1}(K). This contradicts the fact that H is maximal, so we conclude that the order of G/H is prime.

Now suppose that G/H has prime order. Suppose that H is not maximal, that is, there is a proper normal subgroup N with H ⊂ N. Then π(N) is a proper normal subgroup of G/H, which contradicts G/H has prime order. Thus, H is maxima.

3 Results

We prove the following main result:

Theorem 3.1 *Let H be a S-quasinormal subgroup of a group G such that* $[G : H]$ *is a prime number, then H is a normal subgroup of G.*

Proof Let H s-quasinormal subgroup of a group G such that $[G : H]$ is a prime number. Then for every Sylow subgroup P of a group G HP = PH.

Suppose $H \ntrianglelefteq G, then\, H_1 = g^{-1}Hg \neq H, for\, some\, g \in G$, let $K = HH_1 = H_1H$, since $[G : H]$ is a prime number, and $H \subset K \subset G, soK = G$ as H is the maximal subgroup of a group G by lemma 2.10. In particular,= $hh_1,\ for\, some\, h \in H, h_1 \in H_1$. Hence g = $hg^{-1}h_2g\ for\, some\, h, h_2 \in H$, this implies that $g \in HsoH = H_1$, contradicting the assumption. This completes the proof.

Example 3.2 In S_3 group, the Alternating group A_3 is S-quasinormal subgroup, as A_3 is permute with all Sylow subgroups in S_3, and $[S_3 : A_3] = 2$ is a prime number, then by Theorem 3.1 A_3 is a normal subgroup of S_3.

Theorem 3.3 *A S-quasinormal subgroup H of G such that* $[G : H] = 4$ *is a normal subgroup of G.*

Proof Suppose this is false. Then there is a conjugate $H' = g^{-1}$ H g of H such that $H' \neq$ H. Let K = H′H = H′H. Since $H \subset K \subset G$ and $[G : H] = 4$, it follows that K is H or G or else $[K : H] = 2$. 1f K = H, $H' \subseteq K = H, so\, H' = H$ a contradiction.

If K = G, then, as in the proof of Theorem 3.1, H is normal in G. Thus, $[K : H] = 2$ and H is normal in K, also $[G : K] = 2$, and K is normal in G.

We conclude that there are exactly two conjugates of H, namely, H and H'. Let $N = H \cap H'$. By definition, N is the core of H in G and therefore is a normal subgroup of G. Moreover,

$$[K : H] = \left[HH' : H\right] = \left[H' : N\right] = [H : N] = 2.$$

Since$N \subset H \subset G, [G : H] = 4, [H : N] = 2$, and N is normal in G, the group G /N has order 8, H /N is S-quasinormal in G /N and has index 4. We know that every S-quasinormal subgroup G of order 8 is normal in G. Thus, H /N is normal in G/N, so H is normal in G, contradicting the initial assumption. From this, it follows that H is normal in G.

Example 3.4 In The Quaternion group Q_8, Q_8 is a unique 2-Sylow subgroup of order 8 as$|Q_8| = 8 = 2^3$. Let$H = \{+1, -1\}$, H is *S*-quasinormal group in Q_8 as $HQ_8 = Q_8H$ and as$[Q_8 : H] = 4$, Then by Theorem 3.3 H is a normal subgroup of Q_8.

Theorem 3.5 *If H is a S-quasinormal subgroup of a group G and* $[G : H] = 2^r m$, *where r = 1, 2, and m is an odd square-free number, then H is a normal subgroup of G.*

Proof We will argue by induction on$n = 2^r m$. If n = 1, the result is obvious for$n = 2^r$, where r = 1, 2 it follows from Theorems 3.2. Consider any element g ∈ G. Since H is S-quasinormal in G, $H\langle g\rangle$ is a subgroup of G and H is S-quasinormal in $H\langle g\rangle$. If $H\langle g\rangle \neq G$, then by induction hypothesis, H is normal in $H\langle g\rangle$, so Hg = gH. If $H\langle g\rangle$ = G, then $[H\langle g\rangle, H] = n$. This implies that n is the least positive integer k such that g^k ∈ H. Let x = g^p and$y = g^m$, where p is prime (p ≠ 2) and does not divide$m_1 (m_1 = n \setminus p)$. Then the least positive integer k such that $x^k \in H$ is $n/p = m_1$.

So $[H\langle x\rangle, H] = m_1$. Similarly, $[H\langle y\rangle, H] = p$. Since H is S-quasinormal in both $H\langle x\rangle$ and $H\langle y\rangle$, the inductive hypothesis shows that Hx = xH and $Hy = yH$. The fact that (p,m_1) =l implies that ∈ $\langle x, y\rangle$, hence, Hg = gH.

Example 3.6 Let G = S_4, |S_4| = 24 = $3 \cdot 2^3$, Then ∃ a unique Sylow 2-subgroups of order 8, is H_1 = {e, (12)(34), (13)(24), (14), (14)(23), (23), (1243), (1342)},

and ∃ a 4 Sylow 3-subgroups of order 3, are H_2 = {e, (123), (132)}, H_3 = {e, (124), (142)}, H_4 = {e, (134), (143)},

H_5 = {e, (234), (243)}, now.

H_6 = {e, (12)(34), (13)(24), (14)(23)} ≤ S_4, then H_6 it permutes with each Sylow subgroup of S_4, since:

$$H_6 H_1 = H_6 H_1, H_6 \subset H_1$$

$$H_6 H_2 = H_2 H_6 = \{e, (123), (132), (12)(34), (134), (234),$$

$$(13)(24), (243), (124), (14)(23), (142), (143)\}.$$

$$H_6 H_3 = H_3 H_6 = \{e, (124), (142), (12)(34), (143), (243),$$

$$(13)(24), (132), (134), (14)(23), (234), (123)\}.$$

$$H_6 H_4 = H_4 H_6 = \{e, (134), (143), (12)(34), (142), (243),$$

$$(13)(24), (123), (132), (14)(23), (124), (234)\}.$$

$$H_6 H_5 = H_5 H_6 = \{e, (234), (243), (12)(34), (124), (123),$$

$$(13)(24), (143), (123), (14)(23), (134), (142)\}.$$

Then H_6 is S-quasinormal group in S_4, and$[S_4 : H_6] = 6 = 2^1.3$, so by Theorem 3.5, H_6 is a normal subgroup of S_4.

Corollary 3.7 ([4]) *A quasinormal subgroup H of a group G such that [G:H] is prime is a normal subgroup of G.*

Corollary 3.8 ([4]) *If H is a quasinormal subgroup of a group G and* $[G:H] = n$ *is a square-free integer or twice a square-free integer, then H is a normal subgroup of G.*

References

1. Ore, O.: Contributions to the theory of groups of finite order. Duke Math. J. **5**(2), 431–460 (1939)
2. Kegel, O.H.: Sylow-gruppen und subnormalteiler endlicher gruppen. Math. Z. **78**(1), 205–221 (1962)
3. Buckley, J.: Finite groups whose minimal subgroups are normal. Math. Z. **116**(1), 15–17 (1970)
4. Hickerson, D., Stein, S., Yamaoka, K.: When quasinormal implies normal. Am. Math. Mon. **97**(6), 514–518 (1990)
5. Stonehewer, S.E.: Old, recent and new results on quasinormal subgroups. Irish Math. Soc. Bull **56**, 125–133 (2005)

Two-Sided Clifford Wavelet Function in $Cl(p, q)$

Shabnam Jahan Ansari and V. R. Lakshmi Gorty

Abstract Two-sided Clifford wavelet function (CWF) is defined in $Cl_{(p,q)}$ with orthonormal vector basis. The properties like left linearity, right linearity, shifting, modulation dilations and power factor are established. The inversion formula for CWF is also constructed using Fourier transform. Parseval and Plancherel identities are also studied. The study is supported by certain examples related to Mathematical Physics.

Keywords Clifford wavelet function · Inversion formula · Parseval's theorem · Fourier transform

1 Introduction

In [1] author has given the development and construction of Geometric algebra. [15] Geometric algebra is a finite-dimensional vector space over real scalar applications in Physics and Engineering field. The author developed geometric calculus whose fundamental theorem include the generalized Stokes theorem, Residue theorem, and new integral theorems. Quaternion Fourier transform on quaternion field and generalization were introduced in [2]. In [11] author has inspired the straightforward definition of a general geometric Fourier transform covering most versions in the literature.

In [4] introduced basic concepts of the multivector function, vector differential, and vector derivative in geometric algebra. Generalized real Fourier transform on Clifford multivector-valued functions, differentiation properties, and Plancherel theorem were proved. In [6] author has discussed the development of wavelet transform.

S. J. Ansari · V. R. Lakshmi Gorty (✉)
NMIMS University, MPSTME, Basic Science and Humanities, Mumbai 400056, India
e-mail: vr.lakshmigorty@nmims.edu
URL: https://nmims.edu/

S. J. Ansari
e-mail: shabnam.ansari@nmims.edu
URL: https://nmims.edu/

D. Zeidan et al. (eds.), *Mathematics and Computation*, Springer Proceedings in Mathematics & Statistics 418, https://doi.org/10.1007/978-981-99-0447-1_25

In [7] Clifford algebra offers a geometric interpretation for square roots of -1 in the form of blades that square to minus -1, which extends to a geometric interpretation of quaternions as represented as bivectors of a unit cube. In [9] author has described a non-commutative generalization of the complex Fourier-Mellin transform to Clifford algebra valued signal functions over the domain $\mathbb{R}^{p,q}$ taking values in $Cl(p,q)$, $p+q=2$.

In [8] author explains the orthogonal plane split with of quaternion based on the arbitrary choice of one or two linearly independent pure unit quaternion f, g. Also, author has generalized the quaternionic Fourier transform applied to the quaternion field to determine by quaternion f, g and established inverse transformation.

In the present study, authors have developed a two-sided Clifford wavelet function in $Cl(p,q)$. The properties like the inversion formula, Parseval, and convolution theorem are established. Applications in Mathematical Physics are demonstrated.

2 Fourier of CWF in $Cl(p,q)$

Consider a multivector-valued function $f(\mathbf{x})$ in $Cl(p,q)$, i.e., $f : \mathbb{R}^n \to Cl(p,q)$, where $\mathbf{x}$ is a vector variable. $f(\mathbf{x})$ can be decomposed as in [14, 16]:

$$f(\mathbf{x}) = \sum_A f_A(\mathbf{x}) e_A = \langle f(\mathbf{x}) \rangle + \langle f(\mathbf{x}) \rangle_1 + + \langle f(\mathbf{x}) \rangle_2 + \cdots + + \langle f(\mathbf{x}) \rangle_n . \quad (1)$$

$$\mathbf{x} = x_1 e_1 + x_2 e_2 + \cdots + x_p e_p + x_{p+1} e_{p+1} + x_{p+2} e_{p+2} + \cdots + x_q e_q . \quad (2)$$

$$\omega = \omega_1 e_1 + \omega_2 e_2 + \cdots + \omega_p e_p + \omega_{p+1} e_{p+1} + \omega_{p+2} e_{p+2} + \cdots + \omega_q e_q . \quad (3)$$

CWF with respect to mother Clifford wavelet $\psi \in L^2\left(\mathbb{R}^n; Cl_{(p,q)}\right)$ as analogous to [5, 6, 16]:

$$U_{a,\Theta,\mathbf{b}} : L^2\left(\mathbb{R}^n; Cl_{(p,q)}\right) \to L^2\left(G; Cl_{(p,q)}\right). \quad (4)$$

$$\psi(\mathbf{x}) \to U_{a,\Theta,\mathbf{b}}\, \psi(\mathbf{x}) = \psi_{a,\Theta,\mathbf{b}}(\mathbf{x}). \quad (5)$$

$$\psi_{a,\Theta,\mathbf{b}}(\mathbf{x}) = \frac{1}{a^{n/2}} \psi\left(r_\theta^{-1}\left(\frac{\mathbf{x}-\mathbf{b}}{a}\right)\right). \quad (6)$$

The family of wavelets $\psi_{a,\Theta,\mathbf{b}}$ are called daughter Clifford wavelets [3] with $a \in \mathbb{R}^+$ as dilation parameter, $\mathbf{b} \in \mathbb{R}^n$ as the translation vector parameter and Θ as the SO(n) rotation parameter, where SIM(n) is denoted by $Cl_{(p,q)}$ a subgroup of the affine group of motion on $\mathbb{R}^n$ associated with wavelets as follows:

$$G = \mathbb{R}^+ \times \mathrm{SO}(n) \otimes \mathbb{R}^n = \left\{(a, r_\theta(\mathbf{x}), \mathbf{b}) : a \in \mathbb{R}^+, r_\theta(\mathbf{x}) \in \mathrm{SO}(n), \mathbf{b} \in \mathbb{R}^n\right\} \quad (7)$$

where SO(n) is a special orthogonal group of $\mathbb{R}^n$.

For establishing an inversion formula, Parseval and Plancherel identities, certain assumptions about the phase functions $u(\mathbf{x}, \omega)$, $v(\mathbf{x}, \omega)$ need to be made. One possibility is to arbitrarily partition the scalar product

$$\mathbf{x} * \tilde{\omega} = \sum_{l=1}^{n} \mathbf{x}_l \omega_l = u(\mathbf{x}, \omega) + v(\mathbf{x}, \omega)$$

with

$$u(\mathbf{x}, \omega) = \sum_{l=1}^{k} \mathbf{x}_l \omega_l \tag{8}$$

and

$$v(\mathbf{x}, \omega) = \sum_{l=k+1}^{n} \mathbf{x}_l \omega_l. \tag{9}$$

3 Main Results

Definition 1 Fourier of CWF with respect to two square roots of -1 are considered as $e_t, e_t^* \in Cl(p, q)$, of -1, $e_t^2 = e_t^{*2} = -1$, every multivector $A \in Cl(p, q)$ can be split into commuting and anti-commuting parts as in [10]

Lemma 1 *Every multivector $A \in Cl(p, q)$ with respect to the square root $e_t, e_t^* \in Cl(p, q)$ of -1, i.e., $e_t{}^{-1} = -e_t$, then the unique decomposition is as* [13].

$$A_{+e_t} = \frac{1}{2}\left(A + e_t^{-1} A e_t\right).$$

$$A_{-e_t} = \frac{1}{2}\left(A - e_t^{-1} A e_t\right).$$

$$A = A_{+e_t} + A_{-e_t}.$$

$$A_{+e_t} e_t = e_t A_{-e_t}.$$

$$A_{-e_t} e_t = -e_t A_{-e_t}.$$

Definition 2 Fourier of CWF with respective to square root of -1.

Let $e_t, e_t^* \in Cl(p, q)$, $e_t{}^2 = e_t^{*2} = -1$ be any square root of -1. By the construction of the operators of CWF and $\pm$ split produces e_t, e_t^* from [2].

The general two-sided Fourier of CWF $\psi \in L^1(\mathbb{R}^{p,q}, Cl(p, q))$ analogous to [13]

$$F^{e_t, e_t^*}\{\psi\}(\omega) = \int_{\mathbb{R}^{p,q}} e^{-e_t u(\mathbf{x},\omega)} \psi(\mathbf{x}) e^{-e_t^* v(\mathbf{x},\omega)} d^n\mathbf{x} \tag{10}$$

where $d^n\mathbf{x} = dx_1 \cdots dx_n$, $x, \omega \in \mathbb{R}^{p,q}$ and $u, v : \mathbb{R}^{p,q} \times \mathbb{R}^{p,q} \to \mathbb{R}$.

Remark 1 From [7], $\psi = \psi_- + \psi_+$ is obtained from the property of split linearity.

4 Properties of Fourier of CWF

Proposition 1 (Left and right linearity) *For* $\psi_1, \psi_2 \in L^1(\mathbb{R}^{p,q}, Cl(p,q))$ *and constants* $\alpha, \beta \in Cl(p,q)$*, then*

$$F^{e_t,e_t^*}\{\alpha\psi_1 + \beta\psi_2\}(w) = \alpha_{+e_t} F^{e_t,e_t^*}\{\psi_1\}(\omega) + \alpha_{-e_t} F^{-e_t,e_t^*}\{\psi_1\}(\omega) + \beta_{+e_t} F^{e_t,e_t^*}\{\psi_2\}(\omega) + \beta_{-e_t} F^{-e_t,e_t^*}\{\psi_2\}(\omega) \quad (11)$$

and

$$F^{e_t,e_t^*}\{\psi_1\alpha + \psi_2\beta\}(w) = F^{e_t,e_t^*}\{\psi_1\}(\omega)\,\alpha_{+e_t^*} + F^{-e_t,e_t^*}\{\psi_1\}(\omega)\,\alpha_{-e_t^*} + F^{e_t,e_t^*}\{\psi_2\}(\omega)\,\beta_{+e_t^*} + F^{-e_t,e_t^*}\{\psi_2\}(\omega)\,\beta_{-e_t^*}. \quad (12)$$

Proof Considering from remark (3.4) using split linearity, we get

$$\alpha = \alpha_{+e_t} + \alpha_{-e_t};\ e^{-e_t u}\alpha = \alpha_{+e_t} e^{e_t u} + \alpha_{-e_t} e^{-(-e_t)u}. \quad (13)$$

$$\beta = \beta_{+e_t} + \beta_{-e_t};\ e^{-e_t u}\beta = \beta_{+e_t} e^{-e_t u} + \beta_{-e_t} e^{-(-e_t)u}. \quad (14)$$

Also

$$\alpha e^{-e_t^* v} = e^{-e_t^* v}\alpha_{+e_t^*} + e^{-(-e_t^*)v}\alpha_{-e_t^*} \quad (15)$$

$$\beta e^{-e_t^* v} = e^{-e_t^* v}\beta_{+e_t^*} + e^{-(-e_t^*)v}\beta_{-e_t^*}. \quad (16)$$

Using (10) in left-hand-side of (11)

$$\begin{aligned} &F^{e_t,e_t^*}\{\alpha\psi_1 + \beta\psi_2\}(w) \\ &= \int_{\mathbb{R}^{p,q}} e^{-e_t u}\{\alpha\psi_1 + \beta\psi_2\} e^{-e_t^* v} d^n\mathbf{x} \\ &= \int_{\mathbb{R}^{p,q}} \left\{\alpha_{+e_t} e^{-e_t u}\psi_1 + \alpha_{-e_t} e^{-(-e_t)u}\psi_1 + \beta_{+e_t} e^{-e_t u}\psi_2 + \beta_{-f} e^{-(-e_t)u}\psi_2\right\} d^n\mathbf{x} \\ &= \alpha_{+f} F^{e_t,e_t^*}\{\psi_1\}(\omega) + \alpha_{-e_t} F^{-e_t,e_t^*}\{\psi_1\}(\omega) + \beta_{+e_t} F^{e_t,e_t^*}\{\psi_2\}(\omega) + \beta_{-e_t} F^{-e_t,e_t^*}\{\psi_2\}(\omega). \end{aligned}$$

Hence (11) proved.

Similarly right linearity (12) can be proved.

Proposition 2 (Shifting) *For a* **x**-*shift function* $\psi_0(\mathbf{x}) = \psi(\mathbf{x} - \mathbf{x}_0)$, $h \in L^1(\mathbb{R}^{p,q}; Cl(p,q))$, *with constant* $\mathbf{x}_0 \in \mathbb{R}^{p,q}$, *assuming linearity of* $u(\mathbf{x}, \omega)$, $v(\mathbf{x}, \omega)$ *of a vector space with argument* **x**, *we obtain*

$$F^{e_t,e_t^*}\{\psi_0\}(\omega) = e^{-e_t u(\mathbf{x}_0,\omega)} F^{e_t,e_t^*}\{\psi\}(\omega)\ e^{-e_t^* v(\mathbf{x}_0,\omega)}. \quad (17)$$

Proof On substituting $\psi_0(\mathbf{x}) = \psi(\mathbf{x} - \mathbf{x}_0)$ from (10), we get

$$F^{e_l, e_l^*}\{\psi_0\}(\omega)$$

$$= \int_{\mathbb{R}^{p,q}} e^{-e_l u(\mathbf{x},\omega)} \psi(\mathbf{x} - \mathbf{x}_0) e^{-e_l^* v(\mathbf{x},\omega)} d^n\mathbf{x}$$

$$= \int_{\mathbb{R}^{p,q}} e^{-e_l u(\mathbf{y}+\mathbf{x}_0,\omega)} \psi(\mathbf{y}) e^{-e_l^* v(\mathbf{y}+\mathbf{x}_0,\omega)} d^n\mathbf{y}$$

$$= \int_{\mathbb{R}^{p,q}} e^{-e_l u(\mathbf{x}_0,\omega)} e^{-e_l u(\mathbf{y},\omega)} \psi(\mathbf{y}) e^{-e_l^* v(\mathbf{y},\omega)} e^{-e_l^* v(\mathbf{x}_0,\omega)} d^n\mathbf{y}$$

$$= e^{-e_l u(\mathbf{x}_0,\omega)} \int_{\mathbb{R}^{p,q}} e^{-e_l u(\mathbf{y},\omega)} \psi(\mathbf{y}) e^{-e_l^* v(\mathbf{y},\omega)} d^n\mathbf{y}\, e^{-e_l^* v(x_0,\omega)}.$$

Hence the proof (17).

Proposition 3 (Modulation) *Assume that the functions* $u(\boldsymbol{x}, \omega)$, $v(\boldsymbol{x}, \omega)$ *are both linear in their frequency argument* ω *then, for* $F^{e_l, e_l^*}\{\psi_m\}(\omega) = e^{-e_l u(x,\omega_0)} F^{e_l, e_l^*}\psi(\boldsymbol{x})\, e^{-e_l^* v(x,\omega_0)}$ *and constant* $\omega_0 \in Cl(p, q)$ *then the modulation formula is given by*

$$F^{e_l, e_l^*}\{\psi_m\}(\omega) = F^{e_l, e_l^*}\psi(\omega + \omega_0). \tag{18}$$

Proof Assuming that the functions $u(\mathbf{x}, \omega)$ and $v(\mathbf{x}, \omega)$ are both linear in their frequency argument ω. Using (10) in left-hand-side of (18)

$$F^{e_l, e_l^*}\{\psi_m\}(\omega)$$

$$= \int_{\mathbb{R}^{p,q}} e^{-e_l u(\mathbf{x},\omega)} \psi_m(\mathbf{x}) e^{-e_l^* v(\mathbf{x},\omega)} d^n\mathbf{x}$$

$$= \int_{\mathbb{R}^{p,q}} e^{-e_l u(\mathbf{x},\omega)} e^{-e_l u(\mathbf{x},\omega_0)} \psi_m(\mathbf{x}) e^{-e_l^* v(\mathbf{x},\omega_0)} e^{-e_l^* v(\mathbf{x},\omega)} d^n\mathbf{x}$$

$$= \int_{\mathbb{R}^{p,q}} e^{-e_l u(\mathbf{x},\omega+\omega_0)} \psi_m(x) e^{-e_l^* v(\mathbf{x},\omega+\omega_0)} d^n\mathbf{x}$$

$$= F^{e_l, e_l^*}\psi(\omega + \omega_0).$$

Hence (18) is proved.

Proposition 4 (Dilations) *Assume that for constants* $\alpha_1 \ldots \alpha_n \in \mathbb{R}^n \setminus \{0\}$, *and* $x' = \sum_{k=1}^{n} \alpha_k \mathbf{x}_k e_k$ *we have* $u(\mathbf{x}', \omega) = u(\mathbf{x}, \omega')$, *and* $v(\mathbf{x}', \omega) = v(\mathbf{x}, \omega')$ *with* $\omega' = \sum_{k=1}^{n} a_k \omega_k e_k$.

For $\psi_l(\mathbf{x}) = \psi(\mathbf{x}')$, $h \in L^1(\mathbb{R}^{p,q}; Cl(p,q))$ *and* $\omega_l = \sum_{k=1}^{n} \frac{1}{a_k}\omega_k e_k$, *then*

$$F^{e_t,e_t^*}\{\psi_l\}(\omega) = \frac{1}{|\alpha_1 \ldots \alpha_n|} F^{e_t,e_t^*}\{\psi\}(\omega_l). \tag{19}$$

Proof Assume that for constants $\alpha_1 \ldots \alpha_n \in \mathbb{R}^n \setminus \{0\}$ and $\mathbf{x}' = \sum_{k=1}^{n} a_k \mathbf{x}_k e_k$ we have $u(\mathbf{x}', \omega) = u(\mathbf{x}, \omega')$ and $v(\mathbf{x}', \omega) = v(\mathbf{x}, \omega')$ with $\omega' = \sum_{k=1}^{n} a_k \omega_k e_k$.
Using definition (10) in (19)

$$\begin{aligned}
&F^{e_t,e_t^*}\{\psi_d\}(\omega) \\
&= \int_{\mathbb{R}^{p,q}} e^{-e_t u(\mathbf{x},\omega)}\ \psi_d(\mathbf{x})\ e^{-e_t^* v(\mathbf{x},\omega)}\ d^n\mathbf{x} \\
&= \int_{\mathbb{R}^{p,q}} e^{-e_t u(\mathbf{x},\omega)}\ \psi\left(\mathbf{x}'\right)\ e^{-e_t^* v(\mathbf{x},\omega)}\ d^n\mathbf{x} \\
&= \frac{1}{|\alpha_1 \ldots \alpha_n|} \int_{\mathbb{R}^{p,q}} e^{-e_t u(\mathbf{x},\omega)}\ \psi\left(\mathbf{x}'\right)\ e^{-e_t^* v(x,\omega)}\ d^n\mathbf{x}.
\end{aligned}$$

Hence proved.

Proposition 5 (Power factor) *For* e_t, e_t^* *power factors in* $\psi_{p,q}(\mathbf{x}) = {e_t}^p \psi(\mathbf{x})\, e_t^{*q}$, $p, q \in \mathbb{Z}$ *and* $\psi \in L^1(\mathbb{R}^{p,q}; Cl(p,q))$ *from [13], we obtain*

$$F^{p,q}\left\{\psi_{p,q}(\omega)\right\} = {e_t}^p F^{p,q}\{\psi\}(\omega)\, e_t^{*q}. \tag{20}$$

Proof Using (10) in left-hand-site of (20)

$$\begin{aligned}
&F^{e_t,e_t^*}\left\{\psi_{p,q}\right\}(\omega) \\
&= \int_{\mathbb{R}^{p,q}} e^{-e_t u(\mathbf{x},\omega)} \psi_{p,q}(\mathbf{x})\ e^{-e_t^* v(\mathbf{x},\omega)}\ d^n\mathbf{x} \\
&= \int_{\mathbb{R}^{p,q}} e^{-e_t u(\mathbf{x},\omega)} {e_t}^p\ \{\psi\}(\mathbf{x})\, e_t^{*q}\ e^{-e_t^* v(\mathbf{x},\omega)}\ d^n\mathbf{x} \\
&= {e_t}^p \int_{\mathbb{R}^{p,q}} e^{-e_t u(\mathbf{x},\omega)}\ \{\psi\}(\mathbf{x})\, e^{-e_t^* v(\mathbf{x},\omega)}\ d^n\mathbf{x}\, e_t^{*q} \\
&= {e_t}^p F^{p,q}\{\psi\}(\omega)\, e_t^{*q}
\end{aligned}$$

where $e^{-e_t u(\mathbf{x},\omega)} {e_t}^p = {e_t}^p e^{-e_t u(\mathbf{x},\omega)}$ and $e_t^{*q} e^{-e_t^* v(\mathbf{x},\omega)} = e^{-e_t^* v(\mathbf{x},\omega)} e_t^{*q}$.
Thus (20) is proved.

5 Reconstruction for CWF

Theorem 1 *Assuming u, v as in (8) and (9) for $\psi \in L^1(\mathbb{R}^{p,q}; Cl(p,q))$, then the inversion formula is given as*

$$\psi(\mathbf{x}) = \left(F^{e_t,e_t^*}\right)^{-1}\left\{F^{e_t,e_t^*}\{\psi\}\right\}(\mathbf{x}) = \frac{1}{(2\pi)^n}\int_{\mathbb{R}^{p,q}} e^{e_t u(\mathbf{x},\omega)}\, F^{e_t,e_t^*}\psi(\omega)\, e^{e_t^* v(\mathbf{x},\omega)}\, d^n\omega\,. \tag{21}$$

Proof Applying (10) in right-hand-side of (21), we get

$$\begin{aligned}
&\frac{1}{(2\pi)^n}\int_{\mathbb{R}^{p,q}}\int_{\mathbb{R}^{p,q}} e^{e_t u(\mathbf{x},\omega)}\, e^{-e_t u(\mathbf{y},\omega)}\psi(\mathbf{y})\, e^{-e_t^* v(\mathbf{y},\omega)}\, e^{e_t^* v(\mathbf{x},\omega)} d^n\mathbf{y}\, d^n\omega \\
&= \frac{1}{(2\pi)^n}\int_{\mathbb{R}^{p,q}}\int_{\mathbb{R}^{p,q}} e^{e_t u(\mathbf{x}-\mathbf{y},\omega)}\psi(\mathbf{y})\, e^{e_t^* v(\mathbf{x}-\mathbf{y},\omega)} d^n\omega\, d^n\mathbf{y} \\
&= \frac{1}{(2\pi)^n}\int_{\mathbb{R}^{p,q}}\int_{\mathbb{R}^{p,q}} e^{e_t \sum_{l=1}^{k}(\mathbf{x}_l-\mathbf{y}_l)\omega_l}\psi(y)\, e^{e_t^* \sum_{m=k+1}^{n}(\mathbf{x}_m-\mathbf{y}_m)\omega_m}\, d^n\omega\, d^n\mathbf{y} \\
&= \frac{1}{(2\pi)^n}\int_{\mathbb{R}^{p,q}}\int_{\mathbb{R}^{p,q}} \prod_{l=1}^{k} e^{e_t(\mathbf{x}_l-\mathbf{y}_l)\omega_l}\psi(\mathbf{y}) \prod_{m=k+1}^{n} e^{e_t^*(\mathbf{x}_m-\mathbf{y}_m)\omega_m}\, d^n\omega\, d^n\mathbf{y} \\
&= \frac{1}{(2\pi)^n}\int_{\mathbb{R}^{p,q}} \prod_{l=1}^{k}\partial(\mathbf{x}_l-\mathbf{y}_l)\,\psi(y) \prod_{m=k+1}^{n}\partial(\mathbf{x}_m-\mathbf{y}_m) d^n\mathbf{y}
\end{aligned} \tag{22}$$

where

$$\frac{1}{2\pi}\int_{\mathbb{R}} e^{e_t(\mathbf{x}_l-\mathbf{y}_l)\omega_l} d\omega_l = \partial(\mathbf{x}_l-\mathbf{y}_l)\,,\ 1 \le l \le k$$

and

$$\frac{1}{2\pi}\int_{\mathbb{R}} e^{e_t^*(\mathbf{x}_m-\mathbf{y}_m)\omega_m} d\omega_m = \partial(\mathbf{x}_m-\mathbf{y}_m)\,,\ k+1 \le m \le n$$

are used in (22). Hence the proof.

6 Plancherel and Parseval Identities of Fourier of CWF

Theorem 2 *For the function $\psi_1, \psi_2 \in L^2(\mathbb{R}^{p,q}; Cl(p,q))$ and assuming $\tilde{e}_t = -e_t$, $\tilde{e}_t^* = -e_t^*$ then, Plancherel identity is obtained from* [13]

$$\langle \psi_1, \psi_2 \rangle = \frac{1}{(2\pi)^n} \left\langle F^{e_t, e_t^*} \{\psi_1\}, F^{e_t, e_t^*} \{\psi_2\} \right\rangle . \quad (23)$$

Hence Parseval identity is given by

$$\|\psi\| = \frac{1}{(2\pi)^{n/2}} \left\| F^{e_t, e_t^*} \{\psi\} \right\| . \quad (24)$$

Proof On considering

$$\left\langle F^{e_t, e_t^*} \{\psi_1\}, F^{e_t, e_t^*} \{\psi_2\} \right\rangle = \left\langle F^{e_t, e_t^*} \{\psi_1\} (\omega) \left[F^{e_t, e_t^*} \{\psi_2\} (\omega) \right]^{\tilde{}} \right\rangle d^n \omega$$

$$= \int_{\mathbb{R}^{p,q}} \int_{\mathbb{R}^{p,q}} \int_{\mathbb{R}^{p,q}} \left\langle e^{-e_t u(\mathbf{x},\omega)} \psi_1 (\mathbf{x}) \ e^{-e_t^* v(\mathbf{x},\omega)} d^n \mathbf{x} \left[e^{-e_t u(\mathbf{y},\omega)} \psi_2 (\mathbf{y}) \ e^{-e_t^* v(\mathbf{y},\omega)} d^n \mathbf{y} \right]^{\tilde{}} \right\rangle d^n \omega$$

$$= \int_{\mathbb{R}^{p,q}} \int_{\mathbb{R}^{p,q}} \int_{\mathbb{R}^{p,q}} \left\langle e^{-e_t u(\mathbf{x},\omega)} \psi_1 (\mathbf{x}) \ e^{-e_t^* v(\mathbf{x},\omega)} e^{-\tilde{e}_t^* v(\mathbf{y},\omega)} \tilde{\psi}_2 (\mathbf{y}) \, e^{-\tilde{e}_t u(\mathbf{y},\omega)} d^n \mathbf{y} \right\rangle d^n \mathbf{x} d^n \omega$$

$$= \int_{\mathbb{R}^{p,q}} \int_{\mathbb{R}^{p,q}} \int_{\mathbb{R}^{p,q}} \left\langle e^{e_t u(\mathbf{y},\omega)} \, e^{-e_t u(\mathbf{x},\omega)} \psi_1 (\mathbf{x}) \, e^{-e_t^* v(\mathbf{x},\omega)} e^{e_t^* v(\mathbf{y},\omega)} \tilde{\psi}_2 (\mathbf{y}) \, d^n \omega \, d^n \mathbf{y} \right\rangle d^n \mathbf{x}$$

$$= \int_{\mathbb{R}^{p,q}} \int_{\mathbb{R}^{p,q}} \int_{\mathbb{R}^{p,q}} \left\langle e^{-e_t u(\mathbf{x}\text{-}\mathbf{y},\omega)} \psi_1 (\mathbf{x}) \, e^{-e_t^* v(\mathbf{x}\text{-}\mathbf{y},\omega)} \tilde{\psi}_2 (\mathbf{y}) \, d^n \omega \, d^n \mathbf{y} \right\rangle d^n \mathbf{x}$$

$$= (2\pi)^n \int_{\mathbb{R}^{p,q}} \int_{\mathbb{R}^{p,q}} \int_{\mathbb{R}^{p,q}} \left\langle \frac{e^{-e_t \sum_{l=1}^{k} (\mathbf{x}_l - \mathbf{y}_l) \omega_l}}{(2\pi)^k} \psi_1 (\mathbf{x}) \, \frac{e^{-e_t^* \sum_{m=k+1}^{n} (\mathbf{x}_m - \mathbf{y}_m) \omega_m}}{(2\pi)^{n-k}} \tilde{\psi}_2 (\mathbf{y}) \, d^n \omega \, d^n \mathbf{y} \right\rangle d^n \mathbf{x}$$

$$= (2\pi)^n \int_{\mathbb{R}^{p,q}} \int_{\mathbb{R}^{p,q}} \left\langle \prod_{l=1}^{k} \partial \left(\mathbf{x}_l - \mathbf{y}_l \right) \psi_1 (\mathbf{x}) \prod_{m=k+1}^{n} \partial \left(\mathbf{x}_m - \mathbf{y}_m \right) \tilde{\psi}_2 (\mathbf{y}) \ d^n \mathbf{y} \right\rangle d^n \mathbf{x}$$

$$= (2\pi)^n \int_{\mathbb{R}^{p,q}} \left\langle \psi_1 (\mathbf{x}) \ \tilde{\psi}_2 (\mathbf{x}) \right\rangle d^n \mathbf{x}$$

$$= (2\pi)^n \langle \psi_1, \psi_2 \rangle .$$

Parseval identity is obtained by considering $\psi_1 = \psi_2$ in (23) follows as

$$\|\psi\| = \frac{1}{(2\pi)^{n/2}} \left\| F^{e_t, e_t^*} \{\psi\} \right\| .$$

7 Examples

Example 1 Haar wavelet ψ is defined as:

$$\psi(x) = \begin{cases} 1; & 0 \le x < 1/2 \\ -1; & 1/2 \le x < 1 \\ 0; & \text{otherwise.} \end{cases}$$

Find the Fourier of Haar wavelet function in $Cl(p,q)$.

Solution: Applying Fourier of Haar wavelet function in (10), we get

$$\begin{aligned}
&F^{e_t,e_t^*}\{\psi\}(\omega) \\
&= \int_{\mathbb{R}^n} e^{-e_t u(x,\omega)} \psi(x)\, e^{-v e_t^*(x,\omega)} d^n x \\
&= \int_{b_1}^{b_1+1/2} e^{-e_t u(x,\omega)}\, e^{-v e_t^*(x,\omega)} d^n x - \int_{b_1}^{b_1+1/2} e^{-e_t u(x,\omega)}\, e^{-v e_t^*(x,\omega)} d^n x + 0 \\
&= \int_{b_1}^{1/2+b_1} e^{\pm i_n \sum_{l=1}^{n} x_l \omega_l} dx_1 \ldots \int_{b_n}^{1/2+b_n} e^{\pm i_n \sum_{l=1}^{n} x_l \omega_l} dx_n \\
&- \left\{ \int_{1/2+b_1}^{1+b_1} e^{\pm i_n \sum_{l=1}^{n} x_l \omega_l} dx_1 \ldots \int_{1/2+b_n}^{1+b_n} e^{\pm i_n \sum_{l=1}^{n} x_l \omega_l} dx_n \right\} \\
&= \left\{ \left[\frac{e^{\pm i_n \sum_{l=1}^{n} x_l \omega_l}}{(i_n \omega_1 \ldots \omega_n)} \right]_{b_1}^{1/2+b_1} \cdots \left[\frac{e^{\pm i_n \sum_{l=1}^{n} x_l \omega_l}}{(i_n \omega_1 \ldots \omega_n)} \right]_{b_n}^{1/2+b_n} \right\} \\
&- \left\{ \left[\frac{e^{\pm i_1 \sum_{l=1}^{n} x_l \omega_l}}{(i_n \omega_1 \ldots \omega_n)} \right]_{1/2+b_1}^{1+b_1} \cdots \left[\frac{e^{\pm i_n \sum_{l=1}^{n} x_l \omega_l}}{(i_n \omega_1 \ldots \omega_n)} \right]_{b1/2+b_n}^{1+b_n} \right\}.
\end{aligned}$$

Example 2 Maxican hat wavelet: $\psi(x) = \left(1 - x^2\right) e^{x^2/2}$. Find the Fourier of Maxican hat wavelet function in $Cl(p,q)$.

Solution: Using Fourier of Maxican hat wavelet function in (10) gives

$$F^{e_t,e_t^*}\{\psi\}(\omega) = \int_{\mathbb{R}^n} e^{-e_t u(x,\omega)} \left[\left(1 - \sum_{m=1}^{n} x_m^2\right) \prod_{m=1}^{n} e^{-x_m^2/2} \right] e^{-e_t^* v(x,\omega)} d^n x$$

$$= \int_{\mathbb{R}^n} e^{\pm(i_1 x_1 \omega_1 + \ldots i_n x_n \omega_n)} \left[\left(1 - \sum_{m=1}^{n} x_m^2\right) \prod_{m=1}^{n} e^{-x_m^2/2} \right] d^n x \tag{25}$$

considering $m = 1$ in (26) gives

$$e^{-x^2/2+i\omega x}\, i\omega + x e^{-x^2/2+i\omega x} - \frac{1}{2}\left\{\sqrt{2}\omega^2\sqrt{\pi}\, erf_i\left[\frac{\sqrt{2}\omega}{2} + \frac{\sqrt{2}xi}{2}e^{-\omega^2/2}i\right]\right\}.$$

Further $m = 2$ in (26) gives

$$-erf_i\left[-\sqrt{2}\left(\frac{\omega_2 i_2}{2} - \frac{x_2}{2}\right)i\right]\left[\pi erf_i\left\{\sqrt{2}\left(\frac{\omega_1 i_1}{2} - \frac{x_1}{2}\right)i\right\}e^{\sqrt{2}\left(\frac{\omega_2^2 i_2^2}{2} + \frac{\omega_1^2 i_1^2}{2}\right)}\right.$$

$$+\omega_1 i_1\sqrt{2\pi}e^{\left(\frac{\omega_2^2 i_2^2}{2} - \frac{x_1^2}{2} + x_1\omega_1 i_1\right)i} + \sqrt{2\pi}e^{\left(\frac{\omega_2^2 i_2^2}{2} - \frac{x_1^2}{2} \pm x_1\omega_1 i_1\right)i}$$

$$+\frac{\omega_2 i_2 \pi}{2} erf_i\left\{\sqrt{2}\left(\frac{\omega_1 i_1}{2} - \frac{x_1}{2}\right)i\right\}e^{\sqrt{2}\left(\frac{\omega_2^2 i_2^2}{2} - \frac{\omega_1^2 i_1^2}{2}\right)} +$$

$$\left.\frac{\omega_1 i_1 \pi}{2} erf_i\left\{\sqrt{2}\left(\frac{\omega_1 i_1}{2} - \frac{x_1}{2}\right)i\right\}e^{\left\{\sqrt{2}\left(\frac{\omega_2^2 i_2^2}{2} - \frac{\omega_1^2 i_1^2}{2}\right)\right\}}\right]$$

$$+\frac{1}{2}\sqrt{2\pi}\,\omega_2 i_2 erf_i\left\{\sqrt{2}\left(\frac{\omega_1 i_1}{2} - \frac{x_1}{2}\right)i\right\}e^{\left\{\frac{\omega_1^2 i_1^2}{2} - \frac{x_2^2}{2} \pm \omega_2 i_2 x_2\right\}i}$$

$$-\frac{1}{2}\sqrt{2\pi}\,x_2 erf\left\{\sqrt{2}\left(\frac{\omega_1 i_1}{2} - \frac{x_1}{2}\right)i\right\}e^{\left\{\frac{\omega_1^2 i_1^2}{2} - \frac{x_2^2}{2} \pm \omega_2 i_2 x_2\right\}}.$$

Thus for all values of m the Fourier of Maxican hat wavelet in $Cl(p, q)$ is obtained.

8 Application

Apply Fourier on CWF on partial differential equations in Clifford algebra. Considering an initial value problem analogous to [12], we get

$$\frac{\partial\psi}{\partial t} - \Delta^2\psi = 0 \text{ on } \mathbb{R}^{0,q} \times (0, \infty) \tag{26}$$

and $\psi(x_1, \vec{x}) = f(x_1, \vec{x})$ at $t = 0$ where $\mathbb{R}^{0,q} \times (0, \infty)$ is Clifford-Schwatz space and $\Delta^2 = \frac{\partial}{\partial^2 x_1} + \frac{\partial}{\partial^2 \vec{x}}$.

Applying the definition (10) to both sides and from [12], we obtain

$$F^{e_t, e_t^*}\left[\frac{\partial\psi}{\partial t}\right] = e_t^2 F^{e_t, e_t^*}[\psi] + e_t^2 F^{e_t, e_t^*}[\psi]\, e_t^{*2} = -2F^{e_t, e_t^*}[\psi]. \tag{27}$$

General solution of (27) is given by

$$F^{e_t, e_t^*}\left[\frac{\partial \psi}{\partial t}\right] = -Ce^{-t}$$

where C is Clifford constant.

Using initial conditions (27) becomes

$$F^{e_t, e_t^*}\left[\frac{\partial \psi}{\partial t}\right] = -2F^{e_t, e_t^*}[f]. \tag{28}$$

Analogous to [12] and using (28) follows

$$\frac{1}{4\pi t}F^{e_t, e_t^*}\left[e^{-1/4t}\right] = e^{-t}. \tag{29}$$

Applying inversion formula to (29), finally $\psi(\mathbf{x})$ is given as

$$\psi(\mathbf{x}) = (F^{e_t}, {}^{e_t^*})^{-1}\left[\left[e^{-t}\right]F^{e_t, e_t^*}[f]\right].$$

Hence

$$\psi(\mathbf{x}) = \frac{1}{4\pi t}(F^{e_t}, {}^{e_t^*})^{-1}\left[\left[e^{-1/4t}\right]F^{e_t, e_t^*}[f]\right].$$

9 Conclusion

Two-sided Clifford wavelet function (CWF) is defined in $Cl_{(p,q)}$ with orthonormal vector basis. The properties like left linearity, right linearity, shifting, modulation dilations, power factor, inversion formula, Parseval and Plancherel identities are established. Examples are also illustrated in the present work.

References

1. Lounesto, P.: Clifford Algebras and Spinors, 2nd edn. LMS Lecture Note Series, vol. 286, (2001)
2. Hitzer, E.M.: Quaternion fourier transform on quaternion fields and generalizations. Adv. Appl. Clifford Algebr. **17**, 497–517 (2007). August
3. Hitzer, E.M.: "Tutorial on Fourier transformations and wavelet transformations in Clifford geometric algebra. Lecture Notes of the International Workshop for Computational Science with Geometric Algebra (FCSGA2007), pp. 65–87. Nagoya University, Japan (2007)
4. Hitzer, E.M., Mawardi, B.: Clifford fourier transform on multivector fields and uncertainty principles for dimensions $n = 2(\text{mod } 4)$ and $n = 3$ (mod 4). Adv. Appl. Clifford Algebr. **18**, 715–736 (2008). September
5. Pathak, R.S.: The Wavelet Transform, vol. 4. Springer Science Business Media (2009)

6. Chun, L.L.: A Tutorial of the wavelet Transform, p. 22. Taiwan, NTUEE (2010)
7. Hitzer, E.M., Abłamowicz, R.: Geometric roots of -1 in clifford algebras $Cl(p, q)$ with $p + q \leq 4$. Adv. Appl. Clifford Algebr. **21**, 121–144 (2011)
8. Hitzer, E.M.: OPS-QFTs a new type of quaternion fourier transforms based on the orthogonal planes split with one or two general pure quaternions. AIP Conf. Proc. Am. Inst. Phys. **1**, 280–283 (2011). September
9. Hitzer, E.M.: Clifford fourier-mellin transform with two real square roots of -1 in $Cl(p, q)$, $p + q = 2$. AIP Conf. Proc. Am. Inst. Phys. **1493**, 480–485 (2012). November
10. Hitzer, E.M., Helmstetter, J., Abłamowicz, R.: Square Roots of -1 in Real Clifford Algebras in Quaternion and Clifford Fourier Transforms and Wavelets. Birkhäuser, Basel (2013). http://arXiv.org/abs/1204.4576v2
11. Bujack, R., Scheuermann, G., Hitzer, E.M.: A general geometric fourier transform convolution theorem. Adv. Appl. Clifford Algebr. **23**, 15–38 (2013). March
12. Bahri, M., Ashino, R., Vaillancourt, R.: Continuous quaternion Fourier and wavelet transforms. Int. J. Wavelets Multiresolution Inf. Process. **12**, (2014). https://doi.org/10.1142/S0219691314600030
13. Hitzer, E.M.: Two-sided clifford fourier transform with two square roots of -1 in $Cl(p, q)$. Adv. Appl. Clifford Algebr. **24**, 313–332 (2014). June
14. Hitzer, E.M.: New Developments in Clifford Fourier Transforms. In: 2014 Advances in Applied and Pure Mathematics, Proceedings of the International Conference on Pure Mathematics, Applied Mathematics, Computational Methods (PMAMCM 2014), pp. 19–25. Santorini Island, Greece (2014)
15. Bromborsky, A.: An Introduction to Geometric Algebra and Calculus, (2014)
16. Ansari, S.J., Gorty, V.R.L.: Analysis of clifford wavelet transform in $Cl_{(3,1)}$. Commun. (2022)

Generalizations of the Fibonacci Sequence with Zig-Zag Walks

László Németh and László Szalay

Abstract The examination of the recurrence sequences associated with combinatorial constructions has been very extensive in the last decades. One of the most famous recurrence sequences is the Fibonacci sequence. We give two digraph constructions defined on the hyperbolic and on the Euclidean square mosaics, respectively, and we introduce two zig-zag type walks associating to the Fibonacci and its generalized sequences. Then we determine the recurrence relations and we give some examples.

Keywords Recurrence sequence · Hyperbolic Pascal triangle · Zig-zag square graph · Zig-zag sequence · Generalized Fibonacci sequence

1 Introduction

The investigation of the generalizations of Pascal's arithmetic triangle has an extensive literature. One generalization is the hyperbolic Pascal triangle defined on a regular squared grid given by Schläfli's symbol $\{4, q\}$, $q \geq 5$. Belbachir, Németh, and Szalay [1] described some interesting properties. One is that the Fibonacci sequence appears in this structure along a suitable zig-zag walk, considering the directed graph form of the hyperbolic Pascal triangle. Németh and Szalay [2] examined other types of zig-zag walks and presented more recurrence sequences assigned to the hyperbolic Pascal triangle.

The present paper summarizes the studies of diagonal and zig-zag paths on a particular $k+1$ wide, infinite part of the usual Euclidean square lattice as well. Along these paths Németh, and Szalay [3] determined linear recurrence sequences

L. Németh (✉) · L. Szalay
University of Sopron, Sopron 9400 Bajcsy Zs. u. 4,, Hungary
e-mail: nemeth.laszlo@uni-sopron.hu

L. Szalay
e-mail: szalay.laszlo@uni-sopron.hu

L. Szalay
University J. Selye, 945 01 Hradná str. 21, Komárno, Slovakia

D. Zeidan et al. (eds.), *Mathematics and Computation*, Springer Proceedings in Mathematics & Statistics 418, https://doi.org/10.1007/978-981-99-0447-1_26

considering a special type of generalized Fibonacci sequences and that are mostly defined in the *On-Line Encyclopedia of Integer Sequences* (OEIS, [4]). Moreover, we show a new occurrence of the Fibonacci sequence associated with the coefficient of these recurrence sequences.

2 Recurrence Sequences Associated to the Walks on the Hyperbolic Pascal Triangle

In the hyperbolic plane, there are infinite types of square regular mosaics, they are denoted by Schläfli's symbol $\{4, q\}$, where q satisfy $q > 4$. The parameter q means that in one node of the mosaic exactly q regular square meet. Each square regular mosaic induces a hyperbolic Pascal triangle $\mathcal{HPT}$ based on the mosaic $\{4, q\}$, and they can be figured as a digraph, where the vertices and the edges are the vertices and the edges of a well-defined part of the lattice $\{4, q\}$, respectively, further each vertex possesses the value which gives the number of different shortest paths from the base vertex. Fig. 1 illustrates the hyperbolic Pascal triangle $\mathcal{HPT}_{\{4,5\}}$. The values of the most left and most right nodes are 1's and the values inside the triangle are the sums of incoming values (for more details, see Fig. 1 and [1]). In the sequel, we fix the type of $\mathcal{HPT}$ given by mosaic $\{4, 5\}$.

Examining the triangle $\mathcal{HPT}_{\{4,5\}}$ thoroughly in Fig. 1 and starting a walk from the root vertex, continuing with one-step left and one-step right, then again one-step left, one-step right, and so on—shortly $L, R, L, R, L, \ldots$ (follow the red edges), we find that the values of nodes along this zig-zag walk are the terms of the Fibonacci sequence (defined by $F_0 = 0$, $F_1 = 1$ and $F_n = F_{n-1} + F_{n-2}$). Considering a similar

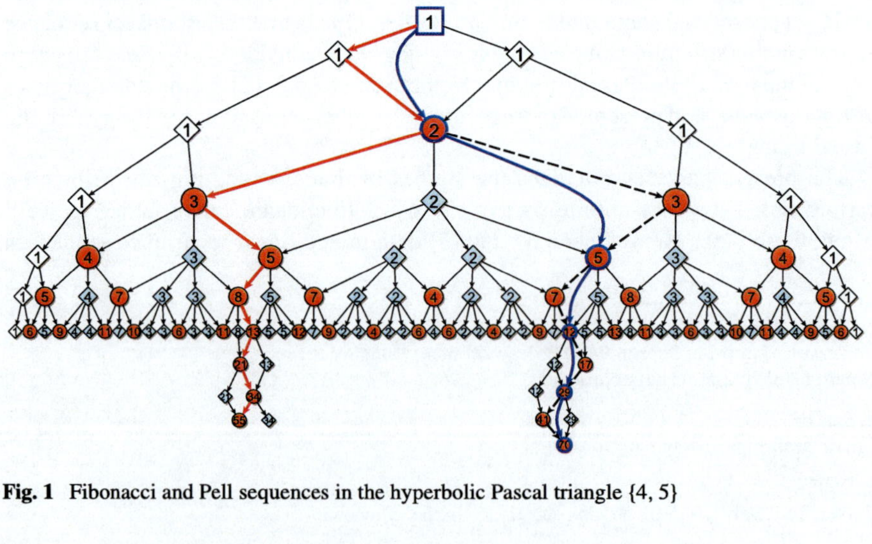

Fig. 1 Fibonacci and Pell sequences in the hyperbolic Pascal triangle $\{4, 5\}$

Table 1 Location of the sequences

	$f_n = \alpha f_{n-1} + f_{n-2}$	$f_n = \alpha f_{n-1} - f_{n-2}$
Condition	$\alpha \geq 1$	$\alpha \geq 2$
Distance of f_n and f_{n-1}	α	$\alpha - 1$
Pattern of steps	$LR^{\alpha-1}$, and $RL^{\alpha-1}$, alternately	$LR^{\alpha-2}$

zig-zag walk, first we step left and right, second right and left, and so on—shortly $LR, RL, LR, RL, \ldots$ (blue walk), the Pell sequence (defined by $P_0 = 0$, $P_1 = 1$ and $P_n = 2P_{n-1} + P_{n-2}$ for $n \geq 2$) appears.

It is easy to show (see [2]) that each pair $i < j \in \mathbb{Z}^+$ can be found next to each other in $\mathcal{HPT}_{\{4,5\}}$.

Theorem 1 ([1, 2]) *Assume $\alpha \in \mathbb{Z}^+$, $f_0 < f_1 \in \mathbb{Z}^+$, and f_0 (in the second case $f_1 - f_0$) is the left neighbor of f_1. Then binary recurrence sequences*

$$f_n = \alpha f_{n-1} \pm f_{n-2}, \quad (n \geq 2)$$

appear in the triangle $\mathcal{HPT}_{\{4,5\}}$ along zig-zag walks.

Table 1 shows the information about the location of the terms of the sequences. Here, for example, $LR^{\alpha-1}$ means that going down from a given element having type red circle, via elements type red circle, first turn left, and then $(\alpha - 1)$-times right. For illustration, see Fig. 2, when the pattern of steps in the zig-zag walk is LRR (or LLR).

3 Recurrence Sequences Associated to the Walks on an Euclidean Square Grid

Consider the Euclidean square lattice and take k consecutive pieces of squares. This is the 0th layer of the k–zig-zag shape. The upper corners are the 1st, 2nd, ..., kth and $(k + 1)$st vertices according to Fig. 3. Extend this by an extra 0th vertex, which is the base vertex. We color it yellow in the figures, and we join it to the 1st vertex by an extra edge. We denote the vertices of the 0th line by small boxes in Fig. 3. Now move the 0the layer to reach the right-down position in the square lattice to obtain the 1st layer, and repeat this procedure with the latest layer infinitely many times. Thus, we define the square k–zig-zag shape or graph, where $k \geq 1$ is the size of the array. Finally, we label the vertices such that a label gives the number of different shortest paths from the base vertex. Figure 4 illustrates the first few layers of the square 4–zig-zag digraph, the vertices are denoted by shaded boxes with their label values and the directed edges are the black arrows. Let $a_{i,j}$ denote the label of the vertex

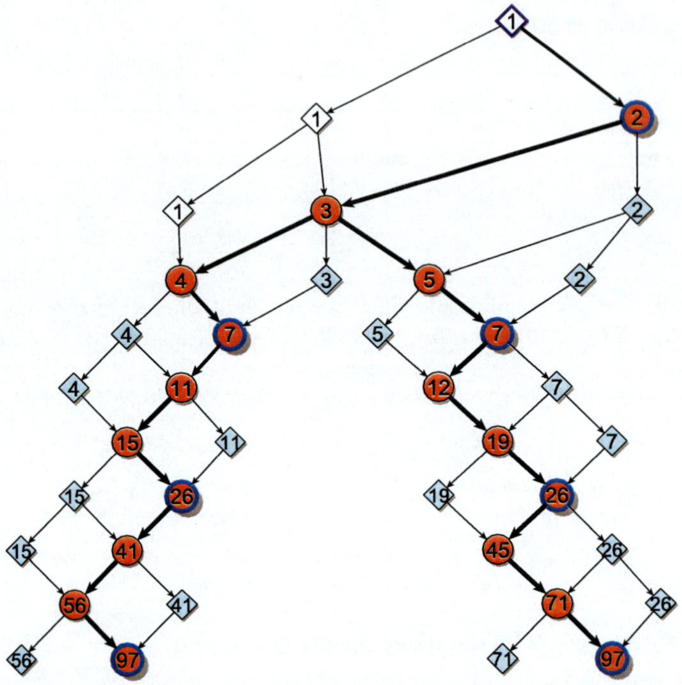

Fig. 2 Appearance of $f_n = 4f_{n-1} - f_{n-2}$, $f_0 = 1$, $f_1 = 2$ in $\mathcal{HPT}_{\{4,5\}}$

Fig. 3 Zig-zag shape

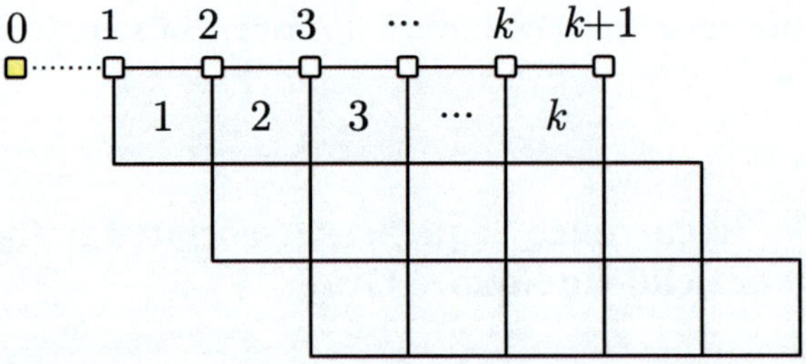

located in ith row and jth position ($0 \leq j \leq k+1, 0 \leq i$). Clearly, the fundamental rule of the construction is given by

$$a_{i,j} = \begin{cases} 1, & \text{if } i = 0; \\ a_{i-1,1}, & \text{if } j = 0, 1 \leq i; \\ a_{i,j-1} + a_{i-1,j+1}, & \text{if } 1 \leq j \leq k, 1 \leq i; \\ a_{i,k}, & \text{if } j = k+1, 1 \leq i. \end{cases} \tag{1}$$

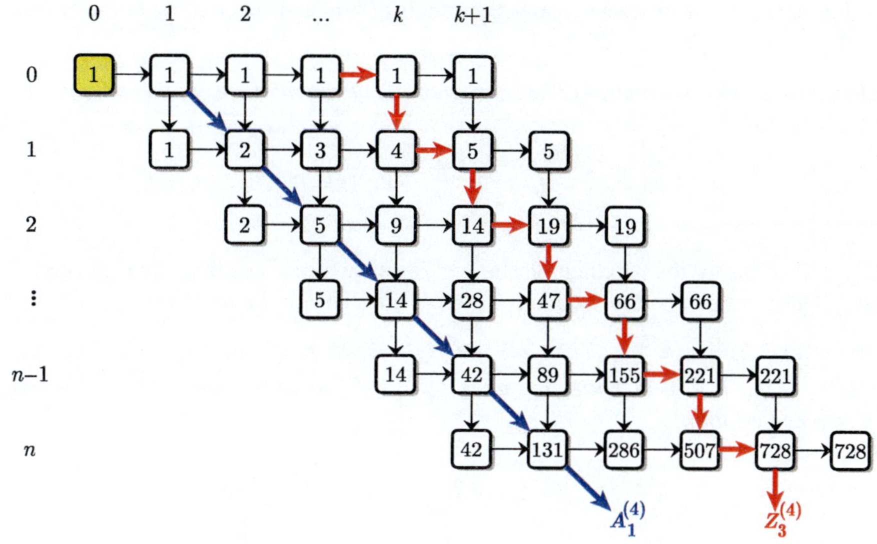

Fig. 4 Square 4–zig-zag digraph ($k = 4$)

For fixed $k \geq 1$ and given $0 \leq j \leq k+1$, let $A_j^{(k)}$ be the sequence defined by $A_j^{(k)} = (a_{i,j})_{i=0}^{\infty}$. The sequence $A_j^{(k)}$ is the jth right-down diagonal sequence of the square k–zig-zag shape. In Fig. 4, the blue arrows represent the sequence $A_1^{(4)}$. We found $A_0^{(k)} = (1, A_1^{(k)})$ and $A_k^{(k)} = A_{k+1}^{(k)}$. These sequences are associated with the zig-zag walks with patterns $RL, RL, RL, \ldots$.

Let $Z_j^{(k)}$, $j \in \{0, 1, \ldots, k\}$ be the jth zig-zag sequence of the square k–zig-zag shape, where $Z_j^{(k)}$ is the merged sequence of $A_j^{(k)}$ and $A_{j+1}^{(k)}$. (In Fig. 4, the red arrows represent the zig-zag sequence $Z_3^{(4)}$). More precisely, $Z_j^{(k)} = (z_{i,j})_{i=0}^{\infty}$, where

$$z_{i,j} = \begin{cases} a_{\ell,j}, & \text{if } i = 2\ell; \\ a_{\ell,j+1}, & \text{if } i = 2\ell + 1. \end{cases} \tag{2}$$

Since $Z_0^{(k)}$ and $Z_k^{(k)}$ are the 'double' of $A_0^{(k)}$ and $A_k^{(k)}$, respectively, usually we examine sequences for $j \in \{1, 2, \ldots, k-1\}$. The $Z_j^{(k)}$ sequences are associated with the zig-zag walks with patterns $R, L, R, L, R, L, \ldots$.

We find that any item $a_{n,j}$, $(n \geq 1)$ is the sum of the certain items of $(n-1)$st row. More precisely, if $0 < j < k+1$, then we obtain the system of recurrence relations

$$a_{n,j} = a_{n-1,j+1} + a_{n,j-1} = a_{n-1,j+1} + a_{n-1,j} + a_{n,j-2} = \cdots = \sum_{\ell=1}^{j+1} a_{n-1,\ell}. \tag{3}$$

Let $p_k(x)$ be the characteristic polynomials of kth recurrence of system (3), then the following theorem holds.

Theorem 2 ([3], Theorem 4) *The characteristic polynomials $p_k(x)$ can be given by*

$$p_k(x) = x^{\lceil \frac{k}{2} \rceil} \sum_{i=0}^{\lfloor \frac{k}{2} \rfloor + 1} (-1)^i \binom{k+2-i}{i} x^{\lfloor \frac{k}{2} \rfloor + 1 - i}, \qquad k \geq 0. \tag{4}$$

Now we record the two theorems of zig-zag sequences. The first one is the corollary of Theorem 2 and the second is a simple corollary of the first one.

Theorem 3 *Given $k \geq 1$. Then all the right-down diagonal sequences $A_j^{(k)}$ for $j \in \{0, 1, \ldots, k, k+1\}$ have the same $(\lfloor \frac{k}{2} \rfloor + 1)$-th order homogeneous linear recurrence relation*

$$a_{n,j} = \sum_{i=0}^{\lfloor \frac{k}{2} \rfloor} (-1)^i \binom{k+1-i}{i+1} a_{n-1-i,j}, \qquad n \geq \left\lfloor \frac{k}{2} \right\rfloor + 1. \tag{5}$$

Theorem 4 *Fixing $k \geq 1$, the zig-zag sequences $Z_j^{(k)}$ for $j \in \{0, 1, \ldots, k\}$ satisfy a $(2\lfloor \frac{k}{2} \rfloor + 2)$-th order homogeneous linear recurrence relation given by*

$$z_{n,j} = \sum_{i=0}^{\lfloor \frac{k}{2} \rfloor} (-1)^i \binom{k+1-i}{i+1} z_{n-1-2i,j}, \qquad n \geq 2\left\lfloor \frac{k}{2} \right\rfloor + 2.$$

Fig. 5 illustrates the 3-zig-zag graph and the zig-zag walk when appeared sequence is the Fibonacci sequence.

Expressing the item $a_{n,j}$ from Eq. (5) we obtain the result of Theorem 3. For example, the first few recurrence relations are

$$\begin{aligned}
k = 0: &\quad a_{n,j} = a_{n-1,j}, \\
k = 1: &\quad a_{n,j} = 2a_{n-1,j}, \\
k = 2: &\quad a_{n,j} = 3a_{n-1,j} - a_{n-2,j}, \\
k = 3: &\quad a_{n,j} = 4a_{n-1,j} - 3a_{n-2,j}, \\
k = 4: &\quad a_{n,j} = 5a_{n-1,j} - 6a_{n-2,j} + a_{n3,j}, \\
k = 5: &\quad a_{n,j} = 6a_{n-1,j} - 10a_{n-2,j} + 4a_{n-3,j}, \\
k = 6: &\quad a_{n,j} = 7a_{n-1,j} - 15a_{n-2,j} + 10a_{n-3,j} - a_{n-4,j}, \\
k = 7: &\quad a_{n,j} = 8a_{n-1,j} - 21a_{n-2,j} + 20a_{n-3,j} - 5a_{n-4,j}, \\
k = 8: &\quad a_{n,j} = 9a_{n-1,j} - 28a_{n-2,j} + 35a_{n-3,j} - 15a_{n-4,j} + a_{n-5,j}, \\
k = 9: &\quad a_{n,j} = 10a_{n-1,j} - 36a_{n-2,j} + 56a_{n-3,j} - 35a_{n-4,j} + 6a_{n-5,j}, \\
k = 10: &\quad a_{n,j} = 11a_{n-1,j} - 45a_{n-2,j} + 84a_{n-3,j} - 70a_{n-4,j} + 21a_{n-5,j} - a_{n-6,j}.
\end{aligned}$$

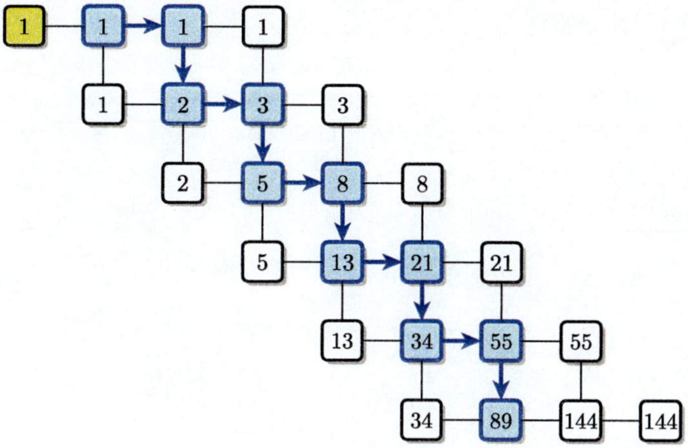

Fig. 5 Fibonacci sequence associated to zig-zag walk in 3-zig-zag graph

We consider the sum of the absolute values of the coefficients of these recurrence sequences (the left-hand side as well), and we gain the first few items of the Fibonacci sequence. Generally, using the well-known properties of Pascal's triangle that the rising-up diagonal sum sequence of Pascal's triangle is the Fibonacci sequence, so

$$\sum_{v=0}^{\lfloor \frac{u}{2} \rfloor} \binom{u-v}{v} = F_{u+1},$$

and from Eq. (5) when $u = k + 2$ and $v = i + 1$ we obtain

$$\sum_{i=0}^{\lfloor \frac{k}{2} \rfloor} \binom{k+1-i}{i+1} = F_{k+3} - 1.$$

Fig. 6 illustrates this coefficient sequence.

Moreover, considering the polynomials $p_k(x)$ of Eq. (4) we find that for non-negative integer $x = m$ the sequence of polynomials $p_k(x)$ becomes integer recurrence sequence with recurrence

$$p_k(m) = mp_{k-1}(m) - p_{k-2}(m), \quad k \geq 1,$$

where the initial values are $p_{-1}(m) = 1$ and $p_0(m) = m$. For the first few m the sequences appear yet in OEIS (see Table 2).

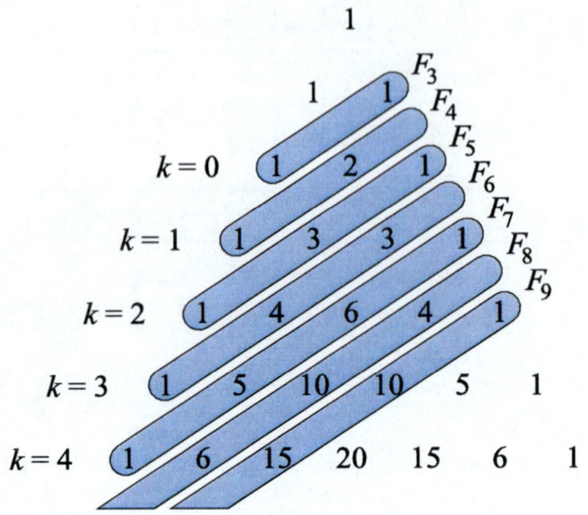

Fig. 6 Fibonacci sequence in Pascal's triangle

Table 2 Sequences in OEIS

x	0	1	2	3	4	5	6	7	8	9	10
$p_k(x)$	A154955	A128834	A090132	A057682	A001792	A039717	A140766	—	A164591	—	A176174

References

1. Belbachir, H., Németh, L., Szalay, l.: Hyperbolic Pascal triangles. Appl. Math. Comput. **273**, 453–464 (2016). https://doi.org/10.1016/j.amc.2015.10.001
2. Németh, L., Szalay, L.: Recurrence sequences in the hyperbolic Pascal triangle corresponding to the regular mosaic {4, 5}. Ann. Math. Inform. **46**, 165–173 (2016)
3. Németh, L., Szalay, L.: Sequences involving square zig-zag shapes. J. Integer Seq. **24**(5), Article 21.5.2 (2021)
4. Sloane, N.J.A., et al.: The on-line encyclopedia of integer sequences, (2021). https://oeis.org/

Mathematics Learning Challenges and Difficulties: A Students' Perspective

David Wafula Waswa and Mowafaq Muhammed Al-kassab

Abstract Mathematics is considered one of the core subjects that all students should learn to a certain level, and a lot of emphasize and importance is bestowed on it by almost all governments. However, students consider it a challenging subject that is difficult to understand. This research therefore seeks to establish what could be the major causes of difficulty in learning mathematics by students in the region. A questionnaire was distributed to 120 university students from five departments of the faculty of Education, and data collected and analyzed using one-sample t-test, two-samples t-test, and analysis of variance through the Minitab software. Results point to three major groups of sources of difficulties encountered by students in learning mathematics: learners' innate cognitive abilities, mathematics problem-solving processes and procedures, and external factors that include overcrowded classes, fear and anxiety, weak foundation, and instructors and instructional materials. Recommendations particular to the Kurdistan region were discussed and they include decongesting classrooms, investing in instructors by offering incentives and providing professional development opportunities, and adequately equipping students with the necessary learning tools and materials.

Keywords Descriptive statistics · t-test · Analysis of variance · Learning challenges · Mathematics competitiveness

1 Introduction

Mathematics has since been viewed as a "difficult" subject, and stereotypes characterizing this label abound. Such stereotypes eventually lead to fear that develops into mathematics phobia among students. The fear and/or phobia negatively impacts on

D. W. Waswa (✉) · M. M. Al-kassab
Department of Mathematics Education, Faculty of Education, Tishk International University-Erbil, Kurdistan Region, Iraq
e-mail: david.wafula@tiu.edu.iq

M. M. Al-kassab
e-mail: mowafaq.mohammed@tiu.edu.iq

D. Zeidan et al. (eds.), *Mathematics and Computation*, Springer Proceedings in Mathematics & Statistics 418, https://doi.org/10.1007/978-981-99-0447-1_27

the students' learning of mathematics leading to poor performance in the subject. The various branches of mathematics present different and sometimes unique challenges to students. In Algebra, for instance, students encounter challenges in converting word problems into mathematical sentences represented by symbols [9], incorrectly substituting and misunderstanding the signs [2] is another common difficulty for students in tackling algebra. Kelanang and Zakaria [3] on the other hand, viewed the origins of the problems that students encounter in learning geometry from 3 perspectives: cognitive and developmental theories, orientation, concept formation, and operative comprehension. In calculus, a course that teachers agree by consensus to be difficult for students [8], students experience challenges with the rate of change concept [6]. In general, [7], generalized sources of difficulties in learning mathematics are categorized in 5 ways; self-factors, teachers, parents, friends, and what they termed as "other" factors. The "other" factors include issues like assessment pressure and the general nature of mathematics. They conceded that other factors like self-esteem, teachers' behaviour, parental cognitive abilities [13], and friends, greatly influence students' disposition towards mathematics learning. Acharya [1] while mostly concurring with the other researchers, he added 2 dimensions, students' prior knowledge and lack of students' labor as possible other sources of difficulty in studying mathematics. Indeed, students find it difficult to understand higher mathematics concepts if their foundation in the subject was weak [4]. It's not enough to just recall previously studied concepts, students must be able to apply them in higher-order mathematics and that is why prior knowledge is key to understanding mathematics. By 'students' labor', Acharya implied students' commitment to the subject. It is obvious mathematics students are required to put in extra hours of study and be committed to "doing" mathematics as opposed to "reading" mathematics, which many students do. But [10] summed the mathematics learning difficulties by putting them in 4 categories: mathematical objects and thinking processes, mathematics teaching processes, students' cognitive processes, and lack of rational attitude towards mathematics. In Kurdistan region of Iraq and in general the larger Iraq and Arab countries, the difficulties talked above are much more general. The general standards of education in the region are relatively low compared to many other places in the world. There are various theories explaining these poor standards of education, perennial unrest in the region and around it, and cultural beliefs that are rooted in family values, such that families would easily sacrifice education on the altar of marriage. For instance, it is common practice for students to repeat grade levels if the teachers deem the performance of the student to be below par. Such practice only serves to discourage students from pursuing learning because it demonstrates to the students that they are not good enough. Mathematics is considered all over the world as an important subject and is made compulsory in many countries for students up to the high school level. Kurdistan region is not an exception to this world trend. It is considered a core subject that all students must take in their early years through high school. Given the importance associated with the subject, many parents, and indeed students themselves also, struggle to continue with it despite the poor outcomes in many instances. This research therefore seeks to establish what could be the major causes of difficulty in learning mathematics by students in the region.

2 Methodology and Research Design

This study is quantitative research with data collected through a questionnaire. The study sample is from a private university in the Kurdistan region of Iraq. Respondents were randomly sampled from the 5 departments in the faculty of education at the university. The questionnaires were handed to respondents by selected responsible students and taken back after they were filled out. Even with this careful handling of the process, out of the 120 questionnaires distributed, 7 were found to be defective and therefore rejected. The remaining 113 were analyzed using the Minitab mathematics software and the results are presented below.

3 Results

Study results from the 5 departments of the faculty of Education as analyzed by the Minitab statistics data analysis software are presented below. First, descriptive statistics for each department were analyzed, Table 1.

Table 1 Descriptive statistics analysis for each department

Departments	Grade level	n	Mean	Median	SE Mean
ELT	1	8	57.13	56.00	3.57
	2	11	52.73	46.00	5.57
	3	16	53.13	54.00	4.15
	4	12	69.42	66.00	7.01
Biology	1	7	56.71	57.00	5.51
	2	7	65.43	61.00	3.77
	3	5	68.20	69.00	3.12
	4	13	58.69	55.00	4.95
Mathematics	1	4	63.50	65.50	3.80
	2	3	66.67	66.00	1.20
	3	4	63.50	63.00	3.38
	4	2	100.00	100.00	1.00
Physics	1	3	53.67	46.00	7.67
	2	3	52.30	62.00	10.70
	3	3	59.33	60.00	3.48
	4	5	62.20	74.00	9.61
Computer	1	4	59.75	61.00	3.57
	2	3	55.30	65.00	11.70

From Table 1, it can clearly be seen that the sampled students from each of the five departments agree with the questionnaire, except for 4th grade mathematics students whose mean was way above 72.

One-Sample t-Test Analysis

After the descriptive statistics, the researchers also carried out a one-sample t-test for individual grade levels in each department. The t-test was based on the hypothesis that the average for individual grade levels in each department is 72 against the alternative hypothesis that the average is not equal to 72.

$$\text{i.e. } H_0 : \mu = 72$$

$$H_1 : \mu \neq 72$$

the results are shown in Table 2.

Table 2 indicates that slightly more than half of the grade levels in the 5 departments had no statistically significant differences in the answering of the questionnaire as can be seen from the P-Values that are greater than 0.05. Therefore, in each the null hypotheses were rejected at the mean $\mu = 72$. The exceptions are grade levels

Table 2 One-sample t-Test analysis for grade levels in each department

Department	Grade level	T-Value	P-Value
ELT	1	−4.17[a]	0.004
	2	−3.46[a]	0.006
	3	−4.54[a]	0.000
	4	−0.37	0.720
Biology	1	−2.78[a]	0.032
	2	−1.74	0.132
	3	−1.22	0.290
	4	−2.69[a]	0.020
Mathematics	1	−2.24	0.111
	2	−4.44[a]	0.047
	3	−2.52	0.087
	4	28.00[a]	0.023
Physics	1	−2.39	0.139
	2	−1.84	0.207
	3	−3.64	0.068
	4	−1.02	0.365
Computer	1	−3.43[a]	0.041
	2	−1.42	0.291

[a] Means significant at 5%

Table 3 Analysis of variance within departments

Department	F-value	P-value
ELT	2.19	0.103
Biology	0.94	0.434
Mathematics	20.31[a]	0.000
Physics	0.28	0.842

[a]Means highly significant at 5%

as can be seen in the table, including 4th grade ELT, 2nd and 3rd grade biology, 1st and 3rd grade mathematics, all grade levels in Physics, and 2nd grade computer.

One-way analysis of variance (ANOVA) was performed for each department to ascertain if there exist any differences in the grade levels within each department. This analysis was performed on 4 departments since the 5th department, Computer Education Department, had only 2-grade levels at the time of this research. Therefore, a Two-Sample t-Test was conducted instead.

Table 3 shows there was a statistically significant difference among the grade levels of mathematics department at $P = 0.000$. This significance is due to grade 4 which has a mean of 100 (see Table 1). The rest of the departments showed no significant differences among the grade levels. The two-sample t-test performed on the computer education department also returned a no statistically significant difference verdict as indicated in Table 4.

A separate descriptive statistics analysis for all departments was performed followed by a One-Sample t-Test and an analysis of variance to check for differences between and among the departments. The results are displayed below (Table 5).

It can be ascertained from the table that since all the means $\mu < 72$, students acting together as departments generally agreed or completely agreed with the questionnaires. On the other hand, the one-sample t-test shown in Table 6 indicates that

Table 4 Two-Sample t-Test for computer education department

Grade level	Means	T-value	P-value
1	59.75	0.41	0.695
2	55.300		

Table 5 Descriptive statistics analysis for departments

Department	n	Mean	Median	SE mean
ELT	47	57.87	54.00	2.81
Biology	32	61.22	60.50	2.56
Mathematics	13	69.85	66.00	4.00
Physics	14	57.64	61.00	4.19
Computer	7	57.86	62.00	4.91

Table 6 One-sample t-Test for departments

Department	n	T-Value	P-Value
ELT	47	−5.04[a]	0.000
Biology	32	−4.22[a]	0.000
Mathematics	13	−0.54	0.600
Physics	14	−3.42[a]	0.005
Computer	7	−2.88[a]	0.028

[a]Means significant at 5%

Table 7 One-way analysis of variance for departments

Source of variation	Degree of freedom	Sum of squares	Mean sum of squares	F-value	P-value
Between departments	4	1627	407 280	1.45	0.221
Error	108	30,200			
Total	112	31,828			

all departments except the mathematics department reject the null hypothesis. The one-sample t-test analysis for departments is shown in the table below.

The One-way analysis of variance for all departments is shown in the Table 7.

The one-way analysis of variance for the departments resulted in no rejection of the null hypothesis indicating that there was no statistically significant difference among the departments.

The last analysis performed was the descriptive analysis for each scale item on the questionnaire, and the results are depicted in the table below.

The descriptive data in Table 8 can be analyzed in three broad ways; there are items associated with the students' own inability to grasp the concepts. As can be seen from the table, 71% of the respondents agree that mathematics is difficult to understand. These agree with more than half, 56%, who accepted that they had difficulties in interpreting mathematics questions. It's obvious that if a student is not able to interpret a question, then such a student will consider the question to be difficult because interpreting is the very beginning of the problem-solving process. Clearly, it's not the mathematics vocabulary that makes it difficult to interpret, and therefore, understand the question. This is evident in the table that more than half of the respondents disagreed with the notion that mathematics vocabulary was difficult to read. This, therefore, leaves us to conclude that it is the mathematics content and or context that is difficult for students to interpret and understand, indeed almost 70% of them expressly said that they are not able to understand the meanings of mathematical expressions. To reinforce this, almost 60% of the respondents confirmed that they do not have the requisite critical thinking abilities for mathematical problem-solving. Critical thinking is necessary for students to recognize quantitative facts and relationships in a mathematical problem in order to solve it, therefore it's not surprising that more than 50% of the are not able to recognize these facts and relationships. Which is not surprising that three quarters of them are not even able to relate lessons to previous ones in order to make sense of the mathematical content. If students are not able to relate two or three lessons, then it would be very difficult for them to relate concepts, hence the difficulties in understanding the subject.

Besides the students' own inability to grasp concepts, the scale items also brought to the fore the difficulties students encounter in understanding the procedural aspect of mathematical problem-solving. As clearly depicted in Table 8, almost 80% of the respondents had difficulties understanding rules and/or methods of problem-solving in mathematics, and almost the same number, 78%, had difficulties in choosing appropriate methods for solving mathematics problems. Indeed, it follows with certainty that such students would find it almost impossible to follow the steps involved in solving mathematical problems, as is seen in the table that about 72% were not able to or find it difficult to do so, and almost 80% find it difficult to understand the rules of mathematical order of operations. As a clear indication, data also shows that over 70% of the respondents confess that they get confused by too many mathematical formulas. It is therefore clear that most students experience some form of difficulties in understanding mathematical procedures and algorithms.

The last aspect of this descriptive data is the external interferences. The data reveals other factors like technology, fear, attitude, teaching materials, and instructors. Table 8 shows that almost 80% of the respondents had a weak foundation in mathematics, and this is clearly displayed by the 75% who accepted that they were unable to continue with mathematics lessons, and the 70% who become uncomfortable or shy away from asking questions in class. The weak foundation may be attributed to the instructors or perhaps instructional materials like textbooks and classroom technology. For instance, 76% of the respondents agree that the textbooks used are not appropriate, and 73% suggest overdependence on technology to solve mathematics problems. This could also have been caused by overcrowded classrooms, 78%, or negative attitude towards the subject as shown by 72% of the respondents. Teachers take a fair share of the cake, 74% agree that teachers are not able to deliver the lesson to students effectively, and 77% think that teachers are unable to use mathematical language that students understand in formulating problems.

4 Discussion and Conclusion

It is evident that mathematics subject presents challenges to learners in various ways and causes of which vary as well.

Table 8 Descriptive statistics analysis for all items

Scale item	Completely agree	Agree	Not sure	Disagree	Completely disagree
Mathematics is difficult to understand	38	33	12	12	5
Difficult to understand rules/methods of problem solving	21	38	18	14	7
Weak foundation in mathematics	23	38	23	12	4
Difficult to interpret mathematics questions	20	36	22	18	4
Too many students in a class affects understanding	32	38	8	16	6
Difficult to read mathematics vocabulary	13	35	16	23	13
Difficult choosing appropriate methods for solving problems	27	28	23	19	3
Difficult to follow all steps in solving a mathematics problem	30	27	15	20	8
Difficult to understand rules of mathematical order of operations	26	34	18	15	7
Dependency on technology to solve mathematical problems	20	27	26	19	8
Lack of critical thinking in solving mathematical problems	20	39	19	16	5
Inability to recognize quantitative facts and relationships in a mathematical problem	22	33	16	23	6

(continued)

Table 8 (continued)

Scale item	Completely agree	Agree	Not sure	Disagree	Completely disagree
Inability to understand the meanings of mathematical expressions	28	40	14	12	6
Fear of failing to solve a mathematical problem	29	23	27	14	7
The teacher is unable to deliver the lesson to the students effectively	25	28	21	22	4
Negative attitude towards mathematics	26	21	25	20	8
The teacher is unable to use mathematical language that the students understand in formulating problems	22	31	23	17	7
Fear of failing to solve a mathematical problem	29	23	27	14	7
Students' inability to relate the lesson to the previous one	19	34	22	22	3
Inability to continue with the lesson	19	34	22	20	4
Feeling uncomfortable when solving a problem	27	24	16	16	15
Feeling shy to ask questions in class	26	28	16	13	17
Repeated absences that lead to lack of understanding of the subject	21	34	23	19	3
The course book is not appropriate for students	26	26	20	19	9
Confused by too many mathematical formulas	31	26	17	13	13

4.1 External Factors

In this research, it was found that students in the Kurdistan region exhibited are affected by external factors that affect their understanding of mathematics concepts.

Weak Foundation

It is evident in some of the results such as a weak foundation in mathematics from the early years in school contributes immensely to the difficulties experienced in later years. This result was also found by the rand group [11] which established that more than two-thirds of Kurdish students have been retained at least one year by the time they reach grade 9, and that in about two-thirds of urban schools, more than 50% of students failed the school's assessment in 2007–2008. This paints a grim situation for educators, more specifically, mathematics educators in the region. One possible explanation for this would be the unrest that has bedeviled the region. After the Iraq war, which lasted for a long time and disrupted all aspects of human development including education, the semi-autonomous Kurdistan region became relatively stable in terms of peace, and this attracted rapid development in terms of education and even physical infrastructure. That's until the ISIS crises that brought uncertainty in the region again. All these cause immense negative consequences on the education sector in the region.

Overcrowded Classes

Overcrowded classes are another reason for poor performance in mathematics. This result is also shared by [11] through the Rand organization found that school infrastructure in the Kurdistan region is not in tandem with the growth pace. They affirmed that all schools at all levels are overcrowded, in poor conditions such that they don't keep up with the rapid growth of student enrollment, leaving no room but double shifts. The lack of sustainable peace in the larger Iraq and war in the neighboring Syria saw an influx of people in the relatively peaceful Kurdistan region [12]. Indeed, many expatriates especially dealing with peace in the larger Iraq and Syria set up office in Erbil, Kurdistan. These large numbers of people exert pressure on the limited resources in the region, and this includes education facilities.

Fear and Anxiety

Fear of and negative attitude towards mathematics creates anxiety in students contributing heavily to lowering standards of mathematics. This research confirmed [5] who found out that mathematics anxiety negatively affects many students' understanding of mathematical concepts. In the Kurdistan region, this anxiety may be attributed to stereotypes that the larger Middle East is good in mathematics and therefore students in Kurdistan should automatically be good in the subject. This puts too much pressure on the students which serve to inhibit creativity in the mathematics classroom.

Instructors and Instructional materials

The researchers also established that instructors and instructional materials are other factors affecting students' understanding of mathematics. In the Kurdistan region, as was established by [11], there is an insufficient number of teachers, and the number is not increasing at a matching pace with the rapidly increasing number of students. This imbalance puts too much pressure on teachers and thereby impacting negatively their delivery of lessons. Compounding the problem is the fact that the few available teachers are ill-trained and ill-equipped to do their job, as concluded by the Rand group. The lack of standard textual materials and classroom technology in many schools exacerbates the problem for teachers, leading to poor performance by students due to a lack of understanding of basic concepts.

4.2 Mathematical Procedures in Problem-Solving

This research established that many students experience difficulties in following and/or understanding the procedures involved in mathematical problem-solving.

Understanding Rules

Many students seem to get confused with mathematics formulas. In the research, participants acknowledged finding it difficult to follow the rules of mathematics such as the order of operation, or even understanding formulas like the quadratic formula. In this case, the possible logical explanation would be a poor foundation in the subject as described above. It is also possible that poor instructions are given by instructors who are not fully equipped or supported appropriately through training and the provision of classroom materials like textbooks.

Choosing Appropriate Methods

This may come as a surprise to many mathematicians, but this research found out that many students are at a loss as to what methods they should use for what problems. Many of them acknowledged being confused by "too many" mathematical formulas. This observation may be attributed to poor instructional strategies by teachers, lack of interest in the subject by students, and the external factors discussed above.

4.3 The Learners Cognitive Abilities

In the research, most respondents acknowledged that they just cannot understand mathematics, saying that they experience difficulties right from the problem interpreting stage. They reinforced this notion by confirming that they were unable to recognize quantitative facts and relationships in a mathematical problem, and even understand the meanings of mathematical expressions. Many agreed that they find it difficult to relate lessons and therefore could not see any flow in concepts. This phenomenon may also be due to poor instructional strategies that lead to disinterest among learners.

5 Recommendations

Based on this study, education stakeholders in the region need to incentivize the teaching profession in order to attract more people. The classrooms need to be decongested and be adequately provided for in terms of classroom materials such as physical manipulatives, classroom technology, and textual materials. More professional development courses, seminars, and/or workshops for instructors to better their instructional strategies and keep up with the latest developments in the industry.

References

1. Acharya, B.R.: Factors affecting difficulties in learning mathematics by mathematics learners. Int. J. Elemen. Educ. **6**(2), 8–15 (2017). https://doi.org/10.11648/J.IJEEDU.20170602.11
2. Gal, H., Linchevski, L.: To see or not to see: analyzing difficulties in geometry from the perspective of visual perception. Educ. Stud. Math. **74**(2), 163–183 (2010). https://doi.org/10.1007/s10649-010-9232-y
3. Kelanang, J.G.P., Zakaria, E.: Mathematics difficulties among primary school students. Adv. Nat. Appl. Sci. **6**(7), 1086–1092 (2012)
4. Marshall, E.M., Staddon, R.V., Wilson, D.A., Mann, V.E.: Addressing maths anxiety and engaging students with maths within the curriculum. MSOR Connetions **15**(3) (2017). https://doi.org/10.21100/msor.v15i3.555
5. Orton, A.: Students' understanding of integration. Educ. Stud. Math. **14**(1), 1–18 (1983)
6. Rameli, M.R.M., Kosnin, A.M.: Challenges in Mathematics Learning: A Study From School Students' Perspective (2016)
7. Robert, A., Speer, N.: Research on the teaching and learning of calculus/elementary analysis. In: The Teaching and Learning of Mathematics at University Level, pp. 283–299. Springer, Dordrecht (2001). https://doi.org/10.1007/0-306-47231-7_26
8. Tambychik, T., Meerah, T.S.M., Aziz, Z.: Mathematics skills difficulties: a mixture of intricacies. Procedia Soc. Behav. Sci. **7**, 171–180 (2010). https://doi.org/10.1016/j.sbspro.2010.10.025
9. Quezada, V.D.: Difficulties and performance in mathematics competences: solving problems with derivatives. Int. J. Eng. Pedagog. **10**(4), 35–53 (2020). https://doi.org/10.3991/ijep.v10i4.12473
10. Vernez, G., Culbertson, S., Constant, L., Karam, R.: Initiatives to improve quality of education in the Kurdistan Region-Iraq. Rand Corporation, RAND Education (2016)
11. Waswa, D.W., D'Cunha, N.: Using Kurdish in preparatory english language classrooms: a replication study at Tishk International University-Erbil, Iraq. Int. J. Soc. Sci. Edu. Stud. **8**(3), 161–169 (2021). https://doi.org/10.23918/ijsses.v8i3p161
12. Waswa, D.W., Al-Kassab, M.M., Alhasoo, A.A.: Dynamics of high school certificate examinations demand: a case study of kurdistan region, Iraq. Computer **10**, 5 (2021). https://doi.org/10.23918/eajse.v7i2p1

13. Waswa, D.W., Al-kassab, M.M., Alhasoo, A.A.: Factors affecting the achievement of demand in high school certificate examination in Erbil, KRG, Iraq. In: AIP Conference Proceedings, vol. 2554, No. 1, p. 030003. AIP Publishing LLC (2023).

Finding Solution to the Initial Value Problem for ODEs First and Second Order by One and the Same Method

V. R. Ibrahimov, G. Yu. Mehdiyeva, and M. N. Imanova

Abstract As is known by using a change of variables, the determination of the solution of ODEs of the second order can be reduced to finding the solution of the system of ODEs of the first order. Therefore, here we have considered a comparison of the multistep methods with the multistep second derivative methods. For this aim suggested here to use the advanced and hybrid methods, which are more exact than the explicit and implicit methods. Some advantages of the proposed methods here have and defined the maximum value of the degree to stable methods. Here for the comparison of these methods with the known ones have defined the disadvantages of the constructed methods and have given the way for the correction mentioned disadvantages of these methods. Constructed, specific methods, which have been applied to solve some simple problems. Note that these methods are not a special case of the known methods. Therefore these methods are independent and they constitute an independent class of methods. For the illustration of the benefits of this method, we have considered the application of some of the suggested methods here to solve some simple problems.

Keywords Initial-value problem for ODE · Multistep second derivate method · Symmetrical multistep methods · Multistep methods of hybrid type · Stability and degree · Bilateral methods

---

V. R. Ibrahimov (✉) · M. N. Imanova
Institue of Control System Named After Academician A. Huseynov, Baku AZ1141, Azerbaijan
e-mail: Ibvag47@mail.ru

M. N. Imanova
e-mail: mehriban.imanova@sdf.gov.az

V. R. Ibrahimov · G. Yu. Mehdiyeva · M. N. Imanova
Computational Mathematics, Baku State University, Baku AZ1148, Azerbaijan
e-mail: imn_bsu@mail.ru

M. N. Imanova
Science Development Foundation, The Republic of Azerbaijan, Baku AZ1025, Azerbaijan

D. Zeidan et al. (eds.), *Mathematics and Computation*, Springer Proceedings in Mathematics & Statistics 418, https://doi.org/10.1007/978-981-99-0447-1_28

1 Introduction

As is known the first direct numerical method for solving ODEs was constructed by Euler. Euler noted that his method coincides with the first two terms of the Taylor expansion. Therefore the error receiving on each point equal to $O(h^2)$. To obtain reliable information about the solution of the investigated problem, Euler suggested to use the calculation of the next terms in the Taylor expansion. By taking into account of this, many scientists tried to construct a method with a second derivative (see for example [1–3]). It is known that one of the popular methods for solving ODEs is the multistep method with constant coefficients, which was fundamentally investigated from the beginning of the 50-th years of the XX century (see for example [4, 5]). This method in a general form can be represented as follows:

$$\sum_{i=0}^{k} \alpha_i y_{n+1} = h \sum_{i=0}^{k} \beta_i y'_{n+i} (n = 0, 1, 2, ...), \tag{1}$$

here the coefficients $\alpha_i, \beta_i (i = 0, 1, ..., k)$ are some real numbers and $\alpha_k \neq 0$. If this method is applied to solve the initial-value problem for ODEs, then receive (see for example [6–14]):

$$\sum_{i=0}^{k} \alpha_i y_{n+1} = h \sum_{i=0}^{k} \beta_i f_{n+i} (n = 0, 1, 2, ...), \tag{2}$$

here $f_m = f(x_m, y_m) (m \geq 0)$, $(m \geq 0)$, but the initial-value problem can be presented in the following form:

$$y' = f(x, y), y(x_0) = y_0, x_0 \leq x \leq X. \tag{3}$$

Suppose that the problem (3) has a unique solution, which is defined in some segments. Let us denote the approximate value of the solution of the problem (3) at the point x_i by y_i and corresponding exact value by the $y(x_i)$. In addition, the mesh-point is denoted in the form $x_{i+1} = x_i + h(i \geq 0)$. And suppose that the function of $f(x, y)$ is defined in some domain in which it has the continuous partial derivatives up to order p, inclusively. As was noted the above method (2) has been investigated by many authors. And for the comparison of the numerical methods primarily used the concepts of stability and the degree (order of accuracy) of the comparison methods. The method (1) by some scientists are called as the finite-difference method. Therefore the quantity of the k are called as the order of the method (1), which takes as the given. By taking this into account, the scientists tried to find some relation between of the order and the degree for the method (2). In 1955, Bakhvalov (see [6]) has investigated method (2) in the case $\alpha_k \neq 0$ and $\beta_i = 0$ prove that for the $k \leq 10$ there are stable methods with the degree $p \leq k$. Method (2) fundamentally investigated by Dahlquist. Dahlquist proved that in the class of the methods (2), there are stable methods with the degree $p \leq 2[k/2] + 2$ if $\alpha_k \neq 0$ and $\beta_k \neq 0$, but with the degree $p \leq k$ for the case $\alpha_k \neq 0$, $\beta_k = 0$. And also have proved that, there are stable methods with degree $p_{\max}$ for all the values of k. Noted that here the conceptions degree and stability define as the following (see for example [4–6]). Definition 1. The method of (2) is called as the stable, if the roots of the polynomial $\rho(\lambda) = \alpha_k\lambda^k + \alpha_{k-1}\lambda^{k-1} + \cdots + \alpha_1\lambda + \alpha_0$ located in the unit circle on the boundary of which there is not multiply root. Definition 2. The method (2) is said to have degree p if the following asymptotic equality is holds:

$$\sum_{i=0}^{k} \left(\alpha_i y(x + ih) - h\beta_i y'(x + ih)\right) = O(h)^{p+1}, h \to 0. \tag{4}$$

It is known that one of the indicators of the effectiveness of the method is its accuracy. Hence, it follows that the method with higher accuracy is effective.

2 Construction Multistep Second Derivative Methods

By taking into account that the stable methods of type (2) has the maximum degree $p_{\max} = 2[k/2] + 2$, we receive that for the construction of the effective methods one can use the methods with higher degrees. For this aim, here proposed to use the multistep second derivative methods with constant coefficients, which can be presented as follows:

$$\sum_{i=0}^{k} \alpha_i y_{n+i} = h \sum_{i=0}^{s} \beta_i y'_{n+i} + h^2 \sum_{i=0}^{l} \gamma_i y''_{n+i}, \tag{5}$$

here the coefficients $\alpha_i, \beta_i, \gamma_i (i = 0, 1, ..., k; j = 0, 1, ..., l)$ are some real numbers and $\alpha_k \neq 0$. In the case $k = s = l$ from the formula (5) it follows the known multistep second derivative methods with the constant coefficients (see for example [16–20]).

Let us note that Euler himself suggested using calculation of them involved in the decomposition of Taylor starting with the third member (see [1, 2]). Notice that in this case the amount of computation will increase. For example in the calculation of the value $y''(x)$ the amount of the computation work will increase by almost two times. Let us consider the following function:

$$y''(x) = f'_x + f'_y \cdot y' (\text{by using problem (3), receive that} y' = f(x, y).$$

It follows from here that calculation the value of the function $y''(s)$ will be more difficult. Let us noted that method (5) successfully used in solving some applied problems. As is known if method (5) stable and has degree of p, then the next one takes place:

$$p \le 2k + 2.$$

It is not difficult to show that by using the next method

$$\sum_{i=0}^{k} \alpha_i y_{n+i} = h \sum_{i=0}^{k} \beta_i f_{n+i+\upsilon_i} (|\upsilon_i| < 1; i = 0, 1, .., k), \tag{6}$$

one can construct stable method with the degree of $p = 2k + 2$. In this case arises some difficulties with the calculation of the values $y_{n+i+\nu_i}$. Note that the method of (6) resembles the Gauss method. One of the popular method of type (6) can presented as follows:

$$y_{n+1} = y_n + h(f_{n+\alpha} + f_{n+1-\alpha})/2, \alpha = \sqrt{3}/6, \tag{7}$$

which has constructed for the $k = 1$ and that has degree $p_{\max} = 4$. The method with the degree $p_{\max} = 6$ constructed for the $k = 2$ and presented as follows:

$$y_{n+2} = y_n + h(5 f_{n+1+\alpha} + 8 f_{n+1} + 5 f_{n+1-\alpha})/9, \ \alpha = \sqrt{15}/5. \tag{8}$$

Note that methods (7) and (8) are stable. If in the method (5) to put $k = s = l = 1$, then from the method (5) can be received the following method, which has the degree $p = 4$:

$$y_{n+1} = y_n + h(f_n + f_{n+1})/2 + h^2(g_n - g_{n+1})/12, \tag{9}$$

here $g(x, y) = f'_x(x, y) + f'_y(x, y) f(x, y)$. By the comparison of the method (7) or (8) with the method (9) receive that in using method (9) it is arises necessity to use some methods for calculation of the predictor values $\hat{y}_{n+1}$ so as method (9) is the implicit. For this aim one can be used some predictor method. It is known that as the predictor method in usually used the explicit method. But in our case the maximum degree for the stable methods satisfies the condition $p \le 2$. Therefore arises necessity to using any implicit method with the degree $p = 3$. Noted that in using method of (7) arises necessity to calculate the values $y_{n+\alpha}$ and $y_{n+1-\alpha}$. It is not difficult to show that by using the value $y_{n+\alpha}$ one can calculate the value $y_{n+1-\alpha}$. Notice, that in application method of (9) arises some difficulties related with the calculation of the function $g(x, y)$. From here, we receive that method (7) has some advantages.

3 Construction Stable Methods with the Degree $p = 3k + 3$

For the construction, more accurate methods let us to consider the following method:

$$\sum_{i=0}^{k}\alpha_i y_{n+i} = h\sum_{i=0}^{k}\beta_i f_{n+i} + h\sum_{i=0}^{k}\gamma_i f_{n+i+\upsilon_i}(|\upsilon_i| < 1;\, i = 0, 1, .., k). \tag{10}$$

Here coefficients $\alpha_i, \beta_i, \gamma_i$ are some real numbers, $\alpha_k \neq 0$. To the proposed method of (10) can be considered as generalizations of the methods (2) and (6). As was shown above, the method can be taken as the better if the method is stable and has the maximum degree. It follows from here that for comparison of the multistep methods it is necessary to define the maximum value for the degree of the method (10). For this aim let us to consider the following Taylor series:

$$y^{(j)}(x+ih) = y^{(j)}(x) + ihy^{(j+1)}(x) + \frac{(ih)^2}{2!}y^{(j+2)}(x) + \cdots + \frac{(ih)^p}{p!}y^{(j+p)}(x) + O(h)^{p+1},\ h \to 0,$$

here $j = 0, 1, 2$ and $y^{(0)}(x) = y(x),\ y^{(1)}(x) = y'(x), y^{(2)}(x) = y''(x)$.

Suppose that method has degree of p. In this case, the following is holds:

$$\sum_{i=0}^{k}\big(\alpha_i y(x+ih) - h\beta_i y'(x+ih) - h\gamma_i y'(x+(i+\nu_i)h)\big) = O(h)^{p+1},\ h \to 0. \tag{11}$$

If in this equality to put $\gamma_i = 0 (i = 0, 1, .., k)$, then from the asymptotic equality of (11), it follows equality of (4). By using the above presented Teylor series in the left hand side of the equality (11), receive:

$$\begin{aligned}&\sum_{i=0}^{k}\big(\alpha_i y(x+ih) - h\beta_i y'(x+ih) - h\gamma_i y'(x+(i+\nu_i)h)\big) = \\ &\sum_{i=0}^{k}\alpha_i y(x) + h\sum_{i=0}^{k}(i\alpha_i - \beta_i - \gamma_i)y'(x) + \cdots \\ &+h^p\sum_{i=0}^{k}\left(\frac{i^p}{p!}\alpha_i - \frac{i^{p-1}}{(p-1)!}\beta_i - \frac{l_i^{p-1}}{(p-1)!}\gamma_i\right)y^{(p)}(x) + O(h)^{p+1}, \\ &l_i = i + \nu_i,\ h \to 0.\end{aligned} \tag{12}$$

By taking asymptotic equality of (11) in the equality of (12), we receive that the following is holds:

$$\sum_{i=0}^{k}\alpha_i y(x) + h\sum_{i=0}^{k}(i\alpha_i - \beta_i - \gamma_i)y'(x) + h^2\sum_{i=0}^{k}\left(\frac{i^2}{2!}\alpha_i - i\beta_i - l_i\gamma_i\right) + \cdots$$
$$h^p\sum_{i=0}^{k}\left(\frac{i^p}{p!}\alpha_i - \frac{i^{p-1}}{(p-1)!}\beta_i - \frac{l_i^{p-1}}{(p-1)!}\gamma_i\right)y^{(p)}(x) = 0. \tag{13}$$

As it is known the system of functions $1, x, x^2, \ldots, x^p$ and the system of functions $y(x), y'(x), \ldots, y^{(p)}(x)$ $y^{(j)}(x) \not\equiv 0$ for all the values of $j (0 \le j \le p)$ are the independent system. By using this, receive that the following must satisfy:

$$\sum_{i=0}^{k}\alpha_i = 0; \sum_{i=0}^{k} i\alpha_i = \sum_{i=0}^{k}(\beta_i + \gamma_i); l_i = i + \nu_i.$$
$$\sum_{i=0}^{k}\frac{i^2}{2!}\alpha_i = \sum_{i=0}^{k}(i\beta_i + l_i\gamma_i); ..; \sum_{i=0}^{k}\frac{i^p}{p!}\alpha_i = \sum_{i=0}^{k}\left(\frac{i^{p-1}}{(p-1)!}\beta_i + \frac{l_i^{p-1}}{(p-2)!}\gamma_i\right), \tag{14}$$

Thus, receive that if the method (10) has the degree p, then its coefficients will satisfy the system (14). In addition, vice versa, for each solution of the system (14) there is a corresponding method with a certain degree. Now let us consider the definitions of some relationship between degree p and order k for the method of (10). The number of equation in the system (14) is equal to $p + 1$ but the number of unknowns is equal to $4k + 4$. It is easy to show that this system has a solution for the $p \le 4k + 2$. It is clear that not all methods with maximum accuracy will be stable or convergent. This question has investigated in the work (see [13–28]) and have proved that there are stable methods of type (10) with the degree $p \le 3k + 3$. As is known the one-step methods are convergent, therefore they can be called as the stable. As was shown above, the $p_{\max} = 4k + 2$ for the method (10). It follows from here, that $p_{\max} = 6$ for the one-step methods. As is known $p_{\max} = 3k + 3$ for the stable methods of type (10). Noted that in the class methods of (6) there are stable method with the degree $p_{\max} = 6$, which can be receive from the method of (6) for the case $k = 2$. Hence, we get that the maximum value of the degree for the stable and instable methods of type (10) coincide for $k = 1$. By taking into account that the system of (14) consists of nonlinear algebraic equations often for solving of which are used the MathCard program. In the case of $k = 2$, have constructed method with the degree $p = 10$. However, the results receiving in the application of that method to solve model problem did not correspond to the theoretical. This is due to the fact that the resulting solution of the system of algebraic equation with some error. Therefore, here recommended using the inequality $p \le 3k + 3$ in the construction of stable methods of type (10). In the investigation method (10), one of the issue is to determine the necessary conditions for the convergence of multistep methods with constant coefficients. For this aim let us to consider the following conditions.

A. The coefficient $\alpha_i, \beta_i, \gamma_i$ are some real numbers, $\alpha_k \neq 0$.

B. Characteristic polynomials $\rho(\lambda) \equiv \sum_{i=0}^{k} \alpha_i \lambda^i$; $\sigma(\lambda) \equiv \sum_{i=0}^{k} \beta_i \lambda^i$; $\gamma(\lambda) \equiv \sum_{i=0}^{k} \gamma_i \lambda^{i+\nu_i}$ have no common factors other than constant.

C. $\sigma(1) + \gamma(1) \neq 0$ and $p \geq 1$ are satisfies.

The condition A is obvious. Therefore let us consider the condition of B. Suppose the contrary and take $\phi(\lambda)$ as the common factor for the polynomials $\rho(\lambda), \sigma(\lambda)$ and $\gamma(\lambda)$. By using the $E^i y(x) = y(x + ih)$ shift operator method of (10) can be presented as follows:

$$\rho(E)y_n - h\sigma(E)y_n' - h\gamma(E)y_n' = 0. \tag{15}$$

By taking into account, that $\phi(\lambda)$ is the common factor, then equality (15) can be presented as follows:

$$\phi(E)(\rho_1(E)y_n - h\sigma_1(E)y_n' - h\gamma_1(E)y_n') = 0.$$

Given than $\phi(\lambda) \neq const$, from this equality get the following:

$$\rho_1(E)y_n - h\sigma_1(E)y_n' - h\gamma_1(E)y_n' = 0, \tag{16}$$

here

$$\rho_1(\lambda) = \rho(\lambda)/\phi(\lambda), \sigma_1(\lambda) = \sigma(\lambda)/\phi(\lambda), \gamma_1(\lambda) = \gamma(\lambda)/\phi(\lambda).$$

By comparison of the equations of (15) and (16), receive that the Eqs. (15) and (16) are equivalent. Noted that equation of (15) is the finite-difference equation with the order of $k_1 < k$. As it is known if are given k_1 initial-values, then the finite-differential equation of (16) will have a unique solution.

Those, receive that the finite-difference equation with the constant coefficients have a unique solution if the number of initial data is less than the order of the difference equation with constant coefficients. It follows from here that our assumption does not hold. Therefore, the conditions B is satisfied. And now let us consider the fulfilment of the condition C. For this aim let us equation of (15) to written as following:

$$\rho(E)y_n = h\sigma(E)y_n' + h\gamma(E)y_n'. \tag{17}$$

By passing to limit for $h \to 0$, receive that

$$\rho(1) = 0, \tag{18}$$

if $x = x_0 + nh$ is fixed point. This condition is called as the necessary condition for the convergence of investigated method. By using condition of (18), the equation can be presented as:

$$\rho_1(E)(E-1)y_n - h\sigma(E)y'_n - h\gamma(E)y'_n = 0$$

or

$$\rho_1(E)(y_{j+1} - y_j) - h\sigma(E)y'_j - h\gamma(E)y'_j = 0, \tag{19}$$

here

$$\rho_1(\lambda) = (\rho(\lambda) - \rho(1))/(\lambda - 1).$$

In the result of the equality, let us change the value of the variable j starting from 0 to n, after summing of the receiving of equalities get the following equality:

$$\rho'(1)(y(x) - y_0) = \sigma(1)\int_{x_0}^{x} y'(s)ds + \gamma(1)\int_{x_0}^{x} y'(s)ds,$$

here $x = x_0 + nh$ fixed point. This equality can be presented as the following form:

$$\rho'(1)(y(x) - y_0) = (\sigma(1) + \gamma(1))\int_{x_0}^{x} y'(s)ds.$$

By using $y(x) - y(x_0) = \int_{x_0}^{x} y'(s)ds$ equality, receive:

$$\rho'(1) = \sigma(1) + \gamma(1). \tag{20}$$

From here receive that if $\sigma(1) + \gamma(1) = 0$ then it follows that $\rho(1) = \rho'(1) = 0$. Consequently, $\lambda = 1$ is the double root. Now will show that in this case the multistep method does not converge. For this, let us consider the following finite-difference equation:

$$\alpha_k y_{n+k} + \alpha_{k-1} y_{n+k-1} + \cdots + \alpha_0 y_n = hc, \tag{21}$$

which is nonhomogeneous finite-difference equation with the constant coefficients. The general solution of the Eq. (21) can be presented as the follows:

$$y_m = \overline{y}_m + y_m^*,$$

Here $\overline{y}_m$ is the general solution of the homogeneous equation, but $y*_m$ is one of partial solution of nonhomogeneous equation, it is easy to prove that $\lim_{h \to 0} y^*_m = 0$. And it is known that $\overline{y}_m$-can be presented as:

$$\overline{y}_m = c_1\lambda_1^m + c_2 m\lambda_1^m + c_3\lambda_2^m + \cdots + c_k\lambda_k^m.$$

As follows from here that if $h \to 0$, then $m \to \infty$. Therefore if $m \to \infty$, then also $\overline{y}_m \to \infty$ and in this case the method does not converge. By this way prove that if $\rho(1) = \rho'(1) = 0$, then method (10) does not converge. Consequently $\rho'(1) \neq 0$, that's why the condition C is satisfied.

And now let us consider construction some concrete methods.

4 Construction of Any Concrete Methods and Application Some of Them to Solve Model Problem

For the construction of the methods of type (10) usually are used Taylor series. By using Taylor series for finding the coefficients $\alpha_i, \beta_i, \gamma_i, \nu_i$ $(i = 0, 1, .., k)$, one can used the nonlinear system of algebraic Eqs. (14).

If $k = 1$ then from the condition $\alpha_1 + \alpha_0 = 0$, receive that $\alpha_1 = 1$ and $\alpha_0 = -1$. In this case, the system of (14) can be written as follows:

$$\beta_1 + \beta_0 + \gamma_1 + \gamma_0 = 1,$$

$$\beta_1 + l_1^j\gamma_1 + l_0^j\gamma_0 = 1/(j+1), l_i = i + \nu_i (i = 0, 1). \tag{22}$$

By solving this system of algebraic equations, one can constructed the numerical methods with the different properties. For example, let us consider the case $\beta_1 = \beta_0 = 0$. In this case, receive the method of (7).

And now let us consider to solving of the system (14) for the case $k = 1$. In this case, the constructed method with the degree $p_{\max} = 6$ can be presented as:

$$y_{n+1} = y_n + h(f_n + f_{n+1})/12 + 5h(f_{n+1/2-\beta} + f_{n+1/2+\beta})/12, \ \beta = \sqrt{5}/10. \tag{23}$$

By the comparison of the methods (7) and (23), we receive that each of them has its advantages and disadvantages. For example, method (23) is implicit, but method (7) is explicit.

Note that by using the solution of the nonlinear system (22) of algebraic equations can been constructed different methods with the different properties. For the illustration of this, let us consider to following methods:

$$y_{n+1} = y_n + h(f_n + 3f_{n+2/3})/4, \tag{24}$$

$$y_{n+1} = y_n + h(f_{n+1} + 3f_{n+1/3})/4. \tag{25}$$

These methods have the order $p = 3$ and stable. Note that methods (7), (24) and (25) are one step methods and method (7) is more exact than the methods (24) and (25). Methods (24) and (25) has maximum order in the considering cases. The maximum value of the degree for these methods equal to $p_{\max} = 3$. As was noted above there are many works dedicated to investigation multistep second derivative methods. Some of them dedicated to construction multistep second derivative methods of hybrid types. For example, in the work [33], problem (3) has investigated by the following method

$$\sum_{i=0}^{k} \alpha_i y_{n+i} = \sum_{j+1}^{l} h^j \sum_{i=0}^{k} \beta_{i,j} y_{n+i}^{(j)} + h \sum_{i=0}^{s} \gamma_i f_{n+\nu_i}, \alpha_k = 1. \tag{26}$$

And have constructed some methods. Noted that methods (30) and (31) (constructed in [33]) have the degree $p = 8$, but they are instable. Methods (19) and (21) constructed in [33] are stable and have the degree $p = 6$ and $p = 5$ respectively which correspond to the above received law. Noted that the method of (26) in a more general form investigated by some authors (see for example [17]). Similar methods were also investigated by different authors (see [18–21]). In this works were used non-classical way for construction their methods. For example method, which constructed in [21] by prof. T.E. Simos in some sense intersects with the Runge–Kutta methods, so it is of some interest with the one step (Runge–Kutta) method. We also have used similar schemes.

For the construction more exact methods, here is suggested to use linear combination of some methods. For this aim let us consider the following equalities:

$$\hat{y}(x_{n+1}) = y(x_n) + hf(x_n y(x_n)) + h^2 y'(x_n)/2! + O(h^3),$$

$$y(x_{n+1}) = y(x_n) + hf(x_{n+1} y(x_{n+1})) - h^2 y''(x_n)/2! + O(h^3),$$

which received by using Euler's methods ($\overline{y}(x_{n+1}) = y(x_{n+1})$). By using these and similar methods proposed here to construct the bilateral methods, which satisfies the following condition:

$$\hat{y}_{n+1} \le y(x_{n+1}) \le y_{n+1} \big(\text{if } y''(x) \ge 0\big).$$

As is known that the following scheme is usually has used for the constructing of bilateral methods

$$\underline{y}_{n+1} \le y(x_{n+1}) \le \overline{y}_{n+1}. \tag{27}$$

Here $\underline{y}_{n+1}$-lower value and $\overline{y}_{n+1}$ upper value.

In the construction of bilateral methods this way is used in cases, when all coefficients are positive. This condition holds for the hybrid methods, since in these methods all coefficients are usually positive. To illustrate the received here theoretical results, let us to consider the following example:

1. $y'(x) = \exp(\lambda x)\cos(x), \; y(0) = 0, \; 0 \le x \le 1,$ (28)

   exact solution for this problem can be presented as the

$$y(x) = (\lambda\cos(x) + \sin(x))\exp(\lambda x)/(\lambda^2 + 1) - \lambda/(\lambda^2 + 1).$$

2. $y'(x) = \lambda y(x), \;\; y(0) = 1, \;\; 0 \le x \le 1$, with the exact solution:

$$y(x) = \exp(\lambda x). \tag{29}$$

In solving this example, some approaches from the following works [28–39] were used.

To solving problem (28) has applied method (23). Since the right hand side of the differential equation independence from the function of $y(x)$, so in finding the solution does not represent any difficulties. Dependences from the values of λ, the solution is other increasing or decreasing. Here, wanted to show that the result obtained correspond to the properties of the solution of problem (28).

From the results tabulated in the Tables 1 and 2, receive that the method behaves in a stable ratio of the error of obtaining when it is used, as well as in changes in the λ-constant, thus receive that method of (8) can be take as the better. And now let us apply method of (7) to solve problem (29), the results of which have tabulated in Tables 3 and 4.

Table 1 Results receiving for $h = 0.1$ by the method (23)

x	$\lambda = 1$	$\lambda = -1$	$\lambda = 5$	$\lambda = -5$
0.1	2.7E-14	2.5E-14	4.9E-10	3.8E-10
0.4	5.4E-13	3.2E-13	1.8E-9	9.9E-10
0.7	1.9E-12	7.9E-13	9.6E-9	1.1E-9
1.0	4.8E-12	1.13E-12	1.3E-7	1.2E-9

Table 2 Results receiving for $h = 0.05$ by the method (23)

x	$\lambda = 1$	$\lambda = -1$	$\lambda = 5$	$\lambda = -5$
0.1	4.0E-16	3.6E-16	7.6E-12	5.9E-12
0.4	8.4E-15	5.0E-15	2.8E-11	1.5E-11
0.7	3.1E-14	1.2E-14	1.5E-10	1.8E-11
1.0	7.5E-14	2.0E-14	2.1E-9	1.9E-11

Table 3 Results receiving by the method (7) for the $h = 0.1$

x	$\lambda = 1$	$\lambda = -1$	$\lambda = 5$	$\lambda = -5$
0.1	4.2E-6	4.0E-6	2.8E-3	2.3E-3
0.4	2.2E-5	1.2E-5	5.1E-2	2.0E-3
0.6	4.2E-5	1.4E-5	2.1E-1	1.1E-3
0.8	6.8E-5	1.6E-5	7.6E-1	5.6E-4
1.0	1.0E-4	1.6E-5	2.5E-0	2.5E-4

Table 4 . results receiving by the method (7) for the $h = 0.05$.

x	$\lambda = 1$	$\lambda = -1$	$\lambda = 5$	$\lambda = -5$
0.1	5.5E-7	4.9E-7	4.3E-4	2.4E-4
0.4	2.8E-6	1.4E-6	4.8E-3	2.1E-4
0.6	5.4E-6	1.7E-6	3.2E-2	1.1E-4
0.8	8.9E-6	1.9E-6	1.1E-1	5.8E-5
1.0	1.3E-5	1.9E-6	3.9E-1	2.6E-5

Method (7) gives the best results for the negative values of the parameter λ i.e. $\lambda < 0$ and the small step-size h.

5 Conclusion

As is known there are basically two large classes of numerical methods for solving initial value problem for the ODEs of first order, which usually called as the one step and multistep methods. Each of these methods has its advantages. It is known that the main advantage of these methods is high degree of accuracy. So, scientists have studied the construction stable methods with the high order of accuracy. Here, for this aim have proposed to use the intersection of those classes methods. And also, have shown that for the construction of the stable methods one can use the multistep multiderivative methods. By using above mentioned, here have compared the above-mentioned class methods and have proved that methods constructed in intersection of named class methods are better, which is confirmed in solving of the numerical example. Noted that the construction of the symmetrical methods and bilateral methods are the promising areas of Computational Mathematics. Hope that these directions will find their followers.

Acknowledgements The authors wishes to express their thanks to academicians Telman Aliyev and Ali Abbasov for their suggestion to investigate the computational aspects of our problem and for their frequent valuable suggestion. This work was supported by the Science Development Foundation under the President of Republic of Azerbaijan—Grant No EIF-MQM-ETS-2020-1(35)-08/01/1-M-01 (for Vagif Ibrahimov and Galina Mehdiyeva).2020–2025, Hubei ChuTian Scholar

Funding, China (For Xiao-Guang Yue). The authors wishes also to thank the anonymous reviewers for their careful reading of the manuscript and their fruitful comments and suggestions.

Conflict of Interests There are no conflict of interests to this work.

References

1. Eyler, Integral calculus, vol. 1, 415p. Moscow, Gostexzdah (1956) (Russian)
2. Krylov A.N.: Lectures on Approximate Calculations, 400 p. Moscow, Gocteh.-izdat (1950) (Russian)
3. Subbotin, M.F.: Celestial Mechanics Course, vol. 2, 404 p. Moscow, ONTI (1937) (Russian)
4. Shura-Bura, M.R.: Error estimates for numerical integration of ordinary differential equations. Prikl. Matem. Mech. № 5, 575–588 (1952) (Russian)
5. Mukhin, I.S.: By the accumulation of errors in the numerical integration of differential-differential equations. Prikl. Mat. Mech. **6**, 752–756 (1952) (Russian)
6. Bakhvalov, N.S.: Some remarks on the question of numerical intefration of differential equation by the finit-difference method. Acad. Sci. Rep. USSA, N3, 1955, 805–808 p., (Russian)
7. Dahlquist, G.: Convergence and stability in the numerical integration of ordinary differential equations. Math. Scand. **4**, 33–53 (1956)
8. Dahlquist, G.: Stability and error bounds in the numerical integration of ODEs, 85 s. Stockholm, K. Tekniska Hofskolans Handlingar, No. 130, pp. 195–987 (1959)
9. Henrici, P.: Discrate Variable Methods in ODE. Wiley, New York, London (1962)
10. Mehdiyeva, G.Y., Imanova, M.N., Ibrahimov, V.R.: Solving Volterra Integro-differential by the Second Derivative Methods Applied Mathematics and Information Sciences, Vol. 9, No. 5, pp. 2521–2527, Sep. 2015
11. Mehdiyeva, G., Ibrahimov, V., Imanova, M.: A Way to Construct a Hybrid Forward jumping method. IOP Conference Series: Materials Science and Engineering, vol. 225 (2017)
12. Ibrahimov, V.R.: ODEs and application proceedings of the report. In: Second International Conference Russia, Bulgaria, One Nonlinear Method for the Numerical Solution of the Koshi Problem for Ordinary Differential Equations, pp. 310–319 (1982)
13. Skvortsov, L.: Explicit two-step runge-kutta methods. Math. Model. **21**, 54–56 (2009)
14. Ibrahimov, V.R.: On a relation between degree and order for the stable advanced formula. J. Comput. Math. Math. Phys. N **7**, 1045–1056 (1990)
15. Mehdiyeva, G.Y., Imanova, M.N., Ibrahimov, V.R.: An application of the hybrid methods to the numerical solution of ordinary differential equations of second order. Vestnik KazNU, Ser. Math, Mech. Inf. No 4 (75), 46–54 (2012)
16. Kobza, J.: Second derivative methods of Adams type. Applikace Mathematicky **20**, 389–405 (1975)
17. Mehdiyeva, G., Ibrahimov, V., Imanova, M.: A way to construct an algorithm that uses hybrid methods. Appl. Math. Sci. HIKARI Ltd **7**(98), 4875–4890 (2013)
18. Simos, T.E.: optimizing a Hybrid Two-step method for the numerical solution of the Schödinger equation and Related problems with respect to Phase-lag. Hindwai Publishing Corporation. J. Appl. Mat. **2012**, article ID 420387, 17 pp. (2012)
19. Fang, T., Liu, C., Hsu, C.-W., Simos, T.E., Tsitouras, C.: Explicit hybrid six-step, six order, fully symmetric methods for solving , Math. Methods Appl. Sci. 1–10 (2019)
20. Monovas, T., Kalogiratov, Z., Rames, H., Simos, T.E.: A new approach on the construction of trigonometrically fitted two step hybrid methods. In: Proceedings of International Conference on Numerical Analysis and Applied Mathematics, AIP Conference Proceedings, vol. 1648, pp. 810009-1–810009-6. AIP Publishing (2015)

21. D'Ambrosio, R., Ferro, M., Paternoster, B.: Two-step hybrid collocation methods for *y"=f(x, y)*. Appl. Math. Lett. **22** ,1076 (2009)
22. .Ibrahimov, V., Mehdiyeva, G., Yue, X.-G., Kaabar, M.K.A., Noeiaghdam, S., Jurayev, D.A.: Novel symmetric mathematical problems. Int. J. Circuits Syst. Signal Process **15**, 1545–1557 (2021)
23. Ibrahimov, V., Imanova, M.: Multistep methods of the hybrid type and their application to solve the second kind Volterra integral equation. Symmetry **6**, 13 (2021)
24. Mehdiyeva, G., Ibrahimov, V., Imanova, M.: On a calculation of definite integrals by using of the calculation of indefinite integrals. UK Oxford, SN Applied Sciences, Springer 118–173 (2019)
25. Ehigie, J.O., Okunuga, S.A., Sofoluwe, A.B., Akanbi, M.A.: On generalized 2-step continuous linear multistep method of hybrid type for the integration of second order ordinary differential equations. Arch. Appl. Res. **2**(6), 362–372 (2010)
26. Imanova, M.N.: On the comparison of Gauss and Hybrid methods and their application to calculation of definite integrals, MMCTSE 2020. J. Phys.: Conf. Ser. (2020). https://doi.org/10.1088/1742-6596/1564/1/012019,1564(2020)012019
27. Fang, T., Liu, C., Hsu, C.-W., Simos, T.E., Tsitouras, C.: Explicit hybrid six-step, six order, fully symmetric methods for solving . Math. Methods Appl. Sci. 1–10 (2019)
28. Mehdiyeva, G., Ibrahimov, V., Imanova, M.: On some comparison of multistep second derivative methods with the multistep hybrid methods and their application to solve integro-differential equations. MMCTSE 2020, 1–9 (2020)
29. Mehdiyeva, G., Ibrahimov, V., Imanova, M.: On a calculation of definite integrals by using of the calculation of indefinite integrals. SN Applied Sciences Springer Nature Sciences (2019)
30. Mehdiyeva, G., Ibrahimov, V., Imanova, M. (2019). On the constraction of the advvanced Hybirid Methods and application to Solving Volterra İntegral Eguations,WSEAS weak transactions on systems and control, vol. 14
31. Butcher, J.: A modified multistep method for the numerical integration of ordinary differential equations. J. Assoc. Comput. Math **12**, 124–135 (1965)
32. Gear, C.: Hybrid methods for initial value problems in ordinary differential equations, SIAM. I. Numer. Anal. **2**, 69–86 (1965)
33. Shokri, A.: The multistep multi derivate methods for the numerical solution of first order initial value problems, TWMS. J. pure Appl. Math. **7**, 88–97 (2016)
34. Burova, I.G.: Application local plynominal and non-polynominal splines of the third order of approximation for the construction of the numerical solution of the Volterra integral. WSEAS Trans. Math. (2021)
35. Imanova, M.: One the multistep method of numerical solution for Volterra integral equation, transactions issue mathematics and mechanics series of physical-technical and mathematical science **I**, 95–104 (2006)
36. Han, H.,Sicheng, L., Lin, H., Nelson, D., Otilia, M., Xiao-Guang, Y.: Risk factor identification of sustainable guarantee net work based on logistic regression algorithm. Sustainability **11**, No 13, 3525 (2019)
37. Kaabar, M.K., Martinez, F., Gomez, I.F., Aguilar, B.G., Kaplan, M.: New apporiximate-analytical solutions for the nonlinear fractional schrödinger equation with second-order spatio-temporal dispersion via dougble laplace transform method. Mathematics
38. Noeiaghdam, S., Jurayev, D.A: Regularization of the III-Posed cauchy problem for matrix factorizations of the helmholtz equation on the plane. Aximos **10**(2), 82 (2021). https://doi.org/10.3390/axioms10020082
39. Jurayev, D.A.: Cauchy problem for matrix factorizations of the Helmholtz equation. Ukr. Math. J. **69**, 1583–1592 (2018)

On Symmetric Matrices with One Positive Eigenvalue and the Interval Property of Some Matrix Classes

Doaa Al-Saafin

Abstract In this paper, totally nonnegative matrices, i.e., matrices having all their minors nonnegative, and matrix intervals with respect to the usual entry-wise partial order are considered. Let $A \in \mathbb{R}^{3\times 3}$ be a symmetric totally nonnegative matrix. Sufficient conditions for the matrix A to be infinitely divisible are presented. Also, it is shown that conditionally positive (negative) and symmetric positive matrices with one positive eigenvalue have the interval property.

Keywords Hadamard power · Hadamard inverse · Matrix interval · Infinitely divisible matrix · Conditionally positive (negative) semidefinite matrix

1 Introduction

Let $A = [a_{ij}]$ and $B = [b_{ij}]$ be real $n \times n$ matrices. Their *Hadamard product* (also called Schur product) $A \circ B$ is defined as the entry-wise product of A and B, i.e., $A \circ B = [a_{ij}b_{ij}]$. A matrix $A = [a_{ij}]$ is *Hadamard invertible* if all its entries are non-zero, and $A^{\circ -1} = [1/a_{ij}]$ is then called the *Hadamard inverse of* A. A is termed *nonnegative* (*positive*), denoted by $A \geq 0\,(A > 0)$, if all its of A are nonnegative (positive). For $A \geq 0$, the *r-th Hadamard power* of A is $A^{\circ r} = [a_{ij}^r]$, $r > 0$. We define the *Hadamard exponential* of A by $e^{\circ A} = [e^{a_{ij}}]$ and if A is positive, the *Hadamard logarithm* of A by $log^{\circ}(A) = [log(a_{ij})]$. Suppose that A is nonnegative and positive semidefinite. We say that A is *infinitely divisible* if the matrix $A^{\circ r}$ is positive semidefinite for every nonnegative r. A square matrix A of order n (≥ 2) is said to be a *PN-matrix* if every principal minor of order k, $2 \leq k \leq n$, is not zero and has the sign of (-1). If A is a positive symmetric matrix and A has exactly one positive eigenvalue, then we say that A is in the class $\mathcal{A}$.

A matrix $A \in \mathbb{R}^{n\times m}$ is called *totally nonnegative* (*positive*), abbreviated TN (TP), if all its minors are nonnegative (positive) real numbers. For $A \in \mathbb{R}^{n\times m}$,

D. Al-Saafin (✉)
Department of Mathematics and Statistics, University of Konstanz, Konstanz, Germany
e-mail: doaa.al-saafin@uni-konstanz.de

D. Zeidan et al. (eds.), *Mathematics and Computation*, Springer Proceedings in Mathematics & Statistics 418, https://doi.org/10.1007/978-981-99-0447-1_29

$\alpha \subseteq \{1, 2, \ldots, n\}$, the principal submatrix of A lying in rows and columns indexed by α will be denoted by $A[\alpha]$.

Let $\mathbf{e} \in \mathbb{R}^n$ be the vector of all ones. A real symmetric $n \times n$ matrix A is said to be *conditionally positive (negative) semidefinite* if $\mathbf{x}^T A\mathbf{x} \geq 0$ (≤ 0) for all $\mathbf{x} \in \mathbb{R}^n$ such that $\mathbf{x}^T\mathbf{e} = 0$. If this inequality is strict then A is *conditionally positive (negative) definite*.

Partition the matrix $A \in \mathbb{R}^{n\times n}$ as

$$[a_{ij}] = \begin{bmatrix} B & \mathbf{x} \\ \mathbf{y}^T & a_{nn} \end{bmatrix}, \tag{1}$$

where $B \in \mathbb{R}^{(n-1)\times(n-1)}$ and $\mathbf{x}, \mathbf{y} \in \mathbb{R}^{n-1}$.

If $a_{nn} \neq 0$, we define the matrix $\tilde{A}$ by

$$\tilde{A} := \begin{bmatrix} B - \frac{1}{a_{nn}}\mathbf{x}\mathbf{y}^T & \mathbf{x} \\ \mathbf{y}^T & a_{nn} \end{bmatrix}. \tag{2}$$

Then the matrix $\tilde{A}[1, \ldots, n-1]$ is called the *Schur complement of* a_{nn} *in* A.

Let $R^{n\times n}$ be endowed with the usual partial order $\leq$, i.e., $A \leq B$, if $0 \leq B - A$, for any $A, B \in R^{n\times n}$. A *matrix interval* is denoted by boldface and is defined as

$$\mathbf{A} := [\underline{A}, \overline{A}] = \{A \in R^{n\times n} \mid \underline{A} \leq A \leq \overline{A}\},$$

where $\underline{A} \leq \overline{A}$ holds for the two *bound matrices* $\underline{A}_{ij} = (\underline{a}_{ij})_{i,j=1}^n$, $\overline{A}_{ij} = (\overline{a}_{ij})_{i,j=1}^n$. By

$$A_c := \frac{1}{2}(\overline{A} - \underline{A}), \quad A_\Delta := \frac{1}{2}(\underline{A} + \overline{A}).$$

Let V be a fixed set of vertex matrices. We say that a class S of matrices has the *interval property (with respect to V)*, if $\mathbf{A} \subset S$ whenever $V(\mathbf{A}) \subset S$. For a collection of various classes of matrices which enjoy the interval property see [6].

Let V_1, V_2 be the following sets of vertex matrices:

$$V_1 = A_c - diag(z)A_\Delta diag(z), \quad V_2 = A_c + diag(z)A_\Delta diag(z),$$

where $z \in \{\pm 1\}^n$. Hence, the number of mutually different matrices in each set is at most 2^{n-1}.

In [4], Bialas and Garloff proved that the set of the positive (semi)definite matrices has the interval property with respect to V_1, see also [10]. In [6], we provide the interval property for other classes of matrices.

2 Background and Key Lemmata

We collect here some key facts needed for our main results. The following lemmata are well known.

Lemma 1 (Schur Product Theorem): *Suppose A and B are positive semidefinite matrices of the same order. Then $A \circ B$ is also positive semidefinite. If A and B are positive definite, then $A \circ B$ is positive definite too.*

In passing, we note that the Hadamard product of two conditionally positive semidefinite matrices need not be conditionally positive semidefinite. A counterexample is provided by $A = \begin{bmatrix} 0.1 & 1 \\ 1 & 2 \end{bmatrix}$ and $B = \begin{bmatrix} 4 & 3.5 \\ 3.5 & 3 \end{bmatrix}$. $A \circ B = \begin{bmatrix} 0.4 & 3.5 \\ 3.5 & 6 \end{bmatrix}$, which is not conditionally positive semidefinite (take $\mathbf{x} = (1, -1)^T$).

Lemma 2 ([7, p. 144]): *The symmetric matrix A is conditionally positive semidefinite if and only if its Hadamard exponential $e^{\circ tA}$ is positive semidefinite for all $t \geq 0$.*

Lemma 3 ([8, Corollary 1.6]): *Let A be Hadamard invertible. Then A is infinitely divisible if and only if A is symmetric, positive, and $log^{\circ}(A)$ is conditionally positive semidefinite.*

Lemma 4 ([2, Theorem 4.4.6]): *A positive symmetric matrix A has one positive eigenvalue if and only if, for each $k \times k$ principal submatrix B of A, $(-1)^{k-1} det\ B \geq 0$.*

Lemma 5 ([4, p. 40]): *The set of the positive (semi)definite matrices has the interval property with respect to V_1.*

3 Main Results

Partition the matrix $A = [a_{ij}] \in \mathbb{R}^{n \times n}$ as in (1). It was shown in [5, Proposition 1.5.2]) that if $a_{nn} \neq 0$ is TN, then the Schur complement of a_{nn} in A, see (2), is TN. In the next proposition, we give a representation of the determinant of the Hadamard inverse of the matrix A.

Proposition 1 *Let $A = [a_{ij}] \in \mathbb{R}^{n \times n}$ be Hadamard invertible and portioned as in (1). Then the following equality holds for some $t > 0$:*

$$det A^{\circ -1} = t\ (-1)^{n-1}\ det((A^{\circ -1} \circ \tilde{A})[1, \ldots, n-1]).$$

Proof Let $A = [a_{ij}] = \begin{bmatrix} B & \mathbf{x} \\ \mathbf{y}^T & a_{nn} \end{bmatrix}$, where $B \in \mathbb{R}^{(n-1)\times(n-1)}$, $\mathbf{x}, \mathbf{y} \in \mathbb{R}^{n-1}$, and $a_{ij} \neq 0$, for all $i, j = 1, \ldots, n$. Then

$$det(\frac{1}{a_{nn}}B^{\circ -1} - (\mathbf{x}\mathbf{y}^T)^{\circ -1})$$

$$= det\begin{pmatrix} \frac{1}{a_{nn}a_{11}} - \frac{1}{a_{1n}a_{n1}} & \frac{1}{a_{nn}a_{12}} - \frac{1}{a_{1n}a_{n2}} & \cdots & \frac{1}{a_{nn}a_{1,n-1}} - \frac{1}{a_{1n}a_{n,n-1}} \\ \frac{1}{a_{nn}a_{21}} - \frac{1}{a_{2n}a_{n1}} & \frac{1}{a_{nn}a_{22}} - \frac{1}{a_{2n}a_{n2}} & \cdots & \frac{1}{a_{nn}a_{2,n-1}} - \frac{1}{a_{2n}a_{n,n-1}} \\ \vdots & \vdots & \ddots & \vdots \\ \frac{1}{a_{nn}a_{n-1,1}} - \frac{1}{a_{n-1,n}a_{n1}} & \frac{1}{a_{nn}a_{n-1,2}} - \frac{1}{a_{n-1,n}a_{n2}} & \cdots & \frac{1}{a_{nn}a_{n-1,n-1}} - \frac{1}{a_{n-1,n}a_{n,n-1}} \end{pmatrix}$$

$$= (-1)^{n-1}\, t'\, det\begin{pmatrix} \frac{\begin{vmatrix} a_{11} & a_{1n} \\ a_{n1} & a_{nn} \end{vmatrix}}{a_{11}a_{nn}} & \frac{\begin{vmatrix} a_{12} & a_{1n} \\ a_{n2} & a_{nn} \end{vmatrix}}{a_{12}a_{nn}} & \cdots & \frac{\begin{vmatrix} a_{1,n-1} & a_{1n} \\ a_{n,n-1} & a_{nn} \end{vmatrix}}{a_{1,n-1}a_{nn}} \\ \frac{\begin{vmatrix} a_{21} & a_{2n} \\ a_{n1} & a_{nn} \end{vmatrix}}{a_{12}a_{nn}} & \frac{\begin{vmatrix} a_{22} & a_{2n} \\ a_{n2} & a_{nn} \end{vmatrix}}{a_{22}a_{nn}} & \cdots & \frac{\begin{vmatrix} a_{2,n-1} & a_{2n} \\ a_{n,n-1} & a_{nn} \end{vmatrix}}{a_{2,n-1}a_{nn}} \\ \vdots & \vdots & \ddots & \vdots \\ \frac{\begin{vmatrix} a_{n-1,1} & a_{n-1,n} \\ a_{n1} & a_{nn} \end{vmatrix}}{a_{n-1,1}a_{nn}} & \frac{\begin{vmatrix} a_{n-1,2} & a_{n-1,n} \\ a_{n2} & a_{nn} \end{vmatrix}}{a_{n-1,2}a_{nn}} & \cdots & \frac{\begin{vmatrix} a_{n-1,n-1} & a_{n-1,n} \\ a_{n,n-1} & a_{nn} \end{vmatrix}}{a_{n-1,n-1}a_{nn}} \end{pmatrix}, \quad (3)$$

for some $t' > 0$. Hence,

$$det(A) = (-1)^{n-1}\, t\, det((A^{\circ -1} \circ \tilde{A})[1, \ldots, n-1]),$$

where $t = a_{nn}^{1-n} t'$. □

In [3], Bhatia gave very simple proofs of the infinite divisibility of some interesting TP matrices like the Cauchy and Pascal matrices. In general, a symmetric TP matrix need not be an infinitely divisible. For example, the matrix $A = \begin{pmatrix} 1.1 & 3 & 1 \\ 3 & 9.1 & 3.9 \\ 1 & 3.9 & 2.9 \end{pmatrix}$ is TP, while $det A^{\circ \frac{1}{9}} < 0$. It was shown in [1, p. 471, proof of Lemma 6] that if A is in class $\mathcal{A}$, then its Hadamard inverse is infinitely divisible. The converse need not be true, for example, the Hadamard inverse of the infinitely divisible matrix $\begin{pmatrix} 1 & 3 & 1 \\ 3 & 10 & 4 \\ 1 & 4 & 2.5 \end{pmatrix}$ has two positive eigenvalues. In the next corollary, we give sufficient conditions for symmetric TN matrices of maximum order 3 to be closed under Hadamard powers.

Corollary 1 *Let $A = [a_{ij}] \in \mathbb{R}^{n\times n}$, $n \le 3$, be a positive symmetric TN matrix with entries satisfying at least one of the following relations:*

1. $a_{11}a_{22} = a_{12}^2$,
2. $a_{12}a_{33} = a_{13}a_{23}$.

Then $A^{\circ -1}$ has exactly one positive eigenvalue.

Proof Since the matrix A has positive entries and $det(A^{\circ-1}[1,2]) \leq 0$, by Lemma 4, the statement is true for $n = 2$, and in the case $n = 3$ we only have to show that $det(A^{\circ-1}) \geq 0$. Now, let $n = 3$. (i) Since $A^{\circ-1}[1,2]$ is positive semidefinite, then Lemma 1 ensures that $det((A^{\circ-1} \circ \tilde{A})[1,2]) \geq 0$, and by Proposition 1 it follows that $det(A^{\circ-1}) \geq 0$.
(ii) By (1), we obtain

$$det(A^{\circ-1}) = m \begin{vmatrix} a_{11} & a_{13} \\ a_{31} & a_{33} \end{vmatrix} \begin{vmatrix} a_{22} & a_{23} \\ a_{32} & a_{33} \end{vmatrix},$$

for some $m > 0$, whence,

$$det(A^{\circ-1}) \geq 0.$$

□

The Pascal matrix $\begin{pmatrix} 1 & 1 & 1 \\ 1 & 2 & 3 \\ 1 & 3 & 6 \end{pmatrix}$ shows that the sufficient conditions in the previous corollary are not necessary, see [3, p. 225] and [5, p. 13].

Theorem 1 *Let A be in class $\mathcal{A}$ and let B be any principal submatrix of A, then B belongs to class $\mathcal{A}$ too.*

Proof Let r be any integer with $1 \leq r \leq n$, and let B denote any $r \times r$ principal submatrix of A. Let $\lambda_1 \leq \lambda_2 \leq \cdots \leq \lambda_n$, $\mu_1 \leq \mu_2 \leq \cdots \leq \mu_r$ be the eigenvalues of A and B, respectively. By [9, Theorem 4.3.15], for each integer k such that $1 \leq k \leq r$, we have

$$\lambda_k \leq \mu_k \leq \lambda_{k+n-r}.$$

So, when $k = r - 1$, $\mu_k \leq \lambda_{n-1} \leq 0$. Since B is positive, then B has one positive eigenvalue. □

Remark 1 Theorem 1 provides a generalization for Lemma 4; a positive symmetric matrix A has one positive eigenvalue if and only if, for each $k \times k$ principal submatrix B of A, $(-1)^{k-1} det B \geq 0$.

Recall that the set of positive semidefinite matrices is subset of the set of conditionally positive semidefinite matrices. In [4], Bialas and Garloff showed that the set of the positive (semi)definite matrices has the interval property. The next theorem gives a generalization of this result.

Theorem 2 ([6, I.P. 4.7]): *The set of the conditionally positive (negative) semidefinite matrices has the interval property with respect to $V_1(V_2)$.*

Proof Let $\mathbf{A} = [\underline{A}, \overline{A}]$ be a matrix interval and assume without loss of generality that $\underline{A}$ and $\overline{A}$ are symmetric. Furthermore, assume that all vertex matrices in V_1 are conditionally positive definite. Let A be a symmetric matrix with $\underline{A} \leq A \leq \overline{A}$. Then it follows that

$$e^{\circ t\underline{A}} \leq e^{\circ tA} \leq e^{\circ t\overline{A}},$$

for all $t \geq 0$. By Lemma 2, $e^{\circ tC}$ is positive semidefinite for all $C \in V_1$. Since there is a one-to-one correspondence between the vertex matrices of A and those of $[e^{\circ t\underline{A}}, e^{\circ t\overline{A}}]$, it follows by Lemma 5 that $e^{\circ tA}$ is positive semidefinite matrix and hence by Lemma 2, A is conditionally positive semidefinite. The statement for conditionally negative semidefinite matrices follows now by using $[-\overline{A}, -\underline{A}]$ and the set V_2 instead of V_1. □

Theorem 3 ([6, I.P. 4.8]): *The set of the positive infinitely divisible matrices has the interval property with respect to* V_1.

Proof Let $\mathbf{A} = [\underline{A}, \overline{A}]$ be a matrix interval and assume without loss of generality that $\underline{A}, \overline{A}$ are symmetric and $\underline{A}$ is positive. Furthermore, assume that all vertex matrices in V_1 are infinitely divisible. Assume that A is a symmetric matrix with $\underline{A} \leq A \leq \overline{A}$. Then it follows that

$$log^{\circ}(\underline{A}) \leq log^{\circ}(A) \leq log^{\circ}(\overline{A}).$$

By Lemma 3, $log^{\circ}(C)$ is conditionally positive semidefinite matrices for all $C \in V_1$. Thus, by Theorem 2, $log^{\circ}(A)$ is conditionally positive semidefinite and hence by Lemma 3, A is infinitely divisible. □

References

1. Bapat, R.B.: Multinomial probabilities, permanents and a conjecture of karlin and rinott. Proc. Am. Math. Soc. **102**(3), 467–472 (1988)
2. Bapat, R.B., Raghavan, T.E.S.: Nonnegative Matrices and Applications (Encyclopedia of Mathematics Science), vol. 64. Cambridge University Press, Cambridge, UK (1997)
3. Bhatia, R.: Infinitely divisible matrices. Amer. Math. Mon. **113**, 221–235 (2006)
4. Bialas, S., Garloff, J.: Intervals of P-matrices and related matrices. Linear Algebr. Appl. **58**, 33–41 (1984)
5. Fallat, M.S., Johnson, C.R.: Totally Nonnegative Matrices, vol. 64. Princeton University Press (2011)
6. Garloff, J., Al-Saafin, D., Adm, M.: Further matrix classes possessing the interval property. Reliab. Comput. **28**, 56–70 (2021)
7. Horn, R.A.: The hadamard product. Proc. Sympos. Appl. Math. **40**, 87–169 (1990)
8. Horn, R.A.: The theory of infinitely divisible matrices and kernels. Trans. Am. Math. Soc. **136**, 269–286 (1969)
9. Horn, R.A., Johnson, C.R.: Matrix Analysis, 1st edn. Cambridge University, New York (1990)
10. Rohn, J.: Positive definiteness and stability of interval matrices. SIAM J. Matrix Anal. Appl. **15**(1), 175–184 (1994)

Global Asymptotic Stability for Discrete-Time SEI Reaction-Diffusion Model

Nidal Anakira, Amel Hioual, Adel Ouannas, Taki-Eddine Oussaeif, and Iqbal M. Batiha

Abstract The global stability of solutions for a discrete-time globally dispersed reaction-diffusion SEI epidemic model with individual immigration is investigated in this work. The global stability is addressed using the Lyapunov functional after giving a discrete form of the reaction-diffusion SEI epidemic model. As in the continuous case, the unique steady-state is proven to be globally stable in the presence of diffusion. To validate the findings of this study, some numerical simulations are provided.

Keywords Discrete-time reaction-diffusion SEI epidemic model · Global asymptotic stability · Lyapunov functional · Numerical simulations

---

N. Anakira (✉)
Department of Mathematics, Faculty of Science and Technology, Irbid National University, Irbid 2600, Jordan
e-mail: dr.nidal@inu.edu.jo

A. Hioual · A. Ouannas · T.-E. Oussaeif
Department of Mathematics and Computer Science, University of Larbi Ben M'hidi, Oum El Bouaghi, Algeria
e-mail: amel.hioual@univ-oeb.dz

A. Ouannas
e-mail: Ouannas.adel@univ-oeb.dz

T.-E. Oussaeif
e-mail: taki_maths@live.fr

I. M. Batiha
Department of Mathematics, Al Zaytoonah University of Jordan, Amman 11733, Jordan
e-mail: i.batiha@zuj.edu.jo

Nonlinear Dynamics Research Center (NDRC), Ajman University, Ajman, UAE

D. Zeidan et al. (eds.), *Mathematics and Computation*, Springer Proceedings in Mathematics & Statistics 418, https://doi.org/10.1007/978-981-99-0447-1_30

1 Introduction

Epidemic infections are the most common cause of death in all living things. The study of epidemiology has drawn the attention of a large number of researchers with the goal of improving the treatment of these diseases via disease planning and predictions, consequently lowering death rates. Infectious illness epidemics are modeled using well-known reaction-diffusion systems. Many of these models are dependent on the groups of individuals considered as well as the disease's transmission characteristics.

In this work, we intend to concern with an infectious illness like tuberculosis that can be described with the help of formulating its states in view of a new version of a discrete-time SEI model that takes into consideration three classes of individuals: Susceptible (S), Exposed (E) and Infectious (I). In other words, the illness has an exposed or latent phase. Because of this time of latency, some migrants will display no symptoms and hence be unaware that they have been infected. In addition, contagious individuals may also relocate from states to others. As a result, even if the disease transmission is successfully prevented at the target place, new cases will join the community, making the decimation of such disease impossible. The work in [3] gave an excellent study of such a model in the situation of Ordinary Differential Equations (ODEs) without spatial diffusion. For the purpose of predicting how the illness may spread in light of integrating spatial diffusion, a new 3-component SEI model was explored in [1].

In several mathematical literature, it has been declared that the discrete-time models defined by certain difference equations are deemed more accurate than continuous models in describing many phenomena (see [5–9]). In other words, the discrete-time models can provide more effective computational numerical simulations than that of the continuous models. As a result, it is appropriate to investigate the discrete-time models that can be represented by certain difference equations. These models have been extensively utilized within many research applications conducted on the endurance, permanence, and global stability of various discrete-time nonlinear systems when the influence of spatial factors is ignored [10, 12–14].

To the best of our knowledge, there are little research papers on the global features of the discrete-time models. The diffusion terms (discrete Laplace operators) are not, nevertheless, included in any epidemical model. Actually, there have been some works on the global stability of some discrete-time diffusion systems [15, 16]. In such works, the positivity, boundedness, and the global stability of the equilibria were obtained, and the discretized models were deduced from the respective continuous models by performing non-standard finite difference schemes, but the Laplace operators have not been addressed in their contents. The diffusion gained from discrete-time models would result in high rich complexity and very complicated dynamical behaviors. This however can be confirmed by studying the stability of these models using appropriate Lyapunov functions, which is deemed an essential topic that should be addressed. From this perspective, we intend to study the global asymptotic stability of the discrete-time SEI model based on employing the continuous-time model investigated in [1].

The remainder of this paper is arranged as follows. In Sect. 2, some useful preliminaries are provided. The discrete-time analog of the continuous reaction-diffusion SEI model is developed in Sect. 3. In Sect. 4, we investigate the dynamical behavior of the considered discrete-time system in terms of the global stability of its equilibria. Section 5 includes some numerical simulations that will verify all findings achieved in this work.

2 Preliminarily

In order to develop the continuous reaction-diffusion SEI model reported in [1] to be formulated to the discrete-time model, we will present some basic facts and preliminaries associated with the forward difference operator.

Definition 1 ([17]) Let $x : \mathbb{N}_a \rightarrow \mathbb{R}$ where $\mathbb{N}_a = \{a, a+1, a+2, ...\}$ and $a \in \mathbb{R}$. The forward difference operator can be outlined as:

$$\Delta x(t) = x(t+1) - x(t).$$

In this connection, we will consider here the following nonlinear integer-order difference system:

$$\begin{cases} \Delta x(n) = f(x(n)), & n \in \mathbb{N} \\ x(0) = x_0, & x_0 \in \mathbb{R}^n, \end{cases} \tag{1}$$

where $x(n) \in \mathbb{R}^n$ is the state-vector of the system and $f : \mathbb{R}^n \rightarrow \mathbb{R}^n$ is a continuous differentiable function. In what follows, we consider that $f(0) = 0$, and hence $x^* = 0$ represents an equilibrium point of system (1).

Theorem 1 ([17]) *If there exists a continuous function* $V : \mathbb{R}^n \rightarrow \mathbb{R}^+$ *such that:*

$$V(0) = 0 \quad and \quad V(x(n)) > 0, \quad \forall x(n) = 0$$
$$\Delta V(x(n)) = V(x(n+1)) - V(x(n)) \leq 0, \quad \forall n \in \mathbb{N}.$$

Then, the trivial solution of system (1) is stable. Moreover, if

$$\Delta V(x(n)) = V(x(n+1)) - V(x(n)) < 0, \quad \forall n \in \mathbb{N},$$

then the trivial solution of the same system is asymptotically stable.

3 Discrete-time SEI Reaction-Diffusion System

In this section, we aim to formulate the discrete-time model by considering the continuous reaction-diffusion SEI model reported in [1]. In particular, the SEI reaction-diffusion epidemic model with the immigration of individuals was given in [1] by:

$$\begin{cases} \partial_t u = d_1 \Delta u + \Lambda_1 - uf(w) - \mu_1 u, & \text{in } \Omega \text{ x } \mathbb{R}^+ \\ \partial_t v = d_2 \Delta v + \Lambda_2 + uf(w) - (\mu_2 + \beta)v, & \text{in } \Omega \text{ x } \mathbb{R}^+ \\ \partial_t w = d_3 \Delta w + \Lambda_3 + \beta v - \mu_3 w, & \text{in } \Omega \text{ x } \mathbb{R}^+, \end{cases} \tag{2}$$

subject to the initial conditions:

$$u(x, 0) = u_0(x), \quad v(x, 0) = v_0(x), \quad w(x, 0) = w_0(x),$$

and to the following homogeneous Neumann boundary conditions:

$$\frac{\partial u}{\partial \nu} = \frac{\partial v}{\partial \nu} = \frac{\partial w}{\partial \nu} = 0, \quad \text{on } \partial\Omega \text{ x } \mathbb{R}^+,$$

where Ω is an open bounded subset of $\mathbb{R}^n$ with piecewise smooth boundary $\partial\Omega$.

As a matter of fact, they can be classified any population into three major groups; vulnerable (or suspectable), exposed, and infectious. The size of these classes are represented by u, v, and w, respectively. The ratios of these classes occurs at the rates of $\Lambda_1 > 0$, $\Lambda_2 \geq 0$, and $\Lambda_3 \geq 0$, respectively. In this regard, we will allow for a nonlinear response to the size of the infectious population in the incidence, and therefore the incidence rate is $uf(w)$ in which f denotes a twice differentiable function fulfilled the following hypotheses:

$$f(w) \geq 0, \ f'(w) \geq 0, \ f''(w) \leq 0.$$

In model (2), the individuals in class v migrate to the class w at rate βv. At the same time, the individuals leave the vulnerable, exposed, and infected classes with per capita death rates of μ_1, μ_2 and μ_3 respectively. In this connection, we suppose that $\mu_1, \mu_2, \mu_3 > 0$, d_1, d_2 and d_3 are the diffusion parameters. From this point of view, we aim in this work to study the discrete version of model (2) which can be expressed, without diffusion, as follows:

$$\begin{cases} \dfrac{du}{dt} = \Lambda_1 - uf(w) - \mu_1 u \\ \dfrac{dv}{dt} = \Lambda_2 + uf(w) - (\mu_2 + \beta)v \\ \dfrac{dw}{dt} = \Lambda_3 + \beta v - \mu_3 w. \end{cases} \tag{3}$$

In what follows, we use the forward Euler discretization scheme according to the following term:

$$\frac{du}{dt}\left(\frac{dw}{dt}, \frac{dw}{dt}\right),$$

which can be represented by:

$$[u(t+h)-u(t)]\ ([v(t+h)-v(t)], [w(t+h)-w(t)]),$$

where h is the step size of the numerical method. With $t = nh$, $u(t) = u(nh)$ and $h = 1$, the Euler's method yields the following system:

$$\begin{cases} u(n+1)-u(n) = \Lambda_1 - u(n)f(w(n)) - \mu_1 u(n) \\ v(n+1)-v(n) = \Lambda_2 + u(n)f(w(n)) - (\mu_2+\beta)v(n) \\ w(n+1)-w(n) = \Lambda_3 + \beta v(n) - \mu_3 w(n). \end{cases} \tag{4}$$

Actually, system (4) is equivalent to the following form:

$$\begin{cases} \Delta u(n) = \Lambda_1 - u(n)f(w(n)) - \mu_1 u(n) \\ \Delta v(n) = \Lambda_2 + u(n)f(w(n)) - (\mu_2+\beta)v(n) \\ \Delta w(n) = \Lambda_3 + \beta v(n) - \mu_3 w(n). \end{cases} \tag{5}$$

Now, in order to move on to the discrete-time version of the model (3), the space factor represented by diffusion can be taken into account in all fundamental aspects. This would generate a one-dimensional discrete-time reaction-diffusion model, which can be outlined as follows:

$$\begin{cases} \Delta u_i^n = d_1\nabla^2 u_i^n + \Lambda_1 - u_i^n f(w_i^n) - \mu_1 u_i^n \\ \Delta v_i^n = d_2\nabla^2 v_i^n + \Lambda_2 + u_i^n f(w(_i^n)) - (\mu_2+\beta)v_i^n \\ \Delta w_i^n = d_3\nabla^2 w_i^n + \Lambda_3 + \beta v_i^n - \mu_3 w_i^n, \ \forall i = \{1, 2, ..., m\}, \end{cases} \tag{6}$$

where m, n are positive integers, d_1, d_2, d_3 are diffusion parameters, and $\nabla^2(\cdot)$ are defined as follows:

$$\begin{aligned} \nabla^2 u_i^n &= u_{i+1}^n - 2u_i^n + u_{i-1}^n \\ \nabla^2 v_i^n &= v_{i+1}^n - 2v_i^n + v_{i-1}^n \\ \nabla^2 w_i^n &= w_{i+1}^n - 2w_i^n + w_{i-1}^n. \end{aligned}$$

In the same regard, the initial conditions u_i^0, v_i^0, w_i^0 as well as the following periodic boundary conditions are considered:

$$\begin{cases} u_0^n = u_m^n, & u_1^n = u_{m+1}^n \\ v_0^n = v_m^n, & v_1^n = v_{m+1}^n \\ w_0^n = w_m^n, & w_1^n = w_{m+1}^n. \end{cases} \tag{7}$$

As indicated in [1], there exists a unique equilibrium (u^*, v^*, w^*) of model (6). This actually means that the exposed and infected classes do not both go to zero at the same time. Anyhow, the equilibrium point of the system at hand can be outlined algebraically to be as follows:

$$\begin{cases} \Lambda_1 = u^* f(w^*) + (\mu_1 + \alpha)u^* \\ \Lambda_2 = -u^* f(w^*) + (\mu_2 + \beta)v^* \\ \mu_3 = \dfrac{\Lambda_3 + \beta v^*}{w^*}. \end{cases} \tag{8}$$

4 Global Stability

In this section, we intend to concern with the global asymptotic stability of the unique positive equilibrium (u^*, v^*, w^*). To this aim, we construct the required conditions for the positive equilibrium to be globally asymptotically stable using the global Lyapunov function.

Theorem 2 *The equilibrium point* (u^*, v^*, w^*) *is globally asymptotically stable of system (6).*

Proof To prove this result, we use the same Lyapunov function that was previously used in [1–3]. For this purpose, we consider the following function:

$$h(y) = y - 1 - \ln y,$$

where $h : \mathbb{R}_*^+ \to \mathbb{R}_*^+$. This function has a strict global minimum, i.e., $h(1) = 0$. Now, consider the following non-negative function:

$$V^n = V_1^n + V_2^n + V_3^n,$$

where

$$V_1^n = u^* \sum_{i=1}^{m} h\left(\frac{u_i^n}{u^*}\right), \quad V_2^n = v^* \sum_{i=1}^{m} h\left(\frac{v_i^n}{v^*}\right), \quad V_3^n = Cw^* \sum_{i=1}^{m} h\left(\frac{w_i^n}{w^*}\right),$$

and

$$C = \frac{u^* f(w^*)}{\beta v^*}.$$

To move forward in this proof, we will use the same proof's procedure given in [2, 4]. For this purpose, we first calculate V_1^n as follows:

$$\begin{aligned}
\Delta V_1^n = V_1^{n+1} - V_1^n &= \sum_{i=1}^{m}\left(u_i^{n+1} - u^n - u^* \ln \frac{u_i^{n+1}}{u_i^n}\right)\\
&= \sum_{i=1}^{m}\left(u_i^{n+1} - u^n - u^* \frac{u_i^{n+1} - u_i^n}{u_i^n}\right) + o(1)\\
&= \sum_{i=1}^{m}(u_i^{n+1} - u^n)\left(1 - \frac{u^*}{u_i^*}\right) + o(1)\\
&= \sum_{i=1}^{m}\left(1 - \frac{u^*}{u_i^n}\right)(d_1\nabla^2 u_i^n + \Lambda_1 - u_i^n f(w_i^n) - \mu_1 u_i^n - u_i^n) + o(1)\\
&= \sum_{i=1}^{m}\left(1 - \frac{u^*}{u_i^n}\right)(\Lambda_1 - u_i^n f(w_i^n) - \mu_1 u_i^n - u_i^n) + d_1\left(1 - \frac{u^*}{u_i^n}\right)(u_{i+1}^n - 2u_i^n + u_{i-1}^n).
\end{aligned}$$

Using (8) yields:

$$\begin{aligned}
\Delta V_1^n =& \sum_{i=1}^{m}\mu_1 u^*\left(1 - \frac{u^*}{u_i^n}\right)\left(1 - \left(1 + \frac{1}{\mu_1}\right)\frac{u_i^n}{u^*}\right) + u^* f(w^*)\left(1 - \frac{u^*}{u_i^n}\right)\left(1 - \frac{uf(w)}{u^* f(w^*)}\right)\\
&- d_1 \sum_{i=1}^{m} u^*\left(\frac{u_{i+1}^n}{u_i^n} + \frac{u_{i-1}^n}{u_i^n} - 2\right),
\end{aligned}$$

or

$$\begin{aligned}
\Delta V_1^n =& \sum_{i=1}^{m} -\mu_1 u^*(h\left(\frac{u^*}{u_i^n}\right) + \left(1 + \frac{1}{\mu_1}\right)h\left(\frac{u_i^n}{u^*}\right) + \ln\left(1 + \frac{1}{\mu_1}\right)\\
&+ u^* f(w^*)\left(-h\left(\frac{u^*}{u_i^n}\right) + h\left(\frac{f(w)}{f(w^*)}\right) - h\left(\frac{uf(w)}{u^* f(w^*)}\right)\right)\\
&- d_1 u^* \sum_{i=1}^{m-1}\left(\sqrt{\frac{u_{i+1}^n}{u_i^n}} - \sqrt{\frac{u_{i-1}^n}{u_i^n}}\right)^2 - d_1 u^* \sum_{i=1}^{m-1}\left(\sqrt{\frac{u_m^n}{u_1^n}} - \sqrt{\frac{u_1^n}{u_m^n}}\right)^2 + o(1).
\end{aligned}$$

Consequently, we have

$$\begin{aligned}
\Delta V_2^n =& V_2^{n+1} - V_2^n = \sum_{i=1}^{m}(v_i^{n+1} - v^n)\left(1 - \frac{v^*}{v_i^*}\right) + o(1)\\
&= \sum_{i=1}^{m}\left(1 - \frac{v^*}{v_i^n}\right)(d_2\nabla^2 v_i^n + \Lambda_2 + u_i^n f(w_i^n) - (\mu_2 + \beta)v_i^n - v_i^n) + o(1),
\end{aligned}$$

i.e.,

$$\Delta V_2^n = \sum_{i=1}^{m}\left(1-\frac{v^*}{v_i^n}\right)\left(d_2\nabla^2 v_i^n + \Lambda_2 + u_i^n f(w_i^n) - v_i^n\frac{\Lambda_2 + u^* f(w^*)}{v^*} - v_i^n\right) + o(1).$$

This implies;

$$\Delta V_2^n = \sum_{i=1}^{m}\Lambda_2 v^*\left(1-\frac{v^*}{v_i^n}\right)\left(1-\left(1+\frac{1}{\Lambda_2}\right)\frac{v_i^n}{v^*}\right) + u^* f(w^*)\left(1-\frac{v^*}{v_i^n}\right)\left(\frac{uf(w)}{u^* f(w^*)} - \frac{v_i^n}{v^*}\right)$$
$$+\sum_{i=1}^{m}\left(1-\frac{v^*}{v_i^n}\right)(d_2\nabla^2 v_i^n) + o(1).$$

That is;

$$\Delta V_2^n = -\Lambda_2 v^*\left(h\left(\frac{v^*}{v_i^n}\right) + \left(1+\frac{1}{\Lambda_2}\right)h\left(\frac{v_i^n}{v^*}\right) + \ln\left(1+\frac{1}{\Lambda_2}\right)\right)$$
$$+ u^* f(w^*)\left(-h(\frac{v^*}{v_i^n}) + h(\frac{u_i^n f(w_i^n)}{u^* f(w^*)}) - h(\frac{v^* u_i^n f(w_i^n)}{v_i^n u^* f(w^*)})\right)$$
$$- d_2 v^* \sum_{i=1}^{m-1}\left(\sqrt{\frac{v_{i+1}^n}{v_i^n}} - \sqrt{\frac{v_{i-1}^n}{v_i^n}}\right)^2 - d_2 v^* \sum_{i=1}^{m-1}\left(\sqrt{\frac{v_m^n}{v_1^n}} - \sqrt{\frac{v_1^n}{v_m^n}}\right)^2 + o(1).$$

In a similar manner, we can have

$$\Delta V_3^n = V_3^{n+1} - V_3^n = \sum_{i=1}^{m} C(w_i^{n+1} - w^n)\left(1-\frac{w^*}{w_i^*}\right) + o(1)$$
$$= \sum_{i=1}^{m} C\left(1-\frac{w^*}{w_i^n}\right)(d_3\nabla^2 w_i^n + \Lambda_3 + \beta v_i^n - \mu_3 w_i^n - w_i^n) + o(1)$$
$$= \sum_{i=1}^{m} C\left(1-\frac{w^*}{w_i^n}\right)\left(d_3\nabla^2 w_i^n + \Lambda_3 + \beta v_i^n - w_i^n\frac{\Lambda_3 + \beta v^*}{w^*} - w_i^n\right) + o(1)$$
$$= \sum_{i=1}^{m} C\Lambda_3\left(1-\frac{w^*}{w_i^n}\right)\left(1-(1-\frac{1}{\Lambda_3})\frac{w_i^n}{w^*}\right) + C\beta v^*\left(1-\frac{w^*}{w_i^n}\right)\left(\frac{v_i^n}{v^*} - \frac{w_i^n}{w^*}\right)$$
$$+ C\sum_{i=1}^{m}\left(1-\frac{w^*}{w_i^n}\right)(d_3\nabla^2 w_i^n) + o(1)$$
$$= -C\Lambda_3 w^*\left(h\left(\frac{w^*}{w_i^n}\right) + \left(1+\frac{1}{\Lambda_3}\right)h\left(\frac{w_i^n}{w^*}\right) + \ln\left(1+\frac{1}{\Lambda_3}\right)\right)$$
$$+ u^* f(w^*)\left(-h\frac{w_i^n}{w^*}) + h\left(\frac{v^*}{v_i^n}\right) - h\left(\frac{w^n v^*}{w^* v_i^n}\right)\right) - d_3 w^* \sum_{i=1}^{m-1}\left(\sqrt{\frac{w_{i+1}^n}{w_i^n}} - \sqrt{\frac{w_{i-1}^n}{w_i^n}}\right)^2$$
$$- d_3 w^* \sum_{i=1}^{m-1}\left(\sqrt{\frac{w_m^n}{w0_1^n}} - \sqrt{\frac{w_1^n}{w_m^n}}\right)^2 + o(1).$$

Now, with the help of using the following proposition [3]:

$$h\left(\frac{f(w_i^n)}{f(w^*)}\right) \le h\left(\frac{w_i^n}{w^*}\right),$$

we can conclude the following inequality:

$$\begin{aligned}
\Delta V^n &= \Delta V_1^n + \Delta V_2^n + \Delta V_3^n \\
&\le \sum_{i=1}^{m} -\mu_1 u^* \left(h\left(\frac{u^*}{u_i^n}\right) + \left(1+\frac{1}{\mu_1}\right) h\left(\frac{u_i^n}{u^*}\right) + \ln\left(1+\frac{1}{\mu_1}\right)\right) \\
&- \Lambda_2 v^* \left(h\left(\frac{v^*}{v_i^n}\right) + \left(1+\frac{1}{\Lambda_2}\right) h\left(\frac{v_i^n}{v^*}\right) + \ln\left(1+\frac{1}{\Lambda_2}\right)\right) \\
&- C\Lambda_3 w^* \left(h\left(\frac{w^*}{w_i^n}\right) + \left(1+\frac{1}{\Lambda_3}\right) h\left(\frac{w_i^n}{w^*}\right) + \ln\left(1+\frac{1}{\Lambda_3}\right)\right) \\
&- u^* f(w^*) \left(h\left(\frac{u^*}{u_i^n}\right) + h\left(\frac{w^*}{w_i^n}\right) + h\left(\frac{v^* u_i^n f(w_i^n)}{v_i^n u^* f(w^*)}\right) + h\left(\frac{w^n v^*}{w^* v_i^n}\right)\right) \\
&- Cd_1 u^* \sum_{i=1}^{m-1} \left(\sqrt{\frac{u_{i+1}^n}{u_i^n}} - \sqrt{\frac{u_{i-1}^n}{u_i^n}}\right)^2 - Cd_1 u^* \sum_{i=1}^{m-1} \left(\sqrt{\frac{u_m^n}{u_1^n}} - \sqrt{\frac{u_1^n}{u_m^n}}\right)^2 \\
&- Cd_2 v^* \sum_{i=1}^{m-1} \left(\sqrt{\frac{v_{i+1}^n}{v_i^n}} - \sqrt{\frac{v_{i-1}^n}{v_i^n}}\right)^2 - Cd_2 v^* \sum_{i=1}^{m-1} \left(\sqrt{\frac{v_m^n}{v_1^n}} - \sqrt{\frac{v_1^n}{v_m^n}}\right)^2 \\
&- Cd_3 w^* \sum_{i=1}^{m-1} \left(\sqrt{\frac{w_{i+1}^n}{w_i^n}} - \sqrt{\frac{w_{i-1}^n}{w_i^n}}\right)^2 - Cd_3 w^* \sum_{i=1}^{m-1} \left(\sqrt{\frac{w_m^n}{w0_1^n}} - \sqrt{\frac{w_1^n}{w_m^n}}\right)^2 + o(1).
\end{aligned}$$

This immediately implies $\Delta V^n \le 0$ which asserts that the equilibrium point (u^*, v^*, w^*) is indeed globally asymptotically stable.

5 Numerical Simulations

In this part, we illustrate a numerical example that aims to show the feasibility of our main results. In system (6), we take $m = 2$ and consider the following function:

$$f(w_i^n) = \frac{w_i^n}{1 + w_i^n}.$$

Thus, we obtain the following discrete-time SEI reaction-diffusion model:

Table 1 Initial data of system (9)

initial conditions	value
u_0^n	25
u_1^n	25
v_0^n	15
v_1^n	15
w_0^n	10
w_1^n	10

Table 2 The parameters of system (9)

parameters	value
Λ_1	1.5
Λ_2	1.2
Λ_3	1.4
μ_1	0.2
μ_2	0.3
μ_3	0.6
β	1.5
d_1	1.4
d_2	1.6
d_3	1

$$\begin{cases} \Delta u_i^n = d_1 \nabla^2 u_i^n + \Lambda_1 - u_i^n \dfrac{w_i^n}{1 + w_i^n} - \mu_1 u_i^n \\ \Delta v_i^n = d_2 \nabla^2 v_i^n + \Lambda_2 + u_i^n \dfrac{w_i^n}{1 + w_i^n} - (\mu_2 + \beta) v_i^n \\ \Delta w_i^n = d_3 \nabla^2 w_i^n + \Lambda_3 + \beta v_i^n - \mu_3 w_i^n, \text{ for} i = 1, 2, \end{cases} \tag{9}$$

subject to the following periodic boundary conditions:

$$\begin{cases} u_0^n = u_2^n, & u_1^n = u_3^n \\ v_0^n = v_2^n, & v_1^n = v_3^n \\ w_0^n = w_2^n, & w_1^n = w_3^n. \end{cases} \tag{10}$$

These conditions are reported in Table 1, whereas the parameters of the above system are listed in Table 2.

After executing a proper MATLAB code, we generate Fig. 1 that illustrates the solutions of the proposed system (9) in three-dimensional case with the initial data set: $(u_i^0, v_i^0, w_i^0) = (50 + 0.05 \left(\sin\left(\frac{x}{5}\right)\right), 15 + 0.05 \left((\cos\left((\frac{x}{5}\right)\right),$

Fig. 1 Numerical solutions of system (9)

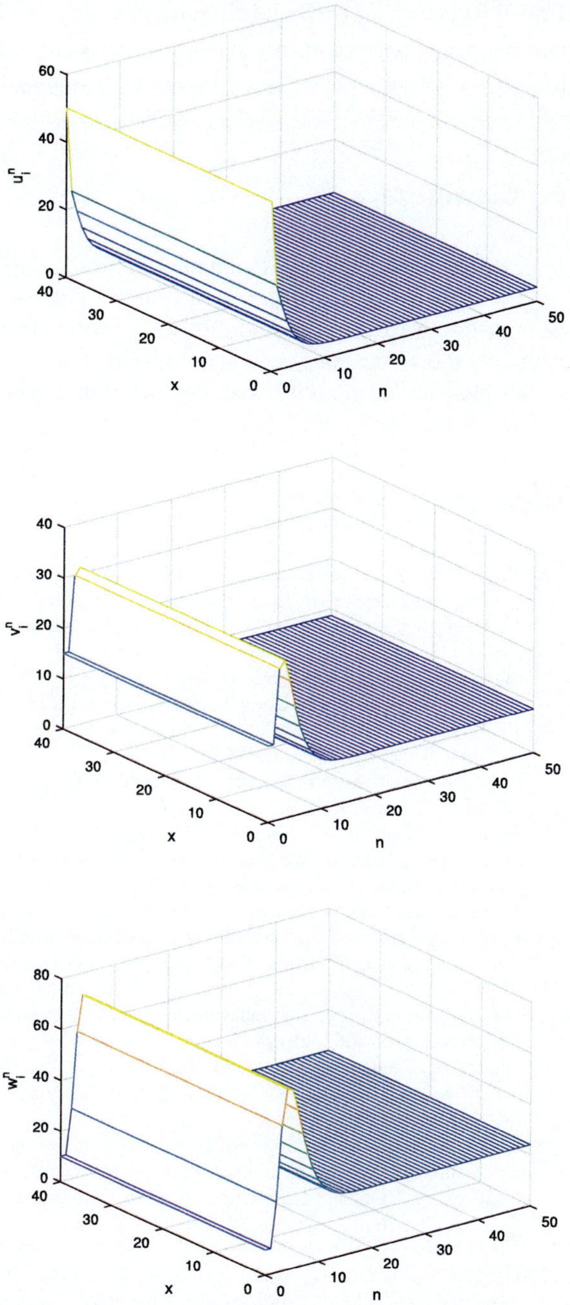

$10 + 0.05\left(\cos\left(\frac{x}{5}\right)\right)$. Obviously, one can observe that the solution of such system is globally asymptotically stable with respect to a fixed steady-state over a given time. This actually reflects the validity of our findings.

6 Conclusion

In this work, the global asymptomatic stability of the unique positive equilibrium of a discrete-time reaction-diffusion SEI model with periodic boundary conditions has been investigated. After providing a new discrete-time version of the aimed model based on the continuous-time reaction-diffusion SEI model, the positive equilibrium's globally asymptotical stability has been demonstrated.

References

1. Abdelmalek, S., Bendoukha, S.: Global asymptotic stability for a SEI reaction-diffusion model of infectious diseases with immigration. Int. J. Biomath. **11**, 1850044 (2018)
2. Abdelmalek, S., Bendoukha, S.: Global asymptotic stability of a diffusive SVIR epidemic model with immigration of individuals. Electron. J. Differ. Equ. **129**(324), 1–14 (2016)
3. Sigdel, R.P., McCluskey, C.C.: Global stability for an SEI model of infectious disease with immigration. Appl. Math. Comput. **243**, 684–689 (2014)
4. Xu, L., Han, R.: Global stability for a discrete space-time Lotka-Volterra system with feedback control. Complexity **2020**, 2960503 (2020)
5. Hioual, A., Ouannas, A.: On fractional variable-order neural networks with time-varying external inputs. Innovat. J. Math. **1**, 52–65 (2022)
6. Wang, B., Ouannas, A., Xia, W.F., Jahanshahi, H., Alotaibi, N.D.: A hybrid approach for synchronizing between two reaction-diffusion systems of integer- and fractional-order applied on certain chemical models. Fractals, preprint (2022)
7. Gasri, A., Ouannas, A., Khennaoui, A.A., Grassi, G.: Chaotic fractional discrete neural networks based on the Caputo h-difference operator: stabilization and linear control laws for synchronization. Eur. Phys. J. Spec. Top. **2022**, 1–15 (2022)
8. Shatnawi, M.T., Djenina, N., Ouannas, A., Batiha, I.M., Grassie, G.: Novel convenient conditions for the stability of nonlinear incommensurate fractional-order difference systems. Alex. Eng. J. **61**, 1655–1663 (2022)
9. Hioual, A., Ouannas, A., Oussaeif, T.E., Grassi, G., Batiha, I.M., Momani, S.: On variable-order fractional discrete neural networks: solvability and stability. Fractal and Fract. **6**, 119 (2022)
10. Abbes, A., Ouannas, A., Shawagfeh, N.: Incommensurate fractional discrete neural network: chaos and complexity. Eur. Phys. J. Plus **137**, 1–15 (2022)
11. Ouannas, A., Batiha, I.M., Bekiros, S., Liu, J., Jahanshahi, H., Aly, A.A., Alghtani, A.H.: Synchronization of the glycolysis reaction-diffusion model via linear control law. Entropy **23**, 1516 (2021)
12. Debbouche, N., Ouannas, A., Batiha, I.M., Grassi, G., Kaabar, M.K.A., Jahanshahi, H., Aly, A.A., Aljuaid, A.M.: Chaotic behavior analysis of a new incommensurate fractional-order Hopfield neural network system. Complexity **2021**, 3394666 (2021)
13. Batiha, I.M., Ouannas, A., Emwas, J.A.: A stabilization approach for a novel chaotic fractional-order discrete neural network. J. Math. Comput. Sci. **11**, 5514–5524 (2021)

14. Mellah, M., Ouannas, A., Khennaoui, A.A.: Fractional discrete neural networks with different dimensions: coexistence of complete synchronization, antiphase synchronization and full state hybrid projective synchronization. Nonlinear Dyn. Syst. Theory **21**, 410 (2021)
15. Zhou, J., Yang, Y., Zhang, T.: Global dynamics of a reaction-diffusion waterborne pathogen model with general incidence rate. J. Math. Anal. Appl. **466**, 835–859 (2018)
16. Yang, Y., Zhou, J.: Global stability of a discrete virus dynamics model with diffusion and general infection function. Int. J. Comput. Math. **96**, 1752–1762 (2019)
17. Elaydi, S.: An Introduction to Difference Equations. Springer, San Antonio, Texas (2015)

Atomic Solution of Euler Equation

Iqbal Jebril, Ghada Eid, Ma'mon Abu Hammad, and Duha AbuJudeh

Abstract In this paper, we find certain solutions of fractional partial differential questions. Tensor product of Banach space is used where separation of variables does not work.

Keywords Atomic solution · Conformable derivative · Tensor product · First section

1 Introduction

There are many definitions available in the literature of fractional derivatives. The main ones are the Riemann Liouville definition and Caputo definition.

Definition 1.1 (*Riemann–Liouville definition*). For $\alpha \in [m-1, m)$, the $\alpha-$ derivative of f is defined by

$$D_c^{\alpha}(f)(w) = \frac{1}{\Gamma(m-\alpha)} \frac{d^m}{dw^m} \int_c^w \frac{f(t)}{(w-t)^{\alpha-m+1}} dt.$$

Definition 1.2 (*Caputo Definition*). For $\alpha \in [m-1, m)$, the $\alpha-$ derivative of f is defined by

$$D_c^{\alpha}(f)(w) = \frac{1}{\Gamma(m-\alpha)} \int_c^w \frac{f^{(m)}(t)}{(w-t)^{\alpha-m+1}} dt.$$

In 2014, [6] introduced a new definition of fractional derivative which is very simple and natural as follows:

I. Jebril (✉) · G. Eid · M. A. Hammad · D. AbuJudeh
Al-Zaytoonah University of Jordan, Queen Alia Airport St. 594 11942, Amman, Jordan
e-mail: i.jebril@zuj.edu.jo

M. A. Hammad
e-mail: m.abuhammad@zuj.edu.jo

D. Zeidan et al. (eds.), *Mathematics and Computation*, Springer Proceedings in Mathematics & Statistics 418, https://doi.org/10.1007/978-981-99-0447-1_31

Given a function $f : [0, \infty) \to \mathbb{R}$, and $w > 0, \alpha \in (0, 1)$. Then for all

$$D^\alpha(f)(w) = \lim_{\varepsilon \to 0} \frac{f\left(w + \varepsilon w^{1-\alpha}\right) - f(w)}{\varepsilon}.$$

If the limit exists, then D^α is called the conformable fractional derivative of order α. Let$T^\alpha(x)$ stands for $T^\alpha(f)(x)$.
Hence,

$$T^\alpha(f)(w) = \lim_{\varepsilon \to 0} \frac{f\left(w + \varepsilon w^{1-\alpha}\right) - f(w)}{\varepsilon}.$$

If f is α-differentiable then define

$$f^\alpha(0) = \lim_{w \to 0} f^\alpha(w).$$

According to this definition, we have the following properties. Let $\alpha \in (0, 1]$, then:

i. $T^\alpha(1) = 0$.
ii. $T^\alpha(w^p) = pw^{p-\alpha}$, for all $p \in \mathbb{R}$.
iii. $T^\alpha(e^{cw}) = cw^{1-\alpha}e^{cw}, c \in \mathbb{R}$.
iv. $T^\alpha(\text{sin}bw) = bw^{1-\alpha}\text{cos}bw, b \in \mathbb{R}$.
v. $T^\alpha(\text{cos}bw) = -bw^{1-\alpha} \sin bw, b \in \mathbb{R}$.
vi. $T^\alpha\left(\frac{1}{\alpha}w^\alpha\right) = 1$.

Further, many functions behave as in the usual derivative. Here are some formulas:

i. $T^\alpha\left(\sin\frac{1}{\alpha}w^\alpha\right) = \cos\frac{1}{\alpha}w^\alpha$.
ii. $T^\alpha\left(\cos\frac{1}{\alpha}w^\alpha\right) = -\sin\frac{1}{\alpha}w^\alpha$.
iii. $T_\alpha\left(e^{\frac{1}{\alpha}w^\alpha}\right) = e^{\frac{1}{\alpha}w^\alpha}$.

Many studies use conformable fractional derivative definition [1, 3–7] and degenerate second-order identification problem in Banach Space [2].

2 Main Result

In this paper, we find certain solutions of fractional partial differential questions. Tensor product of Banach space is used where separation of variables does not work. We want to find an atomic solution of the fractional Euler equation.

$$D_w^{2\alpha} D_r^\beta u + D_r^{2\beta} D_w^\alpha u = u(w, r). \tag{1}$$

With conditions:

$$u(0,r)=1, u^{\alpha}(0,r)=1.$$

$$u(w,0)=1, u^{\beta}(w,0)=1.$$

This is a linear partial differential equation but separation of variables doesn't work. But Theory of tensor product can be used here to get what is called atomic solution.

A solution is called Atomic if it is of the form

$$u(w,r)=W(w)R(r)=W\otimes R, \tag{2}$$

where $W(w)$ and $R(r)$ are not constants. Now, substitute (2) in (1) to get

$$W^{2\alpha}(w)\otimes R^{\beta}(r)+W^{\alpha}(w)\otimes R^{2\beta}(r)=W(w)\otimes R(r). \tag{3}$$

Equation (3) has the tonsorial form:

$$W^{2\alpha}\otimes R^{\beta}+W^{\alpha}\otimes R^{2\beta}=W\otimes R. \tag{4}$$

Since, we have two cases to consider
Case (i)

$$W^{2\alpha}=W^{\alpha}=W.$$

Case (ii)

$$R^{\beta}=R^{2\beta}=R.$$

Let us discuss case (i). $W^{2\alpha}=W^{\alpha}$. This is a linear fractional differential equation. Hence, we get

$$W(w)=a+be^{\frac{w^{\alpha}}{\alpha}}.$$

Using the condition $u(0,r)=1=u^{\alpha}(0,r)$, we get $b=1$ and $a=0$.
Thus,

$$W(w)=\mathrm{e}^{\frac{w^{\alpha}}{\alpha}}.$$

As for $W^{2\alpha}=W$, it gives the same solution.
$W(w)=e^{\frac{w^{\alpha}}{\alpha}}$,
similarly, $W^{\alpha}=W$ it gives

$$W(w)=e^{\frac{w^{\alpha}}{\alpha}}.$$

Thus, we get in all situations $W(w) = e^{\frac{w^\alpha}{\alpha}}$. By Eq. (4)

$$W^{2\alpha} \otimes R^{\beta} + W^{\alpha} \otimes R^{2\beta} = W \otimes R$$

$$e^{\frac{w^\alpha}{\alpha}} \otimes R^{\beta} + e^{\frac{w^\alpha}{\alpha}} \otimes R^{2\beta} = e^{\frac{w^\alpha}{\alpha}} \otimes R$$

$$e^{\frac{w^\alpha}{\alpha}} \left(R^{\beta} + R^{2\beta} - R\right) = 0$$

$$\left(R^{\beta} + R^{2\beta} - R\right) = 0$$

$$\lambda = \frac{-1 + \sqrt{5}}{2}, \lambda_1 = \frac{-1 - \sqrt{5}}{2}.$$

Then

$$R(r) = c_1 e^{\lambda \frac{r^\beta}{\beta}} + c_2 e^{\lambda_1 \frac{r^\beta}{\beta}},$$

and

$u(w, r) = W(w)R(r)$,

where c_1 and c_2 are determined by the initial condition.

Hence

$$u(w, r) = \left(e^{\frac{w^\alpha}{\alpha}}\right)\left(e^{\frac{\lambda r^\beta}{\beta}}\right).$$

Case (ii)

$$R^{\beta} = R^{2\beta} = R.$$

Let us discuss case (ii).$R^{2\beta} = R^{\beta}$. This is a linear fractional differential equation. Hence, we get

$$R(r) = a_1 + b_1 e^{\frac{r^\beta}{\beta}}.$$

Using the condition $u(0, r) = 1 = u^{\alpha}(0, r)$, we get $b_1 = 1$ and $a_1 = 0$.

Thus

$R(r) = e^{\frac{r^\beta}{\beta}}$.

As for $R^{2\beta} = R$, it gives the same solution

$$R(r) = e^{\frac{r^\beta}{\beta}},$$

similarly, $R^{\beta} = R$ it gives

$$R(r) = e^{\frac{r^\beta}{\beta}}.$$

Thus, we get in all situations $R(r) = e^{\frac{r^\beta}{\beta}}$ by Eq. (4)

$$W^{2\alpha} \otimes R^\beta + W^\alpha \otimes R^{2\beta} = W \otimes R$$

$$e^{\frac{r^\beta}{\beta}}\left(W^{2\alpha} + W^\alpha - W\right) = 0$$

$$\left(W^{2\alpha} + W^\alpha - W\right) = 0$$

$$\gamma = \frac{-1+\sqrt{5}}{2}, \gamma_1 = \frac{-1-\sqrt{5}}{2}.$$

Similarly, one can find a solution of the form

$$u(w, r) = W(w)R(r).$$

Conclusion and Future Work

In this research, due to that fact that the method of separation of variables does not sometimes work well in solving several nonlinear fractional partial equations, we find that we should seek about an alternative approach that faces this problem. The Euler equation is deemed as one of these equations. In this work, we have found an atomic solution for Euler equation which has been formulated in view of the conformable fractional derivative definition. This has been performed with the help of using the tensor product technique in the Banach space with some of its properties. In accordance with what we have applied here, we can implement our presented scheme on other nonlinear fractional partial equations formulated in conformable definition or even in other fractional derivative operators. This task has been left to the nearest future for further consideration.

References

1. Abu Hammad, M., Awad, A., Khalil, R.: Properties of conformable fractional chi-square probability distribution. J. Math. Comput. Sci. **10**(4), 1239–1250 (2020)
2. Al Horani, M.H., Favini, A.: Degenerate second-order identification problem in banach spaces. J. Optim. Theory Appl. **120**, 305–326 (2004)
3. Batiha, I., Njadat, N., Batyha, R., Zraiqat, A., Dababneh, A., Momani, S.: Design fractional-order PID controllers for single-joint robot arm model. Int. J. Adv. Soft Comput. Appl. **14**(2), 96–114 (2022)

4. Batiha, I., Oudetallah, J., Ouannas, A., Al-Nana, A., Jebril, I.: Tuning the fractional-order PID-controller for blood glucose level of diabetic patients. Int. J. Adv. Soft Comput. Appl. **13**(2), 1–10 (2021)
5. Bezziou, M., Jebril, I., Dahmani, Z.: A new nonlinear duffing system with sequential fractional derivatives. Chaos, Solitons Fractals **151**, 111247 (2021)
6. Khalil, R., Al Horani, M., Yousef, A., Sababheh M.: A new definition of fractional derivative. J. Comput. Appl. Math. **264**, 65–70 (2014)
7. Khalil, R., Abdullah, L.: Atomic solution of certain inverse problems. Eur. J. Pure Appl. Math. **3**(4), 725–729 (2010)

Solving Non-linear Fractional Coupled Burgers Equation by Sub-equation Method

Worood A. AL-hakim, Maha S. Alsauodi, Gharib M. Gharib, Fatima Alqasem, and May Abu Jalbosh

Abstract In this solution, the real results were reached in solving the nonlinear fractional Coupled Burgers equation, which represents the solution with accuracy, ease and smoothness, which distinguishes it from other solutions, and these results were represented in a clear and expressive graphic for solving partial fractional equations.

Keywords Sub-equation method · Fractional calculus · Coupled Burgers equation · Nonlinear fractional equation · Fractinal equation

1 Introduction

This method is considered one of the most important and modern methods of finding solutions to equilibrium, partial and fractional equations.

Where accurate solutions are obtained in most cases and logical solutions that provide effectiveness for application. Tang, He, Wei, & Fan, Hon [1]. The main topic of this thesis is Solving Non-Linear Fractional Boussinesq—Burgers Equation by Use Sub—Equations Method.

Firstly this contains sections of Fractional Calculus. In the Journal of Computational and Applied Mathematics Abu Hammad and Khalil [2], Abu Hammad and Khalil, Jumarie et al. [3]. Taking conformable fractional derivative and integral and Laplace we take some theorem and property and examples and we take Zhang and

W. A. AL-hakim (✉) · G. M. Gharib · F. Alqasem · M. A. Jalbosh
Department of Mathematics, Zarqa University, Zarqa, Jordan
e-mail: woroodalhakim@yahoo.com

G. M. Gharib
e-mail: ggharib@zu.edu.jo

M. S. Alsauodi
Department of Basic Science, Applied Science Pravate University, Amman, Jordan
e-mail: m_alsoudi@asu.edu.jo

D. Zeidan et al. (eds.), *Mathematics and Computation*, Springer Proceedings in Mathematics & Statistics 418, https://doi.org/10.1007/978-981-99-0447-1_32

Zhang [4] by using five steps, in equation, Successfully obtained. Traveling wave solutions Wang [5], Zhou et al.

So we solve the Non—Linear Fractional fractional derivative Coupled Burgers Equations by sub-equation method.

Equation by Yan and Zhang [6], Singh, Gupta: Exact solutions of a variant Coupled Burgers. In [7] Zeidan et al. [8] and we find the exact solution in the equations by examples.

2 Fractional Calculus

In this work we introduce the famous definition of fractional calculus such that conformable Factional calculus is a field of applied mathematical science that deals with derivatives and integrals of fraction and the idea of fractional derivatives was raised in pints by Lhospital in (1695). Various definitions of fractional calculus were introduced such as Cauchy, Riemann, liouville and cupito; and was recently conformable by Khalil et al. [9].

Fractional calculus is an emerging and interesting branch of applied mathematics, describing theory of derivatives and integrals of any real or complex arbitrary order fractional differential equation (FDE), and has gained much attention due to the fact that the response of the fractional order system ultimately converges with the response of the integer order. Jumarie [10].

Definition 1 Gives a functional f: $[0.\infty) \to$ R Then the conformable fractional derivative f of order α is defined by

$$f^{\alpha} = \lim_{\varepsilon \to 0} \frac{f\left(t + \varepsilon t^{1-\alpha}\right) - f(t)}{\varepsilon}$$

$t > 0, \alpha \in (0, 1]$

Theorem 1 *Let* $\alpha \in (0, 1]$ *and* f *be* $\alpha-$ *differentible at appoint* $t > 0$, *then*:

$$T_{\alpha}(f)(t) = t^{1-\alpha} f^{\backslash}(t)$$

Proof Let $h = \varepsilon t^{1-\alpha} in\ defintion\ ,\ then\ \varepsilon = h t^{\alpha-1}$ as $\varepsilon \to 0, h \to 0$.

$$T_{\alpha}(f)(t) = \lim_{\varepsilon \to 0} \frac{f\left(t + \varepsilon t^{1-\alpha}\right) - f(t)}{\varepsilon} = \lim_{h \to 0} \frac{f(t+h) - f(t)}{h t^{\alpha-1}}$$

$$= t^{1-\alpha} \lim_{h \to 0} \frac{f(t+h) - (t)}{h} = t^{1-\alpha} f^{\backslash}(t)$$

Corollary 1

$$T_{\alpha-1}(f(t)) = t T_{\alpha}(f(t)), \forall \alpha - 1 > 0$$

Theorem 2 *Let* $\alpha \in (0, 1]$ *and f, g be* $\alpha\_differentible$ *at a point* $t > 0$, *then:*

1. $T_\alpha(af + bg) = aT_\alpha(f) + bT_\alpha(g) \quad , \forall a, b \in R$
2. $T_\alpha(t^p) = pt^{p-\alpha}, \quad \forall p \in R$:
3. $T_\alpha(fg) = fT_\alpha(g) + gT_\alpha(f)$
4. $T_\alpha(\frac{f}{g}) = \frac{gT_\alpha(f) - fT_{\alpha(g)}}{g^2}$
5. $T_\alpha\left(\frac{1}{f}\right) = \frac{-T_\alpha(f)}{f^2}$
6. $T_\alpha(\lambda) = 0$

3 Sub-Equation Method

In this section we introduce the main five steps of the fractional Sub-equation method and consider some examples on the proposed method:

A new Numerical technique solves linear and Non-linear Fractional Differential Equations of Order $0 < \alpha < 1$ indicating that Caputo's definition is proposed. The efficiency of this technique will be demonstrated by solving several examples of linear and non-linear differential equation. And non-linear fractional differential equation by Tang et al. [11]. Crompton [12].

Exact solutions. These solitary wave solutions of the perturbation are Nonlinear where an Equation with a power law nonlinearity model can display variety of behaviors:

Step 1:

Suppose that nonlinear FDEs with independent variable $(x_1, x_2, x_3, \ldots x_n, t)$ And dependent variable of *u*:

$$P(u, u_{t,}, u_{x1}, u_{x2}, u_{x3}, \ldots, D_t^\alpha u, D_{x1}^\alpha u, D_{x2}^\alpha u, D_{x3}^\alpha u, \ldots) = 0, \; 0 < a < 1 \tag{1}$$

where $D_t^\alpha au$ and $D_{x1}^\alpha u$,$D_{x2}^\alpha u$, $D_{X3}^\alpha u$, are Jumarie's modified Riemann -Liouville derivatives of u, $= t, x_1, x_2, x_3, \ldots, x_n$ is the unknown function.

Step 2:

Using the traveling wave transformation:

$$U(t, x_1, x_2, x_3, \ldots, x_n) = u(\xi), \xi = ct + k_1x_1 + k_2x_2 + k_3x_3 + \cdots + k_nx_n + \xi_0$$
$$k_nx_n + \xi_o \tag{2}$$

where $k_1, k_2, k_3, \ldots, k_n$ are constants to be determined later, we can turn the FDE (1) into this equation for $u = u\ (\xi)$:

$$P(u, c\,u, K_iu'c^\alpha D_\xi^\alpha u, \; k_\xi^\alpha D_\xi^\alpha, \ldots) = 0, \; where\ i = 1, 2, 3, ... \tag{3}$$

Step 3:

Suppose that the solution of (2) can be done as follows:

$$u\ (\xi) \sum_{x=0}^{n} a_i \varphi_i \tag{4}$$

where a_i(i = 0, 1, 2, 3, … n) and $a_i \neq 0$, the positive integer n can be determined by considering the homogeneous balance between the highest order derivatives and nonlinear terms appearing $\varphi = \varphi(\xi)$ satisfy the following Recite equation:

$$D_{\xi}^{\alpha} = \sigma + \phi^2,\ 0 < a \leq 1 \tag{5}$$

where σ is a constant and the solutions of Eq. (2) are obtained by Zhang using the generalized Exp-function method as follows (5).

4 Fractional Derivative Coupled Burgers Equations by Sub-equation Method

Berger's equation is used in various fields of phenomena and physical experiments such as boundary layer behavior, weather problems, traffic flow and sound, and is used to examine water waves and gas dynamics.

Therefore, the interest in this equation was the focus of attention of many scientists and researchers, and the work was done on analytical solutions and in several ways, including Buckland and others.

Problem 1 Space-Time (1 + 1) fractional derivative Coupled Burgers Equations.

We consider Space-Time fractional coupled Burgers Equations:

$$\begin{cases} D_t^{\alpha}u - D_x^{2\alpha}u + 2uD_x^{\alpha}u + pD_x^{\alpha}(uv) = 0 \\ D_t^{\alpha}v - D_x^{2\alpha}v + 2vD_x^{\alpha}v + qD_x^{\alpha}(uv) = 0 \end{cases},\ 0 < \alpha \leq 1 \tag{6}$$

We make the traveling wave transformation u(x,t) = U(ξ), v(x, t) = V(ξ), and ξ = xk + ct.

Equation (7) reduced to ODE:

$$\begin{cases} C^{\alpha}D_{\xi}^{\alpha}U - K^{2\alpha}D_{\xi}^{2\alpha}U + 2K^{\alpha}uD_{\xi}^{\alpha}U + pK^{\alpha}D_{\xi}^{\alpha}(UV) = 0,\ \ 0 < \alpha \leq 1 \\ C^{\alpha}D_{\xi}^{\alpha}V - K^{2\alpha}D_{\xi}^{2\alpha}V + 2K^{\alpha}VD_{\xi}^{\alpha}V + qK^{\alpha}D_{\xi}^{\alpha}(UV) = 0 \end{cases} \tag{7}$$

We suppose that Eq. (7) has the following general solution:

$$\begin{cases} U(\xi) = \sum_{i=0}^{n} a_i 0^i \\ V(\xi) = \sum_{j=0}^{m} b_j 0^j \end{cases}$$

By balancing the highest order derivative term and nonlinear term in Eq. (7) $D_\xi^{2\alpha}U$, $UD_\xi^{\alpha}U$ and $VD_\xi^{\alpha}V$, $D_\xi^{2\alpha}V$, we have $n + 2 = n + n + 1$ then $n = 1$, $2 + m = m + m + 1$ then $m = n = 1$.

Then Eq. (7) has solution.

$$\begin{cases} U(\xi) = a_0 + a_1\varphi \\ V(\xi) = b_0 + b_1\varphi \end{cases} \tag{8}$$

Substituting (8) along with (4) into (7) and setting the coefficient of $\varphi(\xi)^i$ ($i = 0$, 1, 2, 3) to zero, we can obtain a set of algebraic equation for c, k,b_0, b_1, $a_0$anda_1 as follows:

$$\begin{cases} C^\alpha D_\xi^\alpha\left\{\begin{matrix} a_0+ \\ +a_1\varphi \end{matrix}\right\} - K^{2\alpha}D_\xi^{2\alpha}\left\{\begin{matrix} a_0+ \\ +a_1\varphi \end{matrix}\right\} + 2K^\alpha\left\{\begin{matrix} a_0+ \\ +a_1\varphi \end{matrix}\right\}D_\xi^\alpha\left\{\begin{matrix} a_0+ \\ +a_1\varphi \end{matrix}\right\} + pK^\alpha D_\xi^\alpha\left(\left\{\begin{matrix} a_0+ \\ +a_1\varphi \end{matrix}\right\}\left\{\begin{matrix} b_0+ \\ +b_1\varphi \end{matrix}\right\}\right) = 0 \\ C^\alpha D_\xi^\alpha\left\{\begin{matrix} b_0+ \\ +b_1\varphi \end{matrix}\right\} - K^{2\alpha}D_\xi^{2\alpha}\left\{\begin{matrix} b_0+ \\ +b_1\varphi \end{matrix}\right\} + 2K^\alpha\left\{\begin{matrix} b_0+ \\ +b_1\varphi \end{matrix}\right\}D_\xi^\alpha\left\{\begin{matrix} b_0+ \\ +b_1\varphi \end{matrix}\right\} + qK^\alpha D_\xi^\alpha\left(\left\{\begin{matrix} b_0+ \\ +b_1\varphi \end{matrix}\right\}\left\{\begin{matrix} b_0+ \\ +b_1\varphi \end{matrix}\right\}\right) = 0 \end{cases}$$

$$\begin{cases} C^\alpha a_1\left\{\begin{matrix} \sigma+ \\ +\varphi^2 \end{matrix}\right\} - 2K^{2\alpha}a_1\left\{\begin{matrix} \sigma\varphi+ \\ +\varphi^3 \end{matrix}\right\} + 2K^\alpha\left\{\begin{matrix} a_1\sigma a_0+ \\ +a_1\varphi^2 a_0+ \\ +\sigma a_1^2\varphi+ \\ +a_1^2\varphi^3 \end{matrix}\right\} + pK^\alpha\left\{\begin{matrix} b_0\sigma a_1+ \\ +a_0 b_1\sigma+ \\ +2\sigma b_1\varphi a_1+ \\ +a_0 b_1\varphi^2+ \\ +b_0 a_1\varphi^2+ \\ +2b_1 a_1\varphi^3 \end{matrix}\right\} = 0 \\ C^\alpha b_1\left\{\begin{matrix} \sigma+ \\ +\varphi^2 \end{matrix}\right\} - 2K^{2\alpha}b_1\left\{\begin{matrix} \sigma\varphi+ \\ +\varphi^3 \end{matrix}\right\} + 2K^\alpha\left\{\begin{matrix} b_1\sigma b_0+ \\ +b_1\varphi^2 b_0+ \\ +\sigma b_1^2\varphi+ \\ +b_1^2\varphi^3 \end{matrix}\right\} + qK^\alpha\left\{\begin{matrix} b_0\sigma a_1+ \\ +a_0 b_1\sigma+ \\ +2\sigma b_1\varphi a_1+ \\ +a_0 b_1\varphi^2+ \\ +b_0 a_1\varphi^2+ \\ +2b_1 a_1\varphi^3 \end{matrix}\right\} = 0 \end{cases}$$

From the first equation:

$$\varphi(\xi)^0 : c^\alpha a_1\sigma + 2k^\alpha a_1\sigma a_0 + pk^\alpha a_0 b_1\sigma + pk^\alpha b_0\sigma a_1 = 0$$
$$\varphi(\xi)^1 : -2k^{2\alpha}a_1\sigma + 2k^\alpha\sigma a_1^2 + 2p\sigma k^\alpha b_1 a_1 = 0$$
$$\varphi(\xi)^2 : c^\alpha a_1 + 2k^\alpha a_1 a_0 + pk^\alpha a_0 b_1 + pk^\alpha b_0 a_1 = 0$$
$$\varphi(\xi)^3 : -2k^{2\alpha}a_1 + 2k^\alpha a_1^2 + 2pk^\alpha b_1 a_1 = 0$$

From the second equation:

$$\varphi(\xi)^0{:}c^\alpha b_1\sigma + 2k^\alpha b_1\sigma b_0 + qk^\alpha b_0 a_1\sigma + qk^\alpha a_0\sigma b_1 = 0$$
$$\varphi(\xi)^1{:} - 2\sigma k^{2\alpha} b_1 + 2k^\alpha b_1^2\sigma + 2k^\alpha q b_1 a_1\sigma = 0$$
$$\varphi(\xi)^2{:}c^\alpha b_1 + 2k^\alpha b_0 b_1 + qk^\alpha b_1 a_0 + qk^\alpha b_0 a_1 = 0$$
$$\varphi(\xi)^3{:} - 2k^{2\alpha} b_1 + k^\alpha 2b_1^\wedge 2 + 2k^\alpha q a_1 b_1 = 0$$

By using Mathematica:

$$\text{Solve}\Big[c^\alpha * a_1 * \sigma + 2 * k^\alpha * a_1 * \sigma * a_0 + p * k^\alpha * a_0 * b_1 * \sigma + p * k^\alpha * b_0 * \sigma * a_1 =$$
$$= 0\ \&\&c^\alpha * b_1 * \sigma + 2 * k^\alpha * b_1 * \sigma * b_0 + q * k^\alpha * b_0 * a_1 *^\alpha + q * k^\alpha * a_0 * \sigma * b_1 =$$
$$= 0\ \&\& - 2 * k^{2\alpha} * a_1 * \sigma + 2k^\alpha * \sigma * a_1^2 + 2 * p * \sigma * k^\alpha * b_1 * a_1 =$$
$$= 0\ \&\& - 2\sigma * k^{2\alpha} * b_1 + 2k^\alpha * b_1 \wedge 2 * \sigma + 2k^\alpha * q * b_1 * a_1 * \sigma =$$
$$= 0\ \&\&c^\alpha * a_1 + 2 * k^\alpha * a_1 * a_0 + p * k^\alpha * a_0 * b_1 + p * k^\alpha * b_0 * a_1 =$$
$$= 0\ \&\&c^\alpha * b_1 + 2k^\alpha * b_0 * b_1 + q * k^\alpha * b_1 * a_0 + q * k^\alpha * b_0 * a_1 =$$
$$= 0\ \&\& - 2k^{2\alpha} * a_1 + 2 * k^\alpha * a_1 \wedge 2 + 2 * p * k^\alpha * b_1 * a_1 =$$
$$= 0\ \&\& - 2k^{2\alpha} * b_1 + k^\nu * 2 * b_1 \wedge 2 + 2k^\alpha * q * a_1 * b_1 == 0, \{b_1, \sigma, b_0, a_1, a_0, b_2\}$$

$$\Bigg\{\{b_1 \to 0, a_1 \to 0\}, \Big\{b_1 \to k^\alpha, b_0 \to -\frac{1}{2}c^\alpha k^{-\alpha}, a_1 \to 0, a_0 \to 0\Big\},$$
$$\Big\{b_1 \to 0, b_0 \to 0, a_1 \to k^\alpha, a_0 \to -\frac{1}{2}c^\alpha k^{-\alpha}\Big\},$$
$$\Big\{b_1 \to \frac{k^\alpha(-1+q)}{-1+pq}, b_0 \to -\frac{c^\alpha k^{-\alpha}(-1+q)}{2(-1+pq)}, a_1 \to \frac{k^\alpha(-1+p)}{-1+pq}, a_0 \to -\frac{c^\alpha k^{-\alpha}(-1+p)}{2(-1+pq)}\Big\}\Bigg\}$$

Case 1: $\{b_1 \to 0, a_1 \to 0\}$.

$$\begin{cases} U_1 = a_0 \\ V_1 = b_0 \end{cases} \qquad \blacksquare$$

Case 2: $\left\{b_1 \to k^\alpha, b_0 \to -\frac{1}{2}c^\alpha k^{-\alpha}, a_1 \to 0, a_0 \to 0\right\}$.
where $\sigma < 0.$, $\xi = xk + ct$

$$\begin{cases} U_2 = 0 \\ V_2 = -\frac{1}{2}c^\alpha k^{-\alpha} - k^\alpha\sqrt{-\sigma}\tanh_\alpha\left(\sqrt{-\sigma}\xi\right) \end{cases}$$

$$\begin{cases} U_3 = 0 \\ V_3 = -\frac{1}{2}c^\alpha k^{-\alpha} - k^\alpha\sqrt{-\sigma}\coth_\alpha\left(\sqrt{-\sigma}\xi\right) \end{cases}$$

where $\sigma > 0$, $\xi = x\,k + c\,t$

$$\begin{cases} U_4 = 0 \\ V_4 = -\frac{1}{2}c^\alpha k^{-\alpha} + k^\alpha(\sqrt{\sigma}\tan_\alpha(\sqrt{\sigma}\xi) \end{cases}$$

$$\begin{cases} U_5 = 0 \\ V_5 = -\frac{1}{2}c^\alpha k^{-\alpha} - k^\alpha\sqrt{\sigma}\cot_\alpha(\sqrt{\sigma}\xi) \end{cases}$$

where $\sigma = 0$, $\xi = x\,k + c\,t$, w is constant.

$$\begin{cases} U_6 = 0 \\ V_6 = -\frac{1}{2}c^\alpha k^{-\alpha} - \frac{k^\alpha\Gamma(1+\alpha)}{\xi^\alpha+w} \end{cases} \blacksquare$$

Case 3: $\{b_1 \to 0, b_0 \to 0, a_1 \to k^\alpha, a_0 \to -\frac{1}{2}c^\alpha k^{-\alpha}\}$.
where $\sigma < 0$, $\xi = x\,k + c\,t$

$$\begin{cases} U_7 = -\frac{1}{2}c^\alpha k^{-\alpha} - k^\alpha\sqrt{-\sigma}\tanh_\alpha(\sqrt{-\sigma}\,\xi) \\ V_7 = 0 \end{cases}$$

$$\begin{cases} U_8 = -\frac{1}{2}c^\alpha k^{-\alpha} - k^\alpha\sqrt{-\sigma}\coth_\alpha(\sqrt{-\sigma}\,\xi) \\ V_8 = 0 \end{cases}$$

where $\sigma > 0$, $\xi = x\,k + c\,t$

$$\begin{cases} U_9 = -\frac{1}{2}c^\alpha k^{-\alpha} + k^\alpha(\sqrt{\sigma}\tan_\alpha(\sqrt{\sigma}\,\xi) \\ V_9 = 0 \end{cases}$$

$$\begin{cases} U_{10} = -\frac{1}{2}c^\alpha k^{-\alpha} - k^\alpha\sqrt{\sigma}\cot_\alpha(\sqrt{\sigma}\,\xi) \\ V_{10} = 0 \end{cases}$$

where $\sigma = 0$, $\xi = x\,k + c\,t$, w is constant.

$$\begin{cases} U_{11} = -\frac{1}{2}c^\alpha k^{-\alpha} - k^\alpha\frac{\Gamma(1+\alpha)}{\xi^\alpha+w} \\ V_{11} = 0 \end{cases} \blacksquare$$

Case 4: $\left\{b_1 \to \frac{k^\alpha(-1+q)}{-1+pq}, b_0 \to -\frac{c^\alpha k^{-\alpha}(-1+q)}{2(-1+pq)}, a_1 \to \frac{k^\alpha(-1+p)}{-1+pq}, a_0 \to -\frac{c^\alpha k^{-\alpha}(-1+p)}{2(-1+pq)}\right\}$.
Where $\sigma < 0$, $\xi = x\,k + c\,t$

$$\begin{cases} U_{12} = -\frac{c^\alpha k^{-\alpha}(-1+p)}{2(-1+pq)} - \frac{k^\alpha(-1+p)}{-1+pq}(\sqrt{-\sigma}\tanh_\alpha(\sqrt{-\sigma}\,\xi) \\ V_{12} = -\frac{c^\alpha k^{-\alpha}(-1+q)}{2(-1+pq)} - \frac{k^\alpha(-1+q)}{-1+pq}(\sqrt{-\sigma}\tanh_\alpha(\sqrt{-\sigma}\,\xi)) \end{cases}$$

$$\begin{cases} U_{13} = -\frac{c^\alpha k^{-\alpha}(-1+p)}{2(-1+pq)} - \frac{k^\alpha(-1+p)}{-1+pq}(\sqrt{-\sigma}\coth_\alpha(\sqrt{-\sigma}\,\xi) \\ V_{13} = -\frac{c^\alpha k^{-\alpha}(-1+q)}{2(-1+pq)} - \frac{k^\alpha(-1+q)}{(-1+pq)}(\sqrt{-\sigma}\coth_\alpha(\sqrt{-\sigma}\,\xi) \end{cases}$$

where $\sigma > 0$, $\xi = x\,k + c\,t$

$$\begin{cases} U_{14} = -\frac{c^{\alpha}k^{-\alpha}(-1+p)}{2(-1+pq)} + \frac{k^{\alpha}(-1+p)}{-1+pq}\left(\sqrt{\sigma}\tan_{\alpha}\left(\sqrt{\sigma}\,\xi\right)\right) \\ V_{14} = -\frac{c^{\alpha}k^{-\alpha}(-1+q)}{2(-1+pq)} + \frac{k^{\alpha}(-1+q)}{-1+pq}\left(\sqrt{\sigma}\tan_{\alpha}\left(\sqrt{\sigma}\,\xi\right)\right) \end{cases}$$

$$\begin{cases} U_{15} = -\frac{c^{\alpha}k^{-\alpha}(-1+p)}{2(-1+pq)} - \frac{k^{\alpha}(-1+p)}{-1+pq}\left(\sqrt{\sigma}\cot_{\alpha}\left(\sqrt{\sigma}\,\xi\right)\right) \\ V_{15} = -\frac{c^{\alpha}k^{-\alpha}(-1+q)}{2(-1+pq)} - \frac{k^{\alpha}(-1+q)}{(-1+pq)}\left(\sqrt{\sigma}\cot_{\alpha}\left(\sqrt{\sigma}\,\xi\right)\right) \end{cases}$$

where $w = \text{constant}$, $\sigma = 0$ and $\xi = xk + ct$

$$\begin{cases} U_{16} = -\frac{c^{\alpha}k^{-\alpha}(-1+p)}{2(-1+pq)} - \frac{k^{\alpha}(-1+p)}{-1+pq}\left(\frac{\Gamma(1+\alpha)}{\xi^{\alpha}+w}\right) \\ V_{16} = -\frac{c^{\alpha}k^{-\alpha}(-1+q)}{2(-1+pq)} - \frac{k^{\alpha}(-1+q)}{(-1+pq)}\left(\frac{\Gamma(1+\alpha)}{\xi^{\alpha}+w}\right) \end{cases} \blacksquare$$

4.4: Figures of Space-Time (1+1) fractional derivative Coupled Burgers Equations:

Case 2:

where $\sigma < 0$, Let $\alpha = c = k = 1, \sigma = -1$, then,

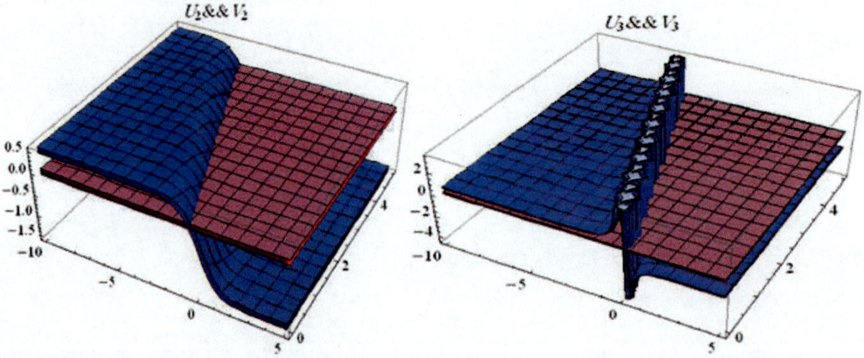

where $\sigma > 0$, Let $\alpha = c = k = \sigma = 1$, then:

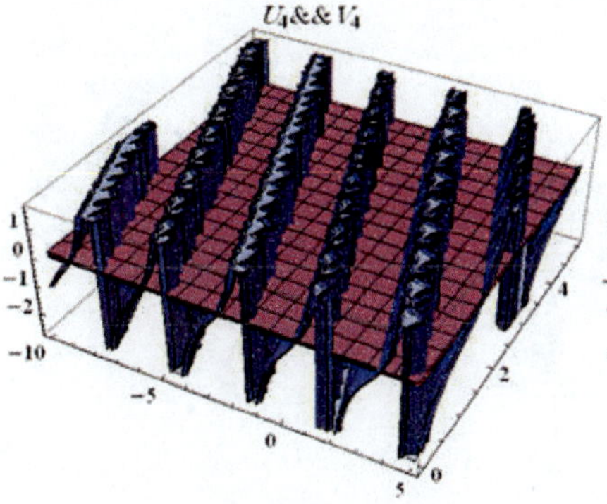

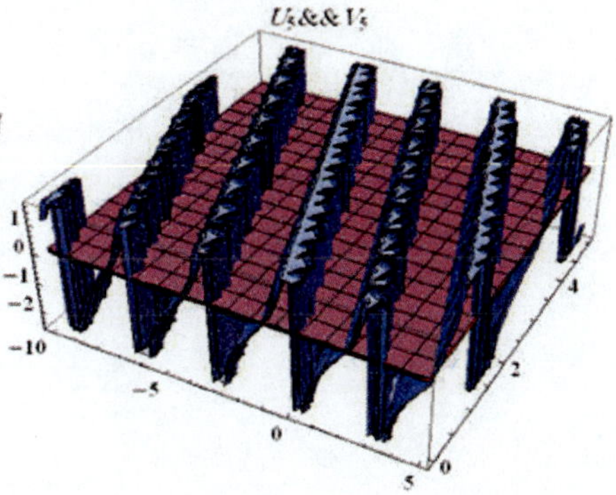

where $\sigma = 0$, Let $\alpha = c = k = 1, w = 0$, then:

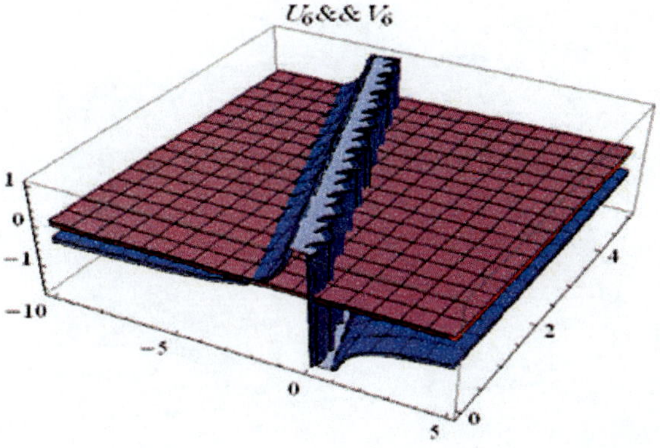

Case 3: Where $\sigma < 0$, Let $\alpha = c = k = 1, \sigma = -1, w = 0$, then:

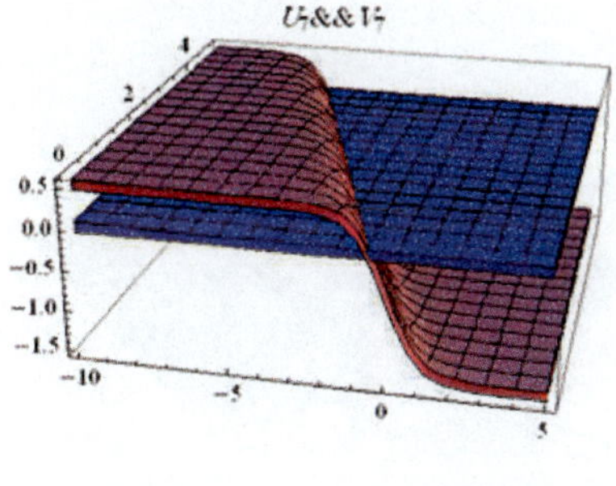

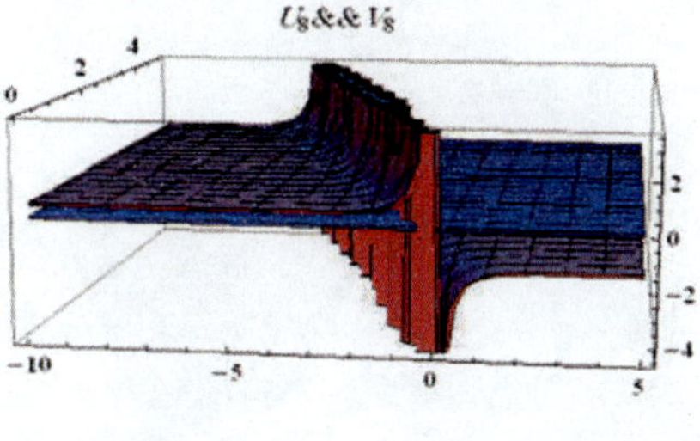

where $\sigma > 0$, Let $\alpha = c = k = \sigma = 1, w = 0$, then:

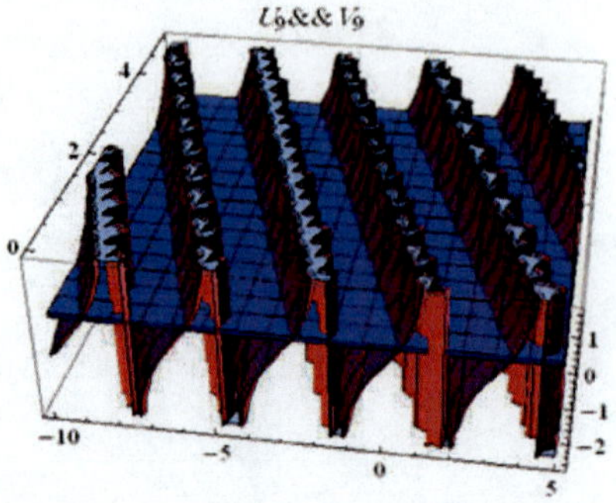

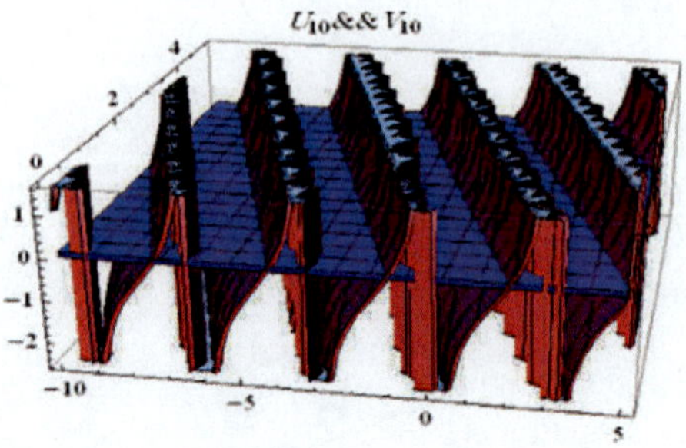

where $\sigma = 0$, Let $\alpha = \mathrm{c} = \mathrm{k} = 1$, $w = 0$, then :

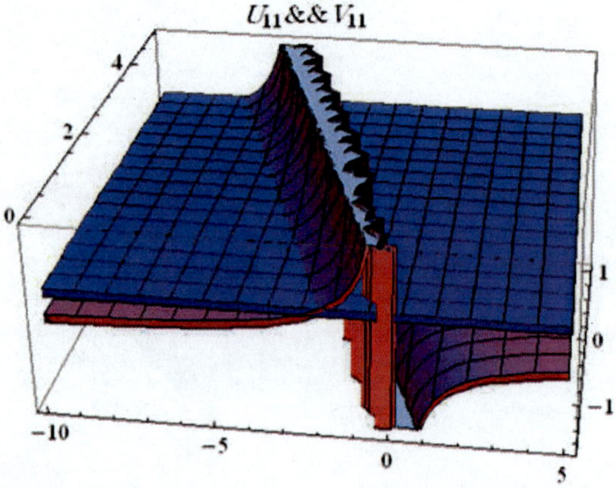

Case 4: Where $\sigma < 0$, Let $\alpha = \mathrm{c} = \mathrm{k} = 1, \sigma = -1, \mathrm{p} = \mathrm{q} = 0.5$&& $pq \neq 1$, then:

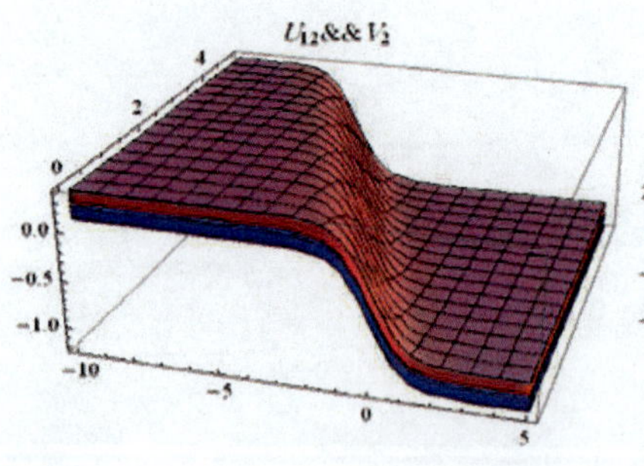

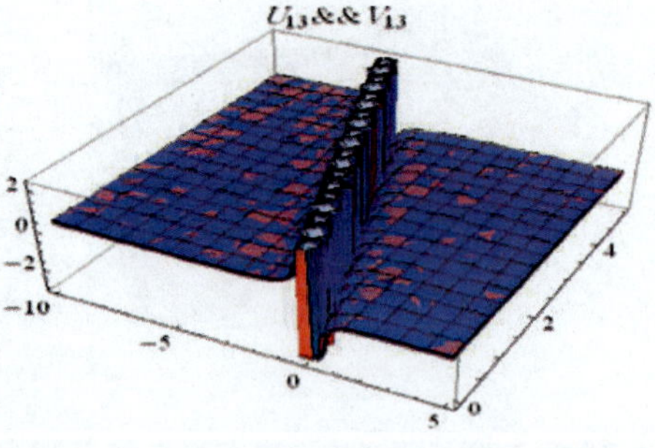

where $\sigma > 0$, $Let \alpha = c = k = \sigma = 1$, &&$p\ q \neq 1$then :

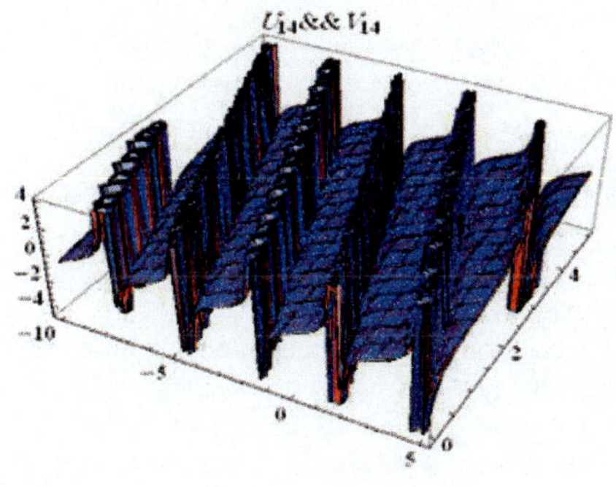

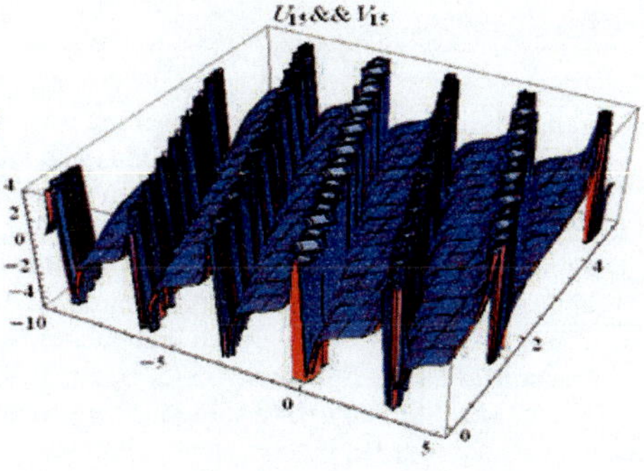

where $\sigma = 0$, Let $\alpha = c = k = 1, w = 0, p = q = 0.5,$ && $pq \neq 1$then :

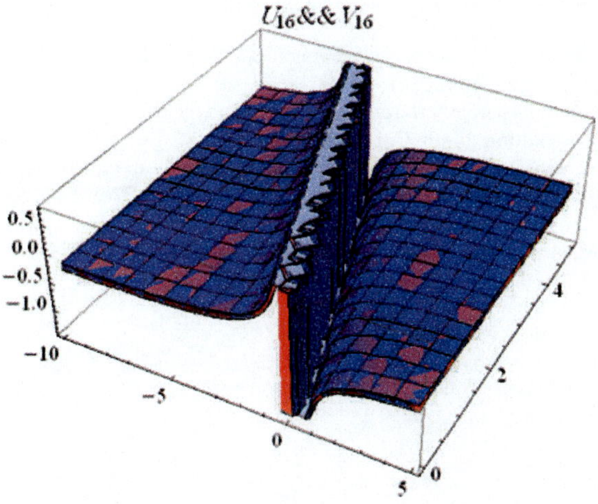

5 Conclusions

In this work, we proposed generalizing the fractional sub- equation method to construct exact solutions of space-time nonlinear fractional derivative systems: coupled Burgers equations. As this method is based on the homogenous balancing principle, it is also applied to other space-time nonlinear fractional derivative systems where the homogeneous balancing principle is satisfied.

References

1. Fan, E., Hon, Y.C.: A series of travelling wave solutions for two variant Boussinesq equations in shallow water waves. Chaos Solitons Fractals **15**, 559–566 (2003)
2. Abu Hammad, M., Khalil, R.: Conformable heat differential equation. Int. J. Pure Appl. Math. (2014)
3. Jumarie, G.: Modified Riemann-Liouville derivative and fractional Taylor series of nondifferentiable functions further results. Comput. Math. Appl. (2006)
4. Zhang, S., Zhang, H.Q.: Fractional sub-equation method and its applications to nonlinear fractional PDEs. Phys. Lett. A **375**(7), 1069–1073 (2011)
5. Wang, M.L.: Solitary wave solutions for variant Boussinesq equations. Phys. Lett. A (1995)
6. Yan, Z., Zhang, H.: New explicit and exact travelling wave solutions for a system of variant Boussinesq equations in mathematical physics. Phys. Lett. A **252**, 291–296 (1999)
7. Singh, K., Gupta, R.K.: Exact solutions of a variant Boussinesq system. Int. J. Eng. Sci. **44**, 1256–1268 (2006)
8. Zeidan, D., Ghau, G.K., Lu, T.-T., Zheng, W.-Q.: Mathematical studies of the solution of Burgers' equations by Adomian decomposition method. Math. Methods Appl. Sci. **43**, 2171–2188 (2020)
9. Khalil, R., Al Horani, M., Yousef, A., Sababheh, M.: A new definition of fractional derivative. J. Comput. Appl. Math. **264** (2014)
10. Jumarie, G.: Fractional partial differential equations and modified Riemann-Liouville derivative new methods for solution. Math. Comput. (2007)
11. Tang, B., He, Y., Wei, L., Zhang, X.: A generalized fractional sub-equation method for fractional differential equations with variable coefficients. Phys. Lett. A **376**(38–39), 2588–2590 (2012)
12. Crompton, B.: An introduction to fractional calculus and the fractional diffusion-wave equation. Unpublished Master Thesis, University of Massachusetts Lowell (2011)

Groups in Which the Commutator Subgroup is Cyclic

Shameseddin Mahmoud Alshorm

Abstract So far, our attention was focused on finite groups in which the commutator is in the center and it's called by the class of CC-groups. Since the center of any group is an abelian group, and the fundamental theorem of finitely generated abelian groups asserts that every abelian group is isomorphic to the direct product of cyclic groups. Then it is reasonable to consider those groups for which the derived subgroup is cyclic. It should be remarked that several authors have investigated particular classes of groups with similar restrictions. For instance, in [1] a bound is obtained for the order of G/G' when G is a $p-group$, $p \neq 2$, with a cyclic commutator subgroup. In this paper, we prove that every finite group which has a cyclic commutator must be supersolvable. Our result makes it possible to apply all properties of supersolvable groups to the so-called Dc-groups.

Keywords Commutators · Cyclic · Supersolvable group

1 Introduction

In 2014, Marcel Herzog, Gil Kaplan, and Arieh Lev reached out to a finite group G, they studied the connections between the sizes of its commutator subgroup $G^{'}$ and its center $Z(G)$. For example, they proved that if G is a solvable group such that $\Phi(G) = 1$ and $|G^{'}| \leq |G|^{1/3}$, $then\ Z(G) \neq 1$. "In a group, the product of two commutators need not be a commutator, consequently the commutator group of a given group cannot be defined as the set of all commutators, but only as the group generated by these. There seems to exist very little in the way of criteria or investigations on the question when all elements of the commutator group are commutators." For more see [2–4].

This is what Oystein Ore says in 1951 in the introduction to his paper "Some remarks on commutators." Since Ore made his comments, numerous contributions

S. M. Alshorm (✉)
Department of Mathematics, Al Zaytoonah University of Jordan, Queen Alia Airport St. 594, Amman 11733, Jordan
e-mail: alshormanshams@gmail.com

D. Zeidan et al. (eds.), *Mathematics and Computation*, Springer Proceedings in Mathematics & Statistics 418, https://doi.org/10.1007/978-981-99-0447-1_33

have been made to this topic and they are widely scattered over the literature. Many results have been rediscovered and republished. A case in point is Ore himself. The main result of Oystein Ore is that the alternating group on n letters, $n \geq 5$, consists entirely of commutators see [5–7]. This was already proved by G. A. Miller in [8] over half a century earlier. A well-known theorem, due to Olga Tausskij, asserts the following: Let G be a non-abelian group of order 2^n, such that $|G/G'| = 4$. Then G is dihedral, generalized quaternion, or semi-dihedral. Moreover, these groups have a cyclic commutator subgroup see [9].

2 A Class of D_c-Group

In this section, we will study the finite groups in which every commutator subgroup is cyclic. For more about the commutator properties see [10–12].

Definition 2.1 If G is finite group with cyclic commutator, then G is called $D_C - group$.

$$D_c = (G : G\, is\, finite\, and\, G^{'}\, is\, cyclic)$$

Note that $D_c - group$ is a not empty class of groups $D_c \neq \emptyset$ as the following example show.

Example 2.2 If G is the abelian group, then $G^{'} = 1 =< e >$ then G' is cyclic. G is $D_c - group$, $G \in D_c$.

Remark 2.3 Every abelian group is $D_c - group$ since the commutator for abelian groups is cyclic, $\mathfrak{A} \leq D_c$, the class of $D_c - groups$ is larger than the class of abelian groups.

Where: denotes the class of finite abelian groups.

Example 2.4 If $G = D_8 \times D_8$ then G is supersolvable since D_8 is supersolvable group and supersolvability is closed undertaking finite direct product. This is an example of supersolvable group which is not $D_c - group$, since the commutator $G' \cong V_4$ and V_4 is not cyclic.

Example 2.5 Let $G = D_{2\times 4} = D_8$ then G is non-abelian group but the commutator $D'_8 \cong \mathbb{Z}_2$ hence $\mathbb{Z}_2$ is cyclic then the commutator is cyclic. This is an example of $D_c - group$ which is not an abelian group.

Note that we can classify the class of $D_c - group$ as this $\mathfrak{A} \leq D_c \leq \mathfrak{U}$.

Where $\mathfrak{U}$: denotes the class of finite supersolvable.

Lemma 2.6 *If G is a $D_c - group$ then every subgroup H from G is a $D_c - group$ ($s - closed$).*

Proof Lemma 2.6 Let H be subgroup $H \leq G$ where $G \in D_c - group$, then G' is cyclic, hence $H' \leq G'$ and every subgroup of cyclic group is cyclic then H' is cyclic then $HisaD_c - group$.

Lemma 2.8 *If G is a $D_c - group$ and N is the normal subgroup of G, $N \trianglelefteq G$ then the quotient group G/N is a $D_c - group$ ($q - closed$).*

Proof Lemma 2.8 $N \trianglelefteq G$, $G\, is\, a\, D_c - group$ and G' is cyclic.

$(G/N)' = G'N/N \cong G'/G' \cap N$, quotient of cyclic group is cyclic.
Then $G'/G' \cap N$ is cyclic that is implies the quotient group $G/N \in D_c - group$.

Remark 2.10 If $G_1, G_2 \in D_c - group$ then $G_1 \times G_2 \notin D_c - group$, we can show that by this counter example.

Example 2.11 Let $G_1 = D_8\, and\, G_2 = D_8$ then $G_1 \times G_2 \notin D_c - group$ since $(G_1 \times G_2)' = \mathbb{Z}_2 \times \mathbb{Z}_2 \cong V_4$ but V_4 is not cyclic.

If we add this restriction that the order for $G_1 and G_2$ must be relatively prime to the previous example, then we have the next lemma.

Lemma 2.12 *If $G_1, G_2 \in D_c - group$ and the order for $G_1 and G_2$ are relatively primes, then $G_1 \times G_2 \in D_c - group$.*

Proof Lemma 2.12 If $G_1, G_2 \in D_c - group$ then G_1', G_2' are cyclic commutators, then $G'_1 \times G'_2$ is cyclic since the direct products of two cyclic groups with relatively prime is cyclic.

Theorem 2.14 *Let G be a group. Then G/G' is abelian.*

Lemma 2.15 Every abelian finite group is supersolvable group.

Lemma 2.16 Let G be a finite group with a normal subgroup N, if N is cyclic and G/N is supersolvable then G is supersolvable.

Lemma 2.17 *If G is a $D_c - group$ then G is supersolvable.*

Proof Lemma 2.17 Assume G' is cyclic then by Theorem 2.14 we have G/G' is abelian and from Lemma 2.15 we get G/G' is supersolvable, since G' is cyclic and G/G' is supersolvable then by Lemma 2.16 we have G is supersolvable.

Remark 2.19 If G is not supersolvable then G' cannot be generated by a single element.

Example 2.20 Take the alternating group A_4 with 4 elements, we know that A_4 is not supersolvable. Therefore $|A_4'| \in \{2, 3\}$ since $A_4' \lhd A_4$ we get $|A_4'| = 4$ and $A_4' \in syl_2(A_4)$.

References

1. van der Waall, R.W.: On finite p-groups whose commutator subgroups are cyclic. Indagationes Mathematicae (Proceedings), **76**(4) 342–345 (1973). https://doi.org/10.1016/1385-7258(73)90030-9
2. Arad, Z., Herzog, M. (eds.): Products of Conjugacy Classes in Groups. Lecture Notes in Mathematics, vol. 1112. Springer, Berlin (1985)
3. Andrea, B.: On the homology of the commutator subgroup of the pure braid group. Proc. Amer. Math. Soc. **149**(6), 2387–2401 (2021)
4. Dummit, D.S., Foote, F.M.: Abstract Algebra, 3rd edn. Wiley (2004)
5. Doerk, K., Hawkes, T.: Finite Soluble Groups, De Gruyter Berlin (1992)
6. Isaacs, I.M.: Finite Group Theory, Graduate studies in mathematics, vol. 92, AMS:350 (2008)
7. Larsen, M., Lu, Z.: Flatness of the commutator map over SLn. Int. Math. Res. Not. IMRN **2021**(8), 5605–5622 (2021)
8. Miller, G.A.: The regular substitution groups whose orders is less than 48. Q. J. Math. **28**, 232–284 (1896)
9. Myers, L.: Math bite: normality of the commutator subgroup. Math. Mag. **68**(1), 49 (1995)
10. Gow, R.: Commutators in fnite simple groups of Lie type. Bull. London Math. Soc. **32**(3), 311–315 (2000)
11. Al-Sharo, K., Alshorman, S.E.: (2021). Finite groups in which every commutator element is central (2021). https://doi.org/10.13140/RG.2.2.33517.10727
12. Alshorman, S.E.: On a class of finite groups with restriction on commutator (2021). https://doi.org/10.13140/RG.2.2.17008.58881/1

On Point Prediction of New Lifetimes Under a Simple Step-Stress Model for Censored Lomax Data

Mohammad A. Amleh

Abstract In this study, we consider a new predictor for future lifetimes of units with Type-II censoring for a simple step-stress model. We assume that the lifetime data of the units are distributed as Lomax distribution with constant shape parameters and scale parameters depending on the stress level. It is also assumed that the stress plan occurs based on a cumulative exposure model. In this context, the best unbiased point predictor is addressed. A data analysis has been performed to compare such new method with the previously obtained point predictors.

Keywords Accelerated life tests · Best unbiased predictor · Cumulative exposure model · Lomax distribution · Maximum likelihood predictor · Step-stress tests

1 Introduction

Accelerated life tests (ALTs) are known in reliability analysis to be used as an evaluation of the lifetime of highly reliable units in a reasonable testing time. In ALTs, the product is tested at higher than usual levels of stress, such as high pressure, vibration, voltage, or temperature to induce early failure times, leading to shorter lifetimes and accelerated damage. Data obtained from such a test need to be analyzed and used to predict new lifetimes based on the model that joins the lifetime distribution to the degree of stress. A special kind of ALT is the step-stress test which allows the experimenter to increase the level of stress at pre-fixed times during the testing experiment. In the basic form of step-stress test, n units are placed on the test at an initial stress level s_1. At the pre-specified time τ_1 the stress level is raised to s_2. Similarly, at the pre-specified time τ_2, the stress level is accelerated from s_2 to s_3, and so on. At the final stage, the stress level is changed from s_{k-1} to s_k at the pre-specified time τ_{k-1}. The experiment stops if all the units tested on the experiment fail, or some termination conditions are used. If only two levels of stress are used, the test is called a simple step-stress test. For more details on these tests, one may refer to Nelson [17]

M. A. Amleh (✉)
Department of Mathematics, Zarqa University, Zarqa, Jordan
e-mail: malamleh@zu.edu.jo

D. Zeidan et al. (eds.), *Mathematics and Computation*, Springer Proceedings in Mathematics & Statistics 418, https://doi.org/10.1007/978-981-99-0447-1_34

and Kundu and Ganguly [13]. In order to analyze the failure times under step-stress tests, we need a model that gives the relationship between the distribution of the lifetimes under various stress levels to the failure times under the step-stress setup. The cumulative exposure model (CEM) is the most popular model in this context, which was proposed by Nelson [16]. In this model, the main assumption is that the remaining lifetime of the experiment units relies only on the cumulative exposure these units have experienced, but without memory on how such exposure has been accumulated. Several authors discussed the step-stress test under CEM and related statistical inferences. Estimation of the parameters in a simple step-stress test under CEM for exponential distribution is addressed by Xiong [19]. Balakrishnan et al. [4] presented a simple step-stress model under Type-I censoring with lifetimes having lognormal distribution. Kateri and Balakrishnan [12] discussed a simple step-stress model with Type-II censoring scheme and Weibull distribution.

For Pareto distribution, Kamal et al. [10] considered the estimation of the parameters of Pareto distribution for a simple step-stress model with complete lifetimes. Chandra and Khan [8] discussed the simple step-stress test and obtained estimators of the parameters for Lomax distribution with a Type-I censoring scheme. Hassan et al. [9] presented the simple step-stress model based on an adaptive Type-II progressive hybrid censoring under Lomax distribution.

The prediction of new order statistics based on the data observed is a fundamental aspect of statistical analysis. It is commonly used in survival analysis and medical studies. More details on point prediction and prediction intervals can be found in Kaminsky and Rhodin [11]. For step-stress tests, Basak and Balakrishnan [5, 6] addressed the problem of prediction of the failure times of units for a simple step-stress test based on exponential distribution under progressive Type-I censoring and Type-II censoring schemes, respectively. Amleh and Raqab [1, 2] presented the prediction problem for step-stress test for Lomax lifetimes under CEM, and for Weibull lifetimes under Khamis-Higgens model, respectively. Recently, Amleh [3] proposed several prediction techniques for Rayleigh distribution under a simple step-stress plan.

In this paper, the simple step-stress for the Lomax distribution according to CEM is considered. It is assumed that failures occur based on the Type-II censoring setup. Based on these assumptions, the aim of the paper is obtaining an explicit form of the best unbiased point predictor and comparing this method with the previously proposed point predictors.

The rest of the paper is designed as follows. The description of the CEM under Lomax distribution and basic model assumptions are discussed in Sect. 2. Point predictors including the best unbiased predictor are presented in Sect. 3. A comparative study among the point predictors is conducted in Sect. 4.

2 Model Description and Related Assumptions

In this section, we provide a brief description of the CEM under a simple step-stress test with Lomax distribution and the related assumptions.

In step-stress tests, a CEM is based on assuming that the failure time distribution of the units at stress level i is related to the failure time distribution of the units at the stress level $i+1$. The test determines that the remaining lifetime of the experimental units relies only on the cumulative exposure the units have seen, without memory of how to accumulate this exposure.

Lomax distribution is a special case of Pareto distribution, it was suggested by Lomax [15]. It is shifted from Pareto distribution which leads the support to begin from zero. It is used widely in economics, engineering, and reliability analysis.

Its probability density function (pdf) is given by

$$f(t, \alpha, \beta) = \frac{\beta}{\alpha}\left(1 + \frac{t}{\alpha}\right)^{-\beta-1}, t > 0, \beta > 0, \alpha > 0, \tag{1}$$

with cumulative distribution function (cdf)

$$F(t, \alpha, \beta) = 1 - \left(1 + \frac{t}{\alpha}\right)^{-\beta}, t > 0, \beta > 0, \alpha > 0, \tag{2}$$

here, β is the shape parameter, while α represents the scale parameter.

The Lomax distribution is featured that its hazard rate function is decreasing in t and given by

$$h(t) = \frac{\beta}{\alpha + t},$$

Accordingly, Lomax distribution is used to represent the lifetime of a decreasing failure rate units. In fact, Bryson [7] argued that Lomax distribution is an alternative to the exponential distribution when the data has heavy tailed distribution.

In the simple step-stress test under Type-II censoring setup, the test is terminated as soon as the rth failure occurs. The experiment is performed as follows. Initially, all n units are put on the lower stress S_1 and continued until time τ. Then, the stress is increased to higher level S_2, and the test runs until a pre-determined r failure times will be observed. Let n_1 denote the random number of failure times before τ, and $n_2 = r - n_1$, denote the number of failure times after τ. If $n_1 \neq r$, the stress level is accelerated to the next step, and the test goes up to the point of r failures.

According to the above situation, the ordered lifetimes that are observed, which are denoted by the vector data **t**, have the following form

$$t_{1:n} < \ldots < t_{n_1:n} < \tau \leq t_{n_1+1:n} < \ldots < t_{r:n}. \tag{3}$$

Here, **t** is the vector of the observed values of the random variable $\mathbf{T} = (T_{1:n}, \ldots, T_{n_1}, T_{n_1+1:n}, \ldots, T_r)$, which represents the Type-II censored lifetimes. The CEM for the simple step-stress plan is given by

$$F(t) = \begin{cases} F_1(t), 0 \leq t < \tau \\ F_2(t - \tau + h), \tau \leq t < \infty, \end{cases} \tag{4}$$

here, h is a solution to the following equation

$$F_1(\tau) = F_2(h).$$

By solving the above equation, we get $h = \frac{\alpha_2}{\alpha_1}\tau$. As a result of that, the Lomax CEM is distributed as

$$F(t) = \begin{cases} 1 - \left(1 + \frac{t}{\alpha_1}\right)^{-\beta}, 0 \leq t < \tau \\ 1 - \left(1 + \frac{\tau}{\alpha_1} + \frac{t-\tau}{\alpha_2}\right)^{-\beta}, \tau \leq t < \infty, \end{cases} \tag{5}$$

with the corresponding pdf

$$f(t) = \begin{cases} \frac{\beta}{\alpha_1}(1 + \frac{t}{\alpha_1})^{-\beta-1}, 0 \leq t < \tau \\ \frac{\beta}{\alpha_2}(1 + \frac{\tau}{\alpha_1} + \frac{t-\tau}{\alpha_2})^{-\beta-1}, \tau \leq t < \infty. \end{cases} \tag{6}$$

3 Point Prediction of New Order Statistics

Now, we discuss the problem of prediction of unobserved lifetimes based on some observed lifetimes for the simple step-stress test under the Lomax CEM. The description of the problem is as follows. Let $T_{1:n} < T_{2:n} < \ldots < T_{r:n}$ be the observed sample, and let$T_{s:n}, s = r + 1, \ldots, n,$ be the unobserved new failure time based on the same test. The point prediction concerns the prediction of the future lifetimes$T_{s:n}$, given the first r observations $T_{i:n}, 0 < i \leq r$.

Based on the Markovian property of the order statistics with Type-II censoring, the conditional density of $Y = T_{s:n}$ given $\boldsymbol{T} = \boldsymbol{t} = (t_{1:n}, \ldots, t_{n_1:n}, t_{n_1+1:n}, \ldots, t_{r:n})$ is the same as the density of $Y = T_{s:n}$ given$T_{r:n} = t_{r:n}$. Therefore, the density of Y given $\boldsymbol{T} = \boldsymbol{t}$ is equivalent to the pdf of the $(s - r)^{th}$ order statistic is taken out of $(n-r)$ units from the truncated distribution of density$\varphi(y) = \frac{f(y)}{1-F(t_{r:n})}$, $y > t_{r:n}$, where $F(y)$ is provided as in Eq. (5). Thus, we have

$$f_{T_{s:n}|T}(y|\boldsymbol{\theta}, data) = c \times \frac{\beta}{\alpha_2}\left(1 + \frac{\tau}{\alpha_1} + \frac{y-\tau}{\alpha_2}\right)^{-\beta(n-s+1)-1}\left(1 + \frac{\tau}{\alpha_1} + \frac{t_{r:n}-\tau}{\alpha_2}\right)^{\beta(n-r)}$$
$$\times\left[\left(1 + \frac{\tau}{\alpha_1} + \frac{t_{r:n}-\tau}{\alpha_2}\right)^{-\beta} - \left(1 + \frac{\tau}{\alpha_1} + \frac{y-\tau}{\alpha_2}\right)^{-\beta}\right]^{s-r-1}, y > t_{r:n}, \quad (7)$$

where $\boldsymbol{\theta} = (\beta, \alpha_1, \alpha_2)$, $c = \frac{(n-r)!}{(s-r-1)!(n-s)!}$.

Amleh and Raqab [1] proposed two point prediction predictors of future lifetimes for Lomax CEM. Now, we give a brief description of such predictors. Further, we present the best unbiased point predictor as an alternative.

3.1 Maximum Likelihood Predictor

The maximum likelihood predictor (MLP) was suggested by Kaminsky and Rhodin [11]. This technique can be used as a prediction of future observations and also an estimation of the unknown parameters in the test. We express the predictive likelihood function (PLF) of $Y = T_{s:n}$ as

$$L(\beta, \alpha_1, \alpha_2, y) = L \propto \prod_{i=1}^{n_1} f_1(t_{i:n}) \prod_{i=n_1+1}^{r} f_2(t_{i:n})[F_2(y) - F_2(t_{r:n})]^{s-r-1} f_2(y)$$
$$[1 - F_2(y)]^{n-s}, 0 \le n_1 \le r, r+1 \le s \le n. \quad (8)$$

The MLP and the predictive maximum likelihood estimators (PMLEs) of the parameters β, α_1 and α_2 are obtained by maximizing the PLF based on the predictive likelihood equations (PLEs) for $\beta, \alpha_1, \alpha_2$ and y. The obtained MLP of Y will be denoted by $\widehat{Y}_M$.

3.2 Conditional Median Predictor

The conditional median predictor (CMP) was developed by Raqab and Nagaraja [18]. A point predictor $\widehat{Y}$ is said to be the CMP of Y, if it is the median of the conditional density of Y given $\boldsymbol{T} = \boldsymbol{t}$, consequently

$$P_\theta\left(Y \le \widehat{Y}|\boldsymbol{T} = \boldsymbol{t}\right) = P_\theta\left(Y \ge \widehat{Y}|\boldsymbol{T} = \boldsymbol{t}\right).$$

Amleh and Raqab [2] obtained the CMP as

$$\widehat{Y}_{CMP} = \left[\alpha_2 + \frac{\alpha_2}{\alpha_1}\tau + (t_{r:n} - \tau)\right][1 - Md_B]^{\frac{-1}{\beta}} - \alpha_2 - \frac{\alpha_2}{\alpha_1}\tau + \tau, \tag{9}$$

where B represents a $Beta(s - r, n - s + 1)$ distribution and Md_B represents its median. The CMP of Y is computed approximately by replacing β, α_1 and α_2 by their corresponding MLEs.

3.3 Best Unbiased Predictor

A point predictor $\widehat{Y}$ of $Y = T_{s:n}$ is said to be a best unbiased predictor (BUP) of Y, if we have

$$E\left(\widehat{Y} - Y\right) = 0,$$

and

$$Var\left(\widehat{Y} - Y\right) \le Var\left(\tilde{Y} - Y\right), \text{ for any unbiased predictor } \tilde{Y} \text{ of } Y.$$

Using the conditional density of Y given $\boldsymbol{T} = \boldsymbol{t}$, as in Eq. (7), the BUP of Y is expressed as

$$\widehat{Y}_{BUP} = E(Y|T) = \int_{t_{r:n}}^{\infty} y g_{T_{s:n}|T}(y|\theta, \text{data}) dy. \tag{10}$$

which can be simplified as

$$\widehat{Y}_{BUP} = \frac{C\beta}{\alpha_2}\left(1 + \frac{\tau}{\alpha_1} + \frac{t_{r:n} - \tau}{\alpha_2}\right)^{\beta(n-r)} \times \int_{t_{r:n}}^{\infty} y\left(1 + \frac{\tau}{\alpha_1} + \frac{y - \tau}{\alpha_2}\right)^{-\beta(n-s+1)-1}$$
$$\left(\left(1 + \frac{\tau}{\alpha_1} + \frac{t_{r:n} - \tau}{\alpha_2}\right)^{-\beta} - \left(1 + \frac{\tau}{\alpha_1} + \frac{y - \tau}{\alpha_2}\right)^{-\beta}\right)^{s-r-1} dy \tag{11}$$

Using the binomial expansion:

$$\left(\left(1 + \frac{\tau}{\alpha_1} + \frac{t_{r:n} - \tau}{\alpha_2}\right)^{-\beta} - \left(1 + \frac{\tau}{\alpha_1} + \frac{y - \tau}{\alpha_2}\right)^{-\beta}\right)^{s-r-1}$$
$$= \sum_{k=0}^{s-r-1}\binom{s-r-1}{k}(-1)^k\left(1 + \frac{\tau}{\alpha_1} + \frac{t_{r:n} - \tau}{\alpha_2}\right)^{-\beta(s-r-k-1)}\left(1 + \frac{\tau}{\alpha_1} + \frac{y - \tau}{\alpha_2}\right)^{-k\beta}, \tag{12}$$

By substituting Eq. (12) in Eq. (11), we obtain

$$\begin{aligned}\widehat{Y}_{BUP} =&\frac{C\beta}{\alpha_2}\sum_{k=0}^{s-r-1}\binom{s-r-1}{k}(-1)^k\left(1+\frac{\tau}{\alpha_1}+\frac{t_{r:n}-\tau}{\alpha_2}\right)^{\beta(n-s-k-1)}\\ &\times\int_{t_{r:n}}^{\infty} y\left(1+\frac{\tau}{\alpha_1}+\frac{y-\tau}{\alpha_2}\right)^{-\beta(n-s+k+1)-1} dy\end{aligned} \tag{13}$$

Using integration by parts and doing some simplifications, the BUP can be obtained precisely as

$$\begin{aligned}\widehat{Y}_{BUP} =&C\sum_{k=0}^{s-r-1}\binom{s-r-1}{k}(-1)^k\\ &\left(\frac{t_{r:n}}{n-s+k+1}-\frac{\alpha_2\left(1+\frac{\tau}{\alpha_1}+\frac{t_{r:n}-\tau}{\alpha_2}\right)}{(n-s-k-1)[1-\beta(n-s-k-1)]}\right).\end{aligned} \tag{14}$$

The BUP of Y can be obtained by replacing β, $\alpha_1$andα_2 by values of the MLEs.

4 Data Analysis

To compare the best unbiased predictor with the other point predictors, we perform a real data analysis. The data is taken from Liu [14] and has been considered by Amleh and Raqab [1]. The data refers to the failure times (in seconds) of nanocrystalline embedded high-k device run under a specific experiment. 40 devices are tested in a step-stress model with stress change time $\tau = 600$ s. 38 lifetimes have been observed before terminating the test. The data are given as follows:

The failure times of the 40 devices

Stress level						Recorded data					
1	8	38	72	97	122	140	163	170	188	198	223
	256	257	265	448							
2	608	611	614	615	616	620	623	623	624	624	631
	636	646	654	660	673	675	680	684	692	693	730
	745										

To make the computations easier, the lifetimes will be divided by 100, and statistical inference will not be affected. Amleh and Raqab [1] showed that Lomax CEM as a suitable model for fitting this data. Moreover, the true CDF of the lifetimes and the corresponding empirical CDF are plotted in Fig. 1.

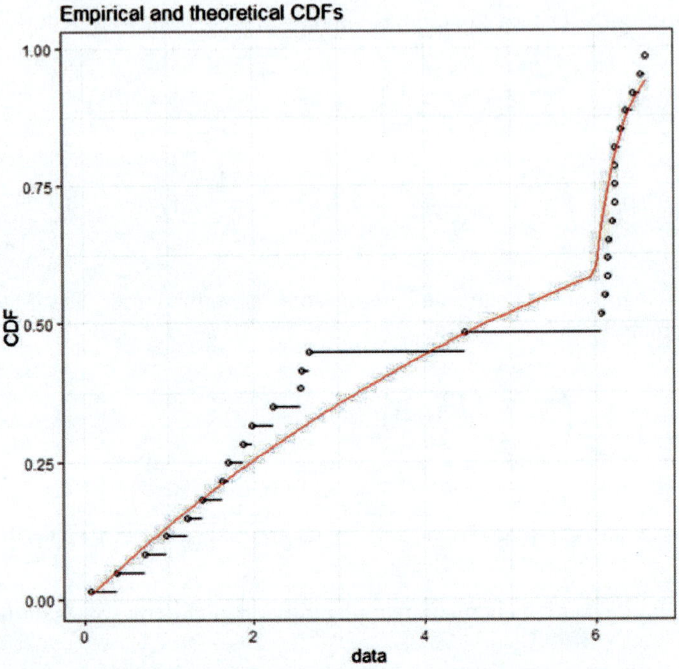

Fig. 1 The estimated CDF based on MLE (solid line) and the empirical CDF (dots)

Assume that the test will be terminated when the 30-th lifetime is observed. So, a Type-II censored sample will be observed, we have $n = 40, r = 30$. The problem is to find the point predictors of the unobserved lifetimes $Y = T_{s:n}$, $s = 31, 32, 33, 34, 35$.

The MLEs of the parameters β, $\alpha_1$andα_2 were obtained to be $\widehat{\beta} = 1.7517$, $\widehat{\alpha}_1 = 17.2768$ and $\widehat{\alpha}_2 = 0.7364$, see Amleh and Raqab [1]. The computations of the point predictors are reported in Table 1. It is noticed that the values of the point predictors are close to the true values. Further, it can be observed that the BUP and the CMP have a clear advantage if s is close to r, while the MLP is closer to the true values when s is close to n. In fact, the BUP and the CMP are computationally attractive when compared to the MLP.

Table 1 Point predictors of future lifetimes $Y = T_{s:n}$

s	True value	MLP	CMP	BUP
31	6.73	6.600	6.664	6.696
32	6.75	6.669	6.769	6.810
33	6.80	6.743	6.896	6.949
34	6.84	6.825	7.053	7.121
35	6.92	6.922	7.251	7.343

References

1. Amleh, M.A., Raqab, M.Z.: Inference in simple step-stress accelerated life tests for Type-II censoring Lomax data. J. Stat. Theory Appl. **20**(2), 364–379 (2021)
2. Amleh, M.A., Raqab, M.Z.: Prediction of censored Weibull lifetimes in a simple step-stress plan with Khamis-Higgins model. Stat. Optim. Inf. Comput. (2021). https://doi.org/10.19139/soic-2310-5070-1069
3. Amleh, M.A.: Prediction of future lifetimes for a simple step-stress model with Type-II censoring and Rayleigh distribution. WSEAS Trans. Math. **21**, 131–143 (2022)
4. Balakrishnan, N., Zhang, L., Xie, Q.: Inference for a simple step-stress model with Type-I censoring and lognormally distributed lifetimes. Commun. Stat.-Theory Methods **38**(10), 1690–1709 (2009)
5. Basak, I.: Prediction of times to failure of censored items for a simple step-stress model with regular and progressive Type-I censoring from the exponential distribution. Commun. Stat.-Theory Methods **43**(10), 2322–2341 (2014)
6. Basak, I., Balakrishnan, N.: Prediction of censored exponential lifetimes in a simple step-stress model under progressive Type-II censoring. Comput. Stat. **32**(4), 1665–1687 (2016)
7. Bryson, M.C.: Heavy-tailed distributions: properties and tests. Technometrics **16**(1), 61–68 (1974)
8. Chandra, N., Khan, M.A.: Optimum plan for step-stress accelerated life testing model under Type-I censored samples. J. Mod. Math. Stat. **7**(5), 58–62 (2013)
9. Hassan, A.S., Assar, S.M., Shelbaia, A.: Optimum step-stress accelerated life test plan for Lomax distribution with an adaptive Type-II progressive hybrid censoring. J. Adv. Math. Comput. Sci. **13**(2), 1–19 (2016)
10. Kamal, M., Zarrin, S., Islam, A.U.: Step stress accelerated life testing for two parameters Pareto distribution. J. Reliab. Theory Appl. **8**, 30–40 (2013)
11. Kaminsky, K.S., Rhodin, L.S.: Maximum likelihood prediction. Ann. Inst. Stat. Math. **37**(3), 507–517 (1985)
12. Kateri, M., Balakrishnan, N.: Inference for a simple step-stress model with Type-II censoring and Weibull distributed lifetimes. IEEE Trans. Reliab. **57**, 616–626 (2008)
13. Kundu, D., Ganguly, A.: Analysis of Step-Stress Models: Existing Results and Some Recent Developments. Academic Press, London (2017)
14. Liu, X.: Bayesian designing and analysis of simple step-stress accelerated life test with Weibull lifetime distribution, Unpublished thesis, the faculty of the Russ College of Engineering and Technology of Ohio University, USA (2010)
15. Lomax, K.S.: Business failures: another example of the analysis of failure data. J. Am. Stat. Assoc. **49**, 847–852 (1954)
16. Nelson, W.: Accelerated life testing-step-stress models and data analyses. IEEE Trans. Reliab. **29**(2), 103–108 (1980)
17. Nelson, W.B.: Accelerated Testing: Statistical Models, Test Plans, Data Analyses. Wiley, New York (1990)
18. Raqab, M.Z., Nagaraja, H.N.: On some predictors of future order statistics. Metron **53**(12), 185–204 (1995)
19. Xiong, C.: Inferences on a simple step-stress model with Type II censored exponential data. IEEE Trans. Reliab. **55**, 67–74 (1998)

Infra Soft β-Open Sets and Their Applications on Infra Soft Topological Spaces

Tareq M. Al-shami and Radwan Abu-Gdairi

Abstract The aim of writing this article is to present the concept of infra soft β-open sets as a new class of generalizations of infra open sets. We first investigate their main properties and study their behaviours under the product of soft spaces and some soft maps. Then, we establish some soft operators such as interior, closure, limit and boundary using infra soft β-open and infra soft β-closed sets. The relationships between them are illustrated and the main features are discussed. Finally, we display some soft maps defined using infra soft β-open and infra soft β-closed sets and scrutinize their master properties.

Keywords Infra soft β-open set · Infra soft β-interior points · Infra soft β-closure points · Infra soft β-continuity

1 Introduction

Molodtsov [57], in 1999, proposed the idea of a soft set as a new mathematical tool to deal with vagueness. He presented some of its applications in some areas. Since the advent of soft sets, they have been applied to address some problems and phenomena in different disciplines such as information system [10], economy [11], linear equations [27], computer science [45] and decision-making problems [48].

Maji et al. [56], in 2003, put forward the main concepts via soft set theory such as the difference, union and intersection operators and a complement of a soft set. To improve these concepts and cancel shortcomings that appeared in these definitions, new versions of these operations and operators were proposed by Ali et al. [3]. Keeping some classical properties via soft set theory was the major goal of [27]. To

T. M. Al-shami (✉)
Department of Mathematics, Sana'a University, P.O. Box 1247, Sanaa, Yemen
e-mail: tareqalshami83@gmail.com

R. Abu-Gdairi
Department of Mathematics, Faculty of Science, Zarqa University, P.O. Box 13110, Zarqa, Jordan
e-mail: rgdairi@zu.edu.jo

D. Zeidan et al. (eds.), *Mathematics and Computation*, Springer Proceedings in Mathematics & Statistics 418, https://doi.org/10.1007/978-981-99-0447-1_35

expand the applications of soft sets, new extensions of soft sets like bipolar soft sets [9] and double-framed soft sets were introduced [34].

As is well known, topology is a novel type of geometry that relies on the neighbourhoods of points instead of measuring the distance between them. Recently, topology has been applied to model some real-life issues as shown in [1, 16, 17, 32, 51, 59]. To study topology via soft set theory, Çağman et al. [46] and Shabir and Naz [60], in 2011, introduced the concept of soft topology. They followed different techniques for studying soft topology. This article follows Shabir and Naz's technique which is defined as a soft topology over the universal set and a fixed set of parameters. The basic concepts and notions of classical topology have been studied in soft topology such as calibre and chain conditions [2], compactness [4, 24, 28, 29, 36, 44], separation axioms [18, 22, 42], fixed point theorem [12, 20], connectedness [49, 52, 54], mappings [21, 26, 30, 53], bioperators [43], covering properties [38, 39, 55], sum of topologies [31, 40] and generalized open sets [5]. Additionally, soft topologies and supra soft topologies were discussed in ordered settings as given in [25]. Al-shami and Kočinac [37] elucidated the conditions under which the soft operators and classical operators of interior and closure are interchangeable. It should be noted that some classical topological properties were generalized to soft topologies without consideration for the divergences between soft topologies and classical topologies, which causes some incorrect forms of some results; so some articles were conducted to put forward the correct frame of these results via soft structures; see [6–8].

The structure of infra soft topologies [13] is one of the recent generalizations of soft topologies. The main advantages of continuously investigating infra soft topologies are the following: (1) many classical topological properties are true in the infra soft topological spaces; (2) easily obtaining the examples that show the interrelations among the different concepts. These advantages are illustrated for the two main topological concepts "compactness and connectedness" in [14, 19]. Also, the concepts of homeomorphisms [15] and separation axioms [33, 35] were introduced in the frame of infra soft topologies. Extensions of infra soft open sets were a goal of some papers. Some types of these extensions were investigated such as infra soft semi-open [23] and infra soft pre-open sets [41]. This manuscript aims to familiarize the notion of infra soft β-open sets as a new extension of infra soft open sets. To confirm that infra soft topology offers a flexible frame to discuss the topological concepts and reveal the relationships between them, we show that several characterizations and properties of soft β-open sets are kept for infra soft β-open sets.

2 Preliminaries

2.1 *Soft Set Theory*

Definition 1 ([57]) The ordered pair $(\mathcal{H}, \mathcal{O})$ is called a soft set over X if $\mathcal{H} : \mathcal{O} \to 2^X$ is a map, where $\mathcal{O}$ denotes a parameter set and 2^X the power set of X.

A soft set is expressed as $(\mathcal{H}, \mathcal{O}) = \{(o, \mathcal{H}(o)) : o \in \mathcal{O} \text{ and } \mathcal{H}(o) \in 2^X\}$.

We used the symbol $C(X_{\mathcal{O}})$ to refer a class of all soft sets over X with $\mathcal{O}$.

Definition 2 ([3]) A soft set $(\mathcal{H}^c, \mathcal{O})$ is called a complement of $(\mathcal{H}, \mathcal{O})$ if $\mathcal{H}^c(o) = X \setminus \mathcal{H}(o)$ for every $o \in \mathcal{O}$.

Definition 3 ([56]) Let $(\mathcal{H}, \mathcal{O})$ be a soft set on X such that $\mathcal{H}(o) = X$ (resp., $\mathcal{H}(o) = \emptyset$) for all $o \in \mathcal{O}$. Then we say that $(\mathcal{H}, \mathcal{O})$ is an absolute (resp., a null) soft set.

The absolute and null soft sets are denoted by Φ and $\widetilde{X}$, respectively.

Definition 4 ([47, 58]) We call $(\mathcal{H}, \mathcal{O})$ a countable (resp. finite) soft set if all components are countable (resp., finite). Otherwise, it is called uncountable (resp. infinite).

Definition 5 ([58]) A soft point on X is a soft set $(\mathcal{H}, \mathcal{O})$ such that $\mathcal{H}(o) = x \in X$ and $\mathcal{H}(o') = \emptyset$ for all $o' \neq o$. It is denoted by δ_o^x.

Definition 6 ([3]) The intersection of soft sets $(\mathcal{H}, \mathcal{O})$ and $(\mathcal{F}, \Delta)$ on X, symbolized by $(\mathcal{H}, \mathcal{O})\widetilde{\bigcap}(\mathcal{F}, \Delta)$, is a soft set $(\mathcal{G}, T)$, where $T = \mathcal{O} \cap \Delta \neq \emptyset$, and a map $\mathcal{G} : T \to 2^X$ is given by $\mathcal{G}(o) = \mathcal{H}(o) \cap \mathcal{F}(o)$ for each $o \in T$.

Definition 7 ([56]) The union of soft sets $(\mathcal{H}, \mathcal{O})$ and $(\mathcal{F}, \Delta)$ on X, symbolized by $(\mathcal{H}, \mathcal{O})\widetilde{\bigcup}(\mathcal{F}, \Delta)$, is a soft set $(\mathcal{G}, T)$, where $T = \mathcal{O} \cup \Delta$ and a map $T : \mathcal{O} \to 2^X$ is given as follows:

$$\mathcal{G}(o) = \begin{cases} \mathcal{H}(o) & : \ o \in \mathcal{O} \setminus \Delta \\ \mathcal{F}(o) & : \ o \in \Delta \setminus \mathcal{O} \\ \mathcal{H}(o) \cup \mathcal{F}(o) & : \ o \in \mathcal{O} \cap \Delta \end{cases}$$

Definition 8 ([50]) A soft set $(\mathcal{H}, \mathcal{O})$ is a subset of a soft set $(\mathcal{F}, \Delta)$, symbolized by $(\mathcal{H}, \mathcal{O})\widetilde{\subseteq}(\mathcal{F}, \Delta)$, if $\mathcal{O} \subseteq \Delta$ and $\mathcal{H}(o) \subseteq \mathcal{F}(o)$ for all $o \in \mathcal{O}$. If $(\mathcal{H}, \mathcal{O})\widetilde{\subseteq}(\mathcal{F}, \Delta)$ and $(\mathcal{F}, \Delta)\widetilde{\subseteq}(\mathcal{H}, \mathcal{O})$, then $(\mathcal{H}, \mathcal{O})$ and $(\mathcal{F}, \Delta)$ are called soft equal.

Definition 9 ([44]) The Cartesian product of $(\mathcal{H}, \mathcal{O})$ and $(\mathcal{F}, \Delta)$, symbolized by $(\mathcal{H} \times \mathcal{F}, \mathcal{O} \times \Delta)$, is defined as $(\mathcal{H} \times \mathcal{F})(o, o') = \mathcal{H}(o) \times \mathcal{F}(o')$ for each $(o, o') \in \mathcal{O} \times \Delta$.

The soft maps definition displayed in [53] was redefined to reduce calculation burden and give a logical explanation for the followed manner of defining some concepts via soft settings such as why we define an injective, or surjective soft map f_ψ according to its classical maps f and ψ.

Definition 10 ([15]) Let $f : X \to \mathcal{S}$ and $\psi : \mathcal{O} \to \Delta$ be two crisp maps. A soft map f_ψ of $C(X_{\mathcal{O}})$ into $C(\mathcal{S}_\Delta)$ is a relation such that each soft point in $C(X_{\mathcal{O}})$ is related to one and only one soft point in $C(\mathcal{S}_\Delta)$ such that

$$f_\psi(\delta_o^x) = \delta_{\psi(o)}^{f(x)} \text{ for each } \delta_o^x \in C(X_{\mathcal{O}}).$$

In addition, $f_\psi^{-1}(\delta_\gamma^y) = \bigsqcup_{\substack{\lambda \in \psi^{-1}(\gamma) \\ x \in f^{-1}(y)}} \delta_\lambda^x$ for each $\delta_\gamma^y \in C(\mathcal{S}_\Delta)$.

Definition 11 ([58]) $f_\psi : C(X_{\mathcal{O}}) \to C(\mathcal{S}_\Delta)$ is called surjective (resp., injective, bijective) if f and ψ are surjective (resp., injective, bijective).

2.2 Infra Soft Topological Spaces

Definition 12 ([13]) We called a subfamily μ of $C(X_{\mathcal{O}})$ an infra soft topology on X if it is closed under finite intersection and Φ is a member of μ.

An infra soft topological space (in short, ISTS) is the triple $(X, \mu, \mathcal{O})$. A soft set is called an infra soft open set if it belongs to μ and the complement of infra soft open is called an infra soft closed set.

Definition 13 ([13]) Let $(\mathcal{H}, \mathcal{O})$ be a subset of $(X, \mu, \mathcal{O})$.

(i) The infra soft closure of $(\mathcal{H}, \mathcal{O})$, denoted by $Cl(\mathcal{H}, \mathcal{O})$, is the intersection of all infra soft closed supersets of $(\mathcal{H}, \mathcal{O})$.
(ii) The infra soft interior of $(\mathcal{H}, \mathcal{O})$, denoted by $Int(\mathcal{H}, \mathcal{O})$, is the union of all infra soft open subsets of $(\mathcal{H}, \mathcal{O})$.

Proposition 1 *([13]) Let $(\mathcal{H}, \mathcal{O})$ and $(\mathcal{F}, \mathcal{O})$ be subsets of an ISTS $(X, \mu, \mathcal{O})$. Then*

(i) $Cl[(\mathcal{H}, \mathcal{O})\widetilde{\bigcup}(\mathcal{F}, \mathcal{O})] = Cl(\mathcal{H}, \mathcal{O})\widetilde{\bigcup}Cl(\mathcal{F}, \mathcal{O})$, *and*
(ii) $Int[(\mathcal{H}, \mathcal{O})\widetilde{\bigcap}(\mathcal{F}, \mathcal{O})] = Int(\mathcal{H}, \mathcal{O})\widetilde{\bigcap}Int(\mathcal{F}, \mathcal{O})$.

Proposition 2 *([13]) Let $(\mathcal{H}, \mathcal{O})$ be an infra soft open set. Then*

$$(\mathcal{H}, \mathcal{O})\widetilde{\bigcap}Cl(\mathcal{F}, \mathcal{O})\widetilde{\subseteq}Cl[(\mathcal{H}, \mathcal{O})\widetilde{\bigcup}(\mathcal{F}, \mathcal{O})] \textit{ for any subset } (\mathcal{F}, \mathcal{O}) \textit{ of } (X, \mu, \mathcal{O})$$

Proposition 3 *([13]) Let $(\mathcal{H}, \mathcal{O})$ be an infra soft closed set. Then*

$$Int[(\mathcal{H}, \mathcal{O})\widetilde{\bigcup}(\mathcal{F}, \mathcal{O})]\widetilde{\subseteq}(\mathcal{H}, \mathcal{O})\widetilde{\bigcup}Int(\mathcal{F}, \mathcal{O}) \textit{ for any subset } (\mathcal{H}, \mathcal{O}) \textit{ of } (X, \mu, \mathcal{O})$$

Definition 14 ([15]) A soft map which is bijective, infra soft continuous and infra soft open is called an infra soft homeomorphism.

We call a property which is kept by any infra soft homeomorphism an infra soft topological property (in short, IST property).

Definition 15 ([15]) Let $f_\psi : (X, \mu, \mathcal{O}) \to (\mathcal{S}, \nu, \Delta)$ be a soft map and $\mathcal{M} \neq \emptyset$ be a subset of X. A soft map $f_{\psi|\mathcal{M}} : (\mathcal{M}, \mu_{\mathcal{M}}, \mathcal{O}) \to (\mathcal{S}, \nu, \Delta)$ which given by $f_{\psi|\mathcal{M}}(\delta_o^m) = f_\psi(\delta_o^m)$ for every $\delta_o^m \in \widetilde{\mathcal{M}}$ is called a restriction soft map of f_ψ on $\mathcal{M}$.

Lemma 1 *Let $f_\psi : (X_1, \mu_1, \mathcal{O}_1) \to (X_2, \mu_2, \mathcal{O}_2)$ be an infra soft homeomorphism map. Then for any subset $(\mathcal{H}, \mathcal{O}_1)$ we have the next two results.*

(i) $f_\psi(Int(\mathcal{H}, \mathcal{O}_1)) = Int(f_\psi(\mathcal{H}, \mathcal{O}_1))$.
(ii) $f_\psi(Cl(\mathcal{H}, \mathcal{O}_1)) = Cl(f_\psi(\mathcal{H}, \mathcal{O}_1))$.

Lemma 2 ([23, 41]) *Consider $(\mathcal{H}_1, \mathcal{O}_1)$ and $(\mathcal{H}_2, \mathcal{O}_2)$ as subsets of $(X_1, \mu_1, \mathcal{O}_1)$ and $(X_2, \mu_2, \mathcal{O}_2)$, respectively. Then*

(i) $Cl[(\mathcal{H}_1, \mathcal{O}_1) \times (\mathcal{H}_2, \mathcal{O}_2)] = Cl(\mathcal{H}_1, \mathcal{O}_1) \times Cl(\mathcal{H}_2, \mathcal{O}_2)$.
(ii) $Int[(\mathcal{H}_1, \mathcal{O}_1) \times (\mathcal{H}_2, \mathcal{O}_2)] = Int(\mathcal{H}_1, \mathcal{O}_1) \times Int(\mathcal{H}_2, \mathcal{O}_2)$.

Proposition 4 ([14]) *Let $\{(X_k, \mu_k, \mathcal{O}_k) : k \in K\}$ be a family of ISTSs. Then $\boldsymbol{\mu} = \{\prod_{k\in K}(o_k, \mathcal{O}_k) : (o_k, \mathcal{O}_k) \in \psi_k\}$ is an infra soft topology on $\boldsymbol{T} = \prod_{k\in K} X_k$ under a set of parameters $\boldsymbol{B} = \prod_{k\in K} \mathcal{O}_k$.*

We call $\boldsymbol{\mu}$, given in proposition above, a product of infra soft topologies, and $(\mathbf{T}, \boldsymbol{\mu}, \mathbf{B})$ a product of infra soft spaces.

3 Main Properties of Infra Soft β-Open Sets

Definition 16 A subset $(\mathcal{H}, \mathcal{O})$ of an ISTS $(X, \mu, \mathcal{O})$ is called infra soft β-open if $(\mathcal{H}, \mathcal{O}) \widetilde{\subseteq} Cl(Int(Cl(\mathcal{H}, \mathcal{O})))$. Its complement is called infra soft β-closed.

Proposition 5 *Every infra soft semi-open (infra soft pre-open) set is an infra soft β-open.*

Proof Let $(\mathcal{H}, \mathcal{O})$ be an infra soft semi-open (resp. infra soft pre-open) set. Then, $(\mathcal{H}, \mathcal{O}) \widetilde{\subseteq} Cl(Int(\mathcal{H}, \mathcal{O}))$ (resp. $(\mathcal{H}, \mathcal{O}) \widetilde{\subseteq} Int(Cl(\mathcal{H}, \mathcal{O})))$. Automatically, we obtain $(\mathcal{H}, \mathcal{O}) \widetilde{\subseteq} Cl(Int(Cl(\mathcal{H}, \mathcal{O})))$, which means that $(\mathcal{H}, \mathcal{O})$ is infra soft β-open.

The next example shows that the converse of the above proposition fails.

Example 1 Let $X = \{x_1, x_2, x_3\}$ and $\mathcal{O} = \{o_1, o_2\}$. Then $\mu = \{\Phi, \widetilde{X}, (\mathcal{H}_1, \mathcal{O}), (\mathcal{H}_2, \mathcal{O})\}$ is an IST on X with $\mathcal{O}$, where

$(\mathcal{H}_1, \mathcal{O}) = \{(o_1, \{x_1\}), (o_2, \{x_2, x_3\})\}$ and
$(\mathcal{H}_2, \mathcal{O}) = \{(o_1, \{x_3\}), (o_2, \{x_1\})\}$.

Let $(\mathcal{H}_5, \mathcal{O}) = \{(o_1, \{x_3\}), (o_2, \{x_3\})\}$ and $(\mathcal{H}_6, \mathcal{O}) = \{(o_1, \{x_1, x_2\}), (o_2, \{x_2, x_3\})\}$. Since $Cl(\mathcal{H}_5, \mathcal{O}) = \widetilde{X}$ and $Cl(Int(Cl(\mathcal{H}_6, \mathcal{O}))) = (\mathcal{H}_6, \mathcal{O})$, then $(\mathcal{H}_5, \mathcal{O})$ and $(\mathcal{H}_6, \mathcal{O})$ are infra soft β-open sets. On the other hand, we have $Int(\mathcal{H}_5, \mathcal{O}) = \Phi$ and $Int(Cl(\mathcal{H}_5, \mathcal{O})) = \{(o_1, \{x_1\}), (o_2, \{x_2, x_3\})\} \widetilde{\not\supseteq} (\mathcal{H}_6, \mathcal{O})$, which means that $(\mathcal{H}_5, \mathcal{O})$ and $(\mathcal{H}_6, \mathcal{O})$ are not infra soft semi-open and infra soft pre-open sets, respectively.

Proposition 6 *The arbitrary unions of infra soft β-open sets is infra soft β-open.*

Proof Let $\{(\mathcal{H}_j, \mathcal{O}) : j \in J\}$ be a class of infra soft β-open sets. Suppose that $J \neq \emptyset$. Then for each $j \in J$, we have $(\mathcal{H}_j, \mathcal{O}) \widetilde{\subseteq} Cl(Int(Cl(\mathcal{H}_j, \mathcal{O})))$. Therefore, $\widetilde{\bigcup}_{j\in J}(\mathcal{H}_j, \mathcal{O}) \widetilde{\subseteq} \widetilde{\bigcup}_{j\in J} Cl(Int(Cl(\mathcal{H}_j, \mathcal{O}))) \widetilde{\subseteq} Cl(Int(Cl(\widetilde{\bigcup}_{j\in J}(\mathcal{H}_j, \mathcal{O}))))$. Hence, $\widetilde{\bigcup}_{j\in J}(\mathcal{H}_j, \mathcal{O})$ is infra soft β-open.

Corollary 1 *The arbitrary intersection of infra soft β-closed sets is infra soft β-closed.*

Proposition 7 *The intersection of infra soft open and infra soft β-open sets is an infra soft β-open set.*

Proof Let $(\mathcal{H}_1, \mathcal{O})$ be an infra soft open set and $(\mathcal{H}_2, \mathcal{O})$ be an infra soft β-open set. Then $(\mathcal{H}_1, \mathcal{O})\widetilde{\bigcap}(\mathcal{H}_2, \mathcal{O})\widetilde{\subseteq}(\mathcal{H}_1, \mathcal{O})\widetilde{\bigcap}Cl(Int(Cl(\mathcal{H}_2, \mathcal{O})))$. It follows from Proposition 2 that $(\mathcal{H}_1, \mathcal{O})\widetilde{\bigcap}Cl(Int(Cl(\mathcal{H}_2, \mathcal{O})))\widetilde{\subseteq}Cl[(\mathcal{H}_1, \mathcal{O})\widetilde{\bigcap}Int(Cl(\mathcal{H}_2, \mathcal{O}))]$ $= Cl(Int[(\mathcal{H}_1, \mathcal{O})\widetilde{\bigcap}Cl(\mathcal{H}_2, \mathcal{O})]\widetilde{\subseteq}$ $Cl(Int(Cl[(\mathcal{H}_1, \mathcal{O})\widetilde{\bigcap}(\mathcal{H}_2, \mathcal{O})]))$. Hence, $(\mathcal{H}_1, \mathcal{O})\widetilde{\bigcap}(\mathcal{H}_2, \mathcal{O})$ is an infra soft β-open set.

Corollary 2 *The union of infra soft closed and infra soft β-closed sets is an infra soft β-closed set.*

Proposition 8 *The image of an infra soft β-open set is infra soft β-open under any infra soft homeomorphism.*

Proof Consider $f_\psi : (X_1, \mu_1, \mathcal{O}_1) \to (X_2, \mu_2, \mathcal{O}_2)$ as an infra soft continuous map and let $(\mathcal{H}, \mathcal{O}_1)$ be an infra soft β-open subset of $(X_1, \mu_1, \mathcal{O}_1)$. Then $f_\psi(\mathcal{H}, \mathcal{O}_1)\widetilde{\subseteq}f_\psi(Cl(Int(Cl(\mathcal{H}, \mathcal{O}_1))))$. It follows from Lemma 1 that $f_\psi(\mathcal{H}, \mathcal{O}_1)\widetilde{\subseteq}$ $Cl(Int(Cl(f_\psi(\mathcal{H}, \mathcal{O}_1))))$. Hence, $f_\psi(\mathcal{H}, \mathcal{O}_1)$ is an infra soft β-open subset of $(X_2, \mu_2, \mathcal{O}_2)$, as required.

Proposition 9 *The product of infra soft β-open sets is an infra soft β-open set.*

Proof Let $(\mathcal{H}_1, \mathcal{O}_1)$ and $(\mathcal{H}_2, \mathcal{O}_2)$ be infra soft β-open subsets of $(X_1, \mu_1, \mathcal{O}_1)$ and $(X_2, \mu_2, \mathcal{O}_2)$, respectively. Then $(\mathcal{H}_1, \mathcal{O}_1) \times (\mathcal{H}_2, \mathcal{O}_2)\widetilde{\subseteq}Cl(Int(Cl(\mathcal{H}_1, \mathcal{O}_1))) \times$ $Cl(Int(Cl(\mathcal{H}_2, \mathcal{O}_2)))$. According to Lemma 2, we obtain $(\mathcal{H}_1, \mathcal{O}_1) \times (\mathcal{H}_2, \mathcal{O}_2)\widetilde{\subseteq}$ $Cl(Int(Cl[(\mathcal{H}_1, \mathcal{O}_1) \times (\mathcal{H}_2, \mathcal{O}_2)]))$ which means that $(\mathcal{H}_1, \mathcal{O}_1) \times (\mathcal{H}_2, \mathcal{O}_2)$ is an infra soft β-open subset of $\widetilde{X_1} \times \widetilde{X_2}$.

4 Infra β-Interior, Infra β-Closure, Infra β-Limit and Infra β-Boundary Soft Points of a Soft Set

Definition 17 Let $(\mathcal{H}, \mathcal{O})$ be a subset of $(X, \mu, \mathcal{O})$. Then

(i) the infra soft β-interior of $(\mathcal{H}, \mathcal{O})$, denoted by $\beta Int(\mathcal{H}, \mathcal{O})$, is the union of all infra soft β-open subsets of $(\mathcal{H}, \mathcal{O})$.
(ii) the infra soft β-closure of $(\mathcal{H}, \mathcal{O})$, denoted by $\beta Cl(\mathcal{H}, \mathcal{O})$, is the intersection of all infra soft β-closed supersets of $(\mathcal{H}, \mathcal{O})$.

Proposition 10 *We have the following properties:*

(i) *$(\mathcal{H}, \mathcal{O})$ is an infra soft β-open subset of $(X, \mu, \mathcal{O})$ iff $\beta Int(\mathcal{H}, \mathcal{O}) = (\mathcal{H}, \mathcal{O})$.*
(ii) *$(\mathcal{H}, \mathcal{O})$ is an infra soft β-closed subset of $(X, \mu, \mathcal{O})$ iff $\beta Cl(\mathcal{H}, \mathcal{O}) = (\mathcal{H}, \mathcal{O})$.*

Proof It comes from Proposition 6 and Corollary 1.

The two characterizations given in the above proposition are generally false for infra soft open and infra soft closed sets.

Proposition 11 *Let $(\mathcal{H}, \mathcal{O})$ be a subset of $(X, \mu, \mathcal{O})$.*

(i) $\delta_o^x \in \beta Int(\mathcal{H}, \mathcal{O})$ *iff there is an infra soft β-open set $(\mathcal{F}, \mathcal{O})$ such that* $\delta_o^x \in (\mathcal{F}, \mathcal{O}) \widetilde{\subseteq} (\mathcal{H}, \mathcal{O})$.
(ii) $\delta_o^x \in \beta Cl(\mathcal{H}, \mathcal{O})$ *iff the intersection of any infra soft β-open set $(\mathcal{F}, \mathcal{O})$ containing δ_o^x and $(\mathcal{H}, \mathcal{O})$ is non-null.*

Proof The proof of (i) is obvious, so we prove (ii).

Let $\delta_o^x \in \beta Cl(\mathcal{H}, \mathcal{O})$. Then every infra soft β-closed set contains $(\mathcal{H}, \mathcal{O})$ contains δ_o^x as well. Suppose that there is an infra soft β-open set $(\mathcal{F}, \mathcal{O})$ containing δ_o^x such that $(\mathcal{H}, \mathcal{O}) \widetilde{\bigcap} (\mathcal{F}, \mathcal{O}) = \Phi$. Therefore, $(\mathcal{H}, \mathcal{O}) \widetilde{\subseteq} (\mathcal{F}^c, \mathcal{O})$ which means that $\delta_o^x \notin \beta Cl(\mathcal{H}, \mathcal{O})$. This is a contradiction. Conversely, suppose that there is an infra soft β-open set $(\mathcal{F}, \mathcal{O})$ containing δ_o^x such that $(\mathcal{H}, \mathcal{O}) \widetilde{\bigcap} (\mathcal{F}, \mathcal{O}) = \Phi$. Therefore, $\beta Cl(\mathcal{H}, \mathcal{O}) \widetilde{\subseteq} (\mathcal{F}^c, \mathcal{O})$ which means that $\delta_o^x \notin \beta Cl(\mathcal{H}, \mathcal{O})$. Hence, we obtain the desired result.

Proposition 12 *Let $(\mathcal{H}, \mathcal{O})$ be a subset of $(X, \mu, \mathcal{O})$. Then*

(i) $(\beta Int(\mathcal{H}, \mathcal{O}))^c = \beta Cl(\mathcal{H}^c, \mathcal{O})$.
(ii) $(\beta Cl(\mathcal{H}, \mathcal{O}))^c = \beta Int(\mathcal{H}^c, \mathcal{O})$.

Proof (i): $(\beta Int(\mathcal{H}, \mathcal{O}))^c = \{\widetilde{\bigcup}_{j \in J} (\mathcal{F}_j, \mathcal{O}) : (\mathcal{F}_j, \mathcal{O})$ is an infra soft β-open set contained in $(\mathcal{H}, \mathcal{O})\}^c = \widetilde{\bigcap}_{j \in J} \{(\mathcal{F}_j^c, \mathcal{O}) : (\mathcal{F}_j^c, \mathcal{O})$ is an infra soft β-closed set containing $(\mathcal{H}^c, \mathcal{O})\} = \beta Cl(\mathcal{H}^c, \mathcal{O})$.

The proof of (ii) is similar to (i).

Proposition 13 *Let $(\mathcal{F}, \mathcal{O})$ be an infra soft open set and $(\Lambda, \mathcal{O})$ be an infra soft closed set in $(X, \mu, \mathcal{O})$. Then*

(i) $(\mathcal{F}, \mathcal{O}) \widetilde{\bigcap} \beta Cl(\mathcal{H}, \mathcal{O}) \widetilde{\subseteq} \beta Cl((\mathcal{F}, \mathcal{O}) \widetilde{\bigcap} (\mathcal{H}, \mathcal{O}))$.
(ii) $\beta Int((\Lambda, \mathcal{O}) \widetilde{\bigcup} (\mathcal{H}, \mathcal{O})) \widetilde{\subseteq} (\Lambda, \mathcal{O}) \widetilde{\bigcup} \beta Int(\mathcal{H}, \mathcal{O})$.

Proof (i): Let $\delta_o^x \in (\mathcal{F}, \mathcal{O}) \widetilde{\bigcap} \beta Cl(\mathcal{H}, \mathcal{O})$. Then $\delta_o^x \in (\mathcal{F}, \mathcal{O})$ and $\delta_o^x \in \beta Cl(\mathcal{H}, \mathcal{O})$. This implies that $(\mathcal{U}, \mathcal{O}) \widetilde{\bigcap} (\mathcal{H}, \mathcal{O}) \neq \Phi$ for every infra soft β-open set $(\mathcal{U}, \mathcal{O})$ containing δ_o^x. It follows from Proposition 7 that $(\mathcal{F}, \mathcal{O}) \widetilde{\bigcap} (\mathcal{U}, \mathcal{O})$ is an infra soft β-open set containing δ_o^x. Therefore, $[(\mathcal{F}, \mathcal{O}) \widetilde{\bigcap} (\mathcal{U}, \mathcal{O})] \widetilde{\bigcap} (\mathcal{H}, \mathcal{O}) \neq \Phi$. Now, $(\mathcal{U}, \mathcal{O}) \widetilde{\bigcap} [(\mathcal{F}, \mathcal{O}) \widetilde{\bigcap} (\mathcal{H}, \mathcal{O})] \neq \Phi$ which means that $\delta_o^x \in \beta Cl((\mathcal{F}, \mathcal{O}) \widetilde{\bigcap} (\mathcal{H}, \mathcal{O}))$. Hence, $(\mathcal{F}, \mathcal{O}) \widetilde{\bigcap} \beta Cl(\mathcal{H}, \mathcal{O}) \widetilde{\subseteq} \beta Cl((\mathcal{F}, \mathcal{O}) \widetilde{\bigcap} (\mathcal{H}, \mathcal{O}))$.

One can prove (ii) following similar arguments.

Theorem 1 *Let $(\mathcal{H}, \mathcal{O})$ and $(\mathcal{F}, \mathcal{O})$ be subsets of $(X, \mu, \mathcal{O})$. Then*

(i) $\beta Int(\widetilde{X}) = \widetilde{X}$.
(ii) $\beta Int(\mathcal{H}, \mathcal{O}) \widetilde{\subseteq} (\mathcal{H}, \mathcal{O})$.

(iii) *If* $(\mathcal{F}, \mathcal{O})\widetilde{\subseteq}(\mathcal{H}, \mathcal{O})$, *then* $\beta Int(\mathcal{F}, \mathcal{O})\widetilde{\subseteq}\beta Int(\mathcal{H}, \mathcal{O})$.
(iv) $\beta Int(\beta Int(\mathcal{H}, \mathcal{O})) = \beta Int(\mathcal{H}, \mathcal{O})$.
(v) $\beta Int(\mathcal{F}, \mathcal{O})\widetilde{\bigcap}\beta Int(\mathcal{H}, \mathcal{O})\widetilde{\subseteq}\beta Int((\mathcal{F}, \mathcal{O})\widetilde{\bigcap}(\mathcal{H}, \mathcal{O}))$.

Proof (i): Since $\widetilde{X}$ is infra soft β-open, $\beta Int(\widetilde{X}) = \widetilde{X}$.

(ii) and (iii) are obvious.

(iv): It is clear that $\beta Int(\beta Int(\mathcal{H}, \mathcal{O}))$ is the largest infra soft β-open set contained in $\beta Int(\mathcal{H}, \mathcal{O})$; however, $\beta Int(\mathcal{H}, \mathcal{O})$ is an infra soft β-open set; hence, $\beta Int(\beta Int(\mathcal{H}, \mathcal{O})) = \beta Int(\mathcal{H}, \mathcal{O})$.

(v): It comes from (iii).

Theorem 2 *Let* $(\mathcal{H}, \mathcal{O})$ *and* $(\mathcal{F}, \mathcal{O})$ *be subsets of* $(X, \mu, \mathcal{O})$. *Then*

(i) $\beta Cl(\Phi) = \Phi$.
(ii) $(\mathcal{H}, \mathcal{O})\widetilde{\subseteq}\beta Cl(\mathcal{H}, \mathcal{O})$.
(iii) *If* $(\mathcal{F}, \mathcal{O})\widetilde{\subseteq}(\mathcal{H}, \mathcal{O})$, *then* $\beta Cl(\mathcal{F}, \mathcal{O})\widetilde{\subseteq}\beta Cl(\mathcal{H}, \mathcal{O})$.
(iv) $\beta Cl(\beta Cl(\mathcal{H}, \mathcal{O}))\widetilde{\subseteq}\beta Cl(\mathcal{H}, \mathcal{O})$.
(v) $\beta Cl((\mathcal{F}, \mathcal{O})\widetilde{\bigcup}(\mathcal{H}, \mathcal{O})) = \beta Cl(\mathcal{F}, \mathcal{O})\widetilde{\bigcup}\beta Cl(\mathcal{H}, \mathcal{O})$.

Proof This is similar to Theorem 1.

Definition 18 A soft point δ_o^x is called an infra soft β-limit point of a subset $(\mathcal{H}, \mathcal{O})$ of $(X, \mu, \mathcal{O})$ provided that $[(\mathcal{F}, \mathcal{O})\backslash\delta_o^x]\widetilde{\bigcap}(\mathcal{H}, \mathcal{O}) \neq \Phi$ for every infra soft β-open set $(\mathcal{F}, \mathcal{O})$ containing δ_o^x.

The soft set of all infra soft β-limit points of $(\mathcal{H}, \mathcal{O})$ is called an infra β-derived soft set. It is denoted by $(\mathcal{H}, \mathcal{O})^{\beta s'}$.

Proposition 14 *Consider* $(\mathcal{F}, \mathcal{O})$ *and* $(\mathcal{H}, \mathcal{O})$ *as subsets of* $(X, \mu, \mathcal{O})$. *Then*

(i) $\Phi^{\beta s'} = \Phi$ *and* $\widetilde{X}^{\beta s'}\widetilde{\subseteq}\widetilde{X}$.
(ii) *If* $(\mathcal{F}, \mathcal{O})\widetilde{\subseteq}(\mathcal{H}, \mathcal{O})$, *then* $(\mathcal{F}, \mathcal{O})^{\beta s'}\widetilde{\subseteq}(\mathcal{H}, \mathcal{O})^{\beta s'}$.
(iii) *If* $\delta_o^x \in (\mathcal{H}, \mathcal{O})^{\beta s'}$, *then* $\delta_o^x \in ((\mathcal{H}, \mathcal{O}) \setminus \delta_o^x)^{\beta s'}$.
(iv) $(\mathcal{F}, \mathcal{O})^{\beta s'}\widetilde{\bigcup}(\mathcal{H}, \mathcal{O})^{\beta s'}\widetilde{\subseteq}((\mathcal{F}, \mathcal{O})\widetilde{\bigcup}(\mathcal{H}, \mathcal{O}))^{\beta s'}$.

Proof It is straightforward.

Theorem 3 *Let* $(\mathcal{H}, \mathcal{O})$ *be a subset of* $(X, \mu, \mathcal{O})$. *Then*

(i) *If* $(\mathcal{H}, \mathcal{O})$ *is an infra soft* β*-closed set, then* $(\mathcal{H}, \mathcal{O})^{\beta s'} \subseteq (\mathcal{H}, \mathcal{O})$.
(ii) $((\mathcal{H}, \mathcal{O})\widetilde{\bigcup}(\mathcal{H}, \mathcal{O})^{\beta s'})^{\beta s'}\widetilde{\subseteq}(\mathcal{H}, \mathcal{O})\widetilde{\bigcup}(\mathcal{H}, \mathcal{O})^{\beta s'}$.
(iii) $\beta Cl(\mathcal{H}, \mathcal{O}) = (\mathcal{H}, \mathcal{O})\widetilde{\bigcup}(\mathcal{H}, \mathcal{O})^{\beta s'}$.

Proof (i) Consider $(\mathcal{H}, \mathcal{O})$ as an infra soft β-closed set such that $\delta_o^x \notin (\mathcal{H}, \mathcal{O})$. Then $\delta_o^x \in (\mathcal{H}^c, \mathcal{O})$. Now, $(\mathcal{H}^c, \mathcal{O})$ is an infra soft β-open set such that $(\mathcal{H}^c, \mathcal{O})\widetilde{\bigcap}(\mathcal{H}, \mathcal{O}) = \Phi$ which means that $\delta_o^x \notin (\mathcal{H}, \mathcal{O})^{\beta s'}$. Thus, $(\mathcal{H}, \mathcal{O})^{\beta s'}\widetilde{\subseteq}(\mathcal{H}, \mathcal{O})$.

(ii) Consider $\delta_o^x \notin (\mathcal{H}, \mathcal{O})\widetilde{\bigcup}(\mathcal{H}, \mathcal{O})^{\beta s'}$. Then $\delta_o^x \notin (\mathcal{H}, \mathcal{O})$ and $\delta_o^x \notin (\mathcal{H}, \mathcal{O})^{\beta s'}$. Therefore, there is an infra soft β-open set $(\mathcal{F}, \mathcal{O})$ such that

$$(\mathcal{F}, \mathcal{O})\widetilde{\bigcap}(\mathcal{H}, \mathcal{O}) = \Phi \tag{1}$$

This implies that

$$(\mathcal{F}, \mathcal{O})\widetilde{\bigcap}(\mathcal{H}, \mathcal{O})^{\beta s'} = \Phi \tag{2}$$

It follows from (1) and (2) that $(\mathcal{F}, \mathcal{O})\widetilde{\bigcap}((\mathcal{H}, \mathcal{O})\widetilde{\bigcup}(\mathcal{H}, \mathcal{O})^{\beta s'}) = \Phi$. Thus, $\delta_o^x \notin ((\mathcal{H}, \mathcal{O})\widetilde{\bigcup}\ (\mathcal{H}, \mathcal{O})^{\beta s'})^{\beta s'}$. Hence, $((\mathcal{H}, \mathcal{O})\widetilde{\bigcup}(\mathcal{H}, \mathcal{O})^{\beta s'})^{\beta s'}\widetilde{\subseteq}((\mathcal{H}, \mathcal{O})\widetilde{\bigcup}(\mathcal{H}, \mathcal{O})^{\beta s'})$, as required.

(iii) It is clear that $(\mathcal{H}, \mathcal{O})\widetilde{\bigcup}(\mathcal{H}, \mathcal{O})^{\beta s'}\widetilde{\subseteq}\beta Cl(\mathcal{H}, \mathcal{O})$. Conversely, let $\delta_o^x \in \beta Cl(\mathcal{H}, \mathcal{O})$. Then for every infra soft β-open set containing δ_o^x, we have $(\mathcal{H}, \mathcal{O})\widetilde{\bigcap}(\mathcal{F}, \mathcal{O}) \neq \Phi$. Without loss of generality, let $\delta_o^x \notin (\mathcal{H}, \mathcal{O})$. Then $[(\mathcal{H}, \mathcal{O})\setminus\delta_o^x]\widetilde{\bigcap}(\mathcal{F}, \mathcal{O}) \neq \Phi$. Consequentially, $\delta_o^x \in (\mathcal{H}, \mathcal{O})^{\beta s'}$. Hence, the proof is complete.

Definition 19 The infra soft β-boundary points of a subset $(\mathcal{H}, \mathcal{O})$ of $(X, \mu, \mathcal{O})$, denoted by $\beta B(\mathcal{H}, \mathcal{O})$, are all the soft points which belong to the complement of $\beta Int(\mathcal{H}, \mathcal{O})\widetilde{\bigcup}\beta Int(\mathcal{H}^c, \mathcal{O})$.

Proposition 15 *Let $(\mathcal{H}, \mathcal{O})$ be a subset of $(X, \mu, \mathcal{O})$. Then*

(i) $\beta B(\mathcal{H}, \mathcal{O}) = \beta Cl(\mathcal{H}, \mathcal{O})\widetilde{\bigcap}\beta Cl((\mathcal{H}^c, \mathcal{O}))$.
(ii) $\beta B(\mathcal{H}, \mathcal{O}) = \beta Cl(\mathcal{H}, \mathcal{O}) \setminus \beta Int(\mathcal{H}, \mathcal{O})$.

Proof (i)
$$\begin{aligned}\beta B(\mathcal{H}, \mathcal{O}) &= \{\delta_o^x \in \widetilde{X} : \delta_o^x \notin \beta Int(\mathcal{H}, \mathcal{O}) \text{ and } \delta_o^x \notin \beta Int((\mathcal{H}^c, \mathcal{O}))\}\\ &= \{\delta_o^x \in \widetilde{X} : \delta_o^x \notin (\beta Cl(\mathcal{H}^c, \mathcal{O}))^c \text{ and } \delta_o^x \notin (\beta Cl(\mathcal{H}, \mathcal{O}))^c\}\\ &= \{\delta_o^x \in \widetilde{X} : \delta_o^x \in \beta Cl(\mathcal{H}^c, \mathcal{O}) \text{ and } \delta_o^x \in \beta Cl(\mathcal{H}, \mathcal{O})\}\\ &= \beta Cl(\mathcal{H}, \mathcal{O})\widetilde{\bigcap}\beta Cl(\mathcal{H}^c, \mathcal{O})\end{aligned}$$

(ii)
$$\begin{aligned}\beta B(\mathcal{H}, \mathcal{O}) &= \beta Cl(\mathcal{H}, \mathcal{O})\widetilde{\bigcap}\beta Cl(\mathcal{H}^c, \mathcal{O})\\ &= \beta Cl(\mathcal{H}, \mathcal{O})\widetilde{\bigcap}(\beta Int(\mathcal{H}, \mathcal{O}))^c\\ &= \beta Cl(\mathcal{H}, \mathcal{O}) \setminus \beta Int(\mathcal{H}, \mathcal{O}).\end{aligned}$$

Corollary 3 *Let $(\mathcal{H}, \mathcal{O})$ be a subset of $(X, \mu, \mathcal{O})$. Then*

(i) $\beta B(\mathcal{H}, \mathcal{O}) = \beta B(\mathcal{H}^c, \mathcal{O})$
(ii) $\beta Cl(\mathcal{H}, \mathcal{O}) = \beta Int(\mathcal{H}, \mathcal{O})\widetilde{\bigcup}\beta B(\mathcal{H}, \mathcal{O})$.

Proposition 16 *Let $(\mathcal{H}, \mathcal{O})$ be a subset of $(X, \mu, \mathcal{O})$. Then*

(i) $(\mathcal{H}, \mathcal{O})$ is infra soft β-open iff $\beta B(\mathcal{H}, \mathcal{O})\widetilde{\bigcap}(\mathcal{H}, \mathcal{O}) = \Phi$.
(ii) $(\mathcal{H}, \mathcal{O})$ is infra soft β-closed iff $\beta B(\mathcal{H}, \mathcal{O})\widetilde{\subseteq}(\mathcal{H}, \mathcal{O})$.

Proof (i) $\beta B(\mathcal{H}, \mathcal{O}) \bigcap (\mathcal{H}, \mathcal{O}) = \beta B(\mathcal{H}, \mathcal{O}) \bigcap \beta Int(\mathcal{H}, \mathcal{O}) = \Phi$. Conversely, let $\delta_o^x \in (\mathcal{H}, \mathcal{O})$. Then $\delta_o^x \in \beta Int(\mathcal{H}, \mathcal{O})$ or $\delta_o^x \in \beta B(\mathcal{H}, \mathcal{O})$. Since $\beta B(\mathcal{H}, \mathcal{O}) \bigcap (\mathcal{H}, \mathcal{O}) = \Phi, \delta_o^x \in \beta Int(\mathcal{H}, \mathcal{O})$. Thus, $(\mathcal{H}, \mathcal{O}) \subseteq \beta Int(\mathcal{H}, \mathcal{O})$ which means that $(\mathcal{H}, \mathcal{O}) = \beta Int(\mathcal{H}, \mathcal{O})$. Hence, $(\mathcal{H}, \mathcal{O})$ is infra soft β-open.

(ii) $(\mathcal{H}, \mathcal{O})$ is infra soft β-closed $\Leftrightarrow$ $(\mathcal{H}^c, \mathcal{O})$ is infra soft β-open $\Leftrightarrow$ $\beta B(\mathcal{H}^c, \mathcal{O}) \bigcap (\mathcal{H}^c, \mathcal{O}) = \Phi \Leftrightarrow \beta B(\mathcal{H}, \mathcal{O}) \bigcap (\mathcal{H}^c, \mathcal{O}) = \Phi \Leftrightarrow \beta B(\mathcal{H}, \mathcal{O}) \subseteq (\mathcal{H}, \mathcal{O})$.

Corollary 4 *A subset $(\mathcal{H}, \mathcal{O})$ of $(X, \mu, \mathcal{O})$ is infra soft β-open and infra soft β-closed iff $\beta B(\mathcal{H}, \mathcal{O}) = \Phi$.*

5 Infra Soft β-Homeomorphism Maps

Definition 20 A soft map $f_\psi : (X, \mu, \mathcal{O}) \to (\mathcal{S}, \nu, \Delta)$ is called infra soft β-continuous at $\delta_o^x \in \widetilde{X}$ if for any infra soft β-open set $(\mathcal{F}, \Delta)$ containing $f_\psi(\delta_o^x)$, there is an infra soft β-open set $(\mathcal{H}, \mathcal{O})$ containing δ_o^x such that $f_\psi(\mathcal{H}, \mathcal{O}) \widetilde{\subseteq} (\mathcal{F}, \Delta)$.

If f_ψ is infra soft β-continuous at all soft points of the domain, then it is called infra soft β-continuous.

Theorem 4 *Let $f_\psi : (X, \mu, \mathcal{O}) \to (\mathcal{S}, \nu, \Delta)$ be an infra soft β-continuous map. Then we have the following five equivalent statements:*

(i) f_ψ is an infra soft β-continuous map;
(ii) The inverse image of each infra soft β-closed set is infra soft β-closed;
(iii) $\beta Cl(f_\psi^{-1}(\mathcal{H}, \Delta)) \widetilde{\subseteq} f_\psi^{-1}(\beta Cl(\mathcal{H}, \Delta))$ for each $(\mathcal{H}, \Delta) \widetilde{\subseteq} \widetilde{\mathcal{S}}$;
(iv) $f_\psi(\beta Cl(\mathcal{F}, \mathcal{O})) \widetilde{\subseteq} \beta Cl(f_\psi(\mathcal{F}, \mathcal{O}))$ for each $(\mathcal{F}, \mathcal{O}) \widetilde{\subseteq} \widetilde{X}$;
(v) $f_\psi^{-1}(\beta Int(\mathcal{H}, \Delta)) \widetilde{\subseteq} \beta Int(f_\psi^{-1}(\mathcal{H}, \Delta))$ for each $(\mathcal{H}, \Delta) \widetilde{\subseteq} \widetilde{\mathcal{S}}$.

Proof $(i) \Rightarrow (ii)$: Let $(\mathcal{H}, \Delta)$ be an infra soft β-closed set in $(\mathcal{S}, \nu, \Delta)$. Then $f_\psi^{-1}(\mathcal{H}^c, \Delta)$ is an infra soft β-open subset of $\widetilde{X}$. Obviously, $f_\psi^{-1}(\mathcal{H}^c, \Delta) = \widetilde{X} - f_\psi^{-1}(\mathcal{H}, \Delta)$; hence, $f_\psi^{-1}(\mathcal{H}, \Delta)$ is an infra soft β-closed subset of $\widetilde{X}$.

$(ii) \Rightarrow (iii)$: According to (ii), $f_\psi^{-1}(\beta Cl(\mathcal{H}, \Delta))$ is an infra soft β-closed subset of $\widetilde{X}$. Then $\beta Cl(f_\psi^{-1}(\mathcal{H}, \Delta)) \widetilde{\subseteq} \beta Cl(f_\psi^{-1}(\beta Cl(\mathcal{H}, \Delta))) = f_\psi^{-1}(\beta Cl(\mathcal{H}, \Delta))$.

$(iii) \Rightarrow (vi)$: According to (iii), $\beta Cl(f_\psi^{-1}(f_\psi(\mathcal{F}, \mathcal{O}))) \widetilde{\subseteq} f_\psi^{-1}(\beta Cl(f_\psi(\mathcal{F}, \mathcal{O})))$. Then
$f_\psi(\beta Cl(\mathcal{F}, \mathcal{O})) \widetilde{\subseteq} f_\psi(f_\psi^{-1}(\beta Cl(f_\psi(\mathcal{F}, \mathcal{O})))) \widetilde{\subseteq} \beta Cl(f_\psi(\mathcal{F}, \mathcal{O}))$.

$(iv) \Rightarrow (v)$: According to (iv), $f_\psi(\beta Cl(\widetilde{X} - f_\psi^{-1}(\mathcal{H}, \Delta))) \widetilde{\subseteq} \beta Cl(f_\psi(\widetilde{X} - f_\psi^{-1}(\mathcal{H}, \Delta)))$. Therefore, $f_\psi(\widetilde{X} - \beta Int(f_\psi^{-1}(\mathcal{H}, \Delta))) = f_\psi(\beta Cl(\widetilde{X} - f_\psi^{-1}(\mathcal{H}, \Delta))) \subseteq \beta Cl(\widetilde{\mathcal{S}} - (\mathcal{H}, \Delta)) = \widetilde{\mathcal{S}} - \beta Int(\mathcal{H}, \Delta)$. Thus, $\widetilde{X} - \beta Int(f_\psi^{-1}(\mathcal{H}, \Delta)) \widetilde{\subseteq} f_\psi^{-1}(\widetilde{\mathcal{S}} - \beta Int(\mathcal{H}, \Delta)) = f_\psi^{-1}(\widetilde{\mathcal{S}}) - f_\psi^{-1}(\beta Int(\mathcal{H}, \Delta))$. Hence, $f_\psi^{-1}(\beta Int(\mathcal{H}, \Delta)) \widetilde{\subseteq} \beta Int(f_\psi^{-1}(\mathcal{H}, \Delta))$.

$(v) \Rightarrow (i)$: Let $(\mathcal{H}, \Delta)$ be an infra soft open subset of $\widetilde{\mathcal{S}}$. According to (v), $f_\psi^{-1}(\mathcal{H}, \Delta) \widetilde{\subseteq} \beta Int(f_\psi^{-1}(\mathcal{H}, \Delta))$. This implies that $f_\psi^{-1}(\mathcal{H}, \Delta) = \beta Int(f_\psi^{-1}(\mathcal{H}, \Delta))$. Hence, f_ψ is infra soft β-continuous.

Theorem 5 *If $f_\psi : (X, \mu, \mathcal{O}) \to (\mathcal{S}, \nu, \Delta)$ is infra soft β-continuous, then the restriction soft map $f_{\psi|\mathcal{M}} : (\mathcal{M}, \mu_\mathcal{M}, \mathcal{O}) \to (\mathcal{S}, \nu, \Delta)$ is infra soft β-continuous provided that $\widetilde{\mathcal{M}}$ is an infra soft open set.*

Proof Consider $(\mathcal{H}, \Delta)$ is an infra soft β-open set in $(\mathcal{S}, \nu, \Delta)$. By hypothesis, $f_{\psi}^{-1}(\mathcal{H}, \Delta)$ is infra soft β-open. Now, $f_{\psi_{|\mathcal{M}}}^{-1}(\mathcal{H}, \Delta) = f_{\psi}^{-1}(\mathcal{H}, \Delta)\widetilde{\bigcap}\widetilde{\mathcal{M}}$. Since $\widetilde{\mathcal{M}}$ is an infra soft open set, it follows from Proposition 7 that $f_{\psi_{|\mathcal{M}}}^{-1}(\mathcal{H}, \Delta)$ is infra soft β-open. Hence, $f_{\psi_{|\mathcal{M}}}$ is an infra soft β-continuous map.

Proposition 17 *Let* $f_{\psi} : (X, \mu, \mathcal{O}) \to (\mathcal{S}, \nu, \Delta)$ *and* $F_{\nu} : (\mathcal{S}, \nu, \Delta) \to (\mathcal{V}, \sigma, \mathcal{U})$ *be infra soft β-continuous. Then* $F_{\nu} \circ f_{\psi}$ *is infra soft β-continuous.*

Proof It is straightforward.

Definition 21 A soft map $f_{\psi} : (X, \mu, \mathcal{O}) \to (\mathcal{S}, \nu, \Delta)$ is called infra soft β-open (resp., infra soft β-closed) if the image of each infra soft β-open (resp., infra soft β-closed) set is infra soft β-open (resp., infra soft β-closed).

Proposition 18 $f_{\psi} : (X, \mu, \mathcal{O}) \to (\mathcal{S}, \nu, \Delta)$ *is an infra soft β-open map iff* $f_{\psi}(\beta Int(\mathcal{H}, \mathcal{O})) \widetilde{\subseteq} \beta Int(f_{\psi}(\mathcal{H}, \mathcal{O}))$ *for each subset of* $(\mathcal{H}, \mathcal{O})$ *of* $\widetilde{X}$.

Proof $\Rightarrow$: Let $(\mathcal{H}, \mathcal{O})$ be a subset of $\widetilde{X}$. Now, $f_{\psi}(\beta Int(\mathcal{H}, \mathcal{O}))\widetilde{\subseteq} f_{\psi}(\mathcal{H}, \mathcal{O})$ and $\beta Int(\mathcal{H}, \mathcal{O})$ is an infra soft β-open set. By hypothesis, $f_{\psi}(\beta Int(\mathcal{H}, \mathcal{O}))$ is infra soft β-open. Therefore, $f_{\psi}(\beta Int(\mathcal{H}, \mathcal{O}))\widetilde{\subseteq}\beta Int(f_{\psi}(\mathcal{H}, \mathcal{O}))$.

$\Leftarrow$: Let $(\Lambda, \mathcal{O})$ be an infra soft open subset of $\widetilde{X}$. Then $f_{\psi}(\mathcal{H}, \mathcal{O})\widetilde{\subseteq}\beta Int(f_{\psi}(\mathcal{H}, \mathcal{O}))$. Therefore, $f_{\psi}(\mathcal{H}, \mathcal{O}) = \beta Int(f_{\psi}(\mathcal{H}, \mathcal{O}))$ which means that f_{ψ} is an infra soft β-open map.

Proposition 19 $f_{\psi} : (X, \mu, \mathcal{O}) \to (\mathcal{S}, \nu, \Delta)$ *is an infra soft β-closed map iff* $\beta Cl(f_{\psi}(\mathcal{H}, \mathcal{O})) \widetilde{\subseteq} f_{\psi}(\beta Cl(\mathcal{H}, \mathcal{O}))$ *for each subset* $(\mathcal{H}, \mathcal{O})$ *of* $\widetilde{X}$.

Proof $\Rightarrow$: Let f_{ψ} be an infra soft β-closed map and $(\mathcal{H}, \mathcal{O})$ be a subset of $\widetilde{X}$. By hypothesis, $f_{\psi}(\beta Cl(\mathcal{H}, \mathcal{O}))$ is infra soft β-closed. Since $f_{\psi}(\mathcal{H}, \mathcal{O})\widetilde{\subseteq} f_{\psi}(\beta Cl(\mathcal{H}, \mathcal{O}))$, $\beta Cl(f_{\psi}(\mathcal{H}, \mathcal{O})) \widetilde{\subseteq} f_{\psi}(\beta Cl(\mathcal{H}, \mathcal{O}))$.

$\Leftarrow$: Suppose that $(\mathcal{H}, \mathcal{O})$ is an infra soft β-closed subset of $\widetilde{X}$. By hypothesis, $f_{\psi}(\mathcal{H}, \mathcal{O})\widetilde{\subseteq} \beta Cl(f_{\psi}(\mathcal{H}, \mathcal{O}))\widetilde{\subseteq} f_{\psi}(\beta Cl(\mathcal{H}, \mathcal{O})) = f_{\psi}(\mathcal{H}, \mathcal{O})$. Therefore, $f_{\psi}(\mathcal{H}, \mathcal{O})$ is infra soft β-closed. Hence, f_{ψ} is an infra soft β-closed map.

Proposition 20 *The concepts of infra soft β-open and infra soft β-closed maps are equivalent under bijectiveness.*

Proof It comes from the fact that a bijective soft map $f_{\psi} : (X, \mu, \mathcal{O}) \to (\mathcal{S}, \nu, \Delta)$ implies that $f_{\psi}(\mathcal{H}^c, \mathcal{O}) = (f_{\psi}(\mathcal{H}, \mathcal{O}))^c$.

Proposition 21 *Let* $f_{\psi} : (X, \mu, \mathcal{O}) \to (\mathcal{S}, \nu, \Delta)$ *and* $F_{\nu} : (\mathcal{S}, \nu, \Delta) \to (\mathcal{V}, \sigma, \mathcal{U})$ *be two soft maps. Then*

(i) If f_{ψ} *and* F_{ν} *are infra soft β-open (resp. infra soft β-closed) maps, then* $F_{\nu} \circ f_{\psi}$ *is an infra soft β-open (resp. infra soft β-closed) map.*

(ii) If $F_{\nu} \circ f_{\psi}$ *is an infra soft β-open (resp. infra soft β-closed) map and* f_{ψ} *is a surjective infra soft β-continuous map, then* F_{ν} *is an infra soft β-open (resp. infra soft β-closed) map.*

(iii) If $F_\nu \circ f_\psi$ is an infra soft β-open (resp. infra soft β-closed) map and F_ν is an injective infra soft β-continuous map, then f_ψ is an infra soft β-open (resp. infra soft β-closed) map.

Proof (i) It is straightforward.

(ii) Consider $(\mathcal{H}, \Delta)$ as an infra soft β-open subset of $(\mathcal{S}, \nu, \Delta)$. By hypothesis, $f_\psi^{-1}(\mathcal{H}, \Delta)$ is an infra soft β-open subset of $(X, \mu, \mathcal{O})$. Again, by hypothesis, $(F_\nu \circ f_\psi)(f_\psi^{-1}(\mathcal{H}, \Delta))$ is an infra soft β-open subset of $(\mathcal{V}, \sigma, \mathcal{U})$. Since f_ψ is surjective, then $(F_\nu \circ f_\psi)(f_\psi^{-1}(\mathcal{H}, \Delta)) = F_\nu(f_\psi(f_\psi^{-1}(\mathcal{H}, \Delta))) = F_\nu(\mathcal{H}, \Delta)$. Hence, F_ν is an infra soft β-open map.

(iii) Consider $(\mathcal{H}, \mathcal{O})$ as an infra soft β-open subset of $(X, \mu, \mathcal{O})$. By hypothesis, $(F_\nu \circ f_\psi)(\mathcal{H}, \mathcal{O})$ is an infra soft β-open subset of $(\mathcal{V}, \sigma, \mathcal{U})$. Again, by hypothesis, $F_\nu^{-1}(F_\nu \circ f_\psi(\mathcal{H}, \mathcal{O}))$ is an infra soft β-open subset of $(\mathcal{S}, \nu, \Delta)$. Since F_ν is injective, then $F_\nu^{-1}(F_\nu \circ f_\psi(\mathcal{H}, \mathcal{O})) = (F_\nu^{-1} F_\nu)(f_\psi(\mathcal{H}, \mathcal{O})) = f_\psi(\mathcal{H}, \mathcal{O})$. Hence, f_ψ is an infra soft β-open map.

Definition 22 A bijective soft map $f_\psi : (X, \mu, \mathcal{O}) \to (\mathcal{S}, \nu, \Delta)$ is called an infra soft β-homeomorphism if it is infra soft β-continuous and infra soft β-open.

Proposition 22 *Let $f_\psi : (X, \mu, \mathcal{O}) \to (\mathcal{S}, \nu, \Delta)$ and $F_\nu : (\mathcal{S}, \nu, \Delta) \to (\mathcal{V}, \sigma, \mathcal{U})$ be infra soft β-homeomorphism maps. Then $F_\nu \circ f_\psi$ is an infra soft β-homeomorphism map.*

Proposition 23 *If $f_\psi : (X, \mu, \mathcal{O}) \to (\mathcal{S}, \nu, \Delta)$ is a bijective soft map, then the following statements are equivalent:*

(i) f_ψ is an infra soft β-homeomorphism.
(ii) f_ψ and f_ψ^{-1} is infra soft β-continuous.
(iii) f_ψ is infra soft β-closed and infra soft β-continuous.

Proposition 24 *If $f_\psi : (X, \mu, \mathcal{O}) \to (\mathcal{S}, \nu, \Delta)$ is an infra soft β-homeomorphism map, then for each $(\mathcal{H}, \mathcal{O}) \in S(X)_A$, we have*

(i) $f_\psi(\beta Int(\mathcal{H}, \mathcal{O})) = \beta Int(f_\psi(\mathcal{H}, \mathcal{O}))$.
(ii) $f_\psi(\beta Cl(\mathcal{H}, \mathcal{O})) = \beta Cl(f_\psi(\mathcal{H}, \mathcal{O}))$.

Proof (i): It comes from Proposition 18 (i) that $f_\psi(\beta Int(\mathcal{H}, \mathcal{O})) \widetilde{\subseteq} \beta Int(f_\psi(\mathcal{H}, \mathcal{O}))$. Conversely, let $\delta_\kappa^s \in \beta Int(f_\psi(\mathcal{H}, \mathcal{O})$. Then there is an infra soft β-open set $(\mathcal{F}, \Delta)$ such that $\delta_\kappa^s \in (\mathcal{F}, \Delta) \widetilde{\subseteq} f_\psi(\mathcal{H}, \mathcal{O})$. By hypothesis, $\delta_o^x = f_\psi^{-1}(\delta_\kappa^s) \in f_\psi^{-1}(\mathcal{F}, \Delta) \widetilde{\subseteq} (\mathcal{H}, \mathcal{O})$ such that $f_\psi^{-1}(\mathcal{F}, \Delta)$ is an infra soft β-open set so that $\delta_o^x \in \beta Int(\mathcal{H}, \mathcal{O})$ which means that $\delta_\kappa^s \in f_\psi(\beta Int(\mathcal{H}, \mathcal{O}))$.

Item (ii) is proved similar to (i).

Theorem 6 *The property of an infra soft β-dense set is an infra soft topological invariant.*

Proof Let $f_{\psi} : (X, \mu, \mathcal{O}) \rightarrow (\mathcal{S}, \nu, \Delta)$ be an infra soft β-homeomorphism and let $(\mathcal{H}, \mathcal{O})$ be an infra soft β-dense set in $(X, \mu, \mathcal{O})$, i.e. $\beta Cl(\mathcal{H}, \mathcal{O}) = \widetilde{X}$. By Proposition 24, (ii) we find $\beta Cl(f_{\psi}(\mathcal{H}, \mathcal{O})) = f_{\psi}(\beta Cl(\mathcal{H}, \mathcal{O})) = f_{\psi}(\widetilde{X}) = \beta Cl(\widetilde{\mathcal{S}}) = \widetilde{\mathcal{S}}$. Thus, $f_{\psi}(\mathcal{H}, \mathcal{O})$ is an infra soft β-dense set in $(\mathcal{S}, \nu, \Delta)$.

Funding: This research has received no external funding.
Conflicts of interest: The authors declare no conflicts of interest.
Availability of data and material: No data were used to support this study.
Code availability: Not applicable.

References

1. Abu-Gdairi, R., El-Gayar, M.A., Al-shami, T.M., Nawar, A.S., El-Bably, M.K.: Some topological approaches for generalized rough sets and their decision-making applications. Symmetry **14**(1), 95 (2022)
2. Alcantud, J.C.R., Al-shami, T.M., Azzam, A.A.: Caliber and chain conditions in soft topologies. Mathematics **9**, 2349 (2021)
3. Ali, M.I., Feng, F., Liu, X., Min, W.K., Shabir, M.: On some new operations in soft set theory. Comput. Math. Appl. **57**, 1547–1553 (2009)
4. Aljarrah, H., Rawshdeh, A., Al-shami, T.M.: On soft compact and soft Lindelöf spaces via soft regular closed sets. Afrika Matematika **33**, 23 (2022). https://doi.org/10.1007/s13370-021-00952-z
5. Al-shami, T.M.: Soft somewhere dense sets on soft topological spaces. Commun. Korean Math. Soc. **33**(4), 1341–1356 (2018)
6. Al-shami, T.M.: Comments on "Soft mappings spaces". Sci. World J. **2019**, Article ID 6903809, 2 pages
7. Al-shami, T.M.: Investigation and corrigendum to some results related to g-soft equality and gf-soft equality relations. Filomat **33**(11), 3375–3383 (2019)
8. Al-shami, T.M.: Comments on some results related to soft separation axioms. Afr. Mat. **31**(7), 1105–1119 (2020)
9. Al-shami, T.M.: Bipolar soft sets: relations between them and ordinary points and their applications. Complexity **2021**, Article ID 6621854, 14 pages
10. Al-shami, T.M.: Compactness on soft topological ordered spaces and its application on the information system. J. Math. **2021**, Article ID 6699092, 12 pages
11. Al-shami, T.M.: On soft separation axioms and their applications on decision-making problem. Math. Problems Eng. **2021**, Article ID 8876978, 12 pages
12. Al-shami, T.M.: Soft separation axioms and fixed soft points using soft semiopen set. J. Appl. Math. **2020**, Article ID 1746103, 11 pages
13. Al-shami, T.M.: New soft structure: infra soft topological spaces. Math. Problems Eng. **2021**, Article ID 3361604, 12 pages
14. Al-shami, T.M.: Infra soft compact spaces and application to fixed point theorem. J. Funct. Spaces **2021**, Article ID 3417096, 9 pages
15. Al-shami, T.M.: Homeomorphism and quotient mappings in infra soft topological spaces. J. Math. **2021**, Article ID 3388288, 10 pages
16. Al-shami, T.M.: Improvement of the approximations and accuracy measure of a rough set using somewhere dense sets. Soft. Comput. **25**(23), 14449–14460 (2021)
17. Al-shami, T.M.: Topological approach to generate new rough set models. Complex & Intell. Syst. (2022). https://doi.org/10.1007/s40747-022-00704-x
18. Al-shami, T.M., Abo-Tabl, E.A.: Soft α-separation axioms and α-fixed soft points. AIMS Math. **6**(6), 5675–5694 (2021)

19. Al-shami, T.M., Abo-Tabl, E.A.: Connectedness and local connectedness on infra soft topological spaces. Mathematics **9**, 1759 (2021)
20. Al-shami, T.M., Abo-Tabl, E.A., Asaad, B.A.: Weak forms of soft separation axioms and fixed soft points. Fuzzy Inf. Eng. **12**(4), 509–528 (2020)
21. Al-shami, T.M., Alshammari, I., Asaad, B.A.: Soft maps via soft somewhere dense sets. Filomat **34**(10), 3429–3440 (2020)
22. Al-shami, T.M., Asaad, B.A., Abo-Tabl, E.A.: Separation axioms and fixed points using total belong and total non-belong relations with respect to soft β-open sets. J. Interdiscip. Math. **24**(4), 1053–1077 (2021)
23. Al-shami, T.M., Azzam, A.A.: Infra soft semiopen sets and infra soft semicontinuity. J. Funct. Spaces **2021**, Article ID 5716876, 11 pages
24. Al-shami, T.M., El-Shafei, M.E.: Two types of separation axioms on supra soft separation spaces. Demonstratio Math. **52**(1), 147–165 (2019)
25. Al-shami, T.M., El-Shafei, M.E.: On supra soft topological ordered spaces. Arab J. Basic Appl. Sci. **26**(1), 433–445 (2019)
26. Al-shami, T.M., El-Shafei, M.E.: Some types of soft ordered maps via soft pre open sets. Appl. Math. & Inf. Sci. **13**(5), 707–715 (2019)
27. Al-shami, T.M., El-Shafei, M.E.: T-soft equality relation. Turk. J. Math. **44**(4), 1427–1441 (2020)
28. Al-shami, T.M., El-Shafei, M.E., Abo-Elhamayel, M.: Almost soft compact and approximately soft Lindelöf spaces. J. Taibah Univ. Sci. **12**(5), 620–630 (2018)
29. Al-shami, T.M., El-Shafei, M.E., Abo-Elhamayel, M.: Seven generalized types of soft semi-compact spaces. Korean J. Math. **27**(3), 661–690 (2019)
30. Al-shami, T.M., El-Shafei, M.E., Asaad, B.A.: Other kinds of soft β mappings via soft topological ordered spaces. Eur. J. Pure Appl. Math. **12**(1), 176–193 (2019)
31. Al-shami, T.M., El-Shafei, M.E., Asaad, B.A.: Sum of soft topological ordered spaces. Adv. Math.: Sci. J. **9**(7), 4695–4710 (2020)
32. Al-shami, T.M., Işk, H., Nawar, A.S., Hosny, R.A.: Some topological approaches for generalized rough sets via ideals. Math. Problems Eng. **2021**, Article ID 5642982, 11 pages
33. Al-shami, T.M., Liu, J.-B.: Two classes of infrasoft separation axioms. J. Math. **2021**, Article ID 4816893, 10 pages
34. Al-shami, T.M., Mhemdi, A.: Belong and nonbelong relations on double-Framed soft sets and their applications. J. Math. **2021**, Article ID 9940301, 12 pages
35. Al-shami, T.M., Mhemdi, A.: Two families of separation axioms on infra soft topological spaces. Filomat **36**(4), 1143–1157 (2022)
36. Al-shami, T.M., Mhemdi, A., Rawshdeh, A., Aljarrah, H.: Soft version of compact and Lindelöf spaces using soft somewhere dense set. AIMS Math. **6**(8), 8064–8077 (2021)
37. Al-shami, T.M., Kočinac, L.D.R.: The equivalence between the enriched and extended soft topologies. Appl. Comput. Math. **18**(2), 149–162 (2019)
38. Al-shami, T.M., Kočinac, L.D.R.: Nearly soft Menger spaces. J. Math. **2020**, Article ID 3807418, 9 pages
39. Al-shami, T.M., Kočinac, L.D.R.: Almost soft Menger and weakly soft Menger spaces. Appl. Comput. Math. **21**(1), 35–51 (2022)
40. Al-shami, T.M., Kočinac, L.D.R., Asaad, B.A.: Sum of soft topological spaces. Mathematics **8**(6), 990 (2020)
41. Al-shami, T.M., Othman, H.A.: Infra pre-open sets and their applications to generate new types of operators and maps. Eur. J. Pure Appl. Math. **15**(1), 261–280 (2022)
42. Al-shami, T.M., Tercan, A., Mhemdi, A.: New soft separation axioms and fixed soft points with respect to total belong andtotal non-belong relations. Demonstratio Math. **54**, 196–211 (2021)
43. Asaad, B.A., Al-shami, T.M., Mhemdi, A.: Bioperators on soft topological spaces. AIMS Math. **6**(11), 12471–12490 (2021)
44. Aygünoğlu, A., Aygün, H.: Some notes on soft topological spaces. Neural Comput. Appl. **21**, 113–119 (2012)

45. Çağman, N., Enginoğlu, S.: Soft matrix theory and its decision making. Comput. Math. Appl. **59**, 3308–3314 (2010)
46. Çağman, N., Karataş, S., Enginoglu, S.: Soft topology. Comput. Math. Appl. **62**, 351–358 (2011)
47. El-Shafei, M.E., Abo-Elhamayel, M., Al-shami, T.M.: Partial soft separation axioms and soft compact spaces. Filomat **32**(13), 4755–4771 (2018)
48. El-Shafei, M.E., Al-shami, T.M.: Applications of partial belong and total non-belong relations on soft separation axioms and decision-making problem. Comput. Appl. Math. **39**(3), 138 (2020)
49. El-Shafei, M.E., Al-shami, T.M.: Some operators of a soft set and soft connected spaces using soft somewhere dense sets. J. Interdiscip. Math. **24**(6), 1471–1495 (2021)
50. Feng, F., Li, Y.M., Davvaz, B., Ali, M.I.: Soft sets combined with fuzzy sets and rough sets: a tentative approach. Soft. Comput. **14**, 899–911 (2010)
51. Hosny, R.A., Asaad, B.A., Azzam, A.A., Al-shami, T.M.: Various topologies generated from E_j-neighbourhoods via ideals. Complexity **2021**, Article ID 4149368, 11 pages
52. Hussain, S.: Binary soft connected spaces and an application of binary soft sets in decision making problem. Fuzzy Inf. Eng. **11**(4), 506–521 (2019)
53. Kharal, A., Ahmed, B.: Mappings on soft classes. New Math. Nat. Comput. **7**(3), 471–481 (2011)
54. Lin, F.: Soft connected spaces and soft paracompact spaces. Int. J. Math. Sci. Eng. **7**(2), 1–7 (2013)
55. Kočinac, L.D.R., Al-shami, T.M., Çetkin, V.: Selection principles in the context of soft sets: Menger spaces. Soft. Comput. **25**, 12693–12702 (2021)
56. Maji, P.K., Biswas, R., Roy, R.: Soft set theory. Comput. & Math. Appl. **45**, 555–562 (2003)
57. Molodtsov, D.: Soft set theory-first results. Comput. & Math. Appl. **37**, 19–31 (1999)
58. Nazmul, S., Samanta, S.K.: Neighbourhood properties of soft topological spaces. Ann. Fuzzy Math. Inf. **6**(1), 1–15 (2013)
59. Salama, A.S., Mhemdi, A., Elbarbary, O.G., Al-shami, T.M.: Topological approaches for rough continuous functions with applications. Complexity **2021**, Article ID 5586187, 12 pages
60. Shabir, M., Naz, M.: On soft topological spaces. Comput. Math. Appl. **61**, 1786–1799 (2011)

An Algorithm of the Prey and Predator Struggle to Survive as a Random Walk Simulation Case Study

Raed M. Khalil and Rania Saadeh

Abstract This paper is an attempt to make use of a mathematical simulation of the tracking problem. Random walk algorithm is used to simulate the interaction over time of hunter and prey in a small rectangular area.

Keywords Hunter · Prey · Random walk · Rugby

1 Introduction

A computer games production company would like to ensure its client's satisfaction and impressions about the game it produces recently. The rugby game is a first-game product. The objective of Rugby is to advance the ball down the field by running it forward in the attempt to score points. Besides that the company needs to simulate the running the ball forward in the attempt to score points. The problem can be solved using algorithms and mathematical simulation of a tracking problem like hunter-prey tracking from a random walk view.

Things in nature often move in complicated ways. You have probably watched the way a butterfly moves. The molecules of the air that you are breathing move in a similar way. This type of motion we call a random walk.

In the next section, the hunter and prey approach is described. The section after contains the simulation study of the rugby game, and then the conclusion section.

R. M. Khalil (✉)
Department of Computer Information Systems, Al Balqa Applied University, Salt, Jordan
e-mail: R.M.Khalil@bau.edu.jo

R. Saadeh
Department of Mathematics, Zarqa University, Zarqa, Jordan
e-mail: rsaadeh@zu.edu.jo

D. Zeidan et al. (eds.), *Mathematics and Computation*, Springer Proceedings in Mathematics & Statistics 418, https://doi.org/10.1007/978-981-99-0447-1_36

2 Hunter and Prey Approach

A hunter in hunter and prey wants to track its prey. The hunter must determine which direction to move in at each stage in order to get as close to the prey as possible. In this simulation, there is no element of surprise. The hunter just notices the prey's movement and tries to catch it.

With this case, a human would be equipped with two primary tools to aid in the endeavor. To begin with, humans intuitively understand that the shortest distance between two sites is a straight line. As a result, if the prey remains static or moves slowly, a human hunter will charge straight at it and catch it [1]. The second tool is that humans can observe the pattern of prior prey movement and forecast where it will go in the future. This allows the human hunter to predict what the prey will do next and act accordingly, not just to reduce the distance between hunter and prey at any one time, but also to reduce the amount of time (and hence effort) required to catch the prey.

In the case of the rugby game, this second tool can be better visualized. Consider the following two rugby players: one is young, swift, and inexperienced, while the other is older, slower, and more experienced. These two players have quite different skill sets, but they can play at the same level and have a lot of fun. International teams frequently include a mixture of these types of players. The younger player has a leg up on the older player in terms of endurance and quickness. This enables him to use the first tracking tool (knowing that the shortest distance is a straight line) to catch his prey on a regular basis (chase down and tackle the member of the opposite team which holds the ball).

Following this strategy, the elder player would become fatigued more quickly. He'd also be more likely to fail because he'd be slightly slower. This athlete will have to rely on his biggest asset, which is his rugby experience [2]. This will allow him to use less energy and move at a slower pace while accomplishing the same goal. He can make accurate predictions about the ball carrier's future moves based on not only his previous match experience, but also the behavior of opposition players during the current encounter. Due to his previous expertise, he will be paying far more attention to the other team's activities than his younger opponent. As a result of his experience, he now has a better ability to forecast.

Why don't we simply educate younger players on how to do this, integrating both advantages in a single player? That is not as simple as it may appear, because the greatest way to learn is through doing.

We'll now look at how to describe this initial tracking tool mathematically, such that a computerized hunter can try to catch a computerized prey in 2D.

The Hunter Algorithm

To put it another way, we assume that the hunter can only move in set steps of a particular length. Before taking each step, the hunter must choose which path is the best for minimizing the gap between it and the prey. This cycle repeats until the prey is apprehended [3].

This program uses the mathematical idea of limited optimization to determine the hunter's stride direction.

To undertake optimization, we'll need an objective function (the distance between the hunter and the prey), whose value we'll strive to minimize by changing the values of some variables (in this case, the step direction) while keeping any relevant constraints in mind (in this case, the length of the step). Pythagoras' theorem [1] is used to characterize the objective function f(x, y) as follows:

$$f(x, y) = \left(x - x_{prey}\right)^2 + \left(y - y_{prey}\right)^2$$

where:

- x = Next x-coordinate of Hunter
- y = Next y-coordinate of Hunter
- xprey = Present x-coordinate of Prey
- yprey = Present y-coordinate of Prey.

The distance between the hunter and the prey is squared, not the actual distance between them. Minimizing the squared distance is the same as minimizing the distance. The primary purpose for using the distance squared is to ensure that the value of f(x, y) is never negative. Numerical operations are simplified as a result of this.

This objective function is to be minimized by changing the values of particular variables, according to the statement. The variables in this situation are x and y, which are the coordinates of the hunter's position. Currently, minimization of this function occurs when x = xprey and y = yprey (the value of f(x, y) is zero, which is the smallest value feasible for a non-negative function).

This, however, implies that the hunter can take arbitrarily lengthy steps, which is unrealistic. This would be the equivalent of the hunter teleporting to the prey and instantaneously grabbing it.

To account for the hunter's limited stepwise mobility, the hunter position must be on a circle with a radius of one step length (the circle's center being the last hunter position). This restriction will be known as h(x, y), and it is written as follows:

$$h(x, y) = (x - x_{hunter})^2 + (y - y_{hunter})^2 - R^2 = 0$$

where:

- x = Next x-coordinate of Hunter
- y = Next y-coordinate of Hunter
- xhunter = Present x-coordinate of Hunter
- yhunter = Present y-coordinate of Hunter
- R = Allowed Hunter step length (radius of circle).

It's worth noting that f(x, y) and h(x, y) are eerily similar. However, the distinctions are significant and should be highlighted. To begin with, f(x, y) is not required to take any precise value (albeit it should be at a minimum), but h(x, y) must always equal zero. h(x, y) is a limitation on the distance between the current hunter position and the next hunter position, whereas f(x, y) is a measure of the distance between the next hunter position and the current prey position.

Finding the derivative of f(x, y) with respect to x and the derivative of f(x, y) with respect to y, setting them both to zero, and solving the system of two equations in two unknowns is the formal method of minimizing a function (f(x, y) This can be written mathematically as

$$\frac{\partial f(x, y)}{\partial x} = 0$$
$$\frac{\partial f(x, y)}{\partial y} = 0$$

This form of attack, on the other hand, would result in the circumstance stated above, in which the hunter would instantaneously switch to the prey position. As a result, we require a formal method for minimizing f(x, y) such that h(x, y) = 0. This can be demonstrated (but not here!) by setting the derivatives of the Lagrangian function (rather than the objective function) for this system to zero. L(x, y, l) is a Lagrangian function that is defined as

$$L(x, y, \lambda) = f(x, y) + \lambda h(x, y)$$

So the criterion for the optimal solution becomes

$$\frac{\partial L(x, y, \lambda)}{\partial x} = \frac{\partial f(x, y)}{\partial x} + \lambda \frac{\partial h(x, y)}{\partial x} = 0$$
$$\frac{\partial L(x, y, \lambda)}{\partial y} = \frac{\partial f(x, y)}{\partial y} + \lambda \frac{\partial h(x, y)}{\partial y} = 0$$
$$\frac{\partial L(x, y, \lambda)}{\partial \lambda} = h(x, y) = 0$$

If we write down the derivatives of these functions the equations above become

$$2\lambda(x - x_{hunter}) + 2\left(x - x_{prey}\right) = 0$$
$$2\lambda(y - y_{hunter}) + 2\left(y - y_{prey}\right) = 0$$
$$(x - x_{hunter})^2 + (y - y_{hunter})^2 - R^2 = 0$$

There are three equations in three unknowns, as can be seen (x, y, l). It's worth noting that R is a user-defined constant and that the prey position (xprey, yprey) can be read (and so can also be treated as constant during each optimization). These equations' answers yield numbers for x and y (the new hunter location) as well as 1. (which can be discarded). It is important to note that numerous solutions are feasible, and the one that minimizes f must be chosen (x, y). This is accomplished by guaranteeing that the Hessian of Lagrangian's eigenvalues are all positive (i.e., that the Hessian is positive definite) at the solution.

So, now that we know how to calculate a plausible, realistic hunter movement at each step, it's only a matter of updating the hunter position and repeating the process until the prey is caught (or flees!). The following pseudo-code illustrates how this could be accomplished:

Define the Prey motion as a function of time (i.e. xprey = xprey(t), yprey = yprey(t)). Note that the Hunter algorithm does not have this information.

Define initial Hunter position (xhunter0, yhunter0)

```
For t = 0 to f(x, y) < R2 OR t > tMax
        Solve
    This gives (x, y)
        Update xhunter = x; yhunter = y
    End
```

Now the hunter and prey movement has been calculated and can be plotted and the time to capture can be seen.

The Seeker Algorithm

The hunter's second tool is more complicated. This is due to the fact that there is a component of forecasting future behavior. Predicting the future is difficult at the best of times, and numerical tracking methods are no exception. The first method of prediction that we shall consider is linear [4]. Making a linear prediction of the prey's mobility based on the current position and a position a predefined time interval previously will be required. If and only if the prey moves in a perfectly straight line, this strategy will produce ideal results (vertically). In this example, perfect means that the prey will be caught as soon as feasible.

However, when the prey motion deviates from linearity, the method's efficiency deteriorates. Furthermore, if the prey motion oscillates often but remains roughly linear, the method's performance will be influenced to some extent by the time interval between the two sample points.

The Tracker Algorithm

So, if linear approximations aren't great, what alternatives are there? It is possible to match a variety of functions to past prey situations that include curvilinear behavior. This fit can also be achieved in a variety of ways [5]. For example, it may be decided to choose the best fit curve through all previous prey placements or a fixed number of the most recent ones. This could be computationally expensive, and the prediction will not be very precise at the end of the day [6]. Making useful forecasts requires making them rapidly based on recent movements and just predicting a little amount ahead.

The optimal quantification of these adjectives (quickly, recently, and usefully) is dependent on the prey's real behavior and the types of functions that are fitted to the motion. It is our responsibility to select an appropriate function and settings that will allow us to follow prey behavior over a wide range of scenarios. This will ensure that the greatest number of prey is captured [7].

Lagrangian interpolation polynomials are a quick and easy approach to fit prey positions. These are functions that yield a polynomial that passes through all of the points in the data set when evaluated [8].

Lagrangian Interpolation Polynomials

A line is defined by two points. Or in other words, a unique line can be drawn through a given pair of points. This line can be expressed as

$$f(x) = \frac{x - x_2}{x_1 - x_2} y_1 + \frac{x - x_1}{x_2 - x_1} y_2$$

A quadratic polynomial can be rendered using three points, a cubic polynomial using four points, and so on. La-formula, grange's which is just an extension of the linear expression above to higher order polynomials, can be used to find these unique functions:

$$f(x) = \sum_{i=1}^{N} \prod_{\substack{j=1 \\ (j \neq i)}}^{N} \frac{(x - x_j)}{(x_i - x_j)} y_i$$

This is simple to develop for any number of points because no optimization is necessary (as there would be with regular curve fitting). If the points do not naturally lie along a low-order polynomial curve, a very high-order polynomial must be fitted to them in order to ensure that the curve passes through all of them. This creates false wave behavior, which can result in wildly erroneous forecasts. To avoid this, it is recommended that the formula be applied to a limited set of current data and that it be used to forecast a little amount ahead of time (which will lessen the effect of any spurious wiggles).

$$\frac{\partial L(x, y, \lambda)}{\partial x} = \frac{\partial f(x, y)}{\partial x} + \lambda \frac{\partial h(x, y)}{\partial x} = 0$$
$$\frac{\partial L(x, y, \lambda)}{\partial y} = \frac{\partial f(x, y)}{\partial y} + \lambda \frac{\partial h(x, y)}{\partial y} = 0$$
$$\frac{\partial L(x, y, \lambda)}{\partial \lambda} = h(x, y) = 0$$

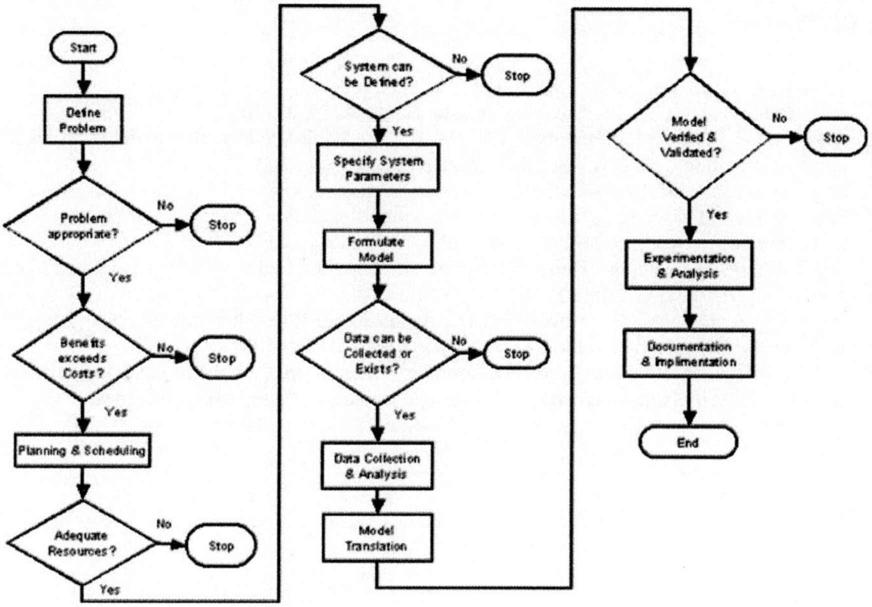

Fig. 1 The simulation study strategy

3 The Simulation Study

The simulation study strategy we follow all over this project is illustrated in Fig. 1.

4 Conclusion

Rugby game is programmed and simulated using the hunter-prey strategy in random walk view, which eventually refreshes our minds and reminds us of a variety of various application ideas that can be implemented using these algorithms. The tracking algorithm is a more powerful technology that has been utilized in game production for a long time and was developed utilizing more mathematical notions.

Last but not least, the company we are considering using this simulation as an advantage point to finish what it started in building a game in the proper human–computer approach.

References

1. Yoshida, T., Jones, L.E., Ellner, S.P., Fussmann, G.F., Hairston, Jr. N.G.: Rapid evolution drives ecological dynamics in a predator-prey system. Nature **424**, 303 (2003)
2. Durney, C.H., Case, S.O., Pleimling, M., Zia, R.K.P.: Stochastic evolution of four species in cyclic competition. J. Stat. Mech. **2012**, P06014 (2012)
3. http://www.pas.rochester.edu/~stte/phy104-F00/n9/notes-9a.html
4. http://mathworld.wolfram.com/RandomWalk1-Dimensional.html
5. https://mathworld.wolfram.com/RandomWalk2-Dimensional.html
6. Von Luxburg, U., Radl, A., Hein, M.: Hitting and commute times in large graphs are often misleading (2010). arXiv:1003.1266
7. Wang, A.Q., Pollock, M., Roberts, G.O., Steinsaltz, D.: Regeneration-enriched Markov processes with application to Monte Carlo (2019). arXiv:1910.05037
8. van der Hofstad, R.: Random Graphs and Complex Networks. In: Cambridge Series in Statistical and Probabilistic Mathematics, vol. 2. Cambridge University Press (2018). To appear

New Modification Methods for Finding Zeros of Nonlinear Functions

Osama Ababneh and Khalid Al-Boureeny

Abstract The objective of this article is to define new efficient iterative methods for finding zeros of nonlinear functions. This procedure is based on Homeier [12] and Newton [12, 20, 27] methods. The proposed methods require only three function evaluations per iteration (only two function evaluations and one first derivative evaluation). The error equations are given theoretically to prove that the suggested methods have third-order convergence. Moreover, the Efficiency Index [20] is 1.4422. Numerical comparisons to demonstrate the exceptional convergence speed of the proposed methods using several types of functions and different initial guesses are included. A comparison with other well-known iterative methods is made. It is observed that our proposed methods are very competitive with the third-order methods.

Keywords Nonlinear functions · Newton methods · Nonlinear equations · Derivative-free methods · Simple roots

1 Introduction

Finding zeros of nonlinear functions by using iterative methods is one of the important problems which have interesting applications in different branches of science, in particular, physics and engineering [20, 22, 27], such as fluid dynamics, nuclear systems, and dynamic economic systems. Also, in mathematics, we do need iterative methods to find rapid solutions for special integrals and differential equations. Recently, there are many numerical iterative methods have been developed to solve these problems, see [1, 5, 6, 14, 16, 19, 27, 30]. These methods have been suggested and analyzed by using a variant of different techniques such as Taylor series. We first looked for the best approximation of which is used in many iterative methods. We obtained this approximation by combining two well-known methods, Potra–Ptak [23] and Weerakon methods [28]. Then, we used Homeier method [12] and the approximation to introduce the first method, which we called the Variant of Homeier

O. Ababneh (✉) · K. Al-Boureeny
Department of Mathematics, Zarqa University, Zarqa, Jordan
e-mail: osababneh@zu.edu.jo

D. Zeidan et al. (eds.), *Mathematics and Computation*, Springer Proceedings in Mathematics & Statistics 418, https://doi.org/10.1007/978-981-99-0447-1_37

Method 1 (VHM1). Finally, we used predictor–corrector technique to improve the first method (VHM1) and we called it Variant of Homeier Method 2 (VHM2). We showed that the new iterative methods are of third order of convergence, Efficiency Index [20] E.I. = 1.4422 and very robust and competitive with other third-order iterative methods.

2 The Established Methods

For the purpose of comparison, three 2-step third-order methods and two 1-step third-order methods are considered. Since these methods are well established, we state the essential formulas used to calculate the simple zero of nonlinear functions and thus compare the effectiveness of the proposed 2-step third-order methods.

Newton Method [3, 4, 9, 20, 22, 24, 27, 29].

One well-known 1-step iterative zero-finding method,

$$x_{n+1} = x_n - \frac{f(x_n)}{f'(x_n)}, n = 0, 1, 2, \ldots \tag{1}$$

Halley Method [7, 8, 10, 11, 15, 24]:

$$x_{n+1} = x_n - \frac{2f(x_n)f'(x_n)}{2f'^2(x_n) - f(x_n)f''(x_n)}, n = 0, 1, 2, \ldots \tag{2}$$

which is widely known Halley's method. It is a cubically converging (p = 3) zero-finding 1-step algorithm. It requires three function evaluations (r = 3) and its E.I. = 1.4422.

Householder method [13, 24]:

$$x_{n+1} = x_n - \frac{f(x_n)}{f'(x_n)}\left\{1 + \frac{f(x_n)f''(x_n)}{2f'^2(x_n)}\right\}, n = 0, 1, 2, \ldots \tag{3}$$

Householder's method is also cubically converging (p = 3) 1-step zero-finding algorithm. It requires three function evaluations (r = 3) and it's E.I. = 1.4422.

Weerakoon and Fernando Method [21, 28]:

$$x_{n+1} = x_n - \frac{2f(x_n)}{f'(x_n) + f'(y_n)}, n = 0, 1, 2, \ldots \text{ where,} \tag{4}$$

$$y_n = x_n - \frac{f(x_n)}{f'(x_n)}$$

Obviously, this is an implicit scheme, which requires having the derivative of the function at the $(n + 1)th$ iterative step to calculate the $(n + 1)th$ iterate itself. They overcome this difficulty by making use of Newton's iterative step to compute the $(n+1)th$ iterate on the right-hand side.

This scheme has also been derived by Ozban by using the arithmetic mean of $f'(xn)$ and $f'(y_n)$ instead of $f'(x_n)$ in Newton's method (1), i.e., $(f'(x_n)+f'(y_n))/2$.

Weerakoon and Fernando method is also a cubically converging (p = 3) 2-step zero-finding algorithm. It requires three function evaluations ($r = 3$) and it's $E.I. =$ 1.4422.

Homeier Method [12, 21]:

$$x_{n+1} = x_n - \left(\frac{f(x_n)}{2}\right)\left\{\frac{1}{f'(x_n)} + \frac{1}{f'(y_n)}\right\}, n = 0, 1, 2, \ldots \tag{5}$$

where $y_n = x_n - \frac{f(x_n)}{f'(x_n)}$.

Homeier's method is also cubically converging (p = 3) 2-step zero-finding algorithm. It requires three function evaluations (r = 3) and its E.I. = 1.4422.

Potra-Ptak Method [23, 25]:

$$x_{n+1} = x_n - \frac{f(x_n) + f(y_n)}{f'(x_n)}, \tag{6}$$

where

$$y_n = x_n - \frac{f(x_n)}{f'(x_n)}$$

Potr-Ptak's method is also a cubically converging (p = 3) 2-step zero-finding algorithm. It requires three function evaluations; two function evaluations and one first derivative (r = 3) and its E.I. = 1.4422.

2.1 Construction of the New Methods

In this section, first we define a new third-order method for finding zeros of a nonlinear function. We do that by combining two well-known methods to obtain a new one. In fact, the new iterative method will be an improvement of the classical Homeier method and this will be our first algorithm.

Secondly, we will improve our first algorithm by assuming a three-step iterative method per full cycle. In order to do that, we perform a Newton iteration at the new third step. We use a third variable Z_n for the third step, which we will approximate lately.

First, we equate (combine) the two methods (4) and (6) to obtain $f'(y_n)$

$$\frac{2f(x_n)}{f'(x_n) + f'(y_n)} \approx \frac{f(x_n) + f(y_n)}{f'(x_n)}$$

$$2f(x_n)f'(x_n) \approx [f(x_n) + f(y_n)]\{f'(x_n) + f'(y_n)\}$$

$$2f(x_n)f'(x_n) \approx [f(x_n) + f(y_n)]f'(x_n) + [f(x_n) + f(y_n)]f'(y_n)$$

$$\frac{2f(x_n)f'(x_n) - [f(x_n) + f(y_n)]f'(x_n)}{[f(x_n) + f(y_n)]} \approx f'(y_n)$$

$$f'(y_n) \approx \frac{f(x_n) - f(y_n)}{f(x_n) + f(y_n)} f'(x_n) \tag{7}$$

Now substituting (7) in (5) in order to get the first algorithm

$$x_{n+1} = x_n - \frac{f(x_n)}{2} * \left\{ \frac{1}{f'(x_n)} + \frac{1}{\frac{f(x_n)-f(y_n)}{f(x_n)+f(y_n)} f'(x_n)} \right\}$$

So, we get the first Algorithm (1) which we will call it a Variant of Homeier Method 1 (VHM1).

For a given $x_\circ$, compute the approximate solution x_n by iterative scheme.

$$y_n = x_n - \frac{f(x_n)}{f'(x_n)}, \ f'(x_n) \neq 0$$

$$x_{n+1} = x_n - \frac{f^2(x_n)}{[f(x_n) - f(y_n)]f'(x_n)}, \text{ for } n = 0, 1, 2, \ldots \tag{8}$$

Now, we need to drive the next algorithm, which will be an improvement of the first algorithm. The main goal is to make the new scheme optimal. We perform a Newton iteration at the new third step which comes next:

$$y_n = x_n - \frac{f(x_n)}{f'(x_n)} \tag{9}$$

$$z_n = x_n - \frac{f^2(x_n)}{[f(x_n) - f(y_n)]f'(x_n)} \tag{10}$$

$$x_{n+1} = z_n - \frac{f(z_n)}{f'(z_n)}. \tag{11}$$

Now, we try to simplify our new scheme to reach the convergence rate three with three function evaluations per full cycle; two function evaluations and one first derivative evaluation. Obviously, $f(z_n)$ and $f'(z_n)$ should be approximated. We replace $f'(z_n)$ by $f'(x_n)$ and write the Taylor expansion of $f(z_n)$ about x_n [25].

$$f(z_n) = f(x_n) + f'(x_n)(z_n - x_n) + \frac{1}{2!} f''(x_n)(z_n - x_n)^2 \tag{12}$$

Now, $f''(x_n)$ should be approximated as well. Once again we write the Taylor expansion of $f(y_n)$ about x_n as follows:

$$f(y_n) = f(x_n) + f'(x_n)(y_n - x_n) + \frac{1}{2!} f''(x_n)(y_n - x_n)^2 \tag{13}$$

From (13) and (9), we obtain $f''(x_n)$ as follows:

$$f''(x_n) = \frac{2f(y_n)[f'(x_n)]^2}{[f(x_n)]^2} \tag{14}$$

Now substituting (14) and (10) in (12), we obtain f(zn) as follows:

$$f(z_n) = \frac{f(x_n) f^2(y_n)}{[f(x_n) - f(y_n)]^2}. \tag{15}$$

Now, we substitute (10) and (15) in (11), also $f'(x_n)$ instead of $f'(z_n)$:

$$x_{n+1} = z_n - \frac{f(z_n)}{f'(z_n)}$$

$$X_{n+1} = \left\{ X_n - \frac{f^2(x_n)}{[f(x_n) - f(y_n)] f'(x_n)} \right\} - \frac{\frac{f(x_n) f^2(y_n)}{[f(x_n) - f(y_n)]^2}}{f'(x_n)}$$

After doing some simplifying work, we get a new algorithm.
Algorithm (1): we will call it the Variant Homeier Method 2 (VHM2).
For a given x_0, compute the approximate solution x_{n+1} by an iterative scheme

$$y_n = x_n - \frac{f(x_n)}{f'(x_n)}, \quad f'(x_n) \neq 0.$$

$$x_{n+1} = x_n - \left\{ 1 + \frac{f(x_n) f(y_n)}{[f(x_n) - f(y_n)]^2} \right\} * \frac{f(x_n)}{f'(x_n)}, \quad \text{for n} = 0, 1, 2, \ldots \tag{16}$$

As we can see, both algorithms require only two function evaluations and only one first derivative evaluation per each cycle. When we compare both algorithms and Homeier method, clearly, there is big difference, which is Homeier method requires one function evaluation and two first derivative evaluations.

2.2 Convergence Criteria of the New Methods

Now, we compute the orders of convergences and corresponding error equations of the proposed methods Algorithms (8) and (16).

Theorem 2.1 *Let* $\propto$ ε I *be a simple zero of sufficiently differentiable function* f: I $\subseteq$ R $\rightarrow$ R *for an open interval* I. *If* x_0 *is sufficiently close to* $\propto$, *then the iterative method defined by Algorithm* (1) *is of order three and satisfies the error equation:*

$$e_{n+1} = (2c_2^2 + 2c_3)e_n^3 + o(e_n^4),$$

where
$C_k = \frac{f^{(k)}(\alpha)}{k!f'(\alpha)}$, k = 2, 3, ... and $e_n = x_n - \alpha$.

Proof Let $\propto$ be a simple zero of f, $f'(\alpha) \neq 0$. Using Taylor's series expansion around $\propto$ in the nth iterate results in

$$f(x_n) = f'(\alpha)e_n + \frac{1}{2!}f''(\alpha)e_n^2 + \frac{1}{3!}f'''(\alpha)e_n^3 + O(e_n^4)$$

$$f(x_n) = f'(\alpha)\left[e_n + c_2 e_n^2 + c_3 e_n^3 + O(e_n^4)\right] \tag{17}$$

$$f'(x_n) = f'(\alpha)\left[1 + 2c_2 e_n + 3c_3 e_n^2 + O(e_n^3)\right] \tag{18}$$

From (17) and (18), we have

$$\frac{f(x_n)}{f'(x_n)} = e_n - c_2 e_n^2 + (2c_2^2 - 2c_3)e_n^3 + O(e_n^3) \tag{19}$$

But $y_n = x_n - \frac{f(x_n)}{f'(x_n)}$, $e_n = x_n - \alpha$. Using (19), we get

$$y_n = x_n - \left\{e_n - c_2 e_n^2 + (2c_2^2 - 2c_3)e_n^3 + O(e_n^3)\right\}$$

$$y_n = \alpha + c_2 e_n^2 + (2c_3 - 2c_2^2)e_n^3 + O(e_n^3)$$

$$(y_n - \alpha) = c_2 e_n^2 + (2c_3 - 2c_2^2)e_n^3 + O(e_n^3) \tag{20}$$

Now by Taylor expansion once again $f(y_n)$ about $\propto$ and using (20):
$f(y_n) = f(\alpha) + f'(\alpha)(y_n - \alpha) + \frac{f''(\alpha)}{2!}(y_n - \alpha)^2$ but $f(\alpha) = 0$,
$f(y_n) = f'(\alpha)\left[(y_n - \alpha) + \frac{f''(\alpha)}{2!f'(\alpha)}(y_n - \alpha)^2\right]$ and $\frac{f''(\alpha)}{2!f'(\alpha)} = C_2$

$$f(y_n) = f'(\alpha)\left[c_2 e_n^2 + (2c_3 - 2c_2^2)e_n^3 + O(e_n^4)\right] \tag{21}$$

$$f^2(x_n) = f'^2(\alpha)\left[e_n^2 + 2c_2e_n^3 + c_3e_n^3 + O(e_n^4)\right] \tag{22}$$

From (17) and (21), we get
$f(x_n) - f(y_n) = f'(\alpha)\left[e_n + (2c_2^2 - c_3)e_n^3\right]$ and by using (18), we obtain

$$f'(x_n)(f(x_n) - f(y_n)) = f'^2(\alpha)\left[e_n + 2c_2e_n^2 + (2c_2^2 + 2c_3)e_n^3 + O(e_n^4)\right] \tag{23}$$

Now by (22) and (23), we get

$$\frac{f^2(x_n)}{f'(x_n)(f(x_n) - f(y_n))} = e_n - (2c_2^2 + 2c_3)e_n^3 + O(e_n^4) \tag{24}$$

Putting (24) in the Algorithm (1), Eq. (8), we get
$x_{n+1} = x_n - \left\{e_n - (2c_2^2 + 2c_3)e_n^3 + O(e_n^4)\right\}$, where $e_n = x_n - \alpha$

$$x_{n+1} = \alpha + (2c_2^2 + 2c_3)e_n^3 + O(e_n^4) \tag{25}$$

Now, $e_{n+1} = x_{n+1} - \alpha$, by substituting (25), we get.

$e_{n+1} = (2c_2^2 + 2c_3)e_n^3 + O(e_n^4)$ and the proof is completed.

Theorem 2.2 Let $\alpha \in$ I *be a simple zero of sufficiently differentiable function* f: I $\subseteq$ R $\rightarrow$ R *for an open interval* I. *If* × 0 *is sufficiently close to* α, *then the iterative method defined by* (8) *is of order three and satisfies the error equation:*

$$e_{n+1} = (2c_3 - c_2^2)e_n^3 + o(e_n^4),$$

Proof Let α be a simple zero of f, $f'(\alpha) \neq 0$, once again, we can follow the same procedure provided in Theorem 2.1.

Using (17) and (21), we get

$$f(x_n)f(y_n) = f'^2(\alpha)\left[c_2e_n^3\right] \tag{26}$$

$$f(x_n) - f(y_n) = f'(\alpha)\left[e_n + (2c_2^2 - c_3)e_n^3\right]$$

$$[f(x_n) - f(y_n)]^2 = f'^2(\alpha)\left[e_n^2\right] \tag{27}$$

And then dividing (26) by (27), we get

$$\frac{f(x_n)f(y_n)}{[f(x_n) - f(y_n)]^2} = c_2e_n \tag{28}$$

By using (19) and (28), we obtain:

$$\left\{1+\frac{f(x_n)f(y_n)}{[f(x_n)-f(y_n)]^2}\right\}*\frac{f(x_n)}{f'(x_n)}=e_n+(c_2^2-2c_3)e_n^3 \tag{29}$$

Now, substituting (29) in Algorithm (1.2), we get

$$\begin{aligned}x_{n+1}=&x_n-\left\{1+\frac{f(x_n)f(y_n)}{[f(x_n)-f(y_n)]^2}\right\}*\frac{f(x_n)}{f'(x_n)}\\&=x_n-\left\{e_n+(c_2^2-2c_3)e_n^3\right\},\text{ and by } e_n=x_n-\alpha,\text{ we get :}\\&=\alpha+(2c_3-c_2^2)e_n^3+O(e_n^4)\end{aligned}$$

Now using the previous result in $e_{n+1}=x_{n+1}-\alpha$

$$e_{n+1}=\left\{\alpha+(2c_3-c_2^2)e_n^3+O(e_n^4)\right\}-\alpha=(2c_3-c_2^2)e_n^3+O(e_n^4),$$

the proof is done.

2.3 More Suggestions

In this section, we present new modifications of important methods for solving nonlinear equations of type $f(x)=0$ using the substitution of the formula (7) $f'(y_n)=\frac{f(x_n)-f(y_n)}{f(x_n)+f(y_n)}f'(x_n)$ in well-known methods.

As we will see, this is so helpful that it reduces the number of required derivative evaluations in iteration schemes. We will introduce only two suggestions as examples and we will show their rate of convergences.

Example 1 Consider Noor and Gupta's fourth-order method [17, 18].

$y_n=x_n-\frac{f(x_n)}{f'(x_n)}$,

$$x_{n+1}=y_n-\frac{f(y_n)}{f'(y_n)}-\frac{1}{2}\left(\frac{f(y_n)}{f'(y_n)}\right)^2*\frac{f'(y_n)}{f(x_n)}*\frac{f'(y_n)-f'(x_n)}{f'(y_n)} \tag{30}$$

By substituting (7) in (30), we get

$$x_{n+1}=y_n-\frac{f(x_n)+f(y_n)}{f(x_n)-f(y_n)}*\frac{f(y_n)}{f'(x_n)}\left(1-\frac{f(y_n)}{f(x_n)}*\frac{f(y_n)}{f(x_n)-f(y_n)}\right) \tag{31}$$

or in another form:

$$x_{n+1}=y_n-\frac{f^3(x_n)-2f(x_n)f^2(y_n)-f^3(y_n)}{[f(x_n)-f(y_n)]^2}*\frac{f(y_n)}{f(x_n)f'(x_n)}$$

Theorem 2.3 Let $\alpha \in I$ be a simple zero of sufficiently differentiable function.

f: I ⊆ R → R for an open interval I. If × 0 is sufficiently close to α, then the iterative method introduced in (31) is of order four and satisfies the error equation:

$$e_{n+1} = (5c_2^3 - c_2c_3)e_n^4 + o(e_n^5),$$

Notes:

1. The suggested method requires only two function evaluations and one derivative evaluation (r = 3).
2. Rate of convergence P = 4.
3. Efficiency Index E.I. $= p^{1/r} = 4^{1/3} = 1.5874$.

Example 2 Consider Jarratt's fourth-order method [2].

$y_n = x_n - \frac{2f(x_n)}{3f'(x_n)},$

$x_{n+1} = x_n - \frac{3f'(y_n)+f'(x_n)}{6f'(y_n)-2f'(x_n)} * \frac{f(x_n)}{f'(x_n)}$

By substituting the previous formula (16), we get

$$x_{n+1} = x_n - \frac{2f(x_n) - f(y_n)}{2f(x_n) - 4f(y_n)} * \frac{f(x_n)}{f'(x_n)}, \tag{32}$$

where

$$y_n = x_n - \frac{f(x_n)}{f'(x_n)}$$

with error equation

$$e_{n+1} = \frac{-c_2}{2}e_n^2 + (c_2^2 - c_3)e_n^3 + o(e_n^4).$$

Theorem 2.4 *Let* α ∈ I *be a simple zero of sufficiently differentiable function* f: I ⊆ R → R *for an open interval* I. If × 0 *is sufficiently close to* α, *then the iterative method introduced in* (32) *is of order two and satisfies the error equation:*

$$e_{n+1} = \frac{-c_2}{2}e_n^2 + (c_2^2 - c_3)e_n^3 + o(e_n^4),$$

3 Numerical Examples

In this section, first we present the results of numerical calculations on different functions and initial guesses to demonstrate the efficiency of the suggested methods, Variant Homeier Method 1 (VHM1) and its improvement (VHM2). Also, we compare

Table 1 Different test functions and their approximate zeros (∝)

Functions	Approximate roots (∝)
$F_1(x) = x\,e^{x^{\wedge}2} - \sin^2 x + 3\cos x + 5$	−1.20764782713919
$F_2(x) = \sin^2 x - x^2 + 1$	1.404491648215341
$F_3(x) = e^x - 5x^2 + 7x - 3$	0.300026392366926
$F_4(x) = \mathrm{Ln}(x^2 + x + 2) - x + 1$	4.152590736757158
$F_5(x) = x^{10} - 2x^3 - x + 1$	0.591448093340752
$F_6(x) = \sin(1 + x) - x + 2$	2.070766727142040
$F_7(x) = \tan(\sqrt{x^2 + 1}) - 7x + 1$	0.410901501707263)
$F_8(x) = \mathrm{Ln}(\cos x + 1) + \sqrt{1 - 2x} + 3x$	−0.706338530699419
$F_9(x) = (x^2 - 10x)^3 - 100{,}000$	−3.450792172105985
$F_{10}(x) = \sqrt{x} - \mathrm{Ln}(x) + \sin x - x$	1.747991025989651

these methods with famous methods, such as Halley's, Weerakoon and Potra–Ptak methods. All computations are carried out with 15 decimal places (See Table 1) approximate zeros α found up to 15th decimal place).

All programs and computations were completed using MATLAB, 2009a. Table 2 displays the number of iterations (IT) and the computational order of convergences (COC). Table 3 displays the number of function evaluations (r), convergence order (P), efficiency index (E.I.), the sum of iterations, and average COC's for each method. When we reached the sought zero α after only three iterations, we used the second formula to compute the COC of the iterative method. Furthermore, we assumed COC is zero when the iterative method diverged. Table 4 displays the number of function evaluations and derivative evaluations required for each method.

4 Conclusion

We have developed two of 2-step iterative methods for finding zeros of nonlinear functions, (VHM1) and (VHM2). The main goal is to find and improve iterative schemes which requires less derivative evaluations of the function, whereas more derivative evaluations in a method cost need more time and effort from an industry point of view. So, both new methods require only two function evaluations and one first derivative evaluation. On the contrary, known methods as Halley and Householder require one function evaluation, one first derivative and one second derivative evaluation whereas Weerakoon and Homeier methods require one function evaluation and two first derivative evaluations (See Table 4). Furthermore, we have proved theoretically that both new methods are of order three. It can be observed that the numerical experiment is displayed in Tables 2, 3, and 4.

In addition, based on numerical experiments, the proposed methods are also compared with the previous well-known iterative methods of the same order of

Table 2 Comparison of various iterative methods

FUNC	Guess	VHM1		VHM2		HAL MD		HOSH MD		WRK MD		POT MD		HOM MD	
		IT	COC	IT	COC	IT	COC	IT	COC	IT	COC	IT	COC	IT	COC
$F_1(x)$	−1	5	3.04	4	4.02	4	2.81	5	3.01	5	3.01	5	3.04	4	2.89
	−1.5	5	3.00	4	3.88	4	3.28	5	2.99	5	2.99	5	2.97	5	2.95
$F_2(x)$	1	5	3.03	4	4.94	5	3.01	6	3.05	5	3.04	17	2.85	4	3.67
	2	5	3.02	4	3.51	5	3.00	5	2.97	5	2.99	5	2.97	4	2.97
$F_3(x)$	0	4	2.83	4	3.78	4	2.84	4	2.74	4	2.83	4	2.74	4	3.26
	0.5	4	3.15	4	4.21	4	3.14	5	3.01	4	3.15	5	3.02	4	3.40
$F_4(x)$	0	5	2.96	4	5.11	**div**	0	6	2.96	5	3.18	4	2.59	5	3.08
	5	4	2.94	3	2.93	4	2.93	4	2.90	3	1.65	4	2.91	4	2.95
$F_5(x)$	0.25	5	3.04	4	4.42	4	2.53	5	3.06	5	3.06	6	3.16	4	3.30
	0.8	4	2.53	4	3.64	5	3.02	5	2.99	4	3.14	4	2.42	4	3.76
$F_6(x)$	1	4	4.03	4	5.12	5	3.00	5	3.03	4	3.02	4	4.63	5	3.20
	2.5	3	2.99	3	3.38	4	3.00	4	3.03	4	3.01	4	3.26	4	3.01
$F_7(x)$	0	4	2.71	4	3.60	4	2.53	4	2.49	4	2.96	4	2.62	4	2.6
	0.7	5	3.04	5	3.35	5	3.24	**div**	0	4	3.94	8	2.88	5	2.95
$F_8(x)$	0	4	3.22	4	4.42	4	3.68	5	3.04	4	3.00	4	3.45	4	3.65
	−1	4	2.98	3	2.94	4	2.89	4	2.92	4	3.01	4	2.97	3	2.96
$F_9(x)$	−3	5	3.09	4	4.12	5	3.04	5	3.18	4	3.11	6	3.01	4	2.91
	−4	4	2.90	4	3.87	4	2.95	4	2.85	4	2.88	6	2.50	4	3.03
$F_{10}(x)$	1	4	2.94	3	3.34	5	3.06	5	3.00	4	3.14	5	3.00	4	3.59
	2	4	2.96	3	3.39	4	3.04	4	2.98	4	2.94	4	2.94	3	1.72

Table 3 Summary of the comparison of variant methods

	VHM1	VHM2	HAL MD	HOS MD	WRK MD	POT MD	HOM MD
Number of func Eval.'s required (r)	3	3	3	3	3	3	3
Convergence order (p) (theoretically)	3	3	3	3	3	3	3
Efficiency Index E.I. $= p^{1/r}$	1.4422	1.4422	1.4422	1.4422	1.4422	1.4422	1.4422
Sum of iterations needed	87	76	83	90	85	108	82
Average COC	3.02	3.90	3.00	2.96	3.00	3.00	3.09
Number of (div)'s	0	0	1	1	0	0	0

Table 4 Type of functions required for each method

	VHM1	VHM2	HAL MD	HOS MD	WRK MD	POT MD	HOM MD
Number of function Eval.'s required (r)	3	3	3	3	3	3	3
Number of function Eval.'s. f(x)	2	2	1	1	1	2	1
Number of 1st derivative Eval.'s. f '(x)	1	1	1	1	2	1	2
Number of 2nd derivative Eval.'s. f '' (x)	0	0	1	1	0	0	0

convergence. The performance of the proposed methods can be seen in Tables 2, 3, and 4.

Moreover, it can easily be seen that both new methods are more efficient, robust, and faster convergence than the other methods with respect to the required number of derivative evaluations for each method, IT's and COC results.

Numerical experiments show that the order of convergence of both methods is at least three.

Conflict of Interest Statement: The authors declare no conflict of interest regarding this publication.

References

1. Ababneh, O.Y.: New iterative methods for solving nonlinear equations and their basins of attraction. WSEAS Trans. Math. Link Disabl. **21**, 9–16 (2022)
2. Ahmad, F., Hussain, S., Rafiq, A.: New twelfth-order j-halley method for solving nonlinear equations. Open Sci. J. Math. Appl. **1**(1), 1–4 (2013)
3. Atkinson, K.E.: An Introduction to Numerical Analysis. Wiley, Inc (1989). ISBN 0-471-62489-6
4. Bonnans, J.F., Gilbert, J.C., Lemaréchal, C., Sagastizábal, C.A.: Numerical Optimization: Theoretical and Practical Aspects, University-text (2nd revised edn. of translation of 1997 French ed.) Berlin: Springer-Verlag. pp. Ope xiv+490 (2006). ISBN 3-540-35445-X. MR 2265882
5. Chun, C., Neta, B.: Comparative study of methods of various orders for finding simple roots of nonlinear equations. J. Appl. Anal. Comput. **9**, 400–427 (2019)
6. Eldanfour, H.M.: Modified Newton's methods with seventh or eighth -order convergence. Gen. Lett. Math. **1**(1), 1–10 (2016). https://doi.org/10.31559/glm2016.1.1.1
7. Ezquerro, J.A., Hernandez, M.A.: Unparametric Halley-type iteration with free second derivative. Int. J. Pure Appl. Math. **6**(1), 103–114 (2003)
8. Ezquerro, J.A., Hernandez, M.A.: On Halley-type iterations with free second derivative. J. Comput. Appl. Math. **170**, 455–459 (2004)
9. Gautschi, W.: Numerical Analysis: An Introduction. Birkhauser (1997)
10. Gutierrez, J.M., Hernandez, M.A.: An acceleration of Newton's method: super-Halley method. Appl. Math. Comput. **117**, 223–239 (2001)
11. Halley, E.: A new exact and easy method of finding the roots of equations generally and that without any previous reduction. Philos. Trans. R. Soc. London **18**, 136–148 (1694)
12. Homeier, H.H.H.: On Newton–type methods with cubic convergence. J. Comput. Appl. Math. **176**, 425–432 (2005)
13. Kumar, S., Kanwar, V., Singh, S.: Modified efficient families of two and three-step predictor-corrector iterative methods for solving nonlinear equations. Appl. Math. **1**, 153–158 (2010)
14. Lambers, J.: Error Analysis for Iterative Methods, 2009–10 : lecture 12 notes, Mat 460/560 (2009)
15. Melman, A.: Geometry and convergence of Halley's method. SIAM Rev. **39**(4), 728–735 (1997)
16. Neta, B.: A new derivative-free method to solve nonlinear equations. Mathematics **9**, 583 (2021). https://doi.org/10.3390/math9060583
17. Noor, M.A., Gupta, V.: Modified householder iterative method free from second derivative for nonlinear equations. Appl. Math. Comput. (2007) in press
18. Noor, M.A., Khan, W.A., Noor, K.I., Al-said, E.: Higher order iterative methods free from second derivative for solving nonlinear equations. Int. J. Phy. Sci. **6**(8), 1887–1897 (2011)
19. Ricceri, B.: A class of equations with three solutions. Mathematics **8**, 478 (2020)
20. Ostrouski, A.M.: Solutions of Equations and System of Equations. Academic Press, New York (1960)
21. Ozban, A.Y.: Some new variants of Newton's method. App. Math. Lett. **17**(2004), 677–682 (2004)
22. Petkovic, M.S., Neta, B., Petkovic, L.D., Dzunic, J.: Multipoint Methods for Solving Nonlinear Equations. Elsevier (2012)
23. Potra, F.A., Ptak, V.: Nondiscrete Introduction and Iterative Processes, Research notes in Mathematics, vol. 103. Pitman, Boston (1984)
24. Scavo, T.R., Thoo, J.B.: On the geometry of Halley's method. Am. Math. Mon. (1994)
25. Soleymani, F., Sharma, R., Li, X., Tohidi, E.: An optimized derivative free form of the potra_ptak method. Math. Comput. Mod. **56**, 97–104 (2012)
26. Thukral, R.: New modifications of Newton-type methods with eighth-order convergence for solving nonlinear equations. J. Adv. Math. **10**(3), 3362–3373 (2015)
27. Troub, J.F.: Iterative Methods for Solution of Equations. Chelsea publishing Company, New York (1977)

28. Weerakoon, S.T., Fernando, G.I.: A variant of Newton's method with accelerated third-order convergence. Appl. Math. Lett. **13**, 87–93 (2000)
29. Ypma, J.T.: Historical development of the Newton-Raphson method. SIAM Rev. **37**(4), 531–551 (1995)
30. Zhanlav, T., Otgondorj, K.: Comparison of some optimal derivative-free three-point iterations. J. Numer. Anal. Approx. Theory. **49**, 76–90 (2020)

On Tempered Exponential Trisplitting for Random Semi-dynamical Systems

Ioan-Lucian Popa, Traian Ceauşu, Larisa Elena Biriş, and Akbar Zada

Abstract In the present paper, the concept of tempered exponential trisplitting for random one-sided discrete-time systems is considered. We establish Datko's type result in terms of invariant projections.

Keywords Stochastic dynamical systems · Tempered exponential trisplitting

1 Introduction and Preliminaries

The concept of exponential trisplitting is a generalization of the well-known notion of exponential trichotomy. Important results for the study of exponential trichotomy for linear discrete-time systems were obtained for the deterministic case. See, for example, [1, 6, 10, 18]. It is worth mentioning [7, 15], where the authors studied the connections between uniform exponential trisplitting and uniform exponential trichotomy, and they presented some necessary and sufficient conditions for uniform exponential trisplitting with invariant projectors, respectively strongly invariant projectors. Characterizations of tempered exponential splitting for random semi-dynamical systems are obtained in [17].

I.-L. Popa (✉)
Department of Computing, Mathematics and Electronics, "1 Decembrie 1918" University of Alba Iulia, 510009 Alba Iulia, Romania
e-mail: lucian.popa@uab.ro

Faculty of Mathematics and Computer Science, Transilvania University of Braşov, Iuliu Maniu Street 50, 500091 Braşov, Romania

T. Ceauşu · L. E. Biriş
Department of Mathematics, West University of Timisoara, Timisoara, Romania
e-mail: traian.ceausu@e-uvt.ro

L. E. Biriş
e-mail: larisa.biris@e-uvt.ro

A. Zada
Department of Mathematics, University of Peshawar, Peshawar 25000, Pakistan
e-mail: akbarzada@uop.edu.pk

D. Zeidan et al. (eds.), *Mathematics and Computation*, Springer Proceedings in Mathematics & Statistics 418, https://doi.org/10.1007/978-981-99-0447-1_38

In the present paper, we consider random discrete-time systems which are defined only on semi-axes, the so-called one-sided systems. The aim of this paper is to extend a result of Datko's type from the deterministic case of linear discrete-time skew product over semiflows to the stochastic one-sided discrete-time random dynamical systems. We obtain a necessary and sufficient condition for tempered exponential splitting with the hypothesis that the projectors are invariant. It is worth mentioning that this approach can be extended to the case of strongly invariant projectors as they are considered in Definition 4.

Let $\mathbb{Z}_+$ denote the set of positive integers. $(X, \|\cdot\|)$ denotes a Banach space. By $\mathcal{B}(X)$, we denote the Banach algebra of all bounded linear operators acting from X into X. By $(\Omega, \mathfrak{F}, \mathbb{P})$, we denote a probability space and $\theta : \Omega \to \Omega$ is a measurable map preserving the probability measure $\mathbb{P}$, that is $\mathbb{P}(\theta B) = \mathbb{P}(B)$, for any $B \in \mathfrak{F}$.

Definition 1 *(see, for example,* [3, 8, 9, 17]) A random variable $\varphi : \Omega \to (0, +\infty)$ is called $\theta-$invariant if $\varphi \circ \theta = \varphi$, that is $\varphi(\theta\omega) = \varphi(\omega)$, for all $\omega \in \Omega$. By convention, we have $\theta^0 = I_\Omega$, where I denotes the identity. As a fast property, we have that $\theta^n \circ \theta^m = \theta^{n+m} = \theta^{m+n} = \theta^m \circ \theta^n$, for all $m, n \in \mathbb{Z}_+$.

The application $\mathbb{Z}_+ \times \Omega \ni (n, z) \to \theta^n\omega \in \Omega$ is measurable for all $n \in \mathbb{Z}_+$. We have that $\mathbb{P}(\theta^n B) = \mathbb{P}(B)$, for all $n \in \mathbb{Z}_+$ and all $B \in \mathfrak{F}$. Also, we have that $\varphi(\theta^n\omega) = \varphi(\omega)$, for all $n \in \mathbb{Z}_+$ and $\omega \in \Omega$.

Further, we consider the metric semi-dynamical system $(\Omega, \mathcal{F}, \mathbb{P}, \theta)$, which is a probability space, with $\theta : \Omega \to \Omega$, measurable. The measurable application $\phi : \mathbb{Z}_+ \times \Omega \to \mathcal{B}(X)$ represents a linear random one-sided discrete-time system on X over a measurable semi-dynamical system θ. For more details about these notions, we can point out the references [8, 9, 13]. Of interest in this paper, we have the following properties:

(a) $\phi(0, \omega) = I_X$, for all $\omega \in \Omega$;
(b) $\phi(n + m, \omega) = \phi(n, \theta^m\omega)\phi(m, \omega)$, for all $n, m \in \mathbb{Z}_+$ and $\omega \in \Omega$.

Obvious from relation *(b)*, we have that

$$\phi(n + m, \omega) = \phi(m, \theta^n\omega)\phi(n, \omega), \quad \text{for all } n, m \in \mathbb{Z}_+ \text{ and } \omega \in \Omega.$$

Throughout this work, the notation (θ, ϕ) will be used for a linear random one-sided discrete-time system (RDTS).

Definition 2 An application $P : \Omega \to \mathcal{B}(X)$ is called a projection if

$$P^2(\omega) = P(\omega), \quad \text{for all } \omega \in \Omega.$$

Definition 3 The projection $P : \Omega \to \mathcal{B}(X)$ is called invariant for the RDTS (θ, ϕ) if

$$\phi(n, \omega)P(\omega) = P(\theta^n\omega)\phi(n, \omega) \tag{1}$$

for all $(n, \omega) \in \mathbb{Z}_+ \times \Omega$.

Remark 1 If the projection $P : \Omega \to \mathcal{B}(X)$ is invariant for the RDTS (θ, ϕ) then $Q : \Omega \to \mathcal{B}(X)$ defined by $Q(\omega) = I - P(\omega)$ is also invariant for RDTS (θ, ϕ).

Definition 4 The projection $P : \Omega \to \mathcal{B}(X)$ is called strongly invariant for the RDTS (θ, ϕ) if (1) is satisfied and $\phi(n, \omega) : Ker\ P(\omega) \to Ker\ P(\theta^n\omega)$ is an isomorphism, for all $(n, \omega) \in \mathbb{Z}_+ \times \Omega$.

As a fast remark from Definition 4, we obtain the following.

Remark 2 The invariant projection $P : \Omega \to \mathcal{B}(X)$ is strongly invariant for the RDTS (θ, ϕ) if $\phi(n, \omega) : Q(\omega)X \to Q(\theta^n\omega)X$ is an isomorphism, for all $(n, \omega) \in \mathbb{Z}_+ \times \Omega$.

Definition 5 If $P_1, P_2, P_3 : \Omega \to \mathcal{B}(X)$ are three strongly invariant projections, then the family $\mathcal{P} = \{P_1, P_2, P_3\}$ is

(a) orthogonal if

$$P_1(\omega) + P_2(\omega) + P_3(\omega) = I, \quad \text{for all } \omega \in \Omega;$$

$$P_k(\omega) = P_j(\omega) = 0, \quad \text{for all } \omega \in \Omega, \quad \text{and any } k, j \in \{1, 2, 3\}, \ k \neq j.$$

(b) invariant for the RDTS (θ, ϕ) if P_j is invariant for the RDTS (θ, ϕ), for all $j \in \{1, 2, 3\}$.
(c) strongly invariant for the RDTS (θ, ϕ) if P_j is strongly invariant for the RDTS (θ, ϕ), for all $j \in \{1, 2, 3\}$.

Definition 6 Let $\mathcal{P} = \{P_1, P_2, P_3\}$ be a family of orthogonal and invariant projections for the RDTS (θ, ϕ). We say that the pair $(\theta, \mathcal{P})$ admits a tempered exponential trisplitting if there exists a function $N : \Omega \to [1, +\infty)$ and θ–invariant random variables $\alpha, \beta, \gamma, \delta : \Omega \to (0, +\infty)$, with $\alpha < \beta$ and $\gamma < \delta$ such that

$$\|\phi(n, \omega)P_1(\omega)x\| \leq N(\omega)e^{\alpha(\omega)n}\|P_1(\omega)x\| \tag{2}$$

$$e^{\beta(\omega)n}\|P_2(\omega)x\| \leq N(\omega)\|\phi(n, \omega)P_2(\omega)x\| \tag{3}$$

$$e^{\gamma(\omega)n}\|\phi(n, \omega)P_3(\omega)x\| \leq N(\omega)\|P_3(\omega)x\| \tag{4}$$

$$\|P_3(\omega)x\| \leq N(\omega)e^{\delta(\omega)n}\|\phi(n, \omega)P_3(\omega)x\| \tag{5}$$

for all $(n, \omega, x) \in \mathbb{Z}_+ \times \Omega \times X$.

Proposition 1 *Let* $(\phi, \mathcal{P})$ *and the functions* $N : \Omega \to [1, +\infty)$ *and* $\alpha, \beta, \gamma, \delta : \Omega \to (0, +\infty)$, *as in Definition 6. Let* $(n, \omega, x) \in \mathbb{Z}_+ \times \Omega \times X$. *Then we have that*

(a) The following are equivalent:

(a1) Relation (2) holds

(a2)

$$\|\phi(m+n,\omega)P_1(\omega)x\| \le N(\omega)e^{\alpha(\omega)(m+n)}\|P_1(\omega)x\| \tag{6}$$

(a3)

$$\|\phi(m+n,\omega)P_1(\omega)x\| \le N(\theta^m\omega)e^{\alpha(\omega)n}\|\phi(m,\omega)P_1(\omega)x\| \tag{7}$$

(a4)

$$\|\phi(m+n,\omega)P_1(\omega)x\| \le N(\theta^n\omega)e^{\alpha(\omega)m}\|\phi(n,\omega)P_1(\omega)x\| \tag{8}$$

(b) The following are equivalent:

(b1) Relation (3) holds

(b2)

$$e^{\beta(\omega)(m+n)}\|P_2(\omega)x\| \le N(\omega)\|\phi(m+n,\omega)P_2(\omega)x\| \tag{9}$$

(b3)

$$e^{\beta(\omega)n}\|\phi(m,\omega)P_2(\omega)x\| \le N(\theta^m\omega)\|\phi(m+n,\omega)P_2(\omega)x\| \tag{10}$$

(b4)

$$e^{\beta(\omega)m}\|\phi(n,\omega)P_2(\omega)x\| \le N(\theta^n\omega)\|\phi(m+n,\omega)P_2(\omega)x\| \tag{11}$$

(c) The following are equivalent:

(c1) Relation (4) holds

(c2)

$$e^{\gamma(\omega)(m+n)}\|\phi(m+n,\omega)P_3(\omega)x\| \le N(\omega)\|P_3(\omega)x\| \tag{12}$$

(c3)

$$e^{\gamma(\omega)n}\|\phi(m+n,\omega)P_3(\omega)x\| \le N(\theta^m\omega)\|\phi(m,\omega)P_3(\omega)x\| \tag{13}$$

(c4)

$$e^{\gamma(\omega)m}\|\phi(m+n,\omega)P_3(\omega)x\| \le N(\theta^n\omega)\|\phi(n,\omega)P_3(\omega)x\| \tag{14}$$

(d) The following are equivalent:

(d1) Relation (5) holds

(d2)

$$\|P_3(\omega)x\| \le N(\omega)e^{\delta(\omega)(m+n)}\|\phi(m+n,\omega)P_3(\omega)x\| \tag{15}$$

(d3)

$$\|\phi(m,\omega)P_3(\omega)x\| \leq N(\theta^m\omega)e^{\delta(\omega)n}\|\phi(m+n,\omega)P_3(\omega)x\| \quad (16)$$

(d4)

$$\|\phi(n,\omega)P_3(\omega)x\| \leq N(\theta^n\omega)e^{\delta(\omega)m}\|\phi(m+n,\omega)P_3(\omega)x\|. \quad (17)$$

Proof The proof is straightforward using Definition 6 and therefore it is omitted.

Remark 3 (a) Definition 6 represents a natural generalization of Definition 2.6 from [7].

(b) In Definition 6, if we consider the θ-invariant variables $\alpha, \beta, \gamma, \delta : \Omega \to (0, +\infty)$, satisfying $\alpha < 0 < \beta$ and $\gamma < 0 < \delta$ then for the pair $(\phi, \mathcal{P})$ we obtain the definition of tempered exponential trichotomy. For the deterministic case, we refer the reader to [14].

(c) In Definition 6, if we consider $P_3(\omega) = 0$ then we recover the definition of tempered exponential splitting (see [17]). For the deterministic case, we refer the reader to [2, 15]. Also, if we consider the θ–invariant variables $\alpha, \beta : \Omega \to (0, +\infty)$, satisfying $\alpha < 0 < \beta$ then we obtain the concept of tempered exponential dichotomy (see [19]). For the deterministic case, we refer the reader to [4, 5, 11] for the case of skew-product semiflows, and [12, 16] for the case of difference equations.

2 Datko-Type Criterion

Definition 7 Let $\mathcal{P} = \{P_1, P_2, P_3\}$ be a family of orthogonal and invariant projections for the RDTS (θ, ϕ). We say that the pair $(\theta, \mathcal{P})$ admits a tempered exponential trisplitting of Datko type if there exist the functions $N, D : \Omega \to [1, +\infty)$ and θ–invariant random variables $\mu, \nu, \xi, \eta : \Omega \to (0, +\infty)$, with $\mu < \nu$ and $\xi < \eta$ such that

$$\sum_{k=n}^{+\infty} e^{\mu(\omega)(n-k)}N(\theta^n\omega)^{-1}\|\phi(k,\omega)P_1(\omega)x\| \leq D(\omega)\|\phi(n,\omega)P_1(\omega)x\| \quad (18)$$

$$\sum_{k=0}^{n} e^{\nu(\omega)(n-k)}N(\theta^n\omega)^{-1}\|\phi(k,\omega)P_2(\omega)x\| \leq D(\omega)\|\phi(n,\omega)P_2(\omega)x\| \quad (19)$$

$$\sum_{k=n}^{+\infty} e^{\xi(\omega)(k-n)}N(\theta^n\omega)^{-1}\|\phi(k,\omega)P_3(\omega)x\| \leq D(\omega)\|\phi(n,\omega)P_3(\omega)x\| \quad (20)$$

$$\sum_{k=0}^{n} e^{\eta(\omega)(k-n)} N(\theta^n\omega)^{-1} \|\phi(k,\omega)P_3(\omega)x\| \le D(\omega)\|\phi(n,\omega)P_3(\omega)x\| \tag{21}$$

for all $(n, \omega, x) \in \mathbb{Z}_+ \times \Omega \times X$.

Theorem 1 *The pair $(\theta, \mathcal{P})$ admits a tempered exponential trisplitting if and only if the pair $(\theta, \mathcal{P})$ admits a tempered exponential trisplitting of Datko type.*

Proof *Necessity.* Let $N : \Omega \to [1, +\infty)$ and the $\theta-$invariant random variables $\alpha, \beta, \gamma, \delta : \Omega \to (0, +\infty)$ as in Definition 6. We consider the θ-invariant random variables $\mu, \nu, \xi, \eta : \Omega \to (0, +\infty)$ with $\alpha < \mu < \nu < \beta$ and $\xi < \gamma < \delta < \eta$ and $D : \Omega \to [1, +\infty)$ defined by

$$D(\omega) = 1 + \frac{e^{\mu(\omega)}}{e^{\mu(\omega)} - e^{\alpha(\omega)}} + \frac{e^{\beta(\omega)}}{e^{\beta(\omega)} - e^{\nu(\omega)}} + \frac{e^{\gamma(\omega)}}{e^{\gamma(\omega)} - e^{\xi(\omega)}} + \frac{e^{\eta(\omega)}}{e^{\eta(\omega)} - e^{\delta(\omega)}}$$

for all $\omega \in \Omega$. Let $(n, \omega, x) \in \mathbb{Z}_+ \times \Omega \times X$. Using (8) from Proposition 1, we have that

$$\begin{aligned}
&\sum_{k=n}^{+\infty} e^{\nu(\omega)(n-k)} N(\theta^n\omega)^{-1} \|\phi(k,\omega)P_1(\omega)x\| \\
&\le \|\phi(n,\omega)P_1(\omega)x\| \sum_{k=n}^{+\infty} e^{(\alpha(\omega)-\mu(\omega))(k-n)} \\
&= \frac{e^{\mu(\omega)}}{e^{\mu(\omega)} - e^{\alpha(\omega)}} \|\phi(n,\omega)P_1(\omega)x\| \\
&\le D(\omega)\|\phi(n,\omega)P_1(\omega)x\|,
\end{aligned}$$

hence (18). Similarly, using (11) we obtain

$$\begin{aligned}
&\sum_{k=0}^{n} e^{\nu(\omega)(n-k)} N(\theta^k\omega)^{-1} \|\phi(k,\omega)P_2(\omega)x\| \\
&\le \|\phi(n,\omega)P_2(\omega)x\| \sum_{k=0}^{n} e^{(\nu(\omega)-\beta(\omega))(n-k)} \\
&= \|\phi(n,\omega)P_2(\omega)x\| e^{(\nu(\omega)-\beta(\omega))n} \sum_{k=0}^{n} e^{(\beta(\omega)-\nu(\omega))k} \\
&\le \|\phi(n,\omega)P_2(\omega)x\| e^{(\nu(\omega)-\beta(\omega))n} \frac{e^{\beta(\omega)}}{e^{\beta(\omega)} - e^{\nu(\omega)} - 1} \\
&= \frac{e^{\beta(\omega)}}{e^{\beta(\omega)} - e^{\nu(\omega)}} \|\phi(n,\omega)P_2(\omega)x\| \\
&\le D(\omega)\|\phi(n,\omega)P_2(\omega)x\|
\end{aligned}$$

from where we obtain that (19) is true. Further, (14) implies that

$$\begin{aligned}
&\sum_{k=n}^{+\infty} e^{\xi(\omega)(k-n)} N(\theta^n\omega)^{-1} \|\phi(k,\omega)P_3(\omega)x\| \\
&\leq \|\phi(n,\omega)P_3(\omega)x\| \sum_{k=n}^{+\infty} e^{(\xi(\omega)-\gamma(\omega))(k-n)} \\
&= \frac{e^{\gamma(\omega)}}{e^{\gamma(\omega)}-e^{\xi(\omega)}} \|\phi(n,\omega)P_3(\omega)x\| \\
&\leq D(\omega)\|\phi(n,\omega)P_3(\omega)x\|
\end{aligned}$$

from where we obtain that (20) holds. Finally, using (17) one can check that

$$\begin{aligned}
&\sum_{k=0}^{n} e^{\eta(\omega)(k-n)} N(\theta^k\omega)^{-1} \|\phi(k,\omega)P_3(\omega)x\| \\
&\leq \|\phi(n,\omega)P_3(\omega)x\| \sum_{k=0}^{n} e^{(\eta(\omega)-\delta(\omega))(k-n)} \\
&= \|\phi(n,\omega)P_3(\omega)x\| e^{(\delta(\omega)-\eta(\omega))n} \frac{e^{(\eta(\omega)-\delta(\omega)(n+1)}}{e^{\eta(\omega)-\delta(\omega)}-1} \\
&= \frac{e^{\eta(\omega)}}{e^{\eta(\omega)}-e^{\delta(\omega)}} \|\phi(n,\omega)P_3(\omega)x\| \\
&\leq D(\omega)\|\phi(n,\omega)P_3(\omega)x\|
\end{aligned}$$

which provides that (21) is also satisfied. Hence, the pair $(\theta, \mathcal{P})$ admits a tempered exponential trisplitting of Datko type.

Sufficiency. Let $(k, \omega, x) \in \mathbb{Z}_+ \times \Omega \times X$. From (18) for $n = 0$, we have that

$$e^{-\mu(\omega)k} \|\phi(k,\omega)P_1(\omega)x\| \leq N(\omega)D(\omega)\|P_1(\omega)x\|$$

so (2) is satisfied. In a similar manner using (20), we obtain

$$e^{\xi(\omega)k} \|\phi(k,\omega)P_3(\omega)x\| \leq N(\omega)D(\omega)\|P_3(\omega)x\|$$

hence (4). Now, let $(n, \omega, x) \in \mathbb{Z}_+ \times \Omega \times X$. For $k = 0$ from (19), we deduce that

$$e^{\nu(\omega)n} \|P_2(\omega)x\| \leq N(\omega)D(\omega)\|\phi(n,\omega)P_2(\omega)x\|$$

so (3) is true. Finally, making use of (21) one sees that

$$e^{-\eta(\omega)n} \|P_3(\omega)x\| \leq N(\omega)D(\omega)\|\phi(n,\omega)P_3(\omega)x\|$$

which conclude that (5) holds. Thus, we may conclude that the pair $(\theta, \mathcal{P})$ admits a tempered exponential trisplitting. This completes the proof.

Acknowledgements This research was funded by "1 Decembrie 1918" University of Alba Iulia through scientific research funds.

References

1. Alonso, A.I., Hong, J., Obaya, R.: Exponential dichotomy and trichotomy for difference equations. Comp. Math. Appl. **38**, 41–49 (1999)
2. Aulbach, B., Kalkbrenner, J.: Exponential forward splitting for noninvertible difference equation. Comput. Math. Appl. **42**, 743–754 (2001)
3. Arnold, L.: Random Dynamical Systems. Springer, New York (1998)
4. Chow, S.-N., Leiva, H.: Two definitions of exponential dichotomy for skew-product semiflows in Banach spaces. Proc. Amer. Math. Sc. **124**, 1071–1081 (1996)
5. Chow, S.-N., Leiva, H.: Existence and Roughness of the Exponential Dichotomy for Skew-Product Semiflow in Banach Spaces. J. Diff. Equ. **120**, 429–477 (1995)
6. Ducrot, A., Magal, P., Seydi, O.: Persistence of exponential trichotomy for linear operators: a Lyapunov-Perron approach. J. Diff. Equ. **28**, 93–126 (2016)
7. Biriş, E.L., Ceauşu, T., Mihiţ, C.L., Popa, I.-L.: Uniform exponential trisplitting - a new criterion for discrete skew-product semiflows. Electron. J. Qual. Theory Diff. Equ. **70**, 1–22 (2019)
8. Cong, N.D.: Topological Dynamics of Random Dynamical Systems. Clarendon, Oxford (1997)
9. Doan, T.S.: Lyapunov exponents for random dynamical systems. Ph.D. Technischen Universitat Dresden (2009)
10. Elaydi, S., Janglajev, K.: Dichotomy and trichotomy of difference equations. J. Diff. Equ. Appl. **3**(5–6), 417–448 (1998)
11. Huy, N.T., Phi, H.: Discrete Characterizations of exponential dichotomy of linear skew-product semiflows over semiflows. J. Math. Anal. Appl. **362**, 46–57 (2010)
12. Lupa, N.: Roughness of $(Z+, Z)$-nonuniform exponential dichotomy for difference equations in Banach spaces. Sci. World J. **6**, 1–12 (2014)
13. Mierczyński, J., Shen, W.: Spectral Theory for Random and Nonautonomous Parabolic equations and Applications. Chapman & Hall/CRC (2008)
14. Popa, I.-L., Ceauşu, T., Bagdasar, O., Agarwal, R.P.: Characterizations of generalized exponential trichotomies for linear discrete time systems. Ann. St. Univ. Ovidius Constanta **27**(2), 153–166 (2019)
15. Megan, M., Popa, I.-L.: Exponential splitting for nonautonomous linear discrete-time systems in Banach spaces. J. Comput. Appl. Math. **312**, 181–191 (2017)
16. Popa, I.-L., Megan, M., Ceauşu, T.: Exponential dichotomies for linear discrete-time systems in Banach spaces. Appl. Anal. Discret. Math. **6**, 140–155 (2012)
17. Popa, I.-L.: Lyapunov functions for random semi-dynamical systems in terms of tempered exponential splitting. Math. Methods Appl. Sci. **44**(15), 11923–11932 (2021)
18. Popa, I.-L., Megan, M., Ceauşu, T.: On h-trichotomy of linear discrete-time systems in Banach spaces, Acta Univ. Apulensis Math. Inform. **39**, 329–339 (2014)
19. Zhou, L., Lu, K., Zhang, W.: Roughness of tempered exponential dichotomies for infinite-dimensional random difference equations. J Diff. Equ. **25**, 4024–4046 (2013)

On q-Laplace Transforms

H. El-Metwally, F. M. Masood, Radwan Abu-Gdairi, and Tareq M. Al-shami

Abstract In this paper, we are concerned about q-Laplace transform which is expected to play a similar role in q-difference analysis as the Laplace transform in continuous analysis or Z transform in difference analysis, especially, in solving q-difference equations.

Keywords q-Laplace transforms · q-difference equations

1 Introduction

Quantum calculus, sometimes called calculus without limits, is equivalent to traditional infinitesimal calculus without the notion of limits. It is defined as "q-calculus", where q stands for quantum. In q-calculus, we are looking for q-analogues of mathematical objects that have the original object as limits when q tends to 1.

The subject of q-calculus started appearing in the nineteenth century in intensive works especially by Jackson [13], Carmichael [5], Mason [16], Adams [2], Trjitzinsky [23], and other authors such as Poincare, Picard, and Ramanujan.

The q-difference has many applications in different mathematics, such as orthogonal polynomials [12], fractal geometry [9, 10], statistical physics [24], quantum mechanics, number theory, and other sciences including mechanics, quantum theory, and theory of relativity [4].

H. El-Metwally (✉)
Department of Mathematics, Mansoura University, Mansoura 51931, Egypt
e-mail: eaash69@yahoo.com

F. M. Masood
Department of Mathematics, Sana'a University, P.O. Box 1247, Sana'a, Yemen

R. Abu-Gdairi
Department of Mathematics, Faculty of Science, Zarqa University, P.O. Box 13110, Zarqa, Jordan
e-mail: rgdairi@zu.edu.jo

T. M. Al-shami
Department of Mathematics, Sana'a University, P.O. Box 1247, Sana'a, Yemen

D. Zeidan et al. (eds.), *Mathematics and Computation*, Springer Proceedings
in Mathematics & Statistics 418, https://doi.org/10.1007/978-981-99-0447-1_39

Laplace transforms have been widely used in mathematical physics and applied mathematics. The theory of the Laplace transform is well known by Sneddon [22], and its generalization was considered by many authors such as Zemanian [25], Rao [20], and Saxena [17–19]. Various existence conditions and a detailed study about the range and invertibility were studied by Rooney [21].

In this paper, we present some definitions, theories, and properties of the q-Laplace transform of some elementary functions such that we need to solve some q-difference equations with some examples.

2 Preliminaries

In this section, we recall the main concepts and properties of q-Laplace transforms which represent an extension of their counterparts via classical Laplace transforms. Then, we list some q-Laplace transforms of some elementary functions.

2.1 Fundamental Concepts of q-Calculus

Definition 1 We define the q-Laplace transformation as a function

$$F(s) = \pounds_q\{f(t)\} = \int_0^{+\infty} e_{q^{-1}}^{-st} f(t) d_q t, \quad s = p + i\sigma \in \mathbb{C}, \tag{1}$$

and we denote $f(t) \rightleftharpoons_q F(s)$. Here, $f(t)$ is denoted as q-original of $F(s)$, while $F(s)$ is denoted as the q-formimag of $f(t)$ by the q-Laplace transformation [1].

Let us recall some basic concepts of q-calculus introduced in published literature [4, 6, 8, 11, 13–15].

The shifted factorial $(m)_n$ is defined by

$$\begin{aligned} (m;q)_0 &= 1, \\ (m;q)_n &= (1-m)(1-qm)(1-q^2m)(1-q^3m)\dots(1-q^{n-1}m) \\ &= \prod_{k=0}^{n-1}(1-q^k m), \qquad n \in N. \end{aligned}$$

A complex number m is defined by

$$\begin{aligned} [m]_q &= 1 + q + q^2 + \dots + q^{m-1} \\ &= \frac{1-q^m}{1-q}, \qquad q \in \mathbb{C} - \{1\}; \ m \in \mathbb{C}, \end{aligned}$$

and the factorial function is

$$[m]_q! = [1]_q[2]_q[3]_q \ldots [m]_q$$
$$= \prod_{n=1}^{m}[n]_q, \qquad q \neq 1; \quad n \in \mathbb{N}, \quad 0 \leq q \leq 1.$$

The q-binomial coefficient $\begin{bmatrix} m \\ k \end{bmatrix}_q$ is defined by

$$\begin{bmatrix} m \\ k \end{bmatrix}_q = \frac{[m]_q!}{[r]_q![m-r]_q!}, \quad r = 0, 1, 2, \ldots m.$$

The function $(x + y)^m$ is defined as

$$(x + y)_q^m = \sum_{r=0}^{m} \begin{bmatrix} m \\ r \end{bmatrix}_q q^{r(r-1)/2} x^{m-r} y^r; \; m \in \mathbb{N}.$$

The exponential function is defined as

$$e_q^t = \sum_{k=0}^{\infty} \frac{t^k}{[k]_q!}, \qquad 0 < |q| < 1\,.$$

The functions e_q^t and $e_{q^{-1}}^{-t}$ satisfy

$$e_q^t \, e_{q^{-1}}^{-t} = 1.$$

The q-derivative $D_q f$ is defined as

$$D_q f(t) = \frac{f(qt) - f(t)}{qt - t}, \qquad 0 < |q| < 1,$$

$$D_q(fg)(t) = g(qt)D_q f(t) + f(t)D_q g(t)$$
$$= f(qt)D_q g(t) + g(t)D_q f(t),$$

and

$$D_q\left(\frac{f}{g}\right)(t) = \frac{g(t)D_q f(t) - f(t)D_q g(t)}{g(qt)g(t)}.$$

The q-integral

$$\int_a^t f(u)d_qu = \sum_{k=0}^{\infty}(1-q)q^k\left[tf\left(tq^k\right) - af\left(q^k a\right)\right], \quad t \in [a,b],$$

and for $a = 0$, we obtain

$$I_q f(t) = \int_0^t f(u)d_qu = \sum_{k=0}^{\infty} t(1-q)q^k f\left(tq^k\right),$$

provided the series converges. Also

$$\int_a^b f(t)d_qt = \int_0^b f(t)d_qt - \int_0^a f(t)d_qt, \quad a \in [0,b].$$

Similarly

$$I_q^0 f(t) = f(t), \qquad I_q^n f(t) = I_q I_q^{n-1} f(t), \qquad n \in \mathbb{N}.$$

Integration by parts is given by

$$\int_a^b f(t)D_q g(t)d_qt = [fg]_a^b - \int_a^b g(qt)D_q f(t)d_qt.$$

2.2 Main Properties of q-Laplace Transforms

Herein, we list the ten basic properties of q-Laplace transforms. 1. Scaling:

$$£_q\{\alpha f(t)\} = \alpha £_q\{f(t)\}, \quad \alpha \in \mathbb{R}.$$

2. Linearity:

$$£_q\{\alpha f(t) + \beta g(t)\} = \alpha £_q\{f(t)\} + \beta £_q\{g(t)\}, \qquad \alpha, \beta \in \mathbb{R}.$$

3. Substitution:

$$e_{q^{-1}}^{-st+s_0t} e_q^{st} f(t) \rightleftharpoons_q F(s-s_0). \tag{2}$$

4. Translation: Consider

$$\eta(t) = \begin{cases} 0, & t < 0, \\ 1, & t \geq 0. \end{cases}$$

it is clear that $f(t) = f(t)\eta(t)$ for $t \geq 0$. Hence, we have

$$\pounds_q\{f(t-t_0)\} = \int_{t_0}^{+\infty} e_{q^{-1}}^{-st} f(t-t_0)\,\eta(t-t_0)\,d_q t.$$

Supposing $t - t_0 = t$, we get

$$\pounds_q\{f(t-t_0)\} = \int_0^{+\infty} \frac{t+t_0}{t} e_q^{st} e_{q^{-1}}^{-s(t+t_0)} f(t) d_q t$$
$$= e_{q^{-1}}^{-st_0} \pounds_q \left\{ \frac{t+t_0}{t} e_q^{st_0} e_q^{st} e_{q^{-1}}^{-s(t+t_0)} f(t) \right\}.$$

5. Transform of derivatives

$$D_q^n f(t) \rightleftharpoons_q \frac{s^n}{q^n} F(s) - \sum_{j=0}^{n-1} \left(\frac{s}{q}\right)^{n-1-j} D_q^j f(0). \tag{3}$$

6. Derivative of transforms

$$(-t)^n q^{\frac{-n}{2}(n+3)} f(tq^{-n}) \rightleftharpoons_q D_{q,s}^n F(s). \tag{4}$$

7 . Transform of $t^n f(t)$ is given by
(i) $t^n f(t) \rightleftharpoons_q (-1)^n D_{q^{-1},s}^n (F(s))$,

(ii) $t^n f(t) \rightleftharpoons_q (-1)^n q^{-\frac{n(n+1)}{2}} D_q^n (F(sq^{-n}))$.

8. Transform of integrals

$$\int_0^t f(t) d_q t \rightleftharpoons_q q \frac{F(s)}{s}.$$

9. Integral of transforms.

$$\int_s^\infty F(s) d_q t \rightleftharpoons_q q \frac{f(qt)}{t}. \tag{5}$$

This formula is especially useful in computing infinite integrals. Indeed, let $s \to 0$ in (5), then

$$\int_0^\infty F(s) d_q t = q \int_0^\infty \frac{f(qt)}{t} d_q t = q \int_0^\infty \frac{f(t)}{t} d_q t.$$

10. Product of transforms

$$f(t) *_q g(t) \rightleftharpoons_q \pounds_q\{f(t)\}\pounds_q\{g(t)\} = F(s)G(s). \tag{6}$$

2.3 q-Laplace Transform of Some Elementary Functions

In what follows, we provide some q-Laplace transform of some elementary functions.

1. If $f(t) = 1$, then

$$F(s) \rightleftharpoons_q \frac{q}{s}.$$

2. If $f(t) = t$, then

$$F(s) \rightleftharpoons_q \frac{q^2}{s^2}.$$

3. If $f(t) = t^n$, then

$$t^n \rightleftharpoons_q [n]_q! \left(\frac{q}{s}\right)^{n+1}.$$

4. If $f(t) = e_q^{\alpha t}$, since $e_q^{\alpha t} = \sum_{k=0}^{\infty} \frac{\alpha^n t^n}{[n]_q!}$, then

$$F(s) \rightleftharpoons_q \frac{q}{s - q\alpha}, \qquad |s| > |\alpha q|. \tag{7}$$

5. If $f(t) = e_{q^{-1}}^{\alpha t}$, since $e_{q^{-1}}^{\alpha t} = \sum_{n=0}^{\infty} \frac{\alpha^n t^n}{[n]_{q^{-1}}!}$, then

$$F(s) \rightleftharpoons_q \frac{q}{s} \sum_{n=0}^{\infty} q^{\frac{n}{2}(n-1)} \left(\frac{q\alpha}{s}\right)^n.$$

6. If $f(t) = \cos_q \alpha t = \frac{e_q^{i\alpha t} + e_q^{-i\alpha t}}{2}$, then

$$F(s) \rightleftharpoons_q \frac{qs}{s^2 + q^2\alpha^2}. \tag{8}$$

7. If $f(t) = \sin_q \alpha t = \frac{e_q^{i\alpha t} - e_q^{-i\alpha t}}{2i}$, then

$$F(s) \rightleftharpoons_q \frac{q^2\alpha}{s^2 + q^2\alpha^2}. \tag{9}$$

8. If $f(t) = \cosh_q \alpha t = \frac{e_q^{\alpha t} + e_q^{-\alpha t}}{2}$, then

$$F(s) \rightleftharpoons_q \frac{q^2\alpha}{s^2 + q^2\alpha^2}. \tag{10}$$

9. If $f(t) = \sinh_q \alpha t = \frac{e_q^{\alpha t} - e_q^{-\alpha t}}{2}$, then

$$F(s) \rightleftharpoons_q \frac{q^2\alpha}{s^2 - q^2\alpha^2}. \tag{11}$$

10. If $f(t) = \sum_{k=0}^{\infty} \alpha_k t^k$, then

$$F(s) \rightleftharpoons_q \frac{q}{s} \sum_{k=0}^{\infty} \alpha_k [k]_q! \left(\frac{q}{s}\right)^k.$$

3 Applications of q-Laplace Transforms to Solve Some q-Difference Equations

In most cases, the search for the q-origin of a given q-image is performed using the results of the basic primitive function transform along with the application of the properties of the q-Laplace transform.

(Note: We put $\pounds_q^{-1}\{F(s)\} = f(t)$, and this is an inverse correspondence of q-Laplace transform.)

Theorem 1 *If the q-image of the unknown origin of q could be written in an integer series of powers $\frac{1}{s}$ of the form*

$$F(s) = \sum_{j=0}^{\infty} \alpha_j s^{-j-1}, \tag{12}$$

(this series is convergent to $F(s)$ for $|s| > R$, where $R = \lim_{n\to\infty} \left|\frac{\alpha_{n+1}}{\alpha_n}\right| \neq \infty$), then the q-original $f(t)$ is given by the formula

$$f(t) = \sum_{k=0}^{\infty} \frac{\alpha_k}{q^{k+1}[k]_q!} t^k. \tag{13}$$

Example 1 Find the inverse of $F(s) = \frac{1}{s - s_0}$.

Solution. We have

$$F(s) = \sum_{k=0}^{\infty} s_0^k s^{-k-1}.$$

Hence,

$$f(t) = \sum_{k=0}^{\infty} \frac{s_0^k}{[k]_q! s^{k+1}} t^k = \frac{1}{q} \sum_{k=0}^{\infty} \frac{(s_0 q^{-1} t)^k}{[k]_q!} = q^{-1} e_q^{s_0 q^{-1} t}.$$

As Laplace transform is widely applied in solving differential and difference equations, the q-Laplace transform is expected to play the same role but now in q-difference equations. The principle lying behind is always the same:

1. Consider a $k-$order linear constant coefficient q-difference equation, with initial conditions

$$c_0 D_q^k y(t) + c_1 D_q^{k-1} y(t) + \cdots + c_{k-1} D_q y(t) + c_k y(t) = g(t), \tag{14}$$

$$y(0) = y_0,\ D_q y(0) = y_1, \ldots,\ D_q^{k-1} y(0) = y_{k-1},$$

by using q-Laplace transform on both sides of the equation and then use the inverse q-Laplace transform to find the unknown function $y(t)$.

For example, consider the case of the second order

$$c_0 D_q^2 y(t) + c_1 D_q y(t) + c_2 y(t) = g(t), \tag{15}$$

$$y(0) = y_0, \quad D_q y(0) = y_1.$$

Let $y(t) \rightleftharpoons_q Y(s)$, $g(t) \rightleftharpoons_q G(s)$, and using (3), then

$$\begin{aligned} D_q y(t) &\rightleftharpoons_q \frac{s}{q} Y(s) - y(0), \\ D_q^2 y(t) &\rightleftharpoons_q \left(\frac{s}{q}\right)^2 Y(s) - \frac{s}{q} y(0) - D_q y(0). \end{aligned} \tag{16}$$

Loading (16) in (15), one gets

$$c_0 \left(\left(\frac{s}{q}\right)^2 Y(s) - \frac{s}{q} y(0) - D_q y(0) \right) + c_1 \left(\frac{s}{q} Y(s) - y(0) \right) + c_2 Y(s) = G(s),$$

$$\left(c_0 \left(\frac{s}{q}\right)^2 + c_1 \frac{s}{q} + c_2 \right) Y(s) - c_0 y_0 \frac{s}{q} - c_0 y_1 - c_1 y_0 = G(s),$$

then

$$Y(s)=\frac{G(s)+c_0y_0\frac{s}{q}+c_0y_1+c_1y_0}{c_0\left(\frac{s}{q}\right)^2+c_1\frac{s}{q}+c_2}. \tag{17}$$

The remaining task consists in finding the express version of $y(t)=£_q^{-1}\{Y(s)\}$.

Example 2 Find the q-original of

$$F(s)=\frac{1}{s(s^2-1)(s^2+4)}. \tag{18}$$

Solution. Since

$$\frac{1}{s(s^2-1)(s^2+4)}=\frac{1}{s(s-1)(s+1)(s^2+4)},$$

then

$$F(s)=\frac{1}{s(s^2-1)(s^2+4)}=\frac{\frac{-1}{4}}{s}+\frac{\frac{1}{10}}{s-1}+\frac{\frac{1}{10}}{s+1}+\frac{\frac{-1}{40}}{s-2i}+\frac{\frac{1}{40}}{s+2i},$$

and by using the inverse q-Laplace transform we obtain

$$\begin{aligned}
f(t)&=£_q^{-1}\{F(s)\}\\
&=£_q^{-1}\left\{\frac{\frac{-1}{4}}{s}+\frac{\frac{1}{10}}{s-1}+\frac{\frac{1}{10}}{s+1}+\frac{\frac{-1}{40}}{s-2i}+\frac{\frac{-1}{40}}{s+2i}\right\}\\
&=\frac{-1}{4}£_q^{-1}\left\{\frac{1}{s}\right\}+\frac{1}{10}£_q^{-1}\left\{\frac{1}{s-1}\right\}+\frac{1}{10}£_q^{-1}\left\{\frac{1}{s+1}\right\}-\frac{1}{40}£_q^{-1}\left\{\frac{1}{s-2i}\right\}+\frac{1}{40}£_q^{-1}\left\{\frac{1}{s+2i}\right\}\\
&=\frac{-1}{4}\frac{1}{q}+\frac{1}{10}\frac{1}{q}e_q^{q^{-1}t}+\frac{1}{10}\frac{1}{q}e_q^{-q^{-1}t}-\frac{1}{40}\frac{1}{q}e_q^{2iq^{-1}t}-\frac{1}{40}\frac{1}{q}e_q^{-2iq^{-1}t}.
\end{aligned}$$

Example 3 Using the q-Laplace transform, solve the equations

$$\begin{aligned}(I)\ \ & D_q^2y(t)+y(t)=0,\\ & y(0)=1,\quad D_qy(0)=0.\end{aligned} \tag{19}$$

$$\begin{aligned}(II)\ \ & D_q^2y(t)-y(t)=0,\\ & y(0)=0,\quad D_qy(0)=1.\end{aligned} \tag{20}$$

$$(III)\ D_q^2 y(t) - 3D_q y(t) + 2y(t) = 0, \quad y(0) = 0, \quad D_q y(0) = 1. \tag{21}$$

Solution.

(I) Using (17) and the data in (19), we get

$$Y(s) = \frac{qs}{s^2 + q^2},$$

which by (8) with $w = 1$ gives $y(t) = \cos_q t$.

(II) Similarly, using (17) and the data in (20), we get

$$Y(s) = \frac{q^2}{s^2 - q^2},$$

which by (11) with $w = 1$ gives $y(t) = \sinh_q t$.

(III) Using (17) and the data in (21), we have

$$\begin{aligned} Y(s) &= \frac{1}{\left(\frac{s}{q}\right)^2 - 3\frac{s}{q} + 2} = \frac{1}{\left(\frac{s}{q} - 2\right)\left(\frac{s}{q} - 1\right)} \\ &= \frac{q}{s - 2q} - \frac{q}{s - q}, \end{aligned}$$

which by (7) with $a = 2$ and $a = 1$ gives

$$y(t) = e_q^{2t} - e_q^t.$$

Example 4 Solve the q-difference equation

$$D_q^2 y(t) + D_q y(t) - 2y(t) = e^{-t}, \quad y(0) = 0, \quad D_q y(0) = 1. \tag{22}$$

Solution. Using (17), we have

$$\left(\frac{s}{q}\right)^2 Y(s) - \frac{s}{q} y(0) - D_q y(0) + \frac{s}{q} Y(s) - y(0) - 2Y(s) = \frac{q}{s + q},$$

and by using the data in (22), we get

$$Y(s) = \frac{1}{\left(\frac{s}{q}+1\right)\left(\frac{s}{q}-1\right)} = \frac{\frac{-1}{2}}{\frac{s}{q}+1} + \frac{1}{\frac{s}{q}-1}$$
$$= \frac{\frac{-1}{2}q}{s+q} + \frac{q}{s-q},$$

then

$$y(t) = £_q^{-1}\{Y(s)\} = \frac{-1}{2}£_q^{-1}\left\{\frac{q}{s+q}\right\} + £_q^{-1}\left\{\frac{q}{s-q}\right\}$$
$$y(t) = \frac{-1}{2}e_q^{-t} + e_q^{t}.$$

Example 5 Solve the system of q-difference equations

$$\begin{cases} D_q t(t) = t(t) + 2y(t), \\ D_q y(t) = 2t(t) + y(t) + 1, \\ t(0) = y(0) = 0. \end{cases} \tag{23}$$

Solution. Using (17), we have

$$\begin{cases} \frac{s}{q}T(s) - t(0) = T(s) + 2Y(s), \\ \frac{s}{q}Y(s) - y(0) = 2T(s) + Y(s) + \frac{q}{s}, \end{cases}$$

and by using the data in (23), we get

$$\begin{cases} T(s) = \frac{2q}{s-q}Y(s), \\ Y(s) = \frac{q^2}{s(s-q)} + \frac{2q}{s-q}t(s), \end{cases} \tag{24}$$

Now inputting the first equation in (24) in the second equation, we obtain

$$Y(s) = \frac{q^2}{s(s-q)} + \frac{4q^2}{(s-q)^2}Y(s)$$
$$\Longrightarrow Y(s) = \frac{q^2(s-q)}{s(s^2-2sq+q^2)} = \frac{q^2(s-q)}{s(s-3q)(s+q)}$$
$$Y(s) = \frac{1}{3}\frac{q}{s} + \frac{1}{6}\frac{q}{s-3q} - \frac{1}{2}\frac{q}{s+q},$$
$$y(t) = \frac{1}{3} + \frac{1}{6}e_q^{3t} - \frac{1}{2}e_q^{-t}.$$

Similarly, inputting the first equation in (24) $Y(s)$ in the first equation, we obtain

$$T(s) = \frac{2q^3}{s(s-3q)(s+q)} = \frac{-2}{3}\frac{q}{s} + \frac{1}{6}\frac{q}{s-3q} + \frac{1}{2}\frac{q}{s+q},$$

$$x(t) = \frac{-2}{3} + \frac{1}{6}e_q^{3t} + \frac{1}{2}e_q^{-t},$$

then, the general solution of (23) is

$$x(t) = \frac{-2}{3} + \frac{1}{6}e_q^{3t} + \frac{1}{2}e_q^{-t},$$
$$y(t) = \frac{1}{3} + \frac{1}{6}e_q^{3t} - \frac{1}{2}e_q^{-t}.$$

References

1. Abdi, W.H.: On q-Laplace transform. Proc. Acad. Sci. India **29A**, 389–408 (1960)
2. Adams, C.R.: On the linear ordinary q-difference equation. Amer. Math. Ser. **II**(30), 195–205 (1929)
3. Al-shami, T.M., El-Shafei, M.E.: T-soft equality relation. Turkish J. Math. **44**(4), 1427–1441 (2020)
4. Bangerezako, G.: An Introduction to q-Difference Equations. University of Burundi, Bujumbura (2007)
5. Carmichael, R.D.: The general theory of linear q-difference equations. Amer. J. Math. **34**, 147–168 (1912)
6. Chung, W.S., Kim, T., In Kwon, H.: On the q-analog of the Laplace transform. Rus. J. Math. Phys. **21**(2), 156–168 (2014)
7. El-Shahed, M., Gaber, M.: Two-dimensional q-differential transformation and its application. Appl. Math. Comp. **217**(22), 9165–9172 (2011)
8. Ernst, T.: The History of q-Calculus and a New Method, U. U. D. M. Report 2000:16, 1101-3591, Department of Mathematics, Uppsala University (2000)
9. Erzan, A.: Finite q-differences and the discrete renormalization group Phys. Lett. A **225**(4–6), 235–238 (1997)
10. Erzan, A., Eckmann, J.P.: q-analysis of fractal sets. Phys. Rev. Lett. **17**, 3245–3248 (1997)
11. Gasper, G., Rahman, M.: Basic Hypergeometric Series. Cambridge University Press, Cambridge (1990)
12. Ismail, M.E.H.: Classical and Quantum Orthogonal Polynomials in One Variable. Cambridge University Press, Cambridge, UK (2005)
13. Jackson, H.F.: q-Difference equations. Amer. J. Math. **32**, 305–314 (1910)
14. Kac, V., Cheung, P.: Quantum calculus. Springer, New York (2002)
15. Kobachi, Nobuo: On q-Laplace transformation. Kumamoto Higher School of Tuen Mun Research Summary Kumamoto Higher School of Tuen Mun **3**, 69–76 (2011)
16. Mason, T.E.: On properties of the solution of linear q-difference equations with entire function coefficients. Amer. J. Math. **37**, 439–444 (1915)
17. Saxena, R.K.: Some thoerems on genralized Laplace tramsfom, I. Proc. Nat. Inst. Sci. India, Part A **26**, 400–413 (1960)

18. Saxena, R.K.: Some thoerems on genralized Laplace tramsfom, II. Riv. Mat. Univ. Parma **2**(2), 287–299 (1961)
19. Saxena, R.K.: Some thoerems on generalize Laplace transform. Proc. Camb. Philos. Soc. **62**, 467–471 (1966)
20. Rao, G.L.N.: The generalized Laplace transform of generalized functions. Ranchi Univ. Math. J **5**, 76–88 (1974)
21. Rooney, P.G.: On integral transformations with G-function kernels. Proc. R. Soc. Edinb. Sect. A: Math. **93**(3–4), 265–297 (1983)
22. Sneddon, I.N.: Fourier Transforms. International Series in Pure and Applied Mathematics (1951)
23. Trjitzinsky, W.J.: Analytic theory of linear q-difference equations. Acta Math. **62**(1), 227–237 (1933)
24. Tsallis, C.: Possible generalization of Boltzmann-Gibbs statistics. J. Stat. Phys. **52**, 479–487 (1988)
25. Zemanian, A.H.: Generalized integral transformations (1968)

An Effective Procedure for Solving Volterra Integro-Differential Equations

N. R Anakira, G. F. Bani-Hani, and O. Ababneh

Abstract In this paper, accurate solutions close to the exact one are obtained successfully based on the homotopy perturbation method (HPM), Laplace transformation and Pade approximant to be an effective procedure for solving linear and nonlinear integral equations of Volterra kind. The obtained results reveal that the new hybrid procedure is powerful, effective and reliable for solving this kind of differential equation.

Keywords Integral equations · HPM procedure · Series expansion · Laplace transform · Pade approximant

1 Introduction

Many engineering and physical problems are mathematically formulated in the form of integral equations of Volterra kind, such as diffusion and concrete problems in mechanics, heat conduction, fluid dynamics and so on of other physical applications [1–4].

To obtain the solution of this type of differential equation, several numerical procedures have been used and employed, for example, Maleknejad et al. [5–7] employ a numerical procedure based on a wavelet, modified block plus function and Bernstein's approximation method to obtain a solution of the first kind Volterra integral equation. Babolian and Masouri [8] used a direct procedure to solve Volterra integral equation. Recently, various numerical procedures were being employed and developed to find the solution of integral equations of Volterra kind; for more details, see [8–12].

N. R. Anakira (✉) · G. F. Bani-Hani
Department of Mathematics, Faculty of Science and Technology, Irbid National University, Irbid 2600, Jordan
e-mail: dr.nidal@inu.edu.jo

O. Ababneh
Department of Mathematics, Faculty of Science, Zarqa University, Zarqa, Jordan

D. Zeidan et al. (eds.), *Mathematics and Computation*, Springer Proceedings in Mathematics & Statistics 418, https://doi.org/10.1007/978-981-99-0447-1_40

This study aims to improve the accuracy of the HPM procedure by applying Laplace transformation to the first few terms of the HPM series approximate solution and then converting the transformed series into a meromorphic function by applying the Pade approximants, and lastly by applying the inverse of the Laplace transformation, we get the required solution to the given problem with high performance. This method is effective, and doesn't need big efforts to get accurate results with high precision.

2 Description of Research Methods

In this section, we present the basic idea of the research methods.

2.1 Analysis of the Homotopy Perturbation Method (HPM)

To illustrate the basic idea of HPM procedure [13–17], we consider the following nonlinear integral equation:

$$A(u) - f(r) = 0, \quad r \in \Omega \tag{1}$$

where A is a general integral operator, B is a boundary operator, $f(r)$ is a known analytic function and Γ is the boundary of the domain Ω. The operator A can be generally divided into two parts L and N, where L is linear, whereas N is nonlinear. Therefore, Eq. (1) can be rewritten as follows:

$$L(u) - N(u) - f(r) = 0. \tag{2}$$

He [4] constructed a homotopy $v : \Omega[0, 1] \rightarrow R$ which satisfies

$$H(v; p) = L(v) - L(v_0) + pL(v_0) + p[N(v) - f(r)] = 0 \tag{3}$$

or

$$H(v; p) = (1 - p)[L(v) - L(v_0)] + p[A(v_0) - F(r)] = 0, \tag{4}$$

where $r \in \Omega, \quad p \in [0, 1]$ is called homotopy parameter, and $v_0(x)$ is an initial approximation of Eq. (1). Hence, it is obvious that

$$H(v; 0) = L(u) - L(v_0) = 0, \quad H(v; 1) = A(v) - F(r) = 0, \tag{5}$$

and the changing process of p from 0 to 1 is just that of $H(v, p)$ from $L(u) - L(v_0)$ to $A(v) - F(r)$. In topology, this is called deformation, where $L(u) - L(v_0)$ and

$A(v) - F(r)$ are called homotopic. Applying the perturbation technique [16], due to the fact that $0 \leq p \leq 1$ can be considered as a small parameter, we can assume that the solution of Eq. (4) or (5) can be expressed as a series in p, as follows:

$$v = v_0 + pv_1 + p^2 v_2 + p^3 v_3 + \cdots \tag{6}$$

where, when $p \to 1$, Eq. (4) or Eq. (5) corresponds to Eq. (3) and becomes the approximate solution of Eq. (1). i.e.,

$$u = \lim_{p \to 1} v = v_0 + v_1 + v_2 + v_3 + \cdots \tag{7}$$

2.2 *Padé Approximation*

Padé approximant [18, 19] is the ratio of two polynomials constructed from the coefficients of the Taylor series expansion of a function $y(x)$.

The $[L/M]$ Padé approximants to a function $y(x)$ are given by

$$\left[\frac{L}{M}\right] = \frac{P_L(x)}{Q_M(x)}$$

where $P_L(x)$ is a polynomial of degree at most L and $Q_M(x)$ is a polynomial of degree at most M. The formal power series

$$y(x) = \sum_{i=1}^{\infty} a_i x^i,$$

$$y(x) - \frac{P_L(x)}{Q_M(x)} = \mathrm{O}\left(x^{L+M+1}\right) \tag{8}$$

determine the coefficients of $P_L(x)$ and $Q_M(x)$ by the equation. Since we can clearly multiply the numerator and denominator by a constant and leave $[L/M]$ unchanged, then we impose the normalization condition

$$Q_M(0) = 1. \tag{9}$$

Finally, we require that $P_L(x)$ and $Q_M(x)$ have no common factors. If we write the coefficient of $P_L(x)$ and $Q_M(x)$ as

$$\begin{cases} P_L(x) = p_0 + p_1 x + p_2 x^2 + \cdots + p_L x^L \\ Q_M(x) = q_0 + q_1 x + q_2 x^2 + \cdots + q_M x^M \end{cases} \tag{10}$$

then, by (9) and (10), we may multiply (8) by $Q_M(x)$, which linearizes the coefficient equations. We can write out (8) in more detail as

$$\begin{cases} a_{L+1} + a_L q_1 + \cdots + a_{L-M+1} q_M = 0 \\ a_{L+2} + a_{L+1} q_1 + \cdots + a_{L-M+2} q_M = 0 \\ \quad\quad\quad\quad\quad\quad . \\ \quad\quad\quad\quad\quad\quad . \\ a_{L+M} + a_{L+M-1} q_1 + \cdots + a_L q_M = 0 \end{cases} \tag{11}$$

$$\begin{cases} a_0 = p_0 \\ a_0 + a_0 q_1 = p_1 \\ a_2 + a_1 q_1 + a_0 q_2 = p_2 \\ \quad\quad . \\ \quad\quad . \\ a_L + a_{L-1} q_1 + \cdots + a_0 q_L = p_L. \end{cases} \tag{12}$$

To solve these equations, we start with equation (11), which is a set of linear equations for all the unknown $q's$. Once the $q's$ are known, then Eq. (12) gives an explicit formula for the unknown $p's$, which complete the solution.

If Eqs. (12) and (11) are non-singular, then we can solve them directly and obtain Eq. (13) [20], where Eq. (13) holds, and if the lower index on a sum exceeds the upper, the sum is replaced by zero:

$$\left[\frac{L}{M}\right] = \frac{det\begin{bmatrix} a_{L-M+1} & a_{L-M+2} & \cdots & a_{L+1} \\ . & . & . & . \\ . & . & . & . \\ . & . & . & . \\ a_L & a_{L+1} & \cdots & a_{L+M} \\ \sum_{j=M}^{L} a_{j-M} x^j & \sum_{j=M-1}^{L} a_{j-M+1} x^j & \cdots & \sum_{j=0}^{L} a_j x^j \end{bmatrix}}{det\begin{bmatrix} a_{L-M+1} & a_{L-M+2} & \cdots & a_{L+1} \\ . & . & . & . \\ . & . & . & . \\ . & . & . & . \\ a_L & a_{L+1} & \cdots & a_{L+M} \\ x^M & x^{M-1} & \cdots & 1 \end{bmatrix}}. \tag{13}$$

To obtain diagonal Padé approximants of different order such as [2/2], [4/4] or [6/6], we can use the symbolic calculus software, Mathematica.

Note that typically the Padé approximant, obtained from a partial Taylor sum, is more accurate than the latter. However, the Padé, being a rational expression, has poles, which are not present in the original function. It is a simple algebraic task to expand the form of an $[N, M]$ Padé in a Taylor series and compute the Padé coefficients by matching with the above [20].

3 Numerical Examples

To illustrate the effectiveness of the new HPM procedure, we shall consider some examples of Volterra integral equations.

3.1 Example 1

Consider the following linear Volterra integral equation of first order taken from Wazwaz 2013 [21]:

$$u'(x) = 1 + \frac{1}{x^2} - \int_0^x u(t)dt, \quad u(0) = 1. \tag{14}$$

To solve this problem using the presented procedure constructed in the previous section, we define the linear and nonlinear operators as given below

$$\begin{aligned} L[v(x,p)] &= \frac{dv(x,p)}{dx}, \\ N[v(x;p)] &= \frac{dv(x,p)}{dx} - 1 - \frac{x^2}{2} + \int_0^x v(t;p)dt. \end{aligned} \tag{15}$$

Then, we construct a homotopy equation in the form

$$(1-p)L[v(x;p)] = H(q)\left(\frac{dv(x,p)}{dx} - 1 - \frac{x^2}{2} + \int_0^x v(t;p)dt\right). \tag{16}$$

Then the zeroth-order problem is given by

$$u_0'(x) = 0, \quad u_0(0) = 1 \tag{17}$$

which has the solution

$$u_0(x) = 1. \tag{18}$$

The first-order problem with its initial conditions is given in the following form:

$$u_1'(x) = h\left(-\frac{x^2}{2} + x - 1\right), \quad u_1(0) = 0 \tag{19}$$

which has the following solution:

$$u_1(x) = -\frac{1}{6}hx\left(x^2 - 3x + 6\right). \tag{20}$$

The second-order problem with its initial conditions is given in the following form along with its solution:

$$u_2'(x) = -\frac{1}{24}h^2x^4 + \frac{h^2x^3}{6} - h^2x^2 + h^2x - h^2 - \frac{hx^2}{2} + hx - h \tag{21}$$

which has the following solution:

$$u_2(x) = -\frac{1}{120}hx\left(h\left(x^4 - 5x^3 + 40x^2 - 60x + 120\right) + 20\left(x^2 - 3x + 6\right)\right). \tag{22}$$

The third-order problem with its initial conditions is given in the following form along with its solution:

$$\begin{aligned} u_3'(x) = &-\frac{1}{720}h^3x^6 + \frac{h^3x^5}{120} - \frac{h^3x^4}{8} + \frac{h^3x^3}{3} - \frac{3h^3x^2}{2} + h^3x \\ &-h^3 - \frac{h^2x^4}{12} + \frac{h^2x^3}{3} - 2h^2x^2 \\ &+2h^2x - 2h^2 - \frac{hx^2}{2} + hx - h), \quad u_3(x)' = 0, \end{aligned} \tag{23}$$

which has the following solution:

$$\begin{aligned} u_3(x) = &-\frac{h^3x^7}{5040} + \frac{h^3x^6}{720} - \frac{h^3x^5}{40} + \frac{h^3x^4}{12} - \frac{h^3x^3}{2} + \frac{h^3x^2}{2} - h^3x \\ &-\frac{h^2x^5}{60} + \frac{h^2x^4}{12} - \frac{2h^2x^3}{3} + h^2x^2 - 2h^2x - \frac{hx^3}{6} + \frac{hx^2}{2} - hx. \end{aligned} \tag{24}$$

Following the same procedure up to the sixth-order problem, and by considering $h = -1$, we have the approximate solution

$$\begin{aligned} u\tilde{(x)} = &-\frac{x^{13}}{6227020800} + \frac{x^{12}}{479001600} - \frac{x^{10}}{3628800} + \frac{x^8}{40320} \\ &-\frac{x^6}{720} + \frac{x^4}{24} - \frac{x^2}{2} + x + 1 \end{aligned} \tag{25}$$

which yields the exact solution (28) in the Limit of infinity terms of the HPM approximate solution. Numerical results resulting from 6 terms of HPM approximate solution are presented in Table 1. In order to improve the accuracy and effectiveness of the HPM procedure, we will use the Laplace transform of the first five terms of the HPM approximate series solution (25), which yields

$$Lu\tilde{(x)} = -\frac{1}{s^7} + \frac{1}{s^5} - \frac{1}{s^3} + \frac{1}{s^2} + \frac{1}{s}. \tag{26}$$

Table 1 Comparison of exact solution and OHAM solution for Example 1

x	Exact solution	HPM approximate solution	Absolute error
0.0	1.00000	1.00000	0
0.2	1.18007	1.18007	0
0.4	1.32106	1.32106	1.33×10^{-15}
0.6	1.42534	1.42534	2.00×10^{-13}
0.8	1.49671	1.49671	8.33×10^{-12}
1.0	1.54030	1.54030	1.49×10^{-10}

For simplicity, consider $s = \frac{1}{z}$; we have

$$L\tilde{u(x)} = z + z^2 - z^3 + z^5 - z^7. \tag{27}$$

Now, we use the Pade approximants of $[\frac{4}{4}] = \frac{z+z^2+z^4}{1+z^2}$ and then using $z = \frac{1}{s}$, and applying the inverse Laplace transform we obtain the exact solution

$$u(x) = x + cos(x). \tag{28}$$

3.2 Example 2

The second problem in this section is the following second-order linear Volterra integral equation with it is initial condition taken from Wazwaz 2013 [21]:

$$u''(x) = 1 + \int_0^x (x-t)u(t)dt, \quad u(0) = 1, \quad u'(0) = 0. \tag{29}$$

Following the same procedure applied in the previous example, we obtained the sixth-order of the HPM approximate solution:

$$\begin{aligned} \tilde{u(x)} = 1 &+ \frac{x^2}{2} + \frac{x^4}{24} + \frac{x^6}{720} + \frac{x^8}{40320} + \frac{x^{10}}{3628800} + \frac{x^{12}}{479001600} + \frac{x^{14}}{87178291200} \\ &+ \frac{x^{16}}{20922789888000} + \frac{x^{18}}{6402373705728000} + \frac{x^{20}}{2432902008176640000} \\ &+ \frac{x^{22}}{1124000727777607680000} + \frac{x^{24}}{620448401733239439360000} \end{aligned} \tag{30}$$

which yields the exact solution $u(x) = cosh(x)$, in the Limit of infinity terms of the order of HPM approximate solution. Numerical results resulting from sixth order of

Table 2 Comparison of exact solution and OHAM solution for Example 1

x	Exact solution	HPM approximate solution	Absolute error
0.0	1.02007	1.02007	0
0.2	1.02007	1.02007	0
0.4	1.08107	1.08107	1.33×10^{-15}
0.6	1.18547	1.18547	2.00×10^{-13}
0.8	1.33743	1.33743	8.33×10^{-12}
1.0	1.54308	1.54308	1.49×10^{-10}

HPM approximate solution are presented in Table 2. In order to improve the accuracy and effectiveness of the HPM procedure, we will use the Laplace transform of the first five terms of the HPM approximate series solution (28), which yields

$$Lu(\tilde{x}) = -\frac{1}{s^9} + \frac{1}{s^7} + \frac{1}{s^5} + \frac{1}{s^3} + \frac{1}{s}. \tag{31}$$

For simplicity, consider $s = \frac{1}{z}$; we have

$$Lu(\tilde{x}) = z + z^3 + z^5 + z^7 + z^9. \tag{32}$$

Now, we use the Pade approximants of $[\frac{4}{4}] = \frac{z}{1-z^2}$ and then using $z = \frac{1}{s}$, and applying the inverse Laplace transform we obtain the exact solution $u(x) = x + cos(x)$.

4 Results and Dissections

Numerical results are obtained using several terms of HPM approximate solutions formulated in the tables. The approximate results obtained reveal that the HPM procedure is a powerful and effective procedure for solving this kind of differential equation, and the solution converges to the exact solutions in the limit of infinity terms. To improve the accuracy of the HPM procedure and obtain the exact solution using only a few terms of the HPM approximate solution, we construct an alternative procedure which modifies the series approximate solution by applying Laplace transformation to the truncated series obtained by HPM, then convert the transformed series into a meromorphic function by Pade approximants, and finally apply the inverse Laplace transform to obtain the exact solutions for the given problem.

5 Conclusion

Based on HPM, a new procedure was proposed for different classes of integral equations of Volterra kind. The reliability, effectiveness and power of this procedure were proved throughout obtaining the exact solutions of the given test problems using only a few terms of the HPM truncated series approximate solutions.

References

1. Ding, H.J., Wang, H.M., Chen, W.Q.: Analytical solution for the electroelastic dynamics of a nonhomogeneous spherically isotropic piezoelectric hollow sphere. Arch. Appl. Mech. **73**(1), 49–62 (2003)
2. Yousefi, S.A.: Numerical solution of Abel's integral equation by using Legendre wavelets. Appl. Math. Comput. **175**(1), 574–580 (2006)
3. Baratella, P.: A Nyström interpolant for some weakly singular linear Volterra integral equations. J. Comput. Appl. Math. **231**(2), 725–734 (2009)
4. Bartoshevich, M.A.: A heat-conduction problem. J. Eng. Phys. **28**(2), 240–244 (1975)
5. Maleknejad, K., Mollapourasl, R., Alizadeh, M.: Numerical solution of Volterra type integral equation of the first kind with wavelet basis. Appl. Math. Comput. **194**(2), 400–405 (2007)
6. Maleknejad, K., Rahimi, B.: Modification of block pulse functions and their application to solve numerically Volterra integral equation of the first kind. Commun. Nonlinear Sci. Numer. Simul. **16**(6), 2469–2477 (2011)
7. Qazza, A.M., Hatamleh, R.M., Alodat, N.A.: About the solution stability of volterra integral equation with random kernel. Far East J. Math. Sci. **100**(5), 671 (2016). Al-Shimmary, A.F., Hussain, A.K., Radhi, S.K.: Numerical solution of Volterra integro-differential equation using 6th order Runge-Kutta method. J. Phys.: Conf. Ser. **1818**(1), 012183 (2021). IOP Publishing
8. Cardone, A., Conte, D., D'Ambrosio, R., Paternoster, B.: Collocation methods for Volterra integral and integro-differential equations: a review. Axioms **7**(3), 45 (2018)
9. Rani, D., Mishra, V.: Solutions of Volterra integral and integro-differential equations using modified Laplace Adomian decomposi-tion method. J. Appl. Math. Stat. Inf. **15**(1), 5–18 (2019)
10. Cimen, E., Yatar, S.: Numerical solution of Volterra integro-differential equation with delay. J. Math. Comput. Sci. **20**, 255–263 (2020)
11. Shayanfard, F., Laeli Dastjerdi, H., Maalek Ghaini, F.M.: Collocation method for approximate solution of Volterra integro-differential equations of the third-kind. Appl. Numer. Math. **150**, 139–148 (2020)
12. Babolian, E., Masouri, Z.: Direct method to solve Volterra integral equation of the first kind using operational matrix with block-pulse functions. J. Comput. Appl. Math. **220**(1–2), 51–57 (2008)
13. Anakira, N.R., Jameel, A.F., Alomari, A.K., Saaban. A., Shakhatreh, M.A., Odat, A.: Approximate approach for solving two points fuzzy boundary value problems. Ital. J. Pure Appl. Math. 1 (2019)
14. Ghasemi, M., Tavassoli Kajani, M., Babolian, E.: Application of He's homotopy perturbation method to nonlinear integro-differential equations. Appl. Math. Comput. **188**(1), 538–548 (2007)
15. Saberi-Nadjafi, J., Ghorbani, A.: He's homotopy perturbation method: an effective tool for solving nonlinear integral and integro-differential equations. Comput. & Math. Appl. **58**(11–12), 2379–2390 (2009)
16. He, J.-H.: Homotopy perturbation method: a new nonlinear analytical technique. Appl. Math. Comput. **135**(1), 73–79 (2003)

17. He, J.-H.: Recent development of the homotopy perturbation method. Topol. Methods Nonlinear Anal. **31**(2), 205–209 (2008)
18. Al-Ahmad, S., Mamat, M., Anakira, N., Alahmad, R.: Modified differential transformation method for solving classes of non-linear differential equations (2022)
19. Al-Ahmad, S., Anakira, N.R., Mamat, M., Jameel, A.F., Alahmad, R., Alomari, A.K.: Accurate approximate solution of classes of boundary value problems using modified differential transform method. TWMS J. App. Eng. Math. **12**(4), 1228–1238 (2022)
20. George, A., Jr.: Essentials of Pade Approximants. Elsevier (1975)
21. Wazwaz, A.M.: The variational iteration method for solving linear and nonlinear Volterra integral and integro-differential equations. Int. J. Comput. Math. **87**(5), 1131–1141 (2010)

New Estimations for Zeros of Polynomials Using Numerical Radius and Similarity of Matrices

Saeed Alkhalely, Aliaa Burqan, and Mowafaq Muhammed Al-Kassab

Abstract The applications of the estimation for the zeros of polynomials are important in many areas of sciences such as signal processing, control theory, communication theory, coding theory, cryptography, etc. Finding the exact zeros of polynomials of higher order is not an easy task and there is no standard method to find them. In this article, we introduce some upper bounds for zeros of monic polynomials with numerical coefficients based on some numerical radius inequalities and similarity of matrices, we also introduce some numerical examples to show that our bounds are better than some existing estimations. Also, we give a new upper bound for zeros of polynomials with matrix coefficients by using the similarity of matrices.

Keywords Monic polynomial · Frobenius companion matrix · Similar matrices · Numerical radius

1 Introductions

Finding the exact zeros of polynomials of higher order is not an easy task and there is no closed form or stander method to find them. Therefore, the researchers directed to

S. Alkhalely
Department of Basic Sciences, Middle East University, Amman, Jordan
e-mail: saeedalkhalely@gmail.com

A. Burqan
Department of Mathematics, Faculty of Science, Zarqa University, Zarqa, Jordan
e-mail: aliaaburqan@zu.edu.jo

M. M. Al-Kassab (✉)
Department of Mathematics Education, Faculty of Education, Tishk International University-Erbil, Kurdistan Region, Iraq
e-mail: mowafaq.muhammed@tiu.edu.iq

D. Zeidan et al. (eds.), *Mathematics and Computation*, Springer Proceedings in Mathematics & Statistics 418, https://doi.org/10.1007/978-981-99-0447-1_41

estimate the zeros of polynomials by using the Frobenius companion matrix corresponding to the monic polynomial where the eigenvalues of the companion matrix are the zeros of the polynomial. Over the years, various mathematicians have estimated the zeros of polynomials using various techniques. A few of them are displayed below where λ is a zero of the monic polynomial $p(z) = z^n + a_n z^{n-1} + \cdots + a_2 z + a_1$.

1. Cauchy bound (1985, [6])

$$|\lambda| \leq 1 + \max\{|a_1|, |a_2|, \ldots, |a_n|\}.$$

2. Montel bound (1985, [6])

$$|\lambda| \leq \max\left\{1, \sum_{i=1}^{n} |a_i|\right\}.$$

3. Carmichael and Mason bound (1985, [6])

$$|\lambda| \leq \left(1 + |a_1|^2 + |a_2|^2 + \cdots + |a_n|^2\right)^{\frac{1}{2}}.$$

4. Fujii and Kubo bound (1993, [3])

$$|\lambda| \leq \cos\frac{\pi}{n+1} + \frac{1}{2}\left(\sqrt{\sum_{i=1}^{n} |a_i|^2} + |a_n|\right).$$

5. Kittaneh bounds (2003, [7])

$$1.\ |\lambda| \leq \frac{1}{2}\left(|a_n| + 1 + \sqrt{(|a_n| - 1)^2 + 4\sqrt{\sum_{i=1}^{n-1} |a_i|^2}}\right),$$

$$2.\ |\lambda| \leq \frac{1}{2}(|a_n| + \cos\frac{\pi}{n} + \sqrt{\left(|a_n| - \cos\frac{\pi}{n}\right)^2 + (|a_{n-1}| + 1)^2 + \sum_{i=1}^{n-2} |a_i|^2}.$$

6. Bhunia et al. bounds (2020, [2])

$$1.\ |\lambda| \leq \frac{1}{2}(|a_n| + \cos\frac{\pi}{n} + \sqrt{\left(|a_n| - \cos\frac{\pi}{n}\right)^2 + \sum_{i=1}^{n-1} |a_i|^2 + 1 + \alpha},$$

where $\alpha = |a_{n-1}| + \sqrt{\sum_{i=1}^{n-1} |a_i|^2}$.

$$2.\ |\lambda| \leq \sqrt{|a_n|^2 + \frac{1}{2}\alpha\left(|a_n| + \frac{1}{2}\alpha\right)} + \sqrt{\cos^2\frac{\pi}{n} + \frac{1}{2}\left(\cos\frac{\pi}{n} + \frac{1}{2}\right)},$$

where $\alpha = \sqrt{\sum_{i=1}^{n-1} |a_i|^2}$.

In this paper, we present some numerical radius inequalities to obtain new bounds for zeros of polynomials. Moreover, we utilize the similarity of matrices to get new upper bounds for the zeros of polynomials with numerical and matrix coefficients. The main results are given in Sects. 2 and 3 includes bounds for zeros of polynomials using similarity of matrices, and the bounds for the zeros of polynomials with matrix coefficients using similarity of matrices are presented in Sect. 4.

2 Main Results

Let M_n denote the algebra of all $n \times n$ complex matrices. For $A \in M_n$, let $\sigma(A), r(A), w(A)$ and $\|A\|$ denote the spectrum, the spectral radius, the numerical radius, and the spectral norm of A, respectively. It is well known that $r(A) \leq w(A) \leq \|A\|$. Consider the monic polynomial of degree n (n > 2):

$$p(z) = z^n + a_n z^{n-1} + a_{n-1} z^{n-2} + \cdots + a_2 z + a_1,$$

where the coefficients $a_i \in \mathbb{C}$ for $i = 1, 2, \ldots, n$, and $a_1 \neq 0$.

The Frobenius companion matrix of p, $C(p)$, associated with the polynomial $p(z)$ is given by

$$C(p) = \begin{bmatrix} -a_n & -a_{n-1} & \cdots & -a_2 & -a_1 \\ 1 & 0 & \cdots & 0 & 0 \\ 0 & 1 & \ddots & 0 & 0 \\ \vdots & \vdots & \ddots & \vdots & \vdots \\ 0 & 0 & \ldots & 1 & 0 \end{bmatrix}_{n \times n}.$$

It is well known that the eigenvalues of $C(p)$ are exactly the zeros of $p(z)$. Using the fact $r(C(p)) \leq w(C(p))$ and since $|\lambda| \leq r(C(p))$ for any $\lambda \in \sigma(c(p))$, we get $|\lambda| \leq w(C(p))$.

Now to display our first new upper bounds we need the following two lemmas, see [3, 7].

Lemma 2.1 *Let $a_i \in \mathbb{C}$ for each $i = 1, 2, \ldots, n$. Then.*

$$w\left(\begin{bmatrix} a_1 & a_2 & \cdots & a_n \\ 0 & 0 & \cdots & 0 \\ 0 & 0 & \cdots & 0 \\ \vdots & \vdots & \cdots & \vdots \\ 0 & 0 & \cdots & 0 \end{bmatrix}\right) = \frac{1}{2}\left(|a_1| + \sqrt{\sum_{i=1}^{n} |a_i|^2}\right).$$

Lemma 2.2 *Let $A \in M_n$. Then.*

$$w(A) \le \frac{1}{2}\||A| + |A^*|\|.$$

Theorem 2.1 Let λ be a zero of $p(z) = z^n + a_n z^{n-1} + \cdots + a_2 z + a_1$. Then.

$$|\lambda| \le \max\{|a_1|, 1\} + \frac{1}{2}\left(|a_n| + \sqrt{\sum_{i=2}^{n} |a_i|^2}\right) = S_1$$

Proof Consider the Frobenius companion matrix of $p(z)$,

$$C(p) = \begin{bmatrix} -a_n & -a_{n-1} & \cdots & -a_2 & -a_1 \\ 1 & 0 & \cdots & 0 & 0 \\ 0 & 1 & \ddots & 0 & 0 \\ \vdots & \vdots & \ddots & \vdots & \vdots \\ 0 & 0 & \cdots & 1 & 0 \end{bmatrix}_{n\times n}.$$

Write $C(p)$ as a sum of two matrices such as $C(p) = A + B$,

where $A = \begin{bmatrix} 0 & 0 & \cdots & 0 & -a_1 \\ 1 & 0 & \cdots & 0 & 0 \\ 0 & 1 & \ddots & 0 & 0 \\ \vdots & \vdots & \ddots & \vdots & \vdots \\ 0 & 0 & \cdots & 1 & 0 \end{bmatrix}$ and $B = \begin{bmatrix} -a_n & -a_{n-1} & \cdots & -a_2 & 0 \\ 0 & 0 & \cdots & 0 & 0 \\ 0 & 0 & \ddots & 0 & 0 \\ \vdots & \vdots & \ddots & \vdots & \vdots \\ 0 & 0 & \cdots & 0 & 0 \end{bmatrix}$.

Then using the property of subadditivity of numerical radius, we have

$$w(C(p)) \le w(A) + w(B)$$

$w(A)$ can be bounded as

$$w(A) \leq \|A\| = \max\{|a_1|, 1\},$$

and Lemma 2.1 yields that

$$w(B) = \frac{1}{2}\left(|a_n| + \sqrt{\sum_{i=2}^{n} |a_i|^2}\right).$$

Thus,

$$|\lambda| \leq \max\{|a_1|, 1\} + \frac{1}{2}\left(|a_n| + \sqrt{\sum_{i=2}^{n} |a_i|^2}\right).$$

In the following, we show with a numerical example that our estimation in Theorem 2.1 is better than some of the existing estimations.

Example 2.1 Consider the polynomial $p(z) = z^4 + z^3 + 4z^2 + z + 1$. The upper bounds for the zeros of this polynomial estimated by different mathematicians are as given in the following table.

Cauchy	5
Montel	8
Carmichael	4.4721
Bhunia et al. (2)	3.811175

while our bound $S_1 = 3.62132$. This shows that for this example, our bound obtained in the above theorem is better than all the estimations mentioned above.

Theorem 2.2 Let λ be a zero of $p(z) = z^n + a_n z^{n-1} + \cdots + a_2 z + a_1$. Then

$$|\lambda| \leq w(p(z)) \leq \frac{1}{2}\left(\max\{|a_1| + 1, 2\} + |a_n| + \sqrt{\sum_{i=2}^{n} |a_i|^2}\right) = S_2$$

Proof Consider $C(p)$ associated with $p(z)$

$$C(p) = \begin{bmatrix} -a_n & -a_{n-1} & \cdots & -a_2 & -a_1 \\ 1 & 0 & \cdots & 0 & 0 \\ 0 & 1 & \ddots & 0 & 0 \\ \vdots & \vdots & \ddots & \vdots & \vdots \\ 0 & 0 & \cdots & 1 & 0 \end{bmatrix}_{n \times n}.$$

Now, we can write $C(p)$ as a sum of two matrices as in the previous proof

$$C(p) = A + B,$$

where $A = \begin{bmatrix} -a_n & -a_{n-1} & \cdots & -a_2 & 0 \\ 0 & 0 & \cdots & 0 & 0 \\ 0 & 0 & \ddots & 0 & 0 \\ \vdots & \vdots & \ddots & \vdots & \vdots \\ 0 & 0 & \cdots & 0 & 0 \end{bmatrix}$ and $B = \begin{bmatrix} 0 & 0 & \cdots & 0 & -a_1 \\ 1 & 0 & \cdots & 0 & 0 \\ 0 & 1 & \ddots & 0 & 0 \\ \vdots & \vdots & \ddots & \vdots & \vdots \\ 0 & 0 & \cdots & 1 & 0 \end{bmatrix}$.

Then using the property of subadditivity of numerical radius, we have

$$w(C(p)) \leq w(A) + w(B),$$

where $w(A) = \frac{1}{2}\left(|a_n| + \sqrt{\sum_{i=2}^{n} |a_j|^2}\right)$.

Lemma 2.2 gives

$$w(B) \leq \frac{1}{2}\||B| + |B^*|\|.$$

Now, by the definition of the absolute value of matrices, we have.

$|B| = (B^*B)^{\frac{1}{2}} = \begin{bmatrix} 1 & 0 & \dots & 0 \\ 0 & 1 & \dots & 0 \\ \vdots & \vdots & \ddots & \vdots \\ 0 & 0 & \dots & |a_1| \end{bmatrix}$ and $|B^*| = (BB^*)^{\frac{1}{2}} = \begin{bmatrix} |a_1| & 0 & \dots & 0 \\ 0 & 1 & \dots & 0 \\ \vdots & \vdots & \ddots & \vdots \\ 0 & 0 & \dots & 1 \end{bmatrix}$.

Thus,

$$w(B) \leq \frac{1}{2}\left\| \begin{bmatrix} |a_1|+1 & 0 & 0 & 0 & \dots & 0 \\ 0 & 2 & 0 & 0 & \dots & 0 \\ 0 & 0 & 2 & 0 & \dots & \vdots \\ 0 & 0 & 0 & 2 & \dots & \vdots \\ \vdots & \vdots & 0 & 0 & \ddots & \vdots \\ 0 & 0 & \dots & 0 & 0 & |a_1|+1 \end{bmatrix} \right\| = \frac{1}{2}\max\{|a_1|+1, 2\}.$$

So,

$|\lambda| \leq \frac{1}{2}\left(\max\{|a_1|+1, 2\} + |a_n| + \sqrt{\sum_{i=2}^{n} |a_i|^2}\right)$ ■

In the following example, we show numerically that our estimation in Theorem 2.2 is better than some of the existing estimations.

Example 2.2 Consider the polynomial $p(z) = z^4 + z^3 + 4z^2 + z + 2$. The upper bounds for the zeros of this polynomial estimated by different mathematicians are shown in the following table.

Cauchy	5
Montel	8
Carmichael	4.7958

while our bound $S_2 = 3.6213$. This shows that for this example, our bound obtained in the above theorem is better than all estimations mentioned above.

3 Bounds for Zeros of Polynomials Using Similarity of Matrices

In this section, Consider the monic polynomial of degree n (n > 2):

$$p(z) = z^n + a_n z^{n-1} + a_{n-1} z^{n-2} + \cdots + a_2 z + a_1,$$

where the coefficients $a_i \in \mathbb{C}$ for $i = 1, 2, \ldots, n$, $a_1 \neq 0$ and n is an even number. Using the similarity of matrices, we give an upper bound for the zeros of $p(z)$, we know that if A and B are similar, then the spectrum of A and the spectrum of B are equal, so we consider the matrix (S-compaion matrix) to be similar to the Frobenius companion matrix which implies that the eigenvalues of companion matrix are exactly the same as to the eigenvalues of our matrix.

Let $K = \begin{bmatrix} I & I \\ 0 & I \end{bmatrix} \in M_n$ where $I \in M_m$and $n = 2m$ Then K is an invertible matrix since and $K^{-1} = \begin{bmatrix} I & -I \\ 0 & I \end{bmatrix}$.

Define a matrix S as $S = KC(p)K^{-1}$.We will call the matrix S by the S-companion matrix of $p(z)$ which is equal

$$S=\begin{bmatrix} -a_n & -a_{n-1} & -a_{n-2} & \cdots & 1-a_{\frac{n}{2}+1} & a_n-a_{\frac{n}{2}} & a_{n-1}-a_{\frac{n}{2}} & \cdots & a_{n-\frac{n}{2}+2}-a_2 & a_{\frac{n}{2}+1}-a_1-1 \\ 1 & 0 & 0 & \cdots & 0 & 0 & 0 & \cdots & 0 & 0 \\ 0 & 1 & 0 & \cdots & 0 & 0 & 0 & \cdots & 0 & 0 \\ 0 & 0 & 1 & \cdots & 0 & 0 & 0 & \cdots & 0 & 0 \\ 0 & 0 & 0 & \ddots & 0 & 0 & 0 & \cdots & 0 & 0 \\ \vdots & \vdots & \vdots & \cdots & \ddots & \vdots & \vdots & \cdots & \vdots & \vdots \\ \vdots & \vdots & \vdots & \cdots & \vdots & \ddots & \vdots & \cdots & \vdots & -1 \\ \vdots & \vdots & \vdots & \cdots & \vdots & \vdots & \ddots & \cdots & \vdots & \vdots \\ \vdots & \vdots & \vdots & \cdots & \vdots & \vdots & \vdots & \ddots & \vdots & \vdots \\ 0 & 0 & 0 & \cdots & 0 & 0 & 0 & 0 & 1 & 0 \end{bmatrix},$$

where -1 occurs in $(\frac{n}{2}+1)^{th}$ row.

To establish our new upper bound, we present the following lemmas. The first lemma can be found in [4], and the second lemma can be found in [5].

Lemma 3.1 *Let $T=[T_{ij}]$ be an $n \times n$ block matrix where $T_{ij} \in M_{mi \times mj}$ and $\sum_{i=1}^{n} m_i = n$. Then.*

$$w(T) \le \frac{1}{2}\left(\sum_{i=1}^{n} w(T_{ii}) + \sqrt{w^2(T_{ii}) + \sum_{\substack{j=1 \\ i \ne j}}^{n} \|T_{ij}\|^2} \right).$$

Lemma 3.2 *Let h_n be the $n \times n$ matrix given by* $h_n = \begin{bmatrix} 0\,0 \ldots\ldots 0 \\ 1\,0 \ldots\ldots 0 \\ \vdots\ 1\ \ddots\quad \vdots \\ \vdots\ \vdots\ \ddots\ \ddots\ \vdots \\ 0\,0\,0\ \ 1\ 0 \end{bmatrix}$.

Then.

$$w(h_n) = cos\frac{\pi}{n+1}.$$

Theorem 3.1 *let λ be a zero of $p(z) = z^n + a_n z^{n-1} + \cdots + a_2 z + a_1$.Then.*

$$|\lambda| \le \frac{1}{2}\left(|a_n| + cos\frac{\pi}{n-1} + \sqrt{cos^2\frac{\pi}{n-1} + 2} + \sqrt{|a_n|^2 + \beta + \left|a_{(\frac{n}{2}+1)} - a_1 - 1\right|^2} \right) = S_3,$$

where $\beta = \sum_{i=(\frac{n}{2}+2)}^{n-1} |a_i|^2 + \left|1 - a_{(\frac{n}{2}+1)}\right|^2 + \sum_{k=0}^{(\frac{n}{2}-2)} \left|a_{(n-k)} - a_{(\frac{n}{2}-k)}\right|^2.$

Proof The S-companion matrix can be written as follows:

$$S = \begin{bmatrix} T_{11} & T_{12} & T_{13} \\ T_{21} & T_{22} & T_{23} \\ T_{31} & T_{32} & T_{33} \end{bmatrix} \text{ where, } T_{11} = [-a_n],\ T_{22} = \begin{bmatrix} 0\,0 & \ldots\ldots & 0 \\ 1\,0 & \ldots\ldots & 0 \\ \vdots\ 1 & \ddots & \vdots \\ \vdots\ \vdots & \ddots\ \ddots & \vdots \\ 0\,0\ 0 & 1 & 0 \end{bmatrix},$$

$$T_{33} = [0],\ T_{13} = \left[a_{(\frac{n}{2}+1)} - a_1 - 1\right],\ T_{23} = [0\ldots0-10\ldots0]^t$$

$$T_{12} = \left[-a_{n-1} - a_{n-2} \ldots 1 - a_{(\frac{n}{2}+1)}a_n - a_{\frac{n}{2}}a_{n-1} - a_{\frac{n}{2}} \ldots a_{(n-\frac{n}{2}+2)} - a_2\right]$$

$$T_{21} = [100\ldots0]^t,\ T_{31} = [0],\ T_{32} = [0\ldots001].$$

Now, the definitions of the numerical radius and the spectral norm with Lemma 3.2 yield that

$$w(T_{11}) = |a_n|,\ w(T_{22}) = \cos\frac{\pi}{n-1},\ w(T_{33}) = 0$$

$\|T_{12}\|^2 = \alpha$; where $\alpha = \sum_{i=0}^{(\frac{n}{2}-2)} \left|a_{(n-i)} - a_{(\frac{n}{2}-i)}\right|^2 + \left|1 - a_{(\frac{n}{2}+1)}\right|^2 + \sum_{(i=\frac{n}{2}+2)}^{(n-1)} |a_i|^2$

$$\|T_{13}\|^2 = \left|a_{(\frac{n}{2}+1)} - a_1 - 1\right|^2,\ \|T_{21}\|^2 = \|T_{23}\|^2 = \|T_{32}\|^2 = 1,\ \|T_{31}\|^2 = 0$$

Applying Lemma 3.1 we get

$$w(S) \le \frac{1}{2}\left(|a_n| + \cos\frac{\pi}{n-1} + \sqrt{\cos^2\frac{\pi}{n-1} + 2 + \sqrt{|a_n|^2 + \beta + \left|a_{(\frac{n}{2}+1)} - a_1 - 1\right|^2}}\right),$$

Use the fact $|\lambda| \le r(C(p)) = r(S) \le w(S)$ we have

$$|\lambda| \le \frac{1}{2}\left(|a_n| + \cos\frac{\pi}{n-1} + \sqrt{\cos^2\frac{\pi}{n-1} + 2 + \sqrt{|a_n|^2 + \beta + \left|a_{(\frac{n}{2}+1)} - a_1 - 1\right|^2}}\right).$$

This completes the proof. ■

The following examples show that our estimations in Theorem 3.1 are better than some existing estimations.

Example 3.1 Consider the polynomial.

$p(z) = z^8 - 30z^7 + \frac{1}{64}z^6 + \frac{1}{16}z^5 + z^4 - 30z^3 + \frac{1}{64}z^2 + \frac{1}{16}z + 1.$

The upper bounds for the zeros of this polynomial $p(z)$ estimated by different researchers are shown in the following table.

Carmichael	42.461845
Montel	62.15625
Fujii and Kubo	37.164726
Kittaneh (1)	31.94029
Kittaneh (2)	36.369237
Bhunia et al., (2)	36.54794

while our bound $S_3 = 31.79726$. This shows that for this example, our bound obtained in the above theorem is better than all the estimations mentioned above.

Example 3.2 We consider the following polynomial.

$p(z) = z^8 + \frac{1}{64}z^6 + \frac{1}{16}z^5 + z^4 - 4z^3 + \frac{1}{64}z^2 + \frac{1}{16}z + 1.$

The upper bounds for the zeros of this polynomial $p(z)$ estimated by different researchers are shown in the following table.

Cauchy	5
Carmichael	4.35985
Montel	18.0083

while our bound $S_3 = 3.8507$. This shows that for this example, our bound obtained in the above theorem is better than all the estimations mentioned above.

4 Bounds for the Zeros of Polynomials with Matrix Coefficients by Using Similarity of Matrices

Consider the monic polynomial

$$P(z) = z^m + A_m z^{m-1} + A_{m-1} z^{m-2} + \cdots + A_2 z + A_1$$

where $A_i \in M_n(\mathbb{C}) \forall i = 1, 2, 3, \ldots, m$ where m is an even number. The Frobenius companion matrix of $P(z)$ is the $nm \times nm$ matrix given by

$$C(p) = \begin{bmatrix} -A_m & -A_{m-1} & \cdots & -A_2 & -A_1 \\ I & 0 & \cdots & 0 & 0 \\ 0 & I & \ddots & 0 & 0 \\ \vdots & \vdots & \ddots & \vdots & \vdots \\ 0 & 0 & 0 & I & 0 \end{bmatrix}_{nm\times nm}.$$

To obtain our new bound for the zeros of $P(z)$ consider the matrix

$$Z = \begin{bmatrix} I & 0 & \cdots & 0 & I & 0 & \cdots & 0 \\ 0 & I & \cdots & 0 & 0 & I & & \vdots \\ \vdots & \ddots & \ddots & & & & \ddots & 0 \\ \vdots & & \ddots & \ddots & & & & I \\ \vdots & & & \ddots & \ddots & & & 0 \\ \vdots & & & & \ddots & \ddots & & \vdots \\ 0 & & & & & & \ddots \; I & 0 \\ 0 & \cdots\cdots & 0 & 0 & \cdots & & 0 & I \end{bmatrix}_{nm\times nm},$$

$$\text{and } Z^{-1} = \begin{bmatrix} I & 0 & \cdots & 0 & -I & 0 & \cdots & 0 \\ 0 & I & \cdots & 0 & 0 & -I & & \vdots \\ \vdots & \ddots & \ddots & & & & \ddots & 0 \\ \vdots & & \ddots & \ddots & & & & -I \\ \vdots & & & \ddots & \ddots & & & 0 \\ \vdots & & & & \ddots & \ddots & & \vdots \\ 0 & & & & & & \ddots \; I & 0 \\ 0 & \cdots\cdots & 0 & 0 & \cdots & & 0 & I \end{bmatrix}_{nm\times nm}.$$

Define the matrix B as $B = ZC(p)Z^{-1}$, then B is similar to $C(p)$ which implies that the eigenvalues of B are exactly the same as the eigenvalues of $C(p)$.

We denote B by the B-companion matrix of $P(z)$ which is equal to

$$B = \begin{bmatrix} -A_m & -A_{m-1} & -A_{m-2} & \cdots & I - A_{\frac{m}{2}+1} & A_m - A_{\frac{m}{2}} & A_{m-1} - A_{\frac{m}{2}} & \cdots & A_{m-\frac{m}{2}+2} - A_2 & A_{\frac{m}{2}+1} - A_1 - I \\ I & 0 & 0 & \cdots & 0 & 0 & 0 & \cdots & 0 & 0 \\ 0 & I & 0 & \cdots & 0 & 0 & 0 & \cdots & 0 & 0 \\ 0 & 0 & I & \cdots & 0 & 0 & 0 & \cdots & 0 & 0 \\ 0 & 0 & 0 & \ddots & 0 & 0 & 0 & \cdots & 0 & 0 \\ \vdots & \vdots & \vdots & \cdots & \ddots & \vdots & \vdots & \cdots & \vdots & \vdots \\ \vdots & \vdots & \vdots & \cdots & \vdots & \ddots & \vdots & \cdots & \vdots & -1 \\ \vdots & \vdots & \vdots & \cdots & \vdots & \vdots & \ddots & \cdots & \vdots & \vdots \\ \vdots & \vdots & \vdots & \cdots & \vdots & \vdots & \vdots & \ddots & \vdots & \vdots \\ 0 & 0 & 0 & \cdots & 0 & 0 & 0 & 0 & I & 0 \end{bmatrix},$$

where -1 in occurs in $(\frac{nm}{2}+1)^{th}$ row.

The following lemmas are needed for proving further results involving estimates for $w(C(P))$ ([1, 8]).

Lemma 4.1 *Let $T = [T_{ij}]$ be an $n \times n$ block matrix with $T_{ij} \in M_{m_i \times m_j}(\mathbb{C})$ and $\sum_{i=1}^{n} m_i = n$. Then.*

$w(T) \leq w\big[t_{ij}\big]$ *where* $t_{ij} = \begin{bmatrix} 0 & T_{ij} \\ T_{ji} & 0 \end{bmatrix}$ *in particular* $t_{ii} = w(T_{ii}), i = 1, 2, 3, \ldots, n.$

Lemma 4.2 *Let l_n be the $n \times n$ block matrix given by*

$$l_n = \begin{bmatrix} 0 & \frac{1}{2}I & 0 & \cdots & 0 \\ \frac{1}{2}I & 0 & \frac{1}{2}I & & \ddots \\ 0 & \frac{1}{2}I & 0 & \ddots & 0 \\ \vdots & & \ddots & \ddots & \frac{1}{2}I \\ 0 & 0 & 0 & \frac{1}{2}I & 0 \end{bmatrix}.$$ *Then the eigenvalue of* $l_n = \lambda_i = cos\frac{\pi i}{n+1}$

for $i = 1, 2, 3, \ldots, n$.

Theorem 4.1 *Let λ be a zero of the polynomial.*

$$P(z) = z^m + A_m z^{m-1} + A_{m-1} z^{m-2} + \cdots + A_2 z + A_1.$$

Then

$$|\lambda| \leq \frac{1}{2}\Bigg(w(A_m) + cos\frac{\pi}{m-1} + \sqrt{w^2(A_m) + \alpha + \|I + A_1 - A_{\frac{m}{2}+1}\|^2} + \sqrt{cos^2\frac{\pi}{m-1} + \alpha + 1} + \sqrt{1 + \|I + A_1 - A_{\frac{m}{2}+1}\|^2}\Bigg),$$

where $\alpha = \sum_{i=\frac{mn}{2}+2}^{mn-1} \|A_i\|^2 + \|1 - A_{\frac{m}{2}+1}\|^2 + \sum_{k=0}^{\frac{mn}{2}-2} \|A_{mn-k} - A_{\frac{mn}{2}-k}\|^2.$

Proof for any two matrices $X, Y \in M_m$ let $T_{X,Y} = \begin{bmatrix} 0 & X \\ Y & 0 \end{bmatrix}$.

Applying Lemma 4.1 on the B-companion matrix, we have
$w(B) \leq w(\tilde{B})$,

where

$$\tilde{B} = \begin{bmatrix} w(A_m) & w[T(A_{m-1}, I)] & w[T(A_{m-2}, 0)] & \cdots & w\left[T\left(I - A_{\frac{m}{2}+1}, 0\right)\right] & w\left[T\left(A_m - A_{\frac{m}{2}+1}, 0\right)\right] & \cdots & w\left[T\left(A_{\frac{m}{2}+1} - A_1 - I, 0\right)\right] \\ w[T(A_{m-1}, I)] & w(0) & \cdots & \cdots & \cdots & \cdots & \cdots & w(0) \\ w[T(A_{m-2}, 0)] & w(I) & \ddots & & & & & \vdots \\ \vdots & \vdots & & \ddots & & & & \vdots \\ w\left[T\left(I - A_{\frac{m}{2}+1}, 0\right)\right] & \vdots & & & \ddots & & & w(T(-I, 0)) \\ w\left[T\left(A_m - A_{\frac{m}{2}+1}, 0\right)\right] & \vdots & & & & \ddots & & \vdots \\ \vdots & \vdots & & & & & \ddots & \vdots \\ w\left[T\left(A_{\frac{m}{2}+1} - A_1 - I, 0\right)\right] & w(0) & \cdots & \cdots & \cdots & w(T(-I, 0)) & \cdots & w(0) \end{bmatrix}.$$

Using the fact that $w\left(\begin{bmatrix} 0 & T \\ 0 & 0 \end{bmatrix}\right) = w\left(\begin{bmatrix} 0 & 0 \\ T & 0 \end{bmatrix}\right) = \frac{\|T\|}{2}$, for any matrix $T \in M_n$.

Then partitioning the matrix $\tilde{B}$ as follows:

$$\tilde{B} = \begin{bmatrix} T_{11} & T_{12} & T_{13} \\ T_{21} & T_{22} & T_{23} \\ T_{31} & T_{32} & T_{33} \end{bmatrix},$$

where

$$T_{11} = w(A_m),\ T_{22} = \begin{bmatrix} 0 & \frac{1}{2} & 0 & \cdots & 0 \\ \frac{1}{2} & 0 & \frac{1}{2} & & \vdots \\ 0 & \frac{1}{2} & \ddots & \ddots & 0 \\ \vdots & & \ddots & \ddots & \frac{1}{2} \\ 0 & \cdots & 0 & \frac{1}{2} & 0 \end{bmatrix},\ T_{33} = w(0) = 0,$$

$$T_{12} = T_{21} =$$

$$\left[w[T(A_{m-1}, I)]w[T(A_{m-2}, I)] \cdots w\left[T\left(I - A_{\frac{m}{2}+1}, 0\right)\right]w\left[T\left(A_m - A_{\frac{m}{2}+1}, 0\right)\right] \ldots w\left[T\left(A_{m-\frac{m}{2}+2} - A_2, 0\right)\right]\right],$$

and $T_{13} = T_{31} = w\left(T\left(A_{\frac{m}{2}+1} - A_1 - I, 0\right)\right)$.

Now, by definitions of the numerical radius and spectral norm and Lemma 4.2, we have

$$w(T_{11}) = w(A_m),\ w(T_{22}) = \cos\frac{\pi}{m-1},\ w(T_{33}) = 0,$$

$$\|T_{12}\|^2 = \|T_{21}\|^2 = \sum_{i=\frac{mn}{2}+2}^{mn-1} \|A_i\|^2 + \|1 - A_{\frac{m}{2}+1}\|^2 + \sum_{i=0}^{\frac{mn}{2}-2} \|A_{mn-i} - A_{\frac{mn}{2}-i}\|^2,$$

$$\|T_{31}\|^2 = \|T_{13}\|^2 = \|I + A_1 - A_{\frac{m}{2}+1}\|^2, \|T_{32}\|^2 = \|T_{23}\|^2 = 1.$$

By applying Lemma 3.1 on the partition of the $\tilde{B}$, we have.

$$w\left(\tilde{B}\right) \le \frac{1}{2}\left(w(A_m) + \cos\frac{\pi}{m-1} + \sqrt{w^2(A_m) + \alpha + \|I + A_1 - A_{\frac{m}{2}+1}\|^2} \right.$$
$$\left. + \sqrt{\cos^2\frac{\pi}{m-1} + \alpha + 1 + \sqrt{1 + \|I + A_1 - A_{\frac{m}{2}+1}\|^2}} \right)$$

Since $|z| \le r(C(p)) = r(B) \le w(B) \le w(\tilde{B})$.
The proof is completed. ■

References

1. Abu-Omar, A., Kittaneh, F.: Upper and lower bounds for the numerical radius with an application to involution operators. J. Math. **45**(4), 1055–1064 (2015). https://doi.org/10.1216/RMJ-2015-45-4-1055
2. Bhunia, P., Bag, S., Nayak, R. K., Paul, K.: Estimations of zeros of a polynomial using numerical radius inequalities (2020). https://doi.org/10.48550/arXiv.2001.09706
3. Fujii, M., Kubo, F.: Buzano's inequality and bounds for roots of algebraic equations. Proc. Am. Math. Soc. **117**, 359–361 (1993). https://doi.org/10.2307/2159168
4. Guelfen, H., Kittaneh, F.: On numerical radius inequalities for operator matrices. Numer. Funct. Anal. Optim. **40**(11), 1231–1241 (2019). https://doi.org/10.1080/01630563.2018.1549073
5. Gustafson, K.E., Rao, D.K.M.: Numerical Range, The field of values of linear operators and matrices. Springer, New York (1997). https://doi.org/10.1007/978-1-4613-8498-4
6. Kato, T.: Notes on some inequalities for linear operators. Math. Ann. **125**, 208–212 (1952)
7. Kittaneh, F.: Bounds for the zeros of polynomials from matrix inequalities. Arch. Math. (Basel) **81**, 601–608 (2003). https://doi.org/10.1007/s00013-003-0525-6
8. Kittaneh, F.: A numerical radius inequality and an estimate for the numerical radius of the Frobenius companion matrix. Stud. Math. **158**, 11–17 (2003). https://doi.org/10.4064/sm158-1-2

A New Paranormed Sequence Space and Invariant Means

Ekrem Savaş

Abstract The purpose of this paper is to introduce the new sequence space which emerges naturally from the concept of invariant means and lacunary sequence. Some inclusion relations and matrix transformations have been discussed.

Keywords Infinite matrices · Lacunary sequence · σ-mean · Matrix transformations

1 Introduction

Let s be the set of all sequences real or complex. By l_∞ and c, we denote the Banach spaces of bounded and convergent sequences $x = (x_k)$ normed by $\|x\| = \sup_k |x_k|$, respectively.

Let σ be a one-to-one mapping of the set of positive integers into itself. A continuous linear functional φ on l_∞ is said to be an invariant mean or a σ-mean if and only if

1. $\varphi \geq 0$ when the sequence $x = (x_n)$ has $x_n \geq 0$ for all n.
2. $\varphi(e) = 1$, where $e = (1, 1, \ldots)$.
3. $\varphi\left(x_{\sigma(n)}\right) = \varphi(x)$ for all $x \in l_\infty$.

For a certain kind of mapping σ every invariant mean φ extends the limit functional on space c, in the sense that $\varphi(x) = \lim x$ for all $x \in c$. Consequently, $c \subset V_\sigma$ where V_σ is the bounded sequences all of whose σ-means are equal (see [14]).

If $x = (x_k)$, by setting $Tx = (Tx_k) = \left(x_{\sigma(k)}\right)$ it can be shown that (see Schaefer [14])

$$V_\sigma = \left\{x \in l_\infty : \lim_m t_{mn}(x) = Le \text{ uniformly in n for some } L = \sigma - \lim x\right\} \quad (1)$$

E. Savaş (✉)
Department of Mathematics, Uşak University, Uşak, Turkey
e-mail: ekremsavas@yahoo.com

D. Zeidan et al. (eds.), *Mathematics and Computation*, Springer Proceedings in Mathematics & Statistics 418, https://doi.org/10.1007/978-981-99-0447-1_42

where

$$t_{mn}(x) = \frac{x_n + Tx_n + \cdots + T^m x_n}{m+1} \text{ and } t_{-1,m} = 0.$$

We say that a bounded sequence $x = (x_k)$ is σ-convergent if and only if $x \in V_\sigma$ such that $\sigma^k(n) \neq n$ for all $n \geq 0, k \geq 1$.

The special case of (1) in which $\sigma(n) = n + 1$ was given by Lorentz [3]. Let $\hat{c}$ denote the set of all almost convergent sequences. Lorentz proved that

$$\hat{c} = \left\{x : \lim_m d_{mn}(x) \text{ exists uniformly in } n\right\}$$

where

$$d_{mn}(x) = \frac{x_n + x_{n+1} + \cdots + x_{n+m}}{m+1}.$$

Just as the concept of almost convergence leads naturally to the concept of strong almost convergence, σ-convergence leads naturally to the concept of strong σ-convergence. A sequence $x = (x_k)$ is said to be strongly σ-convergent (see Mursaleen [7]) if there exists a number L such that

$$\frac{1}{k}\sum_{i=1}^{k} \left|x_{\sigma^i(n)} - L\right| \to 0 \tag{2}$$

as $k \to \infty$ uniformly in n. We write $[V_\sigma]$ as the set of all strong σ-convergent sequences. When (2) holds, we write $[V_\sigma] - \lim x = \ell$. Taking $\sigma(n) = n + 1$, we obtain $[V_\sigma] = [\hat{c}]$ (see [4]) so strong σ- convergence generalizes the concept of strong almost convergence. Note that

$$[V_\sigma] \subset V_\sigma \subset l_\infty.$$

σ-convergent sequences are studied by Savas [6, 8–13] and others.

By a lacunary $\theta = (k_r)$, $r = 0, 1, 2, \ldots$ where $k_0 = 0$, we shall mean an increasing sequence of non-negative integers with $k_r - k_{r-1}$ as $r \to \infty$. The intervals determined by θ will be denoted by $I_r = (k_{r-1}, k_r]$ and $h_r = k_r - k_{r-1}$ (see [2]).

Recently, Savas [13] introduced the space $V(\sigma, \theta)$ of lacunary σ-convergent sequences as follows:

$$V(\sigma, \theta) = \left\{x : \lim_r t_{ri}(x - L) \text{ exists uniformly in } i\right\}$$

where

$$t_{ri}(x) = \frac{1}{h_r}\sum_{k \in I_r} x_{\sigma^i(k)}.$$

Note that in the special case where $\theta = 2^r$, we have $V(\sigma, \theta) = V_\sigma$. If we take $\sigma(n) = n + 1$, $V(\sigma, \theta)$ of lacunary σ-convergent sequences reduces to lacunary almost convergence which is defined in [1].

The purpose of this paper is to consider a new sequence $\overline{N}(\sigma, \theta, p)$, which emerges naturally from the concept of σ-convergence and lacunary sequence. We also study the spaces $\overline{N}(\sigma, \theta, p)$, which generalize $\overline{N}(\sigma, \theta)$ in the same way as $l(p)$ generalize l (see [15]). We discuss a related sequence space. Further we characterize some matrix transformations.

Let $\{p_r\}$ be a bounded sequence of positive real numbers. We define

$$\overline{N}(\sigma, \theta, p) = \left\{x : \sum_r |t_{ri}|^{p_r} \text{ converges uniformly in i}\right\}$$

and

$$\overline{\overline{N}}(\sigma, \theta, p) = \left\{x : \sup_n \sum_r |t_{ri}|^{p_r} < \infty\right\}.$$

(Here and afterwards, summation without limits runs from 0 to ∞.) If $p_r = p$ for all r, we write $\overline{N}(\sigma, \theta)_p$ and $\overline{\overline{N}}(\sigma, \theta)_p$ in place of $\overline{N}(\sigma, \theta, p)$ and $\overline{\overline{N}}(\sigma, \theta, p)$, respectively. If $p = 1$, we write $\overline{N}(\sigma, \theta)$, $\overline{\overline{N}}(\sigma, \theta)$ for $\overline{N}(\sigma, \theta, p)$ and $\overline{\overline{N}}(\sigma, \theta, p)$, respectively.

It is now a natural question whether $\overline{N}(\sigma, \theta, p) = \overline{\overline{N}}(\sigma, \theta, p)$. We are only able to prove that $\overline{N}(\sigma, \theta, p) \subset \overline{\overline{N}}(\sigma, \theta, p)$. We have the following theorem.

Theorem 1 $\overline{N}(\sigma, \theta, p) \subset \overline{\overline{N}}(\sigma, \theta, p)$.

Proof Suppose $x \in \overline{N}(\sigma, \theta, p)$. Then there is a constant R such that

$$\sum_{r \geq R+1} |t_{ri}|^{p_r} \leq 1. \tag{3}$$

Hence, it is enough to show that, for fixed r, $|t_{ri}|^{p_r}$ is bounded, or, equivalently, that t_{ri} is bounded. It follows from (3) that $|t_{ri}| \leq 1$ for $r \geq R + 1$ and all i. But if $r \geq 2$,

$$(h_r + 1)t_{ri} - h_r t_{r-1,i} = x_{\sigma^{k_r}(i)}. \tag{4}$$

Applying (4) with any fixed $r \geq R + 1$, we deduce that $x_{\sigma^{k_r}(i)}$ is bounded. Hence, t_{ri} is bounded for all r. Thus the theorem is proved.

Write $M = \max(1, \sup p_r)$. For $x \in \overline{N}(\sigma, \theta, p)$, define

$$g_p(x) = \sup_n (\sum_r |t_{ri}|^{p_r})^{\frac{1}{M}}; \tag{5}$$

this exists because of Theorem 1. We write

Theorem 2 *(i)* $\overline{N}(\sigma, \theta, p)$ *is a complete linear topological space paranormed by* g_p.

(ii) $\overline{N}(\sigma, \theta, p) \subset \overline{N}(\sigma, \theta, q)$ *for* $p_r \leq q_r$.

Proof It can be proved by "standard" arguments that g_p is a paranorm on $\overline{N}(\sigma, \theta, p)$ and also, with the paranorm topology, the space $\overline{N}(\sigma, \theta, p)$ is complete. As one step in the proof, we shall only show that for fixed x, $\lambda x \to 0$ as $\lambda \to 0$. If $x \in \overline{N}(\sigma, \theta, p)$, then given $\epsilon > 0$ there is a R such that, for all r,

$$\sum_{r \geq R} |t_{ri}|^{p_r} < \varepsilon. \tag{6}$$

So if $0 < \lambda \leq 1$, then

$$\sum_{r \geq R} |t_{ri}(\lambda x)|^{p_r} \leq \sum_{r \geq R} |t_{ri}(x)|^{p_r} \leq \varepsilon,$$

and since, for fixed R,

$$\sum_{r=0}^{R-1} |t_{ri}(\lambda x)|^{p_r} \to 0$$

as $\lambda \to 0$, this completes the proof. If $p_r = p$ for all r, then g_p is a norm for $p \geq 1$ and p-norm for $0 \leq p \leq 1$. To prove (ii), let $x \in \overline{N}(\sigma, \theta, p)$. Then there is an integer R such that (3) holds. Hence for $r \geq R$, $|t_{ri}| \leq 1$ so that

$$|t_{ri}|^{q_r} \leq |t_{ri}|^{p_r}$$

and this completes the proof.

Theorem 3 (*i*) *Let* $inf\ p_r > 0$. *Then* $\overline{\overline{N}}(\sigma, \theta, p)$ *is a complete linear topological space paranormed by* g_p.

(*ii*) $\overline{\overline{N}}(\sigma, \theta, p) \subset \overline{\overline{N}}(\sigma, \theta, q)$ *for* $p_r \leq q_r$.

Proof (i) It can be proved by "standard" arguments. It may, however, be noted that there is an essential difference between the proof of Theorem 3(i) and that of Theorem 2(i). If we are given that $x \in \overline{\overline{N}}(\sigma, \theta, p)$, we cannot assert (6). We now use the assumption that $inf p_r > 0$.

Let $\rho = inf p_r > 0$. Then for $|\lambda| \leq 1$, $|\lambda|^{p_r} \leq |\lambda|^{\rho}$, so that $g_p(\lambda x) \leq |\lambda|^{\rho}\, g_p(x)$. The result clearly follows.

(ii) The proof differs from that of Theorem 2(ii), since we cannot assert (3). If $x \in \overline{\overline{N}}(\sigma, \theta, p)$, then $\sum_r |t_{ri}|^{p_r}$ is bounded. So t_{ri} is bounded for all r, i, say $|t_{ri}| \leq K$. We may suppose that $K \geq 1$. Then

$$\sum_r |t_{ri}|^{q_r} \leq \sum_r K^{q_r - p_r} |t_{ri}|^{p_r} \leq R^M \sum_r |t_{ri}|^{p_r}.$$

Hence, the result follows.

2 Matrix Transformations

Let D = (d_{nk}) be an infinite matrix of complex numbers. Let X and Y be any two subsets of space of all sequences of complex numbers. We write Dx = $(D_n(x))$ if $D_n(x) = \sum_k d_{nk} x_k$ converges for each n.

If $x \in X$ implies that $Dx \in Y$, then we say that A defines a matrix transformation from X into Y, and we denote it by $D : X \to Y$. By (X, Y), we mean the class of matrices D such that $D : X \to Y$. If in X and Y there is some notion of limit or sum, then we write (X, Y, P) to denote the subset of (X, Y) which preserves the limit or sum (see Maddox [5]).

We now characterize some matrix transformations connecting $\overline{\overline{N}}(\sigma, \theta)_p$.

We write, for all integer r, i,

$$t_{ri}(Dx) = \frac{1}{h_r} \sum_{i \in I_r} D_{\sigma^n(i)}(x) = \sum_k d(n, k, r) x_k$$

where

$$d(n, k, r) = \frac{1}{h_r} \sum_{i \in I_r} d_{\sigma^i(n), k}.$$

We have the following.

Theorem 4 *Let* $1 \leq p < \infty$. *Then* $A \in (c, \overline{\overline{N}}(\sigma, \theta)_p$ *if and only if*

$$\sup_n (\sum_r (\sum_k (d(n, k, r))) < \infty \tag{7}$$

The proof is easy, so we omit the details.

References

1. Das, G., Mishra, S.K.: Banach limits and lacunary strong almost convergence. J. Orissa Math. Soc. **2**(2), 61–70 (1983)
2. Freedman, A.R., Sember, J.J., Rapheal, M.: Some Cesaro-type summability spacces. Proc. Lond. Math. Soc. **37**(3), 508–520 (1973)
3. Lorentz, G.G.: A contribution to the theory of divergent sequences. Acta Math. **80**, 167–190 (1948)

4. Maddox, I.J.: Spaces of strongly summable sequences. Quart. J. Math. Oxford Ser. **18**(2), 345–55
5. Maddox, I.J.: Elements of Functional Analysis. Cambridge University Press, Cambridge (1970)
6. Mursaleen: Matrix transformation between some new sequence spaces. Houston J. Math. **9**, 505–509 (1993)
7. Mursaleen: On some new invariant matrix methods of summability. Q. J. Math. **34**, 77–86 (1983)
8. Nuray, F., Savaş, E.: Some new sequence spaces defined by a modulus function. Indian J. Pure Appl. Math. **24**(4), 657–663 (1993)
9. Saraswat, S.K., Gupta, S.K.: Spaces of strongly σ-summable sequences. Bull. Cal. Math. Soc. **75**, 179–184 (1983)
10. Savaş, E.: A note on absolute σ -summability. Istanbul Univ. Fac. Sci. Math. J. **50**, 123–128 (1991)
11. Savaş, E.: Invariant means and generalization of a theorem of S. Mishra. Doğa Türk. J. Math. **14**, 8–14 (1989)
12. Savaş, E.: On strong σ-convergence. J. Orissa Math. Soc. **5**(2), 45–53 (1986)
13. Savaş, E.: On lacunary strong σ-convergence. Indian J. Pure Appl. Math. **21**(4), 359–365 (1990)
14. Schaefer, P.: Infinite matrices and invariant means. Proc. Amer. Math. Soc. **36**, 104–110 (1972)
15. Simons, S.: The sequence spaces $l(p_v)$ and $m(p_v)$. Proc. Lond. Math. Soc. **3**(15), 422–436 (1965)